SECOND EDITION

VISUALIZING
ENVIRONMENTAL
SCIENCE

Linda R. Berg, Ph.D.

Mary Catherine Hager, M.S.

WILEY

In collaboration with

THE NATIONAL GEOGRAPHIC SOCIETY

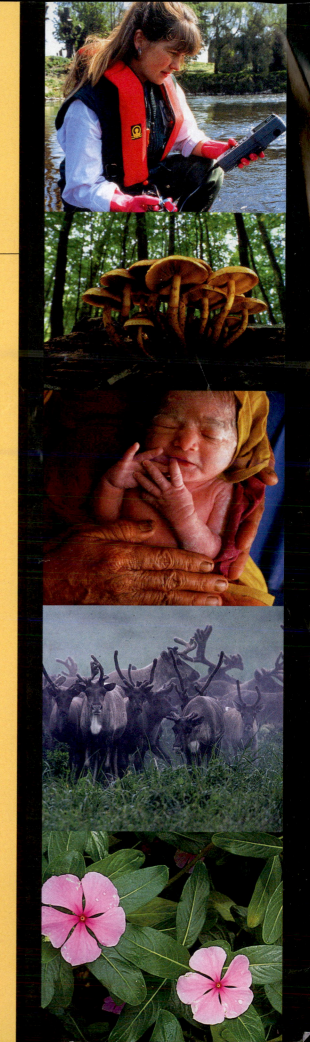

CREDITS

VP AND PUBLISHER Kaye Pace
MANAGING DIRECTOR Helen McInnis
SENIOR EDITOR Rachel Falk
ASSOCIATE EDITOR Merillat Staat
DIRECTOR OF DEVELOPMENT Barbara Heaney
EDITORIAL ASSISTANT Alissa Ertheim
EXECUTIVE MARKETING MANAGER Jeffrey Rucker
PRODUCTION MANAGER Micheline Frederick
SENIOR MEDIA EDITOR Linda Muriello
MEDIA PROJECT MANAGERS Daniela DiMaggio, Bonnie Roth
CREATIVE DIRECTOR Harry Nolan
COVER DESIGNER Harry Nolan
INTERIOR DESIGN Vertigo Design
SENIOR PHOTO EDITOR Elle Wagner
PHOTO RESEARCHER Stacy Gold/
National Geographic Society

This book was set in Times New Roman by GGS Book Services PMG, printed and bound by Quebecor World. The cover was printed by Phoenix Color.

To order books or for customer service please, call 1-800-CALL WILEY (225-5945)

ISBN-13: 978-0-470-11858-0
BRV ISBN: 978-0-470-41808-6

Printed in the United States of America

10 9 8 7 6 5 4 3 2 1

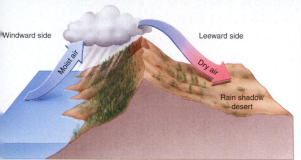

Visualizing Environmental Science 2e is designed to help your students learn effectively. Created in collaboration with the National Geographic Society and our Wiley Visualizing Consulting Editor, Professor Jan Plass of New York University, *Visualizing Environmental Science* 2e integrates rich visuals and media with text to direct student's attention to important information. This approach reduces the cognitive load of complex processes, organizes related pieces of information, and integrates information into clear representations. Beautifully illustrated, *Visualizing Environmental Science* 2e shows your students what the discipline is all about—its main concepts and applications—while also instilling an appreciation and excitement about the richness of the subject.

Visuals, as used throughout this text, are instructional components that display facts, concepts, processes, or principles. They create the foundations for the text and do more than simply support the written or spoken word. The visuals include diagrams, graphs, maps, photographs, illustrations, schematics, animations, and videos.

Why should a textbook based on visuals be effective? Research shows that we learn better from integrated text and visuals than from either medium separately. Beginners in a subject benefit most from reading about the topic, attending class, and studying well-designed and integrated visuals. A visual, with good accompanying discussion, really can be worth a thousand words!

Well-designed visuals can also improve the efficiency with which information is processed by a learner. The more effectively we process information, the more likely it is that we will learn. This process takes place in our working memory. As we learn we integrate new information in our working memory with existing knowledge in our long-term memory.

Have you ever read a paragraph or a page in a book, stopped, and said to yourself: "I don't remember one thing I just read?" This may happen when your working memory has been overloaded, and the text you read was not successfully integrated into long-term memory. Visuals don't automatically solve the problem of overload, but well-designed visuals can reduce the number of elements that working memory must process, thus aiding learning.

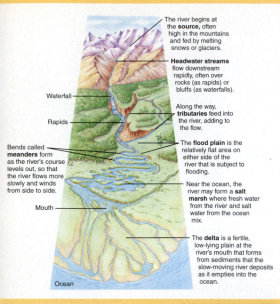

The river begins at the **source**, often high in the mountains and fed by melting snows or glaciers.

Headwater streams flow downstream rapidly, often over rocks (as rapids) or bluffs (as waterfalls).

Waterfall

Rapids

Along the way, **tributaries** feed into the river, adding to the flow.

Bends called **meanders** form as the river's course levels out, so that the river flows more slowly and winds from side to side.

The **flood plain** is the relatively flat area on either side of the river that is subject to flooding.

Near the ocean, the river may form a **salt marsh** where fresh water from the river and salt water from the ocean mix.

Mouth

The **delta** is a fertile, low-lying plain at the river's mouth that forms from sediments that the slow-moving river deposits as it empties into the ocean.

Ocean

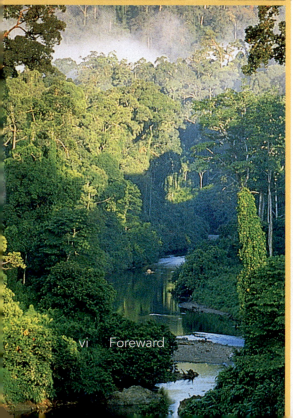

You, as the instructor, facilitate your students' learning. Well-designed visuals, used in class, can help you in that effort. Here are six methods for using the visuals in the *Visualizing Environmental Science* 2e in classroom instruction.

1. **Assign students to study visuals in addition to reading the text.**

   Instead of assigning only one medium of presentation ("Read pages 53 to 69"), it is important to make sure your students know that the visuals are just as essential as the text.

2. **Use visuals during class discussions or presentations.**

   By pointing out important information as the students look at the visuals during class discussions, you can help focus students' attention on key elements of the visuals and help them begin to organize the information and develop an integrated model of understanding. The combination of a verbal explanation of important information with a visual representation can be highly effective.

3. **Use visuals to review content knowledge.**

   Students can review key concepts, principles, processes, vocabulary, and relationships displayed visually. Better understanding results when new information in a student's working memory is linked to prior knowledge.

4. **Use visuals for assignments or when assessing learning.**

   Visuals can be used for comprehension activities or assessments. For example, students could be asked to identify examples of concepts portrayed in visuals. Higher-level thinking activities that require critical thinking, deductive and inductive reasoning, and prediction can also be based on visuals. Visuals can be useful for drawing inferences, predicting, and solving problems.

5. **Use visuals to situate learning in authentic contexts.**

   Learning is made more meaningful when a learner can apply facts, concepts, and principles to realistic situations or examples. Visuals can provide that realistic context.

6. **Use visuals to encourage collaboration.**

   Collaborative groups often are required to practice interactive processes such as giving explanations, asking questions, clarifying ideas, and debating. These interactive, face-to-face processes provide the information needed to build a verbal mental model. Learners also benefit from collaboration in many instances such as decision making or problem solving.

Visualizing Environmental Science 2e not only aids student learning with extraordinary use of visuals, but it also offers an array of remarkable photos, media, and film from the National Geographic Society collections. National Geographic has also performed an invaluable service in fact-checking *Visualizing Environmental Science* 2e, ensuring the accuracy and currency of the text.

Given all of its strengths and resources, *Visualizing Environmental Science* 2e will immerse your students in the discipline and its main concepts and applications, while also instilling an appreciation and excitement for the richness of the subject area.

Additional information on learning tools and instructional design is provided in a special guide to using this book, *Learning from Visuals: How and Why Visuals Can Help Students Learn*, prepared by Matthew Leavitt of Arizona State University. This article is available at the Wiley web site: **www.wiley.com/college/visualizing**. The online Instructor's Manual also provides guidelines and suggestions on using the text and visuals effectively.

Visualizing Environmental Science 2e also offers a rich selection of visuals in the supplementary materials that accompany the book. The following materials are available: Test Bank with visuals used in assessment, PowerPoints, Image Gallery to provide you with the same visuals used in the text, and web-based learning materials for homework and assessment, including images, video, and media resources from National Geographic.

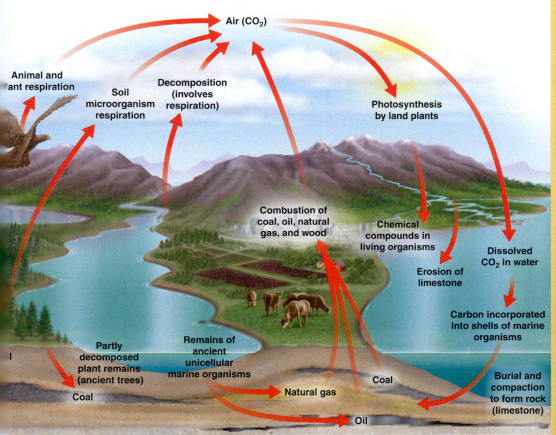

Air (CO$_2$)

Animal and plant respiration

Soil microorganism respiration

Decomposition (involves respiration)

Photosynthesis by land plants

Combustion of coal, oil, natural gas, and wood

Chemical compounds in living organisms

Erosion of limestone

Dissolved CO$_2$ in water

Carbon incorporated into shells of marine organisms

Partly decomposed plant remains (ancient trees)

Remains of ancient unicellular marine organisms

Coal

Natural gas

Coal

Burial and compaction to form rock (limestone)

Oil

PREFACE

The four labels on the diagram read:

④ Some of this heat is transferred back to Earth's surface.

③ Some heat radiated from Earth is absorbed by greenhouse gases.

② Some heat radiated from Earth escapes directly to space.

① Sunlight (radiant energy) is absorbed at surface.

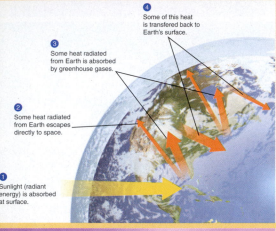

The first edition of *Visualizing Environmental Science* offered students a valuable opportunity to identify and connect the central issues of environmental science through a visual approach. This second edition remains true to that vision, but the visuals have been refined and expanded using insights from research on student learning. *Visualizing Environmental Science* 2e helps students acquire the skills they need to become better learners.

We begin *Visualizing Environmental Science* 2e with an introduction of the environmental dilemmas we face in our world today, emphasizing particularly how unchecked population growth and economic inequity complicate our ability to solve these problems. We stress that solutions rest in creativity and diligence at all levels, from individual commitment to international cooperation. Indeed, a key theme integrated throughout the second edition is the "local to global" scales of environmental science. We offer concrete suggestions that students can adopt to make their own difference in solving environmental problems, and explain the complications that arise when solutions are tackled on a local, regional, national, or global scale.

Yet *Visualizing Environmental Science* 2e is not simply a check-list of "to do" items to save the planet. In the context of an engaging visual presentation, we offer solid discussions of such critical environmental concepts as sustainability, conservation and preservation, and risk analysis. We weave the threads of these concepts throughout our treatment of ecological principles and their application to various ecosystems, the impacts of human population change, and the problems associated with our use of the world's resources. We particularly instruct students in the importance of ecosystem services to a functioning world, and the threats that restrict our planet's ability to provide such services.

This book is intended to serve as an introductory text primarily for nonscience undergraduate students. The accessible format of *Visualizing Environmental Science* 2e, coupled with our assumption that students have little prior knowledge of ecosystem ecology, allows students to easily make the transition from jumping-off points in the early chapters to the more complex concepts they encounter later. With its interdisciplinary presentation, which mirrors the nature of environmental science itself, this book is appropriate for use in one-semester and one-quarter environmental science courses offered by a variety of departments, including environmental studies, biology, agriculture, geography, and geology.

ORGANIZATION

Visualizing Environmental Science 2e is organized around the premise that humans are inextricably linked to the world's environmental dilemmas. We must address these issues as we use Earth's resources and seek to avoid future disasters so often predicted in the media. The book's first four chapters establish the groundwork for understanding what environmental issues we face, how environmental sustainability and human values play a critical role in addressing these issues, how the environmental movement developed over time and was shaped by economics, and how environmental threats from many sources create health hazards that must be evaluated.

Chapters 5, 6, and 7 of *Visualizing Environmental Science* 2e present the intricacies of ecological concepts in a human-dominated world, including energy flow and the cycling of matter through ecosystems, and the various ways that species interact and partition resources. Gaining familiarity with these concepts allows students to better appreciate the variety of terrestrial and aquatic ecosystems that we then introduce, and to develop a richer understanding of the implications to the environment arising from human population change.

The remaining 11 chapters of *Visualizing Environmental Science* 2e deal with the world's resources as we use them today and as we assess their availability and impacts for the future. These issues cover a broad spectrum that includes the sources and effects of air pollution, climate and global atmospheric change, freshwater resources, causes and effects of water pollution, the ocean and fisheries, mineral and soil resources, land resources, agriculture and food resources, biological resources, solid and hazardous waste, and nonrenewable and renewable energy resources. We are particularly pleased to dedicate an entire chapter to a discussion of ocean processes and resources, recognizing the importance of the global ocean to environmental issues.

ILLUSTRATED BOOK TOUR

Many pedagogical features using visuals have been developed specifically for *Visualizing Environmental Science* 2e. Presenting the highly varied and often technical concepts woven throughout environmental science raises challenges for reader and instructor alike. The Illustrated Book Tour on the following pages provides a guide to the diverse features contributing to *Visualizing Environmental Science's* 2e pedagogical plan.

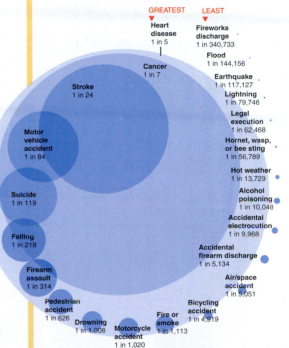

	GREATEST ▼	LEAST ▼
	Heart disease 1 in 5	Fireworks discharge 1 in 340,733
	Cancer 1 in 7	Flood 1 in 144,156
Stroke 1 in 24		Earthquake 1 in 117,127
		Lightning 1 in 79,746
Motor vehicle accident 1 in 84		Legal execution 1 in 62,468
		Hornet, wasp, or bee sting 1 in 56,789
Suicide 1 in 119		Hot weather 1 in 13,729
		Alcohol poisoning 1 in 10,048
Falling 1 in 218		Accidental electrocution 1 in 9,968
Firearm assault 1 in 314		Accidental firearm discharge 1 in 5,134
Pedestrian accident 1 in 626		Air/space accident 1 in 5,051
Drowning 1 in 1,008	Fire or smoke 1 in 1,113	Bicycling accident 1 in 4,919
Motorcycle accident 1 in 1,020		

HELPS STUDENTS VISUALIZE ENVIRONMENTAL SCIENCE

CHAPTER INTRODUCTIONS illustrate certain concepts in the chapter with concise stories about some of today's most pressing environmental issues. These narratives are featured alongside striking accompanying photographs. The chapter openers also include illustrated **CHAPTER OUTLINES** that use thumbnails of illustrations from the chapter to refer visually to the content.

VISUALIZING FEATURES are specially designed multipart visual spreads that focus on a key concept or topic in the chapter, exploring it in detail or in broader context using a combination of photos and figures.

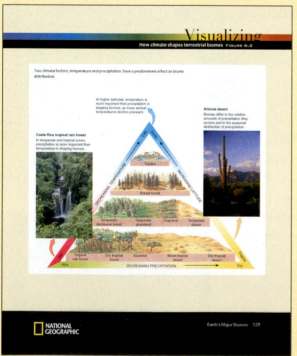

PROCESS DIAGRAMS present a multi-step figure or a combination of figures that describes and depicts a complex process, helping students to observe, follow, and understand the process.

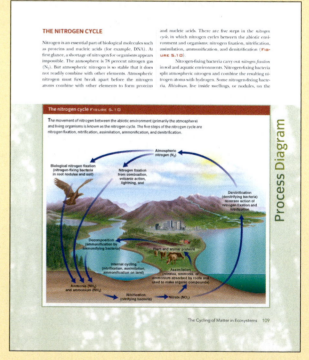

WHAT A SCIENTIST SEES features highlight a concept or phenomenon, using photos and figures that would stand out to a professional in the field, and helping students to develop observational skills.

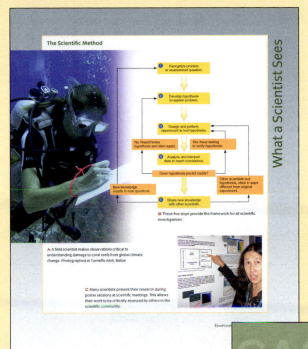

ENVIRODISCOVERY features provide additional topical material about relevant environmental issues.

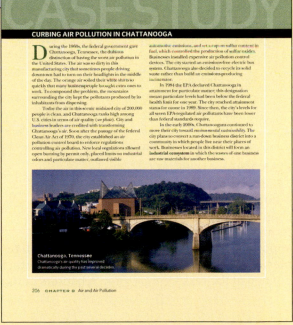

The illustrated **CASE STUDIES** that cap off the text sections of each chapter offer a wide variety of in-depth examinations that address important issues in the field of environmental science.

PROVIDES STUDENTS WITH PROVEN LEARNING TOOLS

Types and Sources of Air Pollution

LEARNING OBJECTIVES

Define air pollution and distinguish between primary and secondary air pollutants.

List the seven major classes of air pollutants and describe their characteristics and sources.

LEARNING OBJECTIVES at the beginning of each section indicate in behavioral terms what the student must be able to do to demonstrate mastery of the material in that section.

CONCEPT CHECK questions at the end of each section give students the opportunity to test their comprehension of the learning objectives.

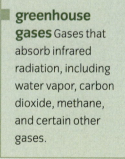

greenhouse gases Gases that absorb infrared radiation, including water vapor, carbon dioxide, methane, and certain other gases.

MARGINAL GLOSSARY TERMS (in green boldface) introduce each chapter's most important terms. The second most important terms appear in **black boldface** and are defined in the text.

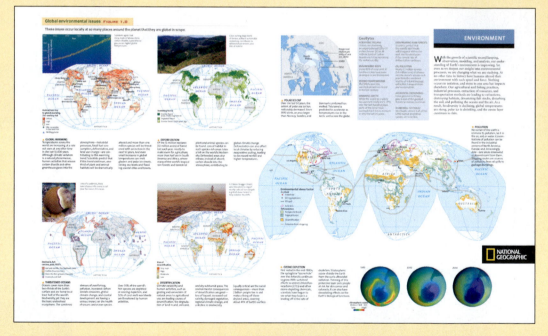

NATIONAL GEOGRAPHIC MAPS, covering Water, Human Population, Land Cover, Energy, Protected Ecosystems, and Environment, are integrated in the appropriate chapters of the book.

A "LOCAL TO GLOBAL" ICON appears in the margins of the text to reinforce the integrated theme that environmental science covers issues from the local to the global scale.

GLOBAL LOCATOR MAPS, prepared by National Geographic, are insets of hemispheric locator maps that help students visualize where the areas shown in particular photographs are situated on a continent.

ILLUSTRATIONS AND PHOTOS support concepts covered in the text, elaborate on relevant issues, and add visual detail. Many of the photos originate from National Geographic's rich sources.

AIR POLLUTION AND HUMAN HEALTH

Generally speaking, exposure to low levels of pollutants irritates the eyes and causes inflammation of the respiratory tract (**TABLE B.1**). Many air pollutants also suppress the immune system, increasing susceptibility to infection. In addition, exposure to air pollution during respiratory illnesses may result in the development later in life of chronic respiratory diseases, such as emphysema and chronic bronchitis. In *emphysema*, the air sacs

...smog was...mining the...tury for the smoky fog prevalent in London because of coal combustion. Traditional London-type smog—that is, smoke pollution—is sometimes called **industrial smog**. The principal pollutants in industrial smog are sulfur oxides and particulate matter. The worst episodes of industrial smog typically occur during winter months, when combustion of household fuel such as heating oil or coal is high. Because of air quality laws and pollution-control devices, industrial smog is generally not a significant problem in highly developed countries today, but it is often serious in many developing countries.

Health effects of several major air pollutants TABLE B.1

Pollutant	Source	Effects
Particulate matter	Industries, motor vehicles	Aggravates respiratory illnesses; long-term exposure may cause chronic conditions such as bronchitis
Sulfur oxides	Electric power plants, industries	Irritate respiratory tract; same effects as particulates
Nitrogen oxides	Motor vehicles, industries, heavily fertilized farmland	Irritate respiratory tract; aggravate respiratory conditions such as asthma and chronic bronchitis
Carbon monoxide	Motor vehicles, industries	Reduces blood's ability to transport oxygen; headache and fatigue at low levels; mental impairment or death at high levels
Ozone	Formed in atmosphere (secondary air pollutant)	Irritates eyes; irritates respiratory tract; produces chest discomfort; aggravates respiratory conditions such as asthma and chronic bronchitis

Effects of Air Pollution   195

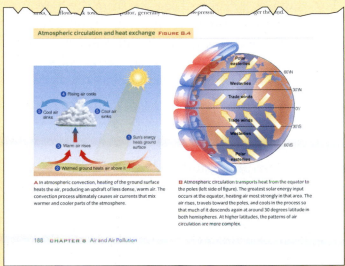

Atmospheric circulation and heat exchange FIGURE B.4

A In atmospheric convection, heating of the ground surface heats the air, producing an updraft of less dense, warm air. The convection process ultimately causes air currents that mix warmer and cooler parts of the atmosphere.

B Atmospheric circulation transports heat from the equator to the poles (left side of figure). The greatest solar energy input occurs at the equator, heating air most strongly in that area. The air rises, travels toward the poles, and cools in the process so that much of it descends again at around 30 degrees latitude in both hemispheres. At higher latitudes, the patterns of air circulation are more complex.

188   **CHAPTER 8**   Air and Air Pollution

TABLES AND GRAPHS, with data sources cited at the end of the text, summarize and organize important information.

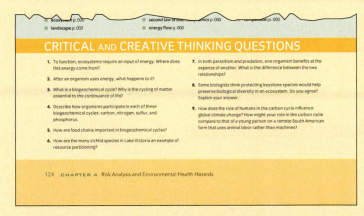

☐ ecosystem p. 000   ☐ second law of thermodynamics p. 000   ☐ competition p. 000
☐ landscape p. 000   ☐ energy flow p. 000

CRITICAL AND CREATIVE THINKING QUESTIONS

1. To function, ecosystems require an input of energy. Where does this energy come from?

2. After an organism uses energy, what happens to it?

3. What is a biogeochemical cycle? Why is the cycling of matter essential to the continuance of life?

4. Describe how organisms participate in each of these biogeochemical cycles: carbon, nitrogen, sulfur, and phosphorus.

5. How are food chains important in biogeochemical cycles?

6. How are the many cichlid species in Lake Victoria an example of resource partitioning?

7. In both parasitism and predation, one organism benefits at the expense of another. What is the difference between the two relationships?

8. Some biologists think protecting keystone species would help preserve biological diversity in an ecosystem. Do you agree? Explain your answer.

9. How does the role of humans in the carbon cycle influence global climate change? How might your role in the carbon cycle compare to that of a young person on a remote South American farm that uses animal labor rather than machines?

124   **CHAPTER 4**   Risk Analysis and Environmental Health Hazards

CRITICAL AND CREATIVE THINKING QUESTIONS encourage critical thinking and highlight each chapter's important concepts and applications.

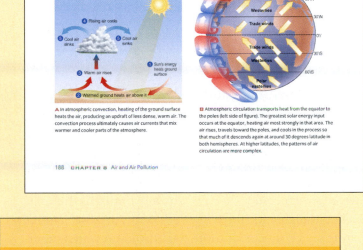

SUMMARY

4 Environmental Economics

1. **Economics** is the study of how people use their limited resources to try to satisfy their unlimited wants. Economies depend on the natural environment as sources for raw materials and sinks for waste products. Both sources and sinks contribute to **natural capital**, which is Earth's resources and processes that sustain living organisms, including humans. Natural capital includes minerals, forests, soils, groundwater, clean air, wildlife, and fisheries.

2. **National income accounts** are a measure of the total income of a nation's goods and services for a given year. An **external cost** is a harmful environmental

or social cost that is borne by people not directly involved in buying or selling a product. National income accounts are incomplete estimates of national economic performance because they do not include both natural resource depletion and the environmental costs of economic activities. Many economists, government planners, and scientists support more comprehensive income accounting that includes these estimates.

3. From an economic point of view, the appropriate amount of pollution is a trade-off between harm to the environment and inhibition of development. The **marginal cost of pollution** is the added cost of an additional unit of pollution. The

marginal cost of pollution abatement is the added cost of reducing one unit of a given type of pollution. Economists think the use of resources for pollution abatement should increase only until the cost of abatement equals the cost of the pollution damage. This results in the **optimum amount of pollution**—the amount of pollution that is economically most desirable.

4. To control pollution, government often uses **command and control regulations**, which are pollution control laws that work by setting limits on levels of pollution. **Incentive-based regulations** are pollution control laws that work by establishing emission targets...industries...ince...

KEY TERMS

☐ utilitarian conservationists p. 000
☐ biocentric preservationists p. 000
☐ full cost accounting p. 000
☐ natural capital p. 000
☐ national income accounts p. 000

☐ external cost p. 000.
☐ marginal cost of pollution p. 000
☐ marginal cost of pollution abatement p. 000
☐ cost-benefit diagram p. 000

☐ optim...
☐ comm...
☐ incen...

CRITICAL AND CREATIVE THINKING QUE...

The **CHAPTER SUMMARY** revisits each learning objective and redefines each marginal glossary term, featured in boldface here, and included in a list of **KEY TERMS**. Students are thus able to study vocabulary words in the context of related concepts.

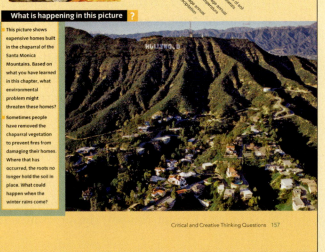

What is happening in this picture ?

☐ This picture shows expensive homes built in the chaparral of the Santa Monica Mountains. Based on what you have learned in this chapter, what environmental problem might threaten these homes?

☐ Sometimes people have removed the chaparral vegetation to prevent fires from damaging their homes. Where that has occurred, the roots no longer hold the soil in place. What could happen when the winter rains come?

Critical and Creative Thinking Questions   157

WHAT IS HAPPENING IN THIS PICTURE? is an end-of-chapter feature that presents students with a photograph relevant to chapter topics but that illustrates a situation students are not likely to have encountered previously. The photograph is paired with questions designed to stimulate creative thinking.

MEDIA AND SUPPLEMENTS

Visualizing Environmental Science 2e is accompanied by a rich array of media and supplements that incorporate the visuals from the textbook extensively to form a pedagogically cohesive package. For example, a Process Diagram from the book may appear in the Instructor's Manual with suggestions on using it as a PowerPoint in the classroom; it may be the subject of a short video or an online animation; and it may also appear with questions in the Test Bank, as part of the chapter review, homework assignment, assessment questions, and other online features.

WILEYPLUS

A powerful online tool that provides instructors and students with an integrated suite of teaching and learning resources in one easy-to-use web site. These include:

VIDEOS

A rich collection of videos have been selected to accompany key topics in the text. Each chapter includes a video clip, available within *WileyPLUS* as digitized streaming video, that illustrates and expands on a concept or topic to aid student understanding. Accompanying each of the videos is contextualized commentary and questions that can further develop student understanding and can be assigned in *WileyPLUS*.

WEB-BASED LEARNING MODULES

A robust suite of multi-media learning resources have been designed for *Visualizing Environmental Science* 2e, again focusing on and using the visuals from the book. Delivered via *WileyPLUS* the content is organized into tutorial animations:

Tutorial Animations and Interactives: Animations visually support the learning of a difficult concept, process, or theory, many of them built around a specific feature such as a Process Diagram, Visualizing feature, or key visual in the chapter. The animations go beyond the content and visuals presented in the book, providing additional visual examples and descriptive narration.

BOOK COMPANION SITE (WWW.WILEY.COM/COLLEGE/BERG)

Instructor Resources on the book companion site include the Test Bank, Instructor's Manual, PowerPoint presentations, and all illustrations and photos in the textbook in jpeg format. Student resources include self quizzes and flashcards.

INSTRUCTOR SUPPLEMENTS

POWERPOINT PRESENTATIONS

A complete set of highly visual PowerPoint presentations by Renee Nerish, Mercer County Community College, is available online and in *WileyPLUS* to enhance classroom presentations. Tailored to the text's topical coverage and learning objectives, these presentations are designed to convey key text concepts, illustrated by embedded text art.

IMAGE GALLERY

All photographs, figures, maps, and other visuals from the text are online and in *WileyPLUS* and can be used as you wish in the classroom. These online electronic files allow you to easily incorporate images into your PowerPoint presentations as you choose, or to create your own overhead transparencies and handouts.

TEST BANK (AVAILABLE IN WILEYPLUS AND ELECTRONIC FORMAT)

The visuals from the textbook are also included in the Test Bank by Vicki Medland, University of Wisconsin–Green Bay. The Test Bank has approximately 1200 test items, with at least 25 percent incorporating visuals from the book. The test items include multiple choice and essay questions testing a variety of comprehension levels. The test bank is available online in MS Word files, as a Computerized Test Bank and within *WileyPLUS*. The easy-to-use test-generation program fully supports graphics, print tests, student answer sheets, and answer keys. The software's advanced features allow you to produce an exam to your exact specifications.

INSTRUCTOR'S MANUAL (AVAILABLE IN ELECTRONIC FORMAT)

For each chapter, materials by Barry Vroginday of DeVry University include suggestions and directions for using web-based learning modules in the classroom and for homework assignments, as well as over 50 creative ideas for in-class activities by David Hassenzahl of the University of Nevada-Las Vegas and Jody Terrell of Texas Woman's University.

ALSO AVAILABLE TO PACKAGE

Environmental Science: Active Learning Laboratories and Applied Problem Sets, 2e by Travis Wagner and Robert Sanford both of University of Southern Maine, is designed to introduce environmental science students to the broad, interdisciplinary field of environmental science by presenting specific labs that use natural and social science concepts to varying degrees and by encouraging a "hands on" approach to understanding the impacts from the environmental/human interface. The laboratory and homework activities are designed to be low-cost and to reflect a sustainability approach in practice and in theory. *Environmental Science: Active Learning Laboratories and Applied Problem Sets, 2e* is available stand -alone or in a package with *Visualizing Environmental Science, 2e.* Contact your Wiley representative for more information.

EarthPulse. Utilizing full-color imagery and National Geographic photographs, *EarthPulse* takes you on a journey of discovery covering topics such as *The Human Condition, Our Relationship with Nature,* and *Our Connected World.* Illustrated by specific examples, each section focuses on trends affecting our world today. Included are extensive full-color world and regional maps for reference. *EarthPulse* is available only in a package with *Visualizing Environmental Science, 2e.* Contact your Wiley representative for more information or visit www.wiley.com/college/earthpulse.

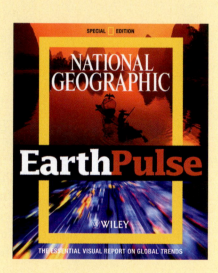

ACKNOWLEDGMENTS

CLASS TESTING AND STUDENT FEEDBACK

To make certain that *Visualizing Environmental Science* met the needs of current students, we asked several instructors to class test a chapter. The feedback that we received from students and instructors confirmed our conviction that the visualizing approach taken in this book is highly effective in helping students to learn. We wish to thank the following instructors and their students who provided us with helpful feedback and suggestions:

Hernan Aubert, Pima Community College; Keith Hench, Kirkwood Community College; Dawn Keller, Hawkeye Community College; Dale Lambert, Tarrant Community College; Janice Padula, Clinton College; and Ashok Malik, Evergreen Valley College.

PROFESSIONAL FEEDBACK

Throughout the process of writing and developing this text and the visual pedagogy, we benefited from the comments and constructive criticism provided by the instructors and colleagues listed below. We offer our sincere appreciation to these individuals for their helpful reviews.

Reviewers of the Second Edition:

Jennifer Andersen, Johnson County Community College
Nancy Bain, Ohio University
Cheryl Berg, Gateway Community College, Phoenix
Stephan Bullard, University of Hartford
Oiyin Pauline Chow, Harrisburg Area Community College
JodyLee Estrada Duek, Pima Community College
Catherine M. Etter, Cape Cod Community College
Peggy Green, Broward Community College
Stelian Grigoras, Northwood University
Carol Hoban, Kennesaw State University
Guang Jin, Illinois State University
Martin Kelly, Genesee Community College
Paul Kramer, Farmingdale State College
Robert S. Mahoney, Johnson & Wales University at Florida
Heidi Marcum, Baylor University

Brian Mooney, Johnson & Wales University at North Carolina
Jacob Napieralski, University of Michigan, Dearborn
Renee Nerish, Mercer County Community College
Neal Phillip, Bronx Community College
Uma Ramakrishnan, Juniata College
Shamili A. Sandiford, College of DuPage
Thomas Sasek, University of Louisiana at Monroe
Richard Shaker, University of Wisconsin, Milwaukee
Charles Shorten, West Chester University
Roy Sofield, Chattanooga State Technical Community College
Charles Venuto, Brevard Community College, Cocoa Campus
Laura J. Vosejpka, Northwood University
John F. Weishampel, University of Central Florida
Karen Wellner, Arizona State University

Reviewers of the First Edition

Mark Anderson, University of Maine
Raymond Beiersdorfer, Youngstown State University
Richard Bowden, Allegheny College
Scott Brame, Clemson University
James A. Brenneman, University of Evansville
Huntting W. Brown, Wright State University
R. Laurence Davis, Northeastern Cave Conservancy, Inc.
Michael S. Dann, Penn State University
Brad C. Fiero, Pima Community College
Michael Freake, Lee University
Jennifer Frick-Ruppert, Brevard College
Todd G. Fritch, Northeastern University
Marcia L. Gillette, Indiana University, Kokomo
Arthur Goldsmith, Hallandale High
Cliff Gottlieb, Shasta College
Syed E. Hasan, University of Missouri—Kansas City
Dawn Keller, Hawkeye Community College
Martin G. Kelly, Buffalo State College
David Kitchen, University of Richmond
Paul Kramer, SUNY Farmingdale
Meredith Gooding Lassiter, Wiona State University
Ernesto Lasso de la Vega, Edison College
Madelyn E. Logan, North Shore Community College
Linda Lyon, Frostburg State University

Timothy F. Lyon, Ball State University
Matthew H. McConeghy, Johnson & Wales University
Rick McDaniel, Henderson State University
Leslie Nesbitt, Niagara University
Ken Nolte, Shasta College
Natalie Osterhoudt, Broward Community College
Barry Perlmutter, Community College of Southern Nevada
Thomas E. Pliske, Florida International University
Katherine Prater, Texas Wesleyan University
Sabine Rech, San Jose State University
Howie Scher, University of Rochester
Nan Schmidt, Pima Community College
Richard B. Schultz, Elmhurst College
Bo Sosnicki, Florida Community College at Jacksonville
Jerry Skinner, Keystone College
Ravi Srinivas, University of St. Thomas;
David Steffy, Jacksonville State University
Andrew Suarez, University of Illinois
Charles Venuto, Brevard Community College
Margaret E. Vorndam, Colorado State University Pueblo
Maud M. Walsh, Louisiana State University
Arlene Westhoven, Ferris State University
Susan M. Whitehead, Becker College
John Wielichowski, Milwaukee Area Technical College

SPECIAL THANKS

We are extremely grateful to the many members of the editorial and production staff at John Wiley and Sons who guided us through the challenging steps of developing this book. Their tireless enthusiasm, professional assistance, and endless patience smoothed the path as we found our way. We thank in particular Senior Editor Rachel Falk, who expertly launched and directed our process; Merillat Staat, Associate Editor, for guiding us through the revision process, coordinating the final stages of development, and providing helpful suggestions during production; and Alissa Etrheim, Editorial Assistant, for her constant attention to detail. We also thank Micheline Frederick, Production Manager, who expertly helped us through production and managed our illustration program. We thank Barbara Heaney, Director of Product and Market Development, who provided valuable suggestions for page layout; Helen McInnis, Managing Director, Wiley Visual Imprint, who managed the concept of the book; Jeof Vita, Art Director, for helping us with difficult layouts; Kaye Pace, Vice President and Executive Publisher, who oversaw the entire project; and Jeffrey Rucker, Executive Marketing Manager for Wiley Visualizing, who adeptly represents the Visualizing imprint.

We also wish to thank those who worked on the media and ancillary materials: Linda Muriello, Senior Media Editor, for her expert work in developing the video and electronic components; and Merillat Staat, Associate Editor, for carefully guiding the completion of the ancillary materials.

We thank Elle Wagner for her unflagging, always swift work in researching and obtaining many of our text images, and Stacy Gold, Research Editor and Account Executive at the National Geographic Image Collection, for her valuable expertise in selecting NGS photos.

Many other individuals at National Geographic offered their expertise and assistance in developing this book: Francis Downey, Vice President and Publisher, and Richard Easby, Supervising Editor, National Geographic School Division; Mimi Dornack, Sales Manager, and Lori Franklin, Assistant Account Executive, National Geographic Image Collection; Dierdre Bevington-Attardi, Project Manager, and Kevin Allen, Director of Map Services, National Geographic Maps; and Devika Levy, Jim Burch, and Michael Garrity of the National Geographic Film Library. We appreciate their contributions and support.

ABOUT THE AUTHORS

Linda R. Berg is an award-winning teacher and textbook author. She received a B.S. in science education, M.S. in botany, and Ph.D. in plant physiology from the University of Maryland. Dr. Berg taught at the University of Maryland-College Park for 17 years and at St. Petersburg College in Florida for 10 years. She has taught introductory courses in environmental science, biology, and botany to thousands of students and has received numerous teaching and service awards. Dr. Berg is also the recipient of many national and regional awards, including the National Science Teachers Association Award for Innovations in College Science Teaching, the Nation's Capital Area Disabled Student Services Award, and the Washington Academy of Sciences Award in University Science Teaching. During her career as a professional science writer, Dr. Berg has authored or co-authored several leading college science textbooks. Her writing reflects her teaching style and love of science.

Mary Catherine Hager is a professional science writer and editor specializing in educational materials for life and earth sciences. She received a double-major B.A. in environmental science and biology from the University of Virginia and a M.S. in zoology from the University of Georgia. Ms. Hager worked as an editor for an environmental consulting firm and as a senior editor for a scientific reference publisher. For the past 15 years, Ms. Hager has published articles in environmental trade magazines, edited federal and state reports addressing wetlands conservation issues, and written components of environmental science and biology textbooks for target audiences ranging from middle school to college. Her writing and editing pursuits are a natural outcome of her scientific training and curiosity coupled with her love of reading and communicating effectively.

CONTENTS *in Brief*

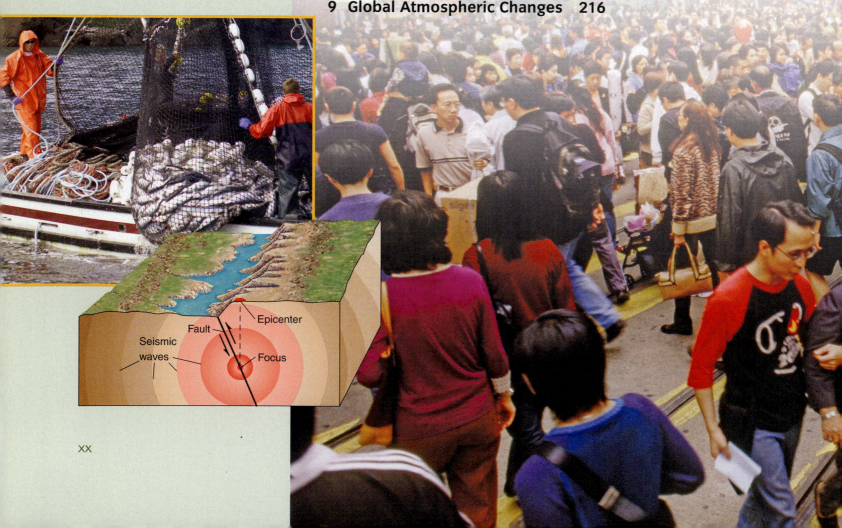

CONTENTS

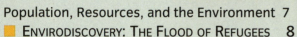

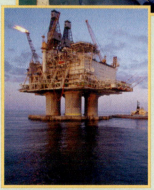

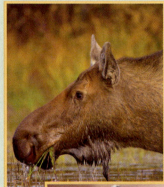

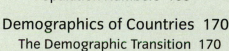

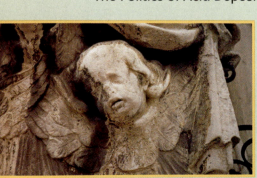

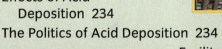

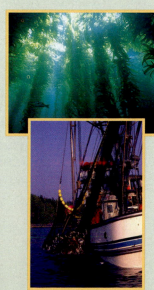

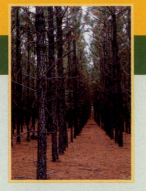
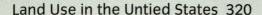

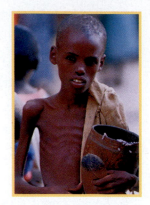

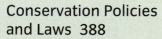

Contents   xxxi

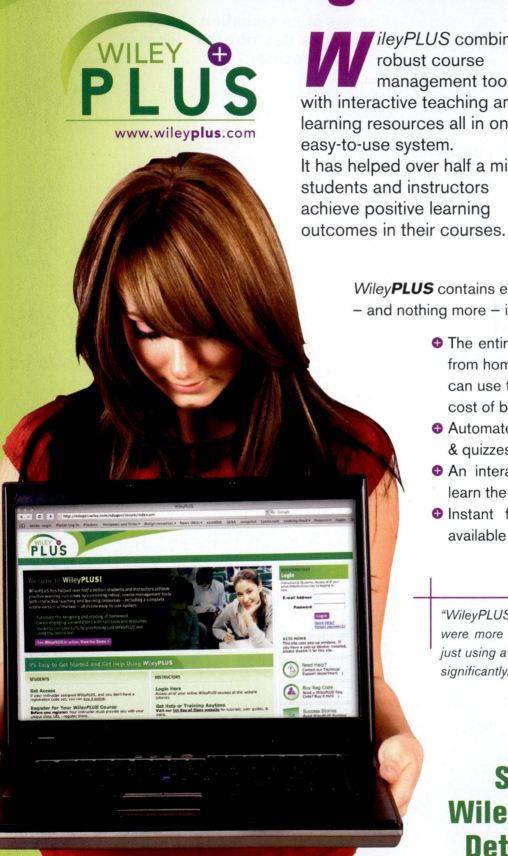

www.wileyplus.com

Wiley is committed to making your entire *WileyPLUS* experience productive & enjoyable by providing the help, resources, and personal support you & your students need, when you need it. It's all here: www.wileyplus.com –

TECHNICAL SUPPORT:

- ⊕ A fully searchable knowledge base of FAQs and help documentation, available 24/7
- ⊕ Live chat with a trained member of our support staff during business hours
- ⊕ A form to fill out and submit online to ask any question and get a quick response
- ⊕ **Instructor-only** phone line during business hours: 1.877.586.0192

FACULTY-LED TRAINING THROUGH THE WILEY FACULTY NETWORK:
Register online: www.wherefacultyconnect.com

Connect with your colleagues in a complimentary virtual seminar, with a personal mentor in your field, or at a live workshop to share best practices for teaching with technology.

1ST DAY OF CLASS...AND BEYOND!
Resources for You & Your Students Need to Get Started & Use WileyPLUS from the first day forward.

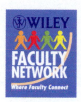

- ⊕ 2-Minute Tutorials on how to set up & maintain your *WileyPLUS* course
- ⊕ User guides, links to technical support & training options
- ⊕ *WileyPLUS for Dummies*: Instructors' quick reference guide to using *WileyPLUS*
- ⊕ Student tutorials & instruction on how to register, buy, and use *WileyPLUS*

YOUR WileyPLUS ACCOUNT MANAGER:

Your personal *WileyPLUS* connection for any assistance you need!

SET UP YOUR WileyPLUS COURSE IN MINUTES!
Select *WileyPLUS* courses with QuickStart contain pre-loaded assignments & presentations created by subject matter experts who are also experienced *WileyPLUS* users.

Interested? See–and Try *WileyPLUS* in action!
Details and Demo: www.wileyplus.com

The Environmental Challenges We Face

A WORLD IN CRISIS

Earth's abundant natural resources have provided the backdrop for a parade of living things to evolve. Although early Earth was inhospitable by modern standards, it provided the raw materials and energy needed for early life forms to arise and adapt. Today, millions of species inhabit the planet.

About 100,000 years ago—a mere blip in Earth's 4.5-billion-year history—an evolutionary milestone occurred, with the appearance of modern humans in Africa. Over time, our population grew, we expanded our range throughout the planet, and we increasingly impacted the environment with our presence and our technologies. Our intellectual capacity has even made it possible for us to venture into space, allowing us a view of the uniqueness of our planet in the solar system (see photograph).

In many ways these technologies have made life better, at least for those of us who live in the United States and other highly developed nations. At the same time, much in our world indicates that we are headed toward serious environmental changes.

Today the human species is the most significant agent of environmental change on our planet. We are overpowering the planet with our burgeoning population; transforming forests, prairies, and deserts to meet our needs and desires; and consuming ever-increasing amounts of Earth's abundant but finite resources—rich topsoil, clean water, and breathable air. We are eradicating thousands upon thousands of unique species as we destroy or alter their habitats. Evidence continues to accumulate that human-induced climate change is putting the natural environment at risk.

This book introduces the major environmental problems that humans have created and considers ways to address these issues. Most importantly, it explains why we must minimize human impact on our planet. We can't afford to ignore the environment because our lives, as well as those of future generations, depend on it. As a wise proverb says, "We have not inherited the world from our ancestors; we have only borrowed it from our children."

NATIONAL GEOGRAPHIC

Human Impacts on the Environment

LEARNING OBJECTIVES

Distinguish among highly developed countries, moderately developed countries, and less developed countries.

Relate human population size to natural resources and resource consumption.

Distinguish between people overpopulation and consumption overpopulation.

Describe the three factors that are most important in determining human impact on the environment.

The satellite photograph in **FIGURE 1.1** is a portrait of about 436 million people. The tiny specks of light represent cities, and the great metropolitan areas, such as New York City along the northeastern seacoast, are ablaze with light.

Earth's central environmental problem, which links all others together, is that there are many people, and the number, both in North America and worldwide, continues to grow. In 1999 the human population as a whole passed a significant milestone: 6 billion individuals. Not only is this number incomprehensibly large, but our population has grown this large in a very brief span of time. In 1960 the human population was only 3 billion (**FIGURE 1.2**). By 1975 there were 4 billion people, and by 1987 there were 5 billion. The 6.6 billion people who currently inhabit our planet consume vast quantities of food and water, use a great deal of energy and raw materials, and produce much waste.

Despite most countries' involvement with family planning, population growth rates won't change overnight. Several billion people will be added to the world in the 21st century, so even if we remain concerned about overpopulation and even if our solutions are very effective, the coming decades may very well see many problems.

poverty A condition in which people are unable to meet their basic needs for food, clothing, shelter, education, or health.

On a global level, nearly one in four people lives in extreme **poverty** (**FIGURE 1.3**). By one measure, living in poverty is defined as having a per person income of less than $2 per day, expressed in U.S. dollars adjusted for purchasing power. About 3.5 billion people—more than half of the world's population—currently live at this level of poverty. Poverty is associated with a short life expectancy, illiteracy, and inadequate access to health services, safe water, and balanced nutrition.

The world population may stabilize toward the end of the 21st century, given the family planning efforts that are currently under way. Population experts at the Population Reference Bureau have noticed a decrease in the fertility rate worldwide to a current average of 2.7 children per woman, and the fertility rate is projected to continue to decline in coming decades. Population experts have made various projections for the world population at the end of the 21st century, from about 7.9 billion to 10.9 billion, based on how fast the fertility rate decreases.

No one knows whether Earth can support so many people indefinitely. Finding ways for it to do so represents one of the greatest challenges of our times. Among the tasks to be accomplished is feeding a world population considerably larger than the present one without destroying the biological communities that support life on our planet. The quality of life available to our children and grandchildren will depend to a large extent on our ability to develop a sustainable system of agriculture to feed the world's people.

A factor as important as population size is a population's level of **consumption**, which is the human use of material and energy. Consumption is intimately connected to a country's **economic growth**, the expansion in output of a nation's goods and services. The world's economy is growing at an enormous rate, yet this growth is unevenly distributed across the nations of the world.

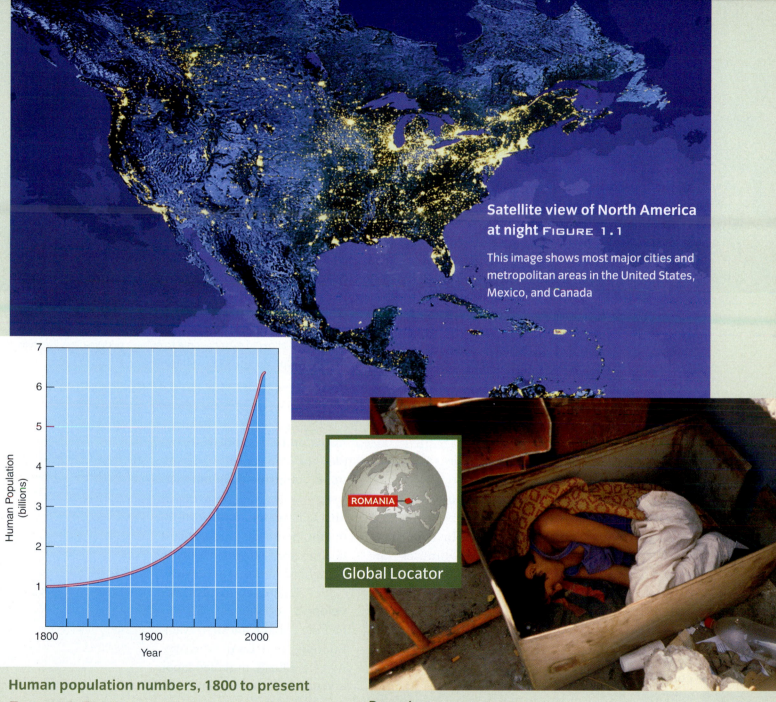

Satellite view of North America at night FIGURE 1.1

This image shows most major cities and metropolitan areas in the United States, Mexico, and Canada

Global Locator

Human population numbers, 1800 to present

FIGURE 1.2

It took thousands of years for the human population to reach 1 billion (in 1800), 130 more years to reach 2 billion (1930), 30 more to reach 3 billion (1960), 15 more to reach 4 billion (1975), 12 more to reach 5 billion (1987), and 12 more to reach 6 billion (1999).

Poverty FIGURE 1.3

An abandoned child sleeps in a box in Bucharest, Romania. The human population problem requires not only a stabilization of population numbers but also an improvement of the economic conditions of people living in extreme poverty.

A A typical Japanese family, from Tokyo, with their possessions. People in highly developed countries consume a disproportionate share of natural resources.

B A typical Mexican family, from Guadalajara, with their possessions. Economic development in this moderately developed country has allowed many people to enjoy a middle-class lifestyle. Other Mexicans live in poverty, however.

THE GAP BETWEEN RICH AND POOR COUNTRIES

highly developed countries Countries with complex industrialized bases, low rates of population growth, and high per person incomes.

moderately developed countries Countries with medium levels of industrialization and per person incomes lower than those of highly developed countries.

Generally speaking, countries are divided into rich (the "haves") and poor (the "have-nots"). Rich countries are known as **highly developed countries**. The United States, Canada, Japan, and most of Europe, which represent about 19 percent of the world's population, are highly developed countries (FIGURE 1.4A).

Poor countries, in which about 81 percent of the world's population live, fall into two subcategories: moderately developed and less developed. Mexico, Turkey, South Africa, and Thailand are examples of **moderately developed countries (MDCs)** (FIGURE 1.4B). People living in MDCs have fewer opportunities for income, education, and health care than people living in highly developed countries.

Examples of **less developed countries** (LDCs) include Bangladesh, Mali, Ethiopia, and Laos (FIGURE 1.4C). Cheap, unskilled labor is abundant in LDCs, but capital for investment is scarce. To improve their economic conditions, many LDCs must borrow money from banks in highly developed countries. Most economies of LDCs are agriculturally based, often on only one or a few crops. As a result, crop failure or a low world market value for that crop is catastrophic to the economy. Hunger, disease, and illiteracy are common in LDCs.

less developed countries Countries with low levels of industrialization, very high rates of population growth, very high infant mortality rates, and very low per person incomes relative to highly developed countries.

⊏ A typical family from Kouakourou, Mali, with all their possessions. The rapidly increasing number of people in less developed countries overwhelms their natural resources, even though individual resource requirements may be low.

POPULATION, RESOURCES, AND THE ENVIRONMENT

Inhabitants of the United States and other highly developed countries consume many more resources per person than do citizens of developing countries. This high rate of resource consumption affects the environment at least as much as the explosion in population that is occurring in other parts of the world.

We can make two useful generalizations about the relationships among population growth, consumption of natural resources, and environmental degradation. One, the amount of resources essential to an individual's survival is small, but rapid population growth (often found in developing countries) tends to overwhelm and deplete a country's soils, forests, and other natural resources. Two, in highly developed nations, individual demands on natural resources are far greater than the requirements for mere survival. To satisfy their desires as well as their basic needs, many people in more affluent nations deplete re-

sources and degrade the global environment through increased consumption of nonessential items such as televisions, jet skis, and cellular phones.

Types of resources When examining the effects of population on the environment, it is important to distinguish between nonrenewable and renewable natural resources. **Nonrenewable resources** include minerals (such as aluminum, tin, and copper) and fossil fuels (coal, oil, and natural gas). Natural processes do not replenish nonrenewable resources within a reasonable duration on the human timescale. Fossil fuels, for example, take millions of years to form.

In addition to a nation's population and its level of resource use, several other factors affect the way nonrenewable resources are used—including how efficiently the resource is extracted and processed and how much of it is required or consumed. Nonetheless, the inescapable fact is that Earth has a finite supply of nonrenewable resources that sooner or later will be exhausted. In time, technological advances may help find or develop substitutes for nonrenewable resources. Slowing the rate of population growth and resource consumption will help us buy time to develop such alternatives.

Some examples of **renewable resources** are trees, fishes, fertile agricultural soil, and fresh water. Nature replaces these resources fairly rapidly, on a scale of days to decades. Forests, fisheries, and agricultural land are particularly important renewable resources in developing countries because they provide food. Indeed, many people in developing countries are subsistence farmers who harvest just enough food for their families to survive.

Rapid population growth can cause renewable resources to be overexploited. For example, large numbers of poor people must grow crops on land—such as mountain slopes or tropical rain forests—that is inappropriate for farming. Although this practice may provide a short-term solution to the need for food, it does not work in the

> **nonrenewable resources** Natural resources that are present in limited supplies and are depleted as they are used.

> **renewable resources** Resources that are replaced by natural processes and that can be used forever, provided they are not overexploited in the short term.

Refugee camp

A young Iraqi girl cries in front of a tent. This photograph, taken in 2007, shows the al-Husseiniya district in Baghdad, where people have fled from the violence in Diyala province.

Global Locator

Sometimes economic, political, or environmental problems force people to leave their homelands en masse. According to the Office of the United Nations High Commissioner for Refugees, in 2007 there were more than 21 million international refugees and internally displaced persons. *International refugees* are people who moved from one country to another, whereas *internally displaced persons* are people who were forced to move but did not cross a border into another country.

Environmental degradation is recognized as a major cause of refugee flight. For the past several decades, deterioration of agricultural lands, coupled with extended droughts and other natural disasters, has caused the displacement of perhaps 30 million people worldwide. The large number of refugees strains the economies and natural resources of the regions into which the refugees migrate.

Many people attempting to flee their homelands are not welcomed with open arms. The industrial world's reluctance to embrace refugees stems from several sources:

- Economic problems at home
- The high costs of processing asylum applications
- Skepticism about whether refugees are actually fleeing political persecution and are not simply "economic migrants"

Failure to address the high demand for asylum leads to a lower quality of life for prospective refugees:

- Refugee camps are growing more violent, with sexual and gender-based violence against women refugees particularly pronounced.
- Human trafficking (charging asylum-seekers money to illegally transport them across borders) is on the rise, with its accompanying financial exploitation, human rights abuses, and violence.
- Less developed nations take their cue from highly developed ones, refusing to accept refugees when more prosperous nations will not.

long run because when these lands are cleared for farming, their agricultural productivity declines rapidly and severe environmental deterioration occurs. Renewable resources, then, are *potentially* renewable. They must be used in a manner that gives them time to replace or replenish themselves.

The effects of population growth on natural resources are particularly critical in developing countries. The economic growth of developing countries is frequently tied to the exploitation of their natural resources, often for export to highly developed countries. Developing countries are faced with the difficult choice of exploiting natural resources to provide for their expanding populations in the short term (that is, to pay for food or to cover debts) or conserving those resources for future generations. It is instructive to note that the economic growth and development of the United States and of other highly developed nations came about through the exploitation—and in some cases the destruction—of their resources. Continued growth and development in highly developed countries now relies significantly on the importation of these resources from less developed countries.

Poverty is tied to the effects of population pressures on natural resources and the environment. Poor people in developing countries find themselves trapped in a vicious cycle of poverty. They use environmental resources unwisely for short-term gain (that is, to survive), but this exploitation degrades the resources and diminishes long-term prospects of economic development.

Population size and resource consumption

Resource issues are clearly related to population size—more people use more resources. An equally important factor is a population's resource consumption. Consumption is both an economic and a social act. Consumption provides the consumer with a sense of identity as well as status among peers. The media, including the advertising industry, promote consumption as a way to achieve happiness. We are encouraged to spend, to consume.

People in highly developed countries are extravagant and wasteful consumers; their use of resources is greatly out of proportion to their numbers. A single child born in a highly developed country such as the United States causes a greater impact on the environment and on resource depletion than perhaps 20 children born in a developing country. Many natural resources are needed to provide the automobiles, air conditioners, disposable diapers, cell phones, DVD players, computers, clothes, newspapers, athletic shoes, furniture, books, and other "comforts" of life in highly developed nations. Thus, the disproportionately large consumption of resources by the United States and other highly developed countries affects natural resources and the environment as much as or more than the population explosion in the developing world.

people overpopulation A situation in which there are too many people in a given geographic area.

consumption overpopulation A situation in which each individual in a population consumes too large a share of resources.

People overpopulation and consumption overpopulation A country is overpopulated if the level of demand on its resource base results in damage to the environment. In comparing human impact on the environment in developing and highly developed countries, we see that a country can be overpopulated in two ways. **People overpopulation** occurs when the environment is worsening because there are too many people, even if those people consume few resources per person. People overpopulation is the current problem in many developing nations. In contrast, **consumption overpopulation** results from the consumption-oriented lifestyles popular in highly developed countries. The effects of consumption overpopulation on the environment are the same as those of people overpopulation—pollution, resource depletion, and degradation of the environment.

Many affluent, highly developed nations, including the United States, suffer from consumption overpopulation. Highly developed nations represent only 20 percent of the world's population, yet they consume significantly more than half of its resources. According to the Worldwatch Institute, highly developed nations account for the lion's share of total resources consumed:

- 86 percent of aluminum used.
- 76 percent of timber harvested.
- 68 percent of energy produced.
- 61 percent of meat eaten.
- 42 percent of the fresh water consumed.

American consumption is actively promoted in Times Square advertisements. Highly developed nations, such as the United States, consume more than 50 percent of the world's resources, produce 75 percent of its pollution and waste, and represent only 19 percent of its total population.

These nations also generate 75 percent of the world's pollution and waste (FIGURE 1.5).

Mathis Wackernagel and colleagues at the Universidad Anáhuac de Xalapa in Mexico compared the ecological impact of individuals living in 52 large nations. In this study, each person has an **ecological footprint**, which is the average amount of productive land and ocean needed to supply that person with food, energy, water, housing, transportation, and waste disposal. In developing nations, such as India and Nigeria, the ecological footprint is about 1 hectare (2.5 acres). In the United States, the ecological footprint is about 9.6 hectares (23.7 acres). If all 6.6 billion people in the world had the same lifestyle and level of consumption as an average American, and assuming no changes in technology, we would need four additional planets the size of Earth (FIGURE 1.6).

New consumers in developing countries As developing nations increase their economic growth and improve their standard of living, more and more people in those countries purchase consumer goods. By the early 2000s, more new cars were sold annually in Asia than in North America and western Europe combined. These new consumers may not consume at the high level of the average consumer in a highly developed nation, but their consumption has increasingly adverse effects on the environment. For example, air pollution from traffic in urban centers in developing countries is bad and getting worse every year. Millions of dollars are lost to health problems caused by air pollution in these cities.

Population, consumption, and environmental impact When you turn on the tap to brush your teeth in the morning, you probably do not think about where the water comes from or about the environmental consequences of removing it from a river or the ground. Likewise, most North Americans do not think about where the energy comes from when they flip on a light switch or start their car, van, or truck. Many of us do not realize that all the materials that make up the products we use every day come from Earth, nor do we grasp that these materials eventually are returned to Earth, mainly in sanitary landfills.

Such human impacts on the environment are difficult to assess. One way to estimate them is to use the three factors most important in determining environmental impact (I):

- The number of people (P).

- The affluence per person, which is a measure of the consumption, or amount of resources used per person (A).

- The environmental effects (resources needed and wastes produced) of the technologies used to obtain and consume the resources (T).

This method of assessment is usually referred to as the IPAT equation: $I = P \times A \times T$.

Ecological footprints FIGURE 1.6

India   United States

A In LDCs such as India, about 1 hectare meets the resource requirements of an average person; the ecological footprint of each individual in a highly developed country such as the United States is almost 10 hectares.

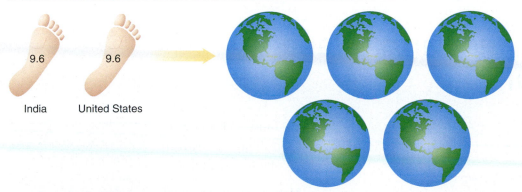

India   United States

B If everyone in the world had the same level of consumption as an average American, it would take the resources and land area of five Earths.

Biologist Paul R. Ehrlich and physicist John P. Holdren first proposed the IPAT model in the 1970s. It shows the mathematical relationship between environmental impacts and the forces that drive them. To determine the environmental impact of carbon dioxide (CO_2) emissions from motor vehicles, for example, multiply the population by the number of cars per person (affluence or consumption per person) by the average annual CO_2 emissions per year (technological impact). This model demonstrates that although improving motor vehicle efficiency and developing cleaner technologies will reduce pollution and environmental degradation, a larger reduction will result if population and per person consumption are also controlled.

Although useful, the IPAT equation must be interpreted with care, in part because we often do not understand all of the environmental impacts of a particular technology. Motor vehicles, for example, are linked not only to global warming from CO_2 emissions but also to air pollution (tailpipe exhaust), water pollution (improperly disposed motor oil and antifreeze), stratospheric ozone depletion (from leakage of air conditioner coolants), and solid waste (disposal of automobiles in

sanitary landfills). There are currently about 550 million cars on the planet, and the number is rising.

The three factors in the IPAT equation are always changing in relation to each other. For example, consumption of a particular resource may increase, but technological advances may decrease the environmental impact of the increased consumption. Consumer trends and choices also affect environmental impact. Because of such trends and uncertainties, the IPAT equation is of limited usefulness when making long-term predictions.

The IPAT equation is valuable in that it helps to identify what we don't know or understand about consumption and its environmental impact. For example, which kinds of consumption have the greatest destructive impact on the environment? Which groups in society are responsible for the greatest environmental disruption? How can we alter the activities of these environmentally disruptive groups? It will take years to address such questions, but the answers should help decision makers in business and government formulate policies that will alter consumption patterns in an environmentally responsible way. The ultimate goal should be to make consumption sustainable so that humanity's current practices do not compromise the ability of future generations to use and enjoy the riches of our planet.

To summarize, as human numbers and consumption increase worldwide, so does humanity's impact on Earth, posing new challenges to us all. Success in achieving sustainability in population size and consumption will require the cooperation of all the world's peoples.

CONCEPT CHECK STOP

How is human population growth related to natural resource depletion and environmental degradation?

What is the difference between people overpopulation and consumption overpopulation?

Sustainability and Earth's Capacity to Support Humans

LEARNING OBJECTIVES

Define environmental sustainability.

Identify human behaviors that threaten environmental sustainability.

One of the most important concepts in this text is **environmental sustainability**. *Sustainability* implies that the environment will function indefinitely without going into a decline from the stresses that human society imposes on natural systems (such as fertile soil, water, and air). Environmental sustainability applies to many levels, including the individual, communal, regional, national, and global levels.

Environmental sustainability is based in part on the following ideas:

- We must consider the effects of our actions on the health and well being of the natural environment, including all living things.

- Earth's resources are not present in infinite supply. We must live within limits that let renewable resources such as fresh water regenerate for future needs.

- We must understand all the costs to the environment and to society of products we consume.

- We must each share responsibility for environmental sustainability.

environmental sustainability The ability to meet humanity's current needs without compromising the ability of future generations to meet their needs.

Many environmental experts think that human society is not operating sustainably because of the following human behaviors (**FIGURES 1.7** and **1.8** [the National Geographic map]):

- We are using nonrenewable resources such as fossil fuels as if they were present in unlimited supplies.

- We are using renewable resources such as fresh water and forests faster than they are replenished naturally.

- We are polluting the environment—the land, rivers, ocean, and atmosphere—with toxins as if the capacity of the environment to absorb them were limitless.

- Our numbers continue to grow despite Earth's finite ability to feed us and to absorb our wastes.

If left unchecked, these activities may threaten the life-support systems of Earth to the extent that recovery is impossible. Our first goal should be to critically evaluate which changes our society is willing to make.

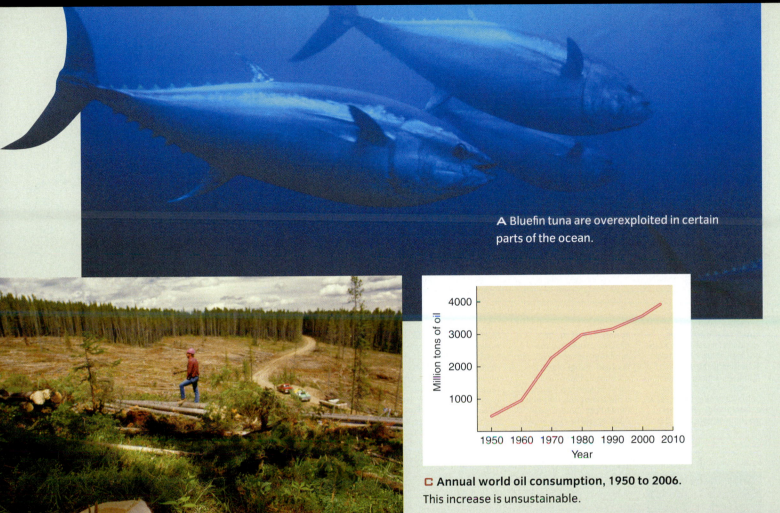

A Bluefin tuna are overexploited in certain parts of the ocean.

C Annual world oil consumption, 1950 to 2006. This increase is unsustainable.

B A clear-cut forest in the Gallatin National Forest in Montana. Forest harvest is unsustainable in many parts of the world.

D An oil refinery produces noxious emissions. Air pollution control devices reduce these toxic emissions but often are not used because they are expensive. Photographed in Baton Rouge, Louisiana.

Global environmental issues FIGURE 1.8

These issues occur locally at so many places around the planet that they are global in scope.

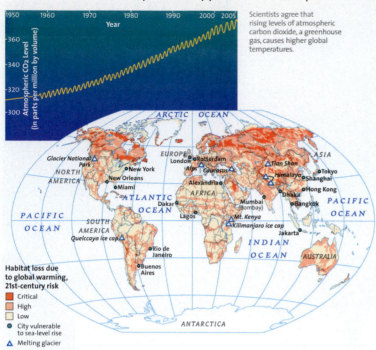

Scientists agree that rising levels of atmospheric carbon dioxide, a greenhouse gas, causes higher global temperatures.

Habitat loss due to global warming, 21st-century risk
- Critical
- High
- Low
- ● City vulnerable to sea-level rise
- △ Melting glacier

Clear-cutting large tracts of timber, without sustainable replanting, contributes to deforestation, erosion, and loss of habitat.

Vanishing forest
- Frontier forest (large, mostly virgin forest)
- Degraded forest
- Frontier forest 8,000 years ago

▲ GLOBAL WARMING

Temperatures across the world are increasing at a rate not seen at any other time in the last 10,000 years. Although climate variation is a natural phenomenon, human activities that release carbon dioxide and other greenhouse gases into the atmosphere—industrial processes, fossil fuel consumption, deforestation, and land use change—are contributing to this warming trend. Scientists predict that if this trend continues, one-third of plant and animal habitats will be dramatically altered and more than one million species will be threatened with extinction in the next 50 years. And even small increases in global temperatures can melt glaciers and polar ice sheets, raising sea levels and flooding coastal cities and towns.

▲ DEFORESTATION

Of the 13 million hectares (32 million acres) of forest lost each year, mostly to make room for agriculture, more than half are in South America and Africa, where many of the world's tropical rain forests and terrestrial plant and animal species can be found. Loss of habitat in such species-rich areas takes a toll on the world's biodiversity. Deforested areas also release, instead of absorb, carbon dioxide into the atmosphere, contributing to global climate change. Deforestation can also affect local climates by reducing evaporative cooling, leading to decreased rainfall and higher temperatures.

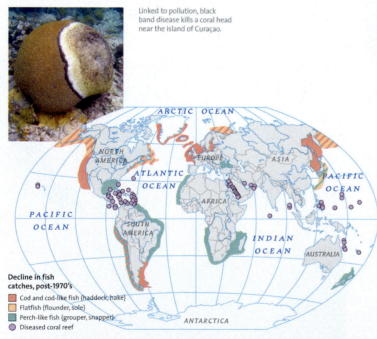

Linked to pollution, black band disease kills a coral head near the island of Curaçao.

Decline in fish catches, post-1970's
- Cod and cod-like fish (haddock, hake)
- Flatfish (flounder, sole)
- Perch-like fish (grouper, snapper)
- ● Diseased coral reef

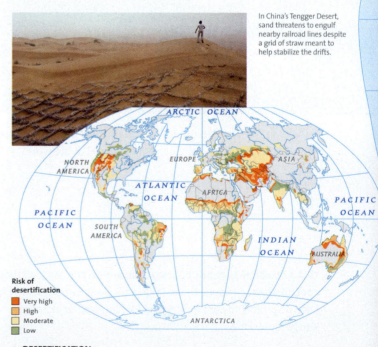

In China's Tengger Desert, sand threatens to engulf nearby railroad lines despite a grid of straw meant to help stabilize the drifts.

Risk of desertification
- Very high
- High
- Moderate
- Low

▲ THREATENED OCEANS

Oceans cover more than two-thirds of the Earth's surface and are home to at least half of the world's biodiversity, yet they are the least understood ecosystems. The combined stresses of overfishing, pollution, increased carbon dioxide emissions, global climate change, and coastal development are having a serious impact on the health of oceans and ocean species. Over 70% of the world's fish species are depleted or nearing depletion, and 50% of coral reefs worldwide are threatened by human activities.

▲ DESERTIFICATION

Climate variability and human activities, such as grazing and conversion of natural areas to agricultural use, are leading causes of desertification, the degradation of land in arid, semiarid, and dry subhumid areas. The environmental consequences of desertification are great—loss of topsoil, increased soil salinity, damaged vegetation, regional climate change, and a decline in biodiversity. Equally critical are the social consequences—more than 2 billion people live in and make a living off these dryland areas, covering about 41% of Earth's surfac

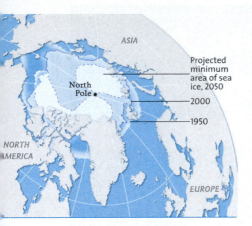

ASIA

NORTH
AMERICA

EUROPE

North
Pole

Projected
minimum
area of sea
ice, 2050

2000

1950

POLAR ICE CAP
...er the last 50 years, the
...tent of polar sea ice has
...ticeably decreased. Since
...70 alone, an area larger
...an Norway, Sweden, and

Denmark combined has
melted. This trend is
predicted to accelerate as
temperatures rise in the
Arctic and across the globe.

GeoBytes

ACIDIFYING OCEANS
Oceans are absorbing
an unprecedented 20 to 25
million tonnes (22 to 28
millions tons) of carbon
dioxide each day, increasing
the water's acidity.

ENDANGERED REEFS
Some 95% of coral reefs in
Southeast Asia have been
destroyed or are threatened.

RECORD TEMPERATURES
The 1990s were the
warmest decade on record
in the last century.

WARMING ARCTIC
While the world as a whole
has warmed nearly 0.6°C (1°F)
over the last hundred years,
parts of the Arctic have
warmed 4 to 5 times as much
in only the last 50 years.

DISAPPEARING RAIN FORESTS
Scientists predict that
the world's rain forests
will disappear within the
next one hundred years
if the current rate of
deforestation continues.

OIL POLLUTION
Nearly 1.3 million tonnes
(1.4 million tons) of oil seep
into the world's oceans each
year from the combined
sources of natural seepage,
extraction, transportation,
and consumption.

ACCIDENTAL DROWNINGS
Entanglement in fishing
gear is one of the greatest
threats to marine mammals.

A FAREWELL TO FROGS?
Worldwide, almost half of the
5,700 named amphibian
species are in decline.

ENVIRONMENT

With the growth of scientific record keeping,
observation, modeling, and analysis, our understanding of Earth's environment is improving. Yet
even as we deepen our insight into environmental
processes, we are changing what we are studying. At
no other time in history have humans altered their
environment with such speed and force. Nothing
occurs in isolation, and stress in one area has impacts
elsewhere. Our agricultural and fishing practices,
industrial processes, extraction of resources, and
transportation methods are leading to extinctions,
destroying habitats, devastating fish stocks, disturbing
the soil, and polluting the oceans and the air. As a
result, biodiversity is declining, global temperatures
are rising, polar ice is shrinking, and the ozone layer
continues to thin.

▼ POLLUTION
No corner of the earth is
immune to pollution, be it in
the air, soil, or water. Concentrations of pollution can be
found in the industrial
centers of North America,
Europe, and, increasingly,
Asia—and areas downwind
or downstream from them.
Shipping routes are sources
of pollution, from oil spills to
garbage dumpings.

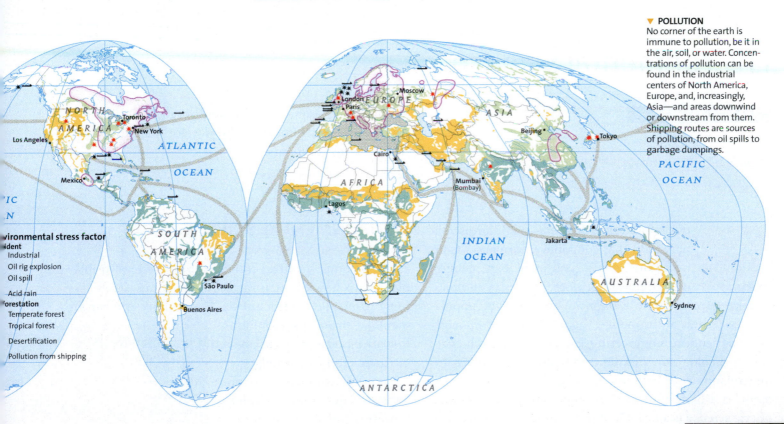

...vironmental stress factor
...ident
Industrial
Oil rig explosion
Oil spill
Acid rain
...orestation
Temperate forest
Tropical forest
Desertification
Pollution from shipping

NATIONAL
GEOGRAPHIC

...OZONE DEPLETION
...t noted in the mid-1980s,
... springtime "ozone hole"
...r the Antarctic continues
... grow. With sustained
...rts to restrict chlorofluo-
...arbons (CFCs) and other
...ne-depleting chemicals,
...ntists have begun to
... what they hope is a
...eling off in the rate of

depletion. Stratospheric
ozone shields the Earth
from the sun's ultraviolet
radiation. Thinning of this
protective layer puts people
at risk for skin cancer and
cataracts. It can also have
devastating effects on the
Earth's biological functions.

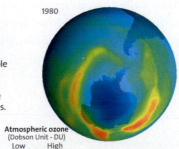

1980

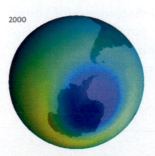

2000

2004

Atmospheric ozone
(Dobson Unit - DU)
Low   High

At first glance, the issues may seem simple. Why don't we just stop the overconsumption, population growth, and pollution? The solutions are more complex than they may initially seem, in part because of various interacting ecological, societal, and economic factors.

Our inadequate scientific understanding of how the environment works and how different human choices affect the environment is a major reason that problems of environmental sustainability are difficult to resolve. Even for established environmental problems, political and social controversy often prevents widespread acceptance that an environmental threat is real. Because the effects of many interactions between the environment and humans are unknown or difficult to predict, we generally don't know if corrective actions should be taken until our scientific understanding is more complete.

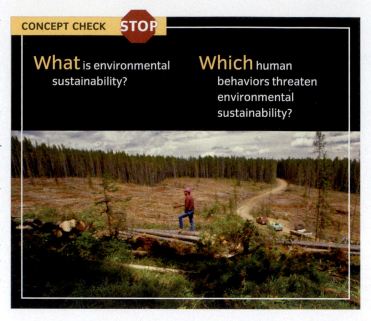

CONCEPT CHECK **STOP**

What is environmental sustainability?

Which human behaviors threaten environmental sustainability?

Environmental Science

LEARNING OBJECTIVES

Define *environmental science*.

Outline the steps of the scientific method.

Environmental science is an interdisciplinary field that combines information from many disciplines, such as biology, geography, chemistry, geology, physics, economics, sociology (particularly *demography*, the study of populations), cultural anthropology, natural resource management, agriculture, engineering, law, politics, and ethics.

Ecology, the discipline of biology that studies the interrelationships between organisms and their environment, is a basic tool of environmental science. **Atmospheric science** is a branch of environmental science that includes the study of weather and climate, greenhouse gases, and other airborne pollutants. **Environmental chemistry** examines chemicals in the environment, including soil and water pollution. **Geosciences**—for example, environmental geology and physical geography—study a wide range of environmental topics, such as soil erosion, groundwater use, ocean pollution, and climate.

Scientists in these sub-disciplines not only evaluate environmental quality but they also develop ways to restore damaged environments.

THE GOALS OF ENVIRONMENTAL SCIENCE

Environmental scientists try to establish general principles about how the natural world functions. They use these principles to develop viable solutions to environmental problems—solutions that are based as much as possible on scientific knowledge.

Environmental problems are generally complex, however, and scientific understanding of them is often less complete than we would like. Environmental scientists are often called on to reach a scientific consensus before the data are complete. As a result, they often make recommendations based on probabilities rather than on precise answers.

environmental science The interdisciplinary study of humanity's relationship with other organisms and the physical environment.

Many of the environmental problems considered in this book are serious ones that we must address. Yet environmental science is not simply a "doom and gloom" listing of problems, coupled with predictions of a bleak future. To the contrary, the focus of environmental science, and our focus as individuals and as world citizens, is on identifying, understanding, and solving problems that we as a society have generated. It is encouraging that individuals, businesses, and governments are already doing a great deal, although more must be done to address the problems of today's world.

SCIENCE AS A PROCESS

The key to successfully solving any environmental problem is rigorous scientific evaluation. It is important to understand clearly just what science is, as well as what it is not. Most people think of **science** as a body of knowledge—a collection of facts about the natural world. However, science is also a dynamic *process*, a systematic way to investigate the natural world. Science seeks to reduce the apparent complexity of our world to general principles, which are then used to make predictions, solve problems, or provide new insights.

Scientists collect objective **data** (singular, *datum*), the information with which science works. Data are collected through observation and experimentation and then analyzed or interpreted (**FIGURE 1.9**). Scientific conclusions are inferred from the available data and are not based on faith, emotion, or intuition. Scientists publish their findings in scientific journals, and other scientists examine and critique their work. A requirement of science is repeatability—that is, observations and experiments must produce consistent data when they are repeated. The scrutiny by other scientists reveals any inconsistencies in results or interpretation. These errors are discussed openly, and ways to eliminate them are developed.

There is no absolute certainty or universal agreement about anything in science. Science is an ongoing enterprise, and scientific concepts must be reevaluated in light of newly discovered data. Thus, scientists never claim to know the final answer about anything because scientific understanding changes.

Several areas of human endeavor are not scientific. Ethical principles often have a religious foundation, and political principles reflect social systems. Some general principles, however, derive not from religion or politics but from the physical world around us. If you drop an apple, it will fall, whether or not you wish it to and despite any laws you may pass forbidding it to do so. Science aims to discover and better understand the general principles that govern the operation of the natural world.

Data collection FIGURE 1.9

A botanist compares a cotton leaf from an experimental plant (*left*) and a control plant (*right*). The two plants were raised under identical conditions, except that the experimental plant was genetically engineered to resist insect pests.

The scientific method The established processes that scientists use to answer questions or solve problems are collectively called the **scientific method**. Although there are many variations of the scientific method, it basically involves five steps:

1. Recognize a question or an unexplained occurrence in the natural world.

2. Develop a *hypothesis*, or an educated guess, to explain the problem.

3. Design and perform an experiment to test the hypothesis.

4. Analyze and interpret the data to reach a conclusion.

5. Share new knowledge with the scientific community.

Although the scientific method is often portrayed as a linear sequence of events, science is rarely as straightforward or tidy as the scientific method implies (see What a Scientist Sees). Good science involves creativity, not only in recognizing questions and developing hypotheses but also in designing experiments. Because scientists try to expand our current knowledge, their work is in the realm of the unknown. Many creative ideas end up as dead ends, and there are often temporary setbacks or reversals of direction as scientific knowledge progresses. Scientific knowledge often expands haphazardly, with the "big picture" emerging slowly from confusing and sometimes contradictory details.

Scientific discoveries are often incorrectly portrayed in the media as "new facts" that have just come to light. At a later time, additional "new facts" that question the validity of the original study are reported. If you were to read the scientific papers on which such media reports are based, however, you would find that all the scientists involved probably made very tentative conclusions based on their data. Science progresses from uncertainty to less uncertainty, not from certainty to greater certainty. Thus, science is self-correcting over time, despite the fact that it never "proves" anything.

> **scientific method**
> The way a scientist approaches a problem, by formulating a hypothesis and then testing it.

The importance of prediction A scientific **hypothesis** needs to be useful—it needs to tell you something you want to know. A hypothesis is most useful when it makes predictions because the predictions provide a way to test the validity of the hypothesis. If your experiment refutes your prediction, then you must carefully recheck the entire experiment. If the prediction is still refuted, then you must reject the hypothesis. The more verifiable predictions a hypothesis makes, the more valid that hypothesis is.

Each of the many factors that influence a process is called a **variable**. To evaluate alternative hypotheses about a specific variable, it is necessary to hold all other variables constant so that they are not misleading or confusing. To test a hypothesis about a variable, we carry out two forms of the experiment in parallel. In the **experimental group**, the chosen variable is altered in a known way. In the **control group**, that variable isn't altered. In all other respects, the experimental group and the control group are the same. We then ask, "What is the difference, if any, between the outcomes for the two groups?" Any difference must be due to the influence of the variable we changed because all other variables remained the same. Much of the challenge of science lies in designing control groups and in successfully isolating a single variable from all other variables.

Theories A **theory** is an integrated explanation of numerous hypotheses, each of which is supported by a large body of observations and experiments. A theory condenses and simplifies many data that previously appeared to be unrelated. Because a theory demonstrates the relationships among different data, it simplifies and clarifies our understanding of the natural world. A good theory grows as additional information becomes known. It predicts new data and suggests new relationships among a range of natural phenomena.

Theories are the solid ground of science, the explanations of which we are most sure. This definition contrasts sharply with the general public's use of the word *theory*, which implies lack of knowledge or a guess. In this book, the word *theory* is always used in its scientific sense, to refer to a broadly conceived, logically coherent, and well-supported explanation.

The Scientific Method

1. Recognize problem or unanswered question.

2. Develop hypothesis to explain problem.

Make predictions based on hypothesis.

3. Design and perform experiment to test hypothesis.

No. Reject/revise hypothesis and start again.

Yes. Keep testing to verify hypothesis.

4. Analyze and interpret data to reach conclusions.

Does hypothesis predict reality?

New knowledge results in new questions.

Other scientists test hypothesis, often in ways different from original experiment.

5. Share new knowledge with other scientists.

B These five steps provide the framework for all scientific investigations.

A A field scientist makes observations critical to understanding damage to coral reefs from global climate change. Photographed at Turneffe Atoll, Belize.

C Many scientists present their research during poster sessions at scientific meetings. This allows their work to be critically assessed by others in the scientific community.

Despite the fact that theories are generally accepted, there is no absolute truth in science, only varying degrees of uncertainty. Science is continually evolving as new evidence comes to light, and therefore its conclusions are always provisional or uncertain. It is always possible that the results of a future experiment will contradict a prevailing theory and show at least one aspect of it to be false.

Uncertainty, however, does not mean that scientific conclusions are invalid. For example, overwhelming evidence links cigarette smoking and incidence of lung cancer. We can't state with absolute certainty that every smoker will be diagnosed with lung cancer, but this uncertainty does not mean that there is no correlation between smoking and lung cancer. On the basis of the available evidence, we say that people who smoke have an increased risk of developing lung cancer.

In conclusion, the aim of science is to increase human comprehension by explaining the processes and events of nature. Scientists work under the assumption that all phenomena in the natural world have natural causes, and they formulate theories to explain these phenomena. The process of science as a human endeavor has shaped the world we live in and transformed our views of the universe and how it works.

CONCEPT CHECK STOP

What is environmental science? What are some of the disciplines involved in environmental science?

What are the five steps of the scientific method? Why is each important?

How We Handle Environmental Problems

LEARNING OBJECTIVE

List and briefly describe the five stages of solving environmental problems.

Before examining the environmental problems discussed in the remaining chapters of this book, let's consider the elements that contribute to solving those problems. How, for example, can we handle water pollution in a river (**FIGURE 1.10**)? At what point are conclusions regarded as certain enough to warrant action? Who makes the decisions, and what are the trade-offs? Viewed

Monitoring water pollution FIGURE 1.10

This pollution control officer is measuring the oxygen level in the Severn River near Shrewsbury, England. When dissolved oxygen levels are high, pollution levels (of sewage, fertilizer, and such) are low.

Addressing environmental problems FIGURE 1.11

These five steps provide a framework for addressing environmental problems.

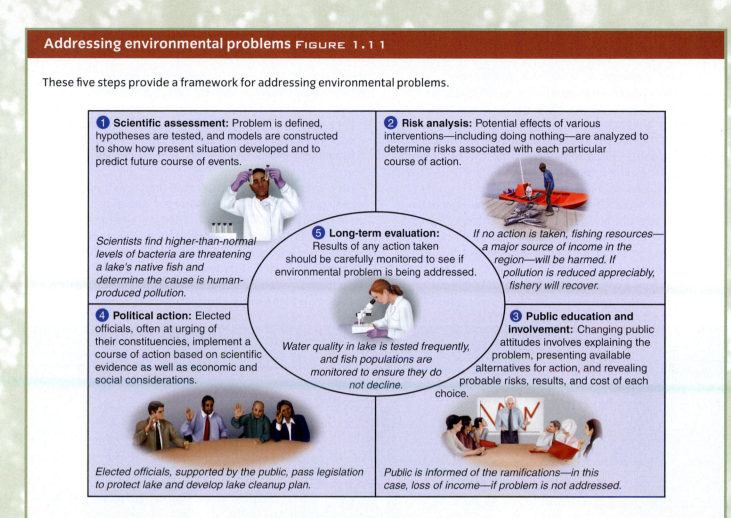

1 Scientific assessment: Problem is defined, hypotheses are tested, and models are constructed to show how present situation developed and to predict future course of events.

Scientists find higher-than-normal levels of bacteria are threatening a lake's native fish and determine the cause is human-produced pollution.

2 Risk analysis: Potential effects of various interventions—including doing nothing—are analyzed to determine risks associated with each particular course of action.

If no action is taken, fishing resources—a major source of income in the region—will be harmed. If pollution is reduced appreciably, fishery will recover.

5 Long-term evaluation: Results of any action taken should be carefully monitored to see if environmental problem is being addressed.

Water quality in lake is tested frequently, and fish populations are monitored to ensure they do not decline.

4 Political action: Elected officials, often at urging of their constituencies, implement a course of action based on scientific evidence as well as economic and social considerations.

Elected officials, supported by the public, pass legislation to protect lake and develop lake cleanup plan.

3 Public education and involvement: Changing public attitudes involves explaining the problem, presenting available alternatives for action, and revealing probable risks, results, and cost of each choice.

Public is informed of the ramifications—in this case, loss of income—if problem is not addressed.

---

simply, there are five stages in addressing an environmental problem (**FIGURE 1.11**):

1. Scientific assessment
2. Risk analysis
3. Public education and involvement
4. Political action
5. Long-term evaluation

These five stages represent an ideal approach to systematically addressing environmental problems. In real life, seeking solutions to environmental problems is rarely so neat and tidy, particularly when the problem is exceedingly complex, of regional or global scale, or has high costs and unclear benefits for the money invested. Quite often, the public becomes aware of a problem, which triggers discussion of remediation before the problem is clearly identified and scientifically assessed.

CONCEPT CHECK STOP

What are the five steps used to solve an environmental problem?

The NIMBY response is an unavoidable complication in addressing many environmental issues. NIMBY stands for "not in my backyard." As soon as people hear that a power plant, an incinerator, or a hazardous waste disposal site may be situated nearby, the NIMBY response rears its head. Part of the reason NIMBYism is so prevalent is that, despite the assurances experts give that a site will be safe, no one can guarantee complete safety and no possibility of an accident.

A sister response to NIMBY is NIMTOO, which stands for "not in my term of office." Politicians who wish to get reelected are sensitive to their constituents' concerns and are not likely to support the construction of power plants or waste disposal sites in their districts.

Given human nature, the NIMBY and NIMTOO responses are not surprising. But emotional reactions, however reasonable they may be, do little to constructively solve complex environmental problems. Consider the disposal of radioactive waste from nuclear power plants. There is universal agreement that we need to safely isolate radioactive waste until it decays enough to cause little danger. But NIMBY and NIMTOO, with their associated demonstrations, lawsuits, and administrative hearings, prevent us from effectively dealing with radioactive waste disposal. Every potential disposal site is near someone's home, in some politician's state. Most people agree that our generation has the responsibility to dispose of hazardous waste. Only we want to put it in some other state, in someone else's backyard. Arguing against any disposal scheme that is proposed will simply result in letting the waste remain where it is now. Although this may be the only politically acceptable solution, it is unacceptable from a safety viewpoint.

Not in my backyard

Steam rises from two of the cooling towers of a nuclear power plant. Many nuclear power plants store highly radioactive spent fuel on site because there is currently no place to safely dispose of it.

THE NEW ORLEANS DISASTER

Hurricane Katrina, which hit the north-central Gulf Coast in August 2005, was one of the most devastating storms in U.S. history. It produced a storm surge that caused severe damage to New Orleans as well as to other coastal cities and towns in the region. The high waters caused levees and canals to fail, flooding 80 percent of New Orleans and many nearby neighborhoods (see photograph).

Most people are aware of the catastrophic loss of life and property caused by Katrina. Here we focus on how humans altered the geography and geology of the New Orleans area in ways that exacerbated the storm damage. This discussion is applicable to many of the world's cities located on river deltas; these cities are also vulnerable to storms and flooding.

The Mississippi River Delta formed over millennia from sediments deposited at the mouth of the river. The location of New Orleans is ideal for industry and for sea and river commerce, but the city's development has disrupted the delta-building processes.

Over the years, engineers have constructed a system of canals to aid navigation and a system of levees to control flooding because the city is at or below sea level. The canals allowed salt water to intrude and kill the freshwater marsh vegetation. The levees prevented the deposition of sediments that remain behind after floodwaters subside (The sediments are now deposited in the Gulf of Mexico.) Under natural conditions, these sediments replenish and maintain the delta, building up coastal wetlands.

As the city has grown, new development has taken place on wetlands—bayous, waterways, and marshes—that were drained and filled in. Before their destruction, these coastal wetlands provided some protection against flooding from storm surges. We are not implying that had Louisiana's wetlands been intact. New Orleans would not have suffered any damage from a hurricane of Katrina's magnitude. However, had these wetlands been largely unaltered, they would have moderated the storm's damage by absorbing much of the water from the storm surge.

Another reason that Katrina devastated New Orleans is that the city has been subsiding (sinking) for many years, primarily because New Orleans is built on unconsolidated sediment (no bedrock underneath). Many wetlands scientists also attribute this subsidence to the extraction of the area's rich supply of underground natural resources—groundwater, oil, and natural gas. As these resources are removed, the land compacts, lowering the city. New Orleans and nearby coastal areas are subsiding an average of 11 mm each year. At the same time, the sea level has been rising an average of 1 mm to 2.5 mm per year due to human-induced changes in climate.

New Orleans in the aftermath of Hurricane Katrina
This photograph was taken about two weeks after the storm. The red object in the foreground is a barge.

SUMMARY

1 Human Impacts on the Environment

1. **Highly developed countries** are countries that have complex industrialized bases, low rates of population growth, and high per person incomes. **Moderately developed countries** are developing countries that have medium levels of industrialization and average per person incomes lower than those of highly developed countries. **Less developed countries (LDCs)** are developing countries with low levels of industrialization, very high rates of population growth, very high infant mortality rates, and very low per person incomes (relative to highly developed countries). **Poverty**, which is common in LDCs, is a condition in which people are unable to meet their basic needs for food, clothing, shelter, education, or health.

2. The increasing global population is placing stresses on the environment, as humans consume ever-increasing quantities of food and water, use more energy and raw materials, and produce enormous amounts of waste and pollution. **Nonrenewable resources** are natural resources that are present in limited supplies and are depleted as they are used. **Renewable resources** are resources that natural processes replace and that therefore can be used forever, provided that they are not exploited in the short term.

3. **People overpopulation** is a situation in which too many people live in a given geographic area. Developing countries have people overpopulation. **Consumption overpopulation** is a situation that occurs when each individual in a population consumes too large a share of resources. Highly developed countries have consumption overpopulation.

4. Environmental impact and the forces that drive it can be modeled by the IPAT equation, $I = P \times A \times T$. Environmental impact (I) has three factors: the number of people (P); the affluence per person (A), which is a measure of the consumption, or amount of resources used per person; and the environmental effect of the technologies used to obtain and consume those resources (T).

2 Sustainability and Earth's Capacity to Support Humans

1. **Environmental sustainability** is the ability to meet humanity's current needs without compromising the ability of future generations to meet their needs. Sustainability implies that the environment can function indefinitely without going into a decline from the stresses that human society imposes on natural systems.

2. Human behaviors that threaten environmental sustainability include overuse of renewable and nonrenewable resources, pollution, and overpopulation.

3 Environmental Science

1. **Environmental science** is the interdisciplinary study of humanity's relationship with other organisms and the nonliving physical environment. Environmental science encompasses many problems involving human numbers, Earth's natural resources, and environmental pollution.

2. The **scientific method** is the way a scientist approaches a problem, by formulating a **hypothesis** and then testing it by means of an experiment. (1) A scientist recognizes and states the problem or unanswered question. (2) The scientist develops a hypothesis, or an educated guess, to explain the problem. (3) An experiment is designed and performed to test the hypothesis. (4) **Data**, the results obtained from the experiment, are analyzed and interpreted to reach a conclusion. (5) The conclusion is shared with the scientific community.

4 How We Handle Environmental Problems

1. Addressing environmental problems ideally requires five stages. (1) Scientific assessment involves identifying a potential environmental problem and collecting data to construct a model. (2) Risk analysis evaluates the potential effects of intervention. (3) Public education and involvement occur when the results of scientific assessment and risk analysis are placed in the public arena. (4) In political action, elected or appointed officials implement a particular risk-management strategy. (5) Evaluation monitors the effects of the action that was taken.

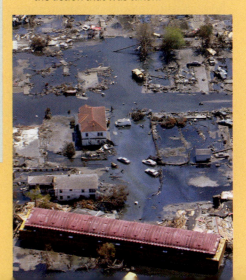

CRITICAL AND CREATIVE THINKING QUESTIONS

1. Criticize the following statement: "Population growth in developing countries is of much more concern than is population growth in highly developed countries."

2. Give at least two examples of things that you can do as an individual to promote environmental sustainability.

3. People want scientists to give them precise, definitive answers to environmental problems. Explain why this is not possible.

4. Write a brief paragraph describing two or three ways your personal quality of life could be improved. Do any of these involve material items? If so, explain how these items would improve your life.

5–7. Your throat feels scratchy, and you think you're coming down with a cold. You take a couple of vitamin C pills and feel better. You conclude that vitamin C helps prevent colds.

5. Is your conclusion valid from a scientific standpoint? Why or why not?

6. Develop a testable hypothesis to answer the question "Does taking vitamin C pills help prevent colds?"

7. Describe how you would conduct a controlled experiment to test your hypothesis.

8–10. Examine the graph, which shows an estimate of the discrepancy between the wealth of the world's poorest countries and that of the richest countries.

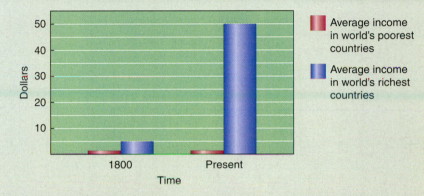

Average income in world's poorest countries

Average income in world's richest countries

8. How has the distribution of wealth changed from the 1800s to the present? How would you explain this difference?

9. Based on the trend evident in this graph, predict what the graph might look like in 100 years.

10. Some economists think that our current path of economic growth is unsustainable. Do the data in this graph support or refute this idea? Explain your answer.

What is happening in this picture ?

- What are these people protesting?

- How is this opposition an example of NIMBY?

- What kinds of incentives might make these people more willing to consider locating a nuclear waste site near them?

THE PEOPLE OF NEW HAMPSHIRE
AGAINST THE NUCLEAR DUMP

Environmental Sustainability and Human Values

2

THE GLOBAL COMMONS

Ecologist Garrett Hardin (1915–2003) is best known for his 1968 essay "The Tragedy of the Commons." In it he contended that our inability to solve complex environmental problems is the result of a struggle between short-term individual welfare and long-term *environmental sustainability* (the ability to meet humanity's current needs without compromising the ability of future generations to meet their needs).

Hardin used the commons to illustrate this struggle. In medieval Europe, the inhabitants of a village shared pastureland, called the commons, and each herder could bring animals onto the commons to graze. The more animals a herder brought onto the commons, the greater the advantage to that individual. When every herder in the village brought as many animals onto the commons as possible, however, the plants were killed from overgrazing.

In today's world, Hardin's parable has particular relevance at the global level. The commons are those parts of our environment that are available to everyone but for which no single individual has responsibility: the atmosphere, water, wildlife, fisheries, and forests (see photograph). These modern-day commons, sometimes collectively called the *global commons*, are experiencing increasing environmental stress. The world needs effective legal and economic policies to prevent the degradation of our global commons. Clearly, all people, businesses, and governments must foster a strong sense of *stewardship*, or shared responsibility, for the sustainable care of our planet.

This chapter examines the role of ethics and values in environmental issues. As you read, keep in mind these words from the *Earth Charter*, formulated in 1992 by representatives from 178 countries: "Let ours be a time remembered for the awakening of a new reverence for life, the firm resolve to achieve sustainability, the quickening of the struggle for justice and peace, and the joyful celebration of life."

NATIONAL GEOGRAPHIC

Human Use of the Earth

Environmental sustainability is a concept that people have discussed for many years. *Our Common Future,* the 1987 report of the U.N. World Commission on Environment and Development, presented the closely related concept of **sustainable development** (FIGURE 2.1). The authors of *Our Common Future* pointed out that sustainable development includes meeting the needs of the world's poor. The report also linked the environment's ability to meet present and future needs to the state of technology and social organization existing at a given time and in a given place. The number of people, their degree of affluence (that is, their level of consumption), and their choices of technology all interact to produce the total effect of a given society, or of society at large, on the sustainability of the environment.

Even using the best technologies imaginable, Earth's productivity still has its limits, and our use of it can't be expanded indefinitely. *Sustainable development can occur only within the limits of the environment.* To live within these limits, population growth must be held at a level that we can sustain, and the wealthy must first stabilize their use of natural resources and then reduce this use to a level that can be maintained. The world does not contain nearly enough resources to sustain everyone at the level of consumption that is enjoyed in the United States, Europe, and Japan. Suitable strategies, however, do exist to reduce these levels of consumption without concurrently reducing the real quality of life.

sustainable development Economic growth that meets the needs of the present without compromising the ability of future generations to meet their needs.

Sustainable development FIGURE 2.1

Three factors—environmentally sound decisions, economically viable decisions, and socially equitable decisions—interact to promote sustainable development.

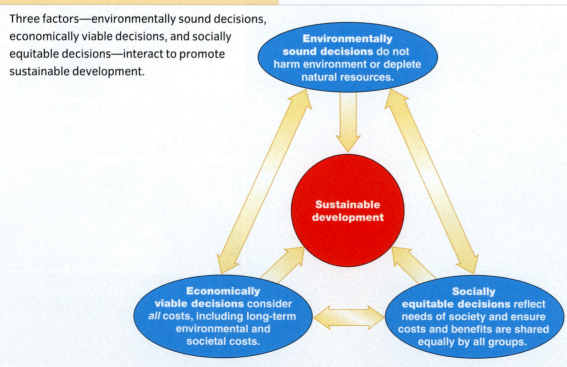

Environmentally sound decisions do not harm environment or deplete natural resources.

Sustainable development

Economically viable decisions consider *all* costs, including long-term environmental and societal costs.

Socially equitable decisions reflect needs of society and ensure costs and benefits are shared equally by all groups.

The challenge of eradicating poverty in developing countries FIGURE 2.2

Global Locator

INDIA

A Squatters live in dilapidated shacks under a water pipe in Delhi, India. Leaks in the pipe provide water for these people.

B Desperately poor pavement people eat, sleep, and raise their families on a street in Calcutta, India.

SUSTAINABLE CONSUMPTION

As you saw in Chapter 1, **consumption overpopulation** is pollution and degradation of the environment that occurs when each individual in a population consumes too large a share of resources. Consumption overpopulation stems from the lifestyles of people living in highly developed nations. *Lifestyle* is interpreted broadly to include goods and services bought for food, clothing, housing, travel, recreation, and entertainment. In evaluating consumption overpopulation, all aspects of the production, use, and disposal of these goods and services are taken into account, including environmental costs. Such an analysis provides a sense of what it means to consume sustainably versus unsustainably.

Sustainable consumption, like sustainable development, forces us to address whether our present actions undermine the long-term ability of the environment to meet

> **sustainable consumption** The use of goods and services that satisfy basic human needs and improve the quality of life but that also minimize resource use.

the needs of future generations. Factors that affect sustainable consumption include population, economic activities, technology choices, social values, and government policies.

At the global level, sustainable consumption requires the eradication of poverty, which in turn requires that poor people in developing countries *increase* their consumption of certain essential resources (FIGURE 2.2). For their increased consumption to be sustainable, however, the consumption patterns of people in highly developed countries must change.

Widespread adoption of sustainable consumption will not be easy. It will require major changes in the consumption patterns and lifestyles of most people in highly developed countries. Some examples of promoting sustainable consumption include switching from motor vehicles to public transport and bicycles and developing durable, repairable, recyclable products.

An increasing number of people in the United States and other highly developed nations have embraced a type of sustainable consumption known as **voluntary simplicity**, which recognizes that individual happiness and quality of life are not necessarily linked to the accumulation of material goods. People who embrace voluntary simplicity recognize that a person's values and character define that individual more than how many things he or she owns. This belief requires a change in behavior as people purchase and use fewer items than they might have formerly. It is a commitment at the individual level to saving the planet for future generations.

One example of voluntary simplicity is car sharing. Car-sharing programs, which are designed for people who use a car occasionally, offer an economical alternative to individual ownership (**FIGURE 2.3**). Car sharing may also reduce the numbers of cars manufactured. Studies show that most car sharers drive significantly fewer miles than they did before they joined the program.

Sustainable consumption and voluntary simplicity are not popular ideas with policymakers, politicians, and economists, or even with consumers such as you and me. They contains an inherent threat to "business as usual," but they also offer new and exciting opportunities. Many scientists and population experts increasingly advocate adoption of sustainable consumption before it is forced on us by an environmentally degraded, resource-depleted world.

As people adopt new lifestyles, they must be educated so that they understand the reasons for changing practices that may be highly ingrained or traditional. Formal education and informal education are both important in bringing about change and in contributing to sustainable consumption. If people understand the way the natural world functions, they can appreciate their own place in it and value sustainable actions.

Any long-term involvement in the condition of the world must start with individuals—our values, attitudes, and practices. Each of us makes a difference, and it is ultimately our collective activities that make the world what it is.

CONCEPT CHECK STOP

What is sustainable development?

What is sustainable consumption? How is it linked to a reduction in world poverty?

Human Values and Environmental Problems

We now shift our attention to the views of different individuals and societies and how those views affect our ability to understand and solve sustainability problems. **Ethics** is the branch of philosophy that is derived through the logical application of human **values**. These values are the principles that an individual or a society considers important or worthwhile. Values are not static entities but change as societal, cultural, political, and economic priorities change. Ethics helps us determine which forms of conduct are morally acceptable and unacceptable, right and wrong. Ethics plays a role in any types of human activities that involve intelligent judgment and voluntary action. Whenever alternative, conflicting values occur, ethics helps us choose which value is better, or worthier, than other values.

Environmental ethics examines moral values to determine how humans should relate to the natural environment. Environmental ethicists consider such issues as what role we should play in determining the fate of Earth's resources, including other species, or how we might develop an environmental ethic that is acceptable in the short term for us as individuals and also in the long term for our species and the planet. These issues and others like them are difficult intellectual questions that involve political, economic, societal, and individual trade-offs.

environmental ethics A field of applied ethics that considers the moral basis of environmental responsibility.

Environmental ethics considers not only the rights of people living today, both individually and collectively, but also the rights of future generations (**FIGURE 2.4**). This aspect of environmental ethics is critical because the impacts of today's activities and technologies are changing the environment. In some cases these impacts may be felt for hundreds or even thousands of years. Addressing issues of environmental ethics puts us in a better position to use science and technology for long-term environmental sustainability.

Tomorrow's generation FIGURE 2.4

The choices made today will determine whether future generations, such as these students from Bailey Elementary School in Falls Church, Virginia, will inherit a sustainable world.

WORLDVIEWS

Each of us has a particular **worldview**—that is, a commonly shared perspective based on a collection of our basic values that helps us make sense of the world, understand our place and purpose in it, and determine right and wrong behaviors. These worldviews lead to behaviors and lifestyles that may or may not be compatible with environmental sustainability.

Two extreme, competing **environmental worldviews** are the Western worldview and the deep ecology worldview. These two worldviews, admittedly broad generalizations, are at nearly opposite ends of a spectrum of worldviews relevant to global sustainability problems, and each

> **environmental worldview** A worldview based on how the environment works, our place in the environment, and right and wrong environmental behaviors.

> **Western worldview** A worldview based on human superiority over nature, the unrestricted use of natural resources, and economic growth to manage an expanding industrial base.

approaches environmental responsibility in a radically different way.

The traditional **Western worldview**, also known as the *expansionist worldview*, is human centered and utilitarian. It mirrors the beliefs of the 19th-century **frontier attitude**, a desire to conquer and exploit nature as quickly as possible (**FIGURE 2.5**). The Western worldview also advocates the inherent rights of individuals, accumulation of wealth, and unlimited consumption of goods and services to provide material comforts. According to the Western worldview, humans have a primary obligation to humans and are therefore responsible for managing natural resources to benefit human society. Thus, any concerns about the environment are derived from human interests.

Western worldview FIGURE 2.5

A Logging operations in 1884. This huge logjam occurred on the St. Croix River near Taylors Falls, Minnesota.

Global Locator

B The Western worldview in action today. These logs were obtained illegally from a protected Indonesian forest.

The **deep ecology worldview** is a diverse set of viewpoints that dates from the 1970s and is based on the work of Arne Naess, a Norwegian philosopher, and others, including ecologist Bill Devall and philosopher George Sessions. The principles of deep ecology, as expressed by Naess in *Ecology, Community and Lifestyle* (1989), include:

1. Both human and nonhuman life has intrinsic value. The value of nonhuman life forms is independent of the usefulness they may have for narrow human purposes.

2. Richness and diversity of life forms contribute to the flourishing of human and nonhuman life on Earth (**FIGURE 2.6**).

3. Humans have no right to reduce this richness and diversity except to satisfy vital needs.

4. Present human interference with the nonhuman world is excessive, and the situation is rapidly worsening.

5. The flourishing of human life and cultures is compatible with a substantial decrease in the human population. The flourishing of nonhuman life requires such a decrease.

6. Significant change of life conditions for the better requires changes in economic, technological, and ideological structures.

7. The ideological change is mainly that of appreciating life quality rather than adhering to a high standard of living.

8. Those who subscribe to the foregoing points have an obligation to participate in the attempt to implement the necessary changes.

Preservation of biological diversity is an important part of the deep ecology worldview FIGURE 2.6

A Western hemlock trees, ferns, mosses, and other plants in Olympic National Park, Washington.

B Sea star perched on a hard coral in shallow water along the coast of Vanuatu in the South Pacific Ocean.

C An African black leopard, also known as a black panther, in a tree. Leopards often hide their food in trees to avoid sharing it with lions and hyenas.

Compared to the Western worldview, the deep ecology worldview represents a radical shift in how humans relate themselves to the environment. The deep ecology worldview stresses that all forms of life have the right to exist and that humans are not different or separate from other organisms. Humans have an obligation to themselves and to the environment. The deep ecology worldview advocates sharply curbing human population growth. It does not advocate returning to a society free of today's technological advances but instead proposes a significant rethinking of our use of current technologies and alternatives. It asks individuals and societies to share an inner spirituality connected to the natural world.

Most people today do not fully embrace either the Western worldview or the deep ecology worldview. The Western worldview is **anthropocentric** and emphasizes the importance of humans as the overriding concern in the grand scheme of things. In contrast, the deep ecology worldview is **biocentric** and views humans as one species among others. The planet's natural resources could not support its more than 6.6 billion humans if each consumed the high level of goods and services sanctioned by the Western worldview. On the other hand, the world as envisioned by the deep ecology worldview could support only a fraction of the existing human population (**FIGURE 2.7**).

These worldviews, while not practical for widespread adoption, are useful to keep in mind as you examine various environmental issues in later chapters. In the meantime, you should think about your own worldview and discuss it with others—whose worldviews will probably be different from your own. Thinking leads to actions, and actions lead to consequences. What are the short-term and long-term consequences of your particular worldview? We must develop and incorporate into our culture a long-lasting, environmentally sensitive worldview if the environment is to be sustainable for us, for other living organisms, and for future generations.

Embracing deep ecology FIGURE 2.7

At one time or another, most of us yearn for the simpler life that the tenets of deep ecology advocate. However, there are far too many people and far too little land for us all to embrace this lifestyle. Photographed on Gotland Island, Sweden.

CONCEPT CHECK STOP

What is environmental ethics?

What assumptions are made in the Western worldview?

What assumptions are made in the deep ecology worldview?

Environmental Justice

In the early 1970s, the Board of National Ministries of the American Baptist Churches coined the term *eco-justice* to link social and environmental ethics. At the local level, eco-justice encompasses environmental inequities faced by low-income minority communities. Many studies indicate that low-income communities and/or communities of color are more likely than others to have chemical plants, hazardous waste facilities, sanitary landfills, sewage treatment plants, and incinerators (**FIGURE 2.8**). A 1990 study at Clark Atlanta University, for example, found that six of Houston's eight incinerators were located in predominantly black neighborhoods. Such communities often have limited involvement in the political process and may not even be aware of their exposure to increased levels of pollutants.

Because people in low-income communities frequently lack access to sufficient health care, they may not be treated adequately for exposure to environmental contaminants. The high incidence of asthma in many minority communities may be caused or exacerbated by exposure to environmental pollutants. Few studies, except those documenting lead contamination, have examined how environmental pollutants interact with other socioeconomic factors to cause health problems. Currently, we have little scientific evidence showing to what extent a polluted environment is responsible for the disproportionate health problems of poor and minority communities. For example, a 1997 study of residents in San Francisco's polluted Bayview Hunters Point area found that hospitalization rates for chronic illnesses were the highest in the state. Researchers failed, however, to link these illnesses to an increased exposure to toxic pollutants.

In addition to their increased exposure to pollution, low-income communities may not receive equal benefits from federal cleanup programs. A 1992 paper published in the *National Law Journal* reported that toxic waste sites in white communities were cleaned up faster and better than those in minority communities.

A children's playground overlooks a pulp mill FIGURE 2.8

Poor minority neighborhoods often have the most polluted and degraded environments. Photographed in Kingsport, Tennessee.

ENVIRONMENTAL JUSTICE AND ETHICAL ISSUES

There is an increasing awareness that environmental decisions such as where to locate a hazardous waste landfill have important ethical dimensions. The most basic ethical dilemma centers on the rights of the poor and disenfranchised versus the rights of the rich and powerful. Whose rights should have priority in these decisions? The challenge is to find and adopt solutions that respect all social groups, including those yet to be born. **Environmental justice** is a fundamental human right in an ethical society. Although we may never completely eliminate past environmental injustices, we have a moral imperative to prevent them today so that their nega-

> **environmental justice** The right of every citizen to adequate protection from environmental hazards.

tive effects do not disproportionately affect any particular segment of society.

In response to these concerns, a growing environmental justice movement has emerged at the grassroots level as a strong motivator for change. Advocates are calling for special efforts to clean up hazardous sites in low-income neighborhoods, from inner-city streets to Native American reservations. Many advocates cite the need for more research on human diseases that environmental pollutants may influence.

CONCEPT CHECK **STOP**

What is environmental justice?

Which communities are exposed to a disproportionate share of environmental hazards?

An Overall Plan for Sustainable Living

LEARNING OBJECTIVES

Relate poverty and population growth to carrying capacity and global sustainability.

Discuss problems related to loss of forests and declining biological diversity.

Describe the extent of food insecurity.

Define *enhanced greenhouse effect* and explain how stabilizing climate is related to energy use.

Describe at least two problems in cities in the developing world.

There is no shortage of suggestions for ways to address the world's many environmental problems. We have organized this section around the five recommendations for sustainable living presented in the 2006 book *Plan B 2.0: Rescuing a Planet Under Stress and a Civilization in Trouble* by Lester R. Brown. If we as individuals and collectively as governments were to focus our efforts and financial support on Brown's plan, we think the quality of human life would be much improved. Brown's five recommendations for sustainable living are:

1. Eliminate poverty and stabilize the human population.

2. Protect and restore Earth's resources.

3. Provide adequate food for all people.

4. Mitigate climate change.

5. Design sustainable cities.

Seriously addressing these recommendations offers hope for the kind of future we want for our children and grandchildren (**FIGURE 2.9**).

◀ **Family planning in Egypt** Women at a health clinic learn about family planning and birth control.

RECOMMENDATION 1: ELIMINATE POVERTY AND STABILIZE THE HUMAN POPULATION.

Young trees in ▶ Scotland These evergreens are being cultivated as part of a reforestation project on land unsuitable for growing crops.

RECOMMENDATION 2: PROTECT AND RESTORE EARTH'S RESOURCES.

▼ **Apartment buildings with solar panels in the modern city Orot, Israel.**

RECOMMENDATION 4: MITIGATE CLIMATE CHANGE.

VIEW THIS IN ACTION in your WileyPLUS course

Fish breeding pens off the coast of Norway. ▶ Fish farming is expanding at an unprecedented scale, in part because wild fisheries are being overfished.

RECOMMENDATION 3: PROVIDE ADEQUATE FOOD FOR ALL PEOPLE.

▲ **Bicycles at a train station in Amsterdam.** Each resident in the Netherlands rides a bicycle an average of 573 mi (917 km) per year.

RECOMMENDATION 5: DESIGN SUSTAINABLE CITIES.

RECOMMENDATION 1: ELIMINATE POVERTY AND STABILIZE THE HUMAN POPULATION

The ultimate goal of economic development is to make it possible for humans throughout the world to enjoy long, healthy lives. A serious complication lies in the fact that the distribution of the world's resources is unequal. We who live in the United States are collectively the wealthiest people who have ever existed, with the highest standard of living (shared with a few other rich countries). The United States, with fewer than 5 percent of the world's people, controls about 25 percent of the world's economy but depends on other nations for this prosperity. Yet we often seem unaware of this relationship and tend to underestimate our effects on the environment that supports us.

Failing to confront the problem of poverty around the world makes it impossible to attain global sustainability (**FIGURE 2.10**). For example, most people would find it unacceptable that about 29,000 infants and children under the age of five die each day (2006 data from U.N. Children's Fund). Most of these deaths could have been prevented through access to adequate food and basic medical techniques and supplies. For us to allow so many to go hungry and to live in poverty threatens the global ecosystem that sustains us all. Every-

one must have a reasonable share of Earth's productivity. As U.S. President Franklin Delano Roosevelt said in his second inaugural address in 1937, "The test of our progress is not whether we add more to the abundance of those who have much; it is whether we provide enough for those who have too little."

Raising the standard of living for poor countries requires the universal education of children and the elimination of illiteracy. Improving the status of women is crucial because women are often disproportionately disadvantaged in poor countries. In many developing countries, women have few rights and little legal ability to protect their property, their rights to their children, and their income.

We have entered an era of global trade, within which we must establish guidelines for national, corporate, and individual behaviors. For example, the flow of money from developing countries to highly developed countries has exceeded the flow in the other direction for many years. Former West German Chancellor Willy Brandt termed this phenomenon "a blood transfusion from the sick to the healthy." A world that values social justice and environmental sustainability must reverse this flow. Debts from the poorest countries should be forgiven more readily than they are now, and international development assistance should be enhanced.

Child at work FIGURE 2.10

This child works as a laborer at a quarry in Accra, Ghana. He was blinded from an accident at the quarry.

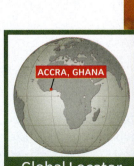

Global Locator

Population growth rates are generally highest where poverty is most intense. If we pay consistent attention to overpopulation and devote the resources necessary to make family planning available for everyone, the human population will stabilize. If we do not continue to emphasize family planning measures, we simply will not achieve population stability.

To stay within Earth's **carrying capacity**, we must reach and sustain a stable population and reduce excessive consumption. These goals must be coupled with educational programs everywhere, so that people understand that Earth's carrying capacity is not unlimited. There is no hope for a peaceful world without overall population stability, and there is no hope for regional economic sustainability without regional population stability.

> ■ **carrying capacity**
> The maximum population that can be sustained by a given environment or by the world as a whole.

RECOMMENDATION 2: PROTECT AND RESTORE EARTH'S RESOURCES

To build a sustainable society, we must preserve the natural systems that support us. The conservation of nonrenewable resources, such as oil and minerals, is obvious, although discoveries of new supplies of nonrenewable resources sometimes give the illusion that they are inexhaustible. Renewable resources such as forests, biodiversity, soils, fresh water, and fisheries must be used in ways that ensure their long-term productivity. Their capacity for renewal must be understood and respected. However, renewable resources have been badly damaged over the past 200 years. Until environmental sustainability becomes a part of economic calculations, susceptible natural resources will continue to be consumed unsustainably, driven by short-term economics.

The world's forests
Many of the world's forests are being cut, burned, or seriously altered at a frightening rate. For example, logging in Oregon, Washington, Alaska, Canada, and Siberia is destroying old-growth forests. Tropical forests are also being overexploited. Many products—hardwoods; foods such as beef, bananas, coffee, and tea; and medicines—come to the industrialized world from the tropics. As trees are destroyed, only a small fraction of them are replanted.

The pressure of rapid population growth and widespread poverty also harms the world's forests. In many developing countries, forests have traditionally served as a "safety valve" for the poor, who, by consuming small tracts of forest on a one-time basis and moving on, find a source of food, shelter, and clothing. But now the numbers of people in developing countries are too great for their forests to support. Tropical rain forests—biologically the world's richest terrestrial areas—have been reduced to less than half their original area. Methods of forest clearing that were suitable when population levels were lower and forests had time to recover from temporary disturbances simply do not work any longer. They convert a potentially renewable resource into an unsustainable one.

Loss of biodiversity
We have a clear interest in protecting Earth's **biological diversity** and managing it sustainably because we obtain from living organisms all our food, most medicines, many building and clothing materials, biomass for energy, and numerous other products. In addition, organisms and the natural environment provide an array of **ecosystem services** without which we would not survive. These services include the protection of watersheds and soils, the development of fertile agricultural lands, the determination of both local climate and global climate, and the maintenance of habitats for animals and plants.

> ■ **biological diversity** The number and variety of Earth's organisms.

Over the next few decades, we can expect human activities to cause the rate of extinction to increase to perhaps hundreds of species a day. How big a loss is this? Unfortunately, we still have a limited knowledge about the world's biological diversity. An estimated five-sixths of all species have not yet been scientifically described. Some 80 percent of the species of plants, animals, fungi, and microorganisms on which we depend are found in developing countries. How will these relatively poor countries sustainably manage and conserve these precious resources? Biological diversity is an intrinsically local problem, and each nation must address it for the sake of its own people's future, as well as for the

Yanomami children in Brazil enjoy a photographer's camera. Intrusion into isolated areas such as the Amazon Basin threatens both biological diversity and the cultures of indigenous people who have lived in harmony with nature for hundreds of generations.

world at large. Like most other challenges of sustainable development, biological diversity can be addressed adequately only if we provide international assistance where needed, including help in training scientists and engineers from developing countries.

Biological diversity and human cultural diversity are intertwined: They are, in fact, two sides of the same coin. **Cultural diversity** is Earth's variety of human communities, each with its individual languages, traditions, and identities (FIGURE 2.11). Cultural diversity enriches the collective human experience. For that reason, the U.N. Educational, Scientific, and Cultural Organization supports the protection of minorities in the context of cultural diversity.

RECOMMENDATION 3: PROVIDE ADEQUATE FOOD FOR ALL PEOPLE

Globally, more than 800 million people lack access to the food needed for healthy, productive lives. This estimate, according to a 2004 report by the U.N. Food and Agriculture Organization, includes a high percentage of chil-

dren. Children are particularly susceptible to food deficiencies because their brains and bodies cannot develop properly without adequate nutrition. Most malnourished people live in rural areas of the poorest developing nations. The link between poverty and **food insecurity** is inescapable.

> **food insecurity**
> The condition in which people live with chronic hunger and malnutrition.

Improving agriculture is one of the highest priorities for achieving global sustainability. In general, grain production per person has kept pace with human population growth over the past 50 years. However, expanded agricultural productivity has taken place at high environmental costs. Moreover, the global population continues to expand, putting additional pressure on food production.

Worldwide, little additional land that is not currently under cultivation is suitable for agriculture. One way to increase the productivity of agricultural land is through **multi-cropping**, or growing more than one crop per year. For example, winter wheat and summer soybean crops are grown in some areas of the United States. However, multi-cropping can be accomplished only in regions where water supplies are adequate for irrigation. Also, care must be taken to prevent a decline in soil fertility from such intensive use.

The negative environmental effects of agriculture, including loss of soil fertility, soil erosion, aquifer depletion, soil and water pollution, and air pollution, must be brought under control (**Figure 2.12**). Many strategies exist to retard the loss of topsoil, conserve water, conserve energy, and reduce the use of agricultural chemicals. For example, in **conservation tillage**, residues from previous crops are left in the soil, partially covering it and helping to hold topsoil in place.

Damage to soil resources Figure 2.12

Extensive withdrawal of water contributed to erosion in this once-fertile area of Kenya. Careful stewardship of the land prevents such damage.

We must develop sustainable agricultural systems that provide improved dietary standards, such as the inclusion of high-quality protein in diets in developing countries. China's expanding use of **aquaculture** is an example of efficient protein production. The carp that are raised in Chinese aquaculture are efficient at converting food into high-quality protein. In China, fish production by aquaculture now exceeds poultry production. However, aquaculture, like all human endeavors, has negative environmental effects that must be addressed for it to be sustainable on a large scale.

RECOMMENDATION 4: MITIGATE CLIMATE CHANGE

A widely discussed human effect on the environment is climate change caused by the **enhanced greenhouse effect**. Both highly developed and developing countries contribute to major increases in CO_2 in the atmosphere, as well as to the increasing amounts of methane, nitrous oxide, tropospheric ozone, and CFCs. The most important greenhouse gas, CO_2, is produced when we burn fossil fuels—coal, oil, and natural gas.

> **enhanced greenhouse effect** The additional warming produced by increased levels of gases that absorb infrared radiation.

Although Earth's climate has been relatively stable during the past 10,000 years, human activities are causing it to change. The average global temperature increased by almost 1° Celsius during the 20th century; more than half of that warming occurred during the past 30 years. Climate scientists generally agree that Earth's climate will continue to change rapidly during the 21st century.

The outcomes of these likely changes are serious, because modern society has evolved and successfully adapted to conditions as they are. Keeping in mind that the change from the last ice age to the present was accompanied by an increase in global temperature of 5° Celsius puts the consequences of the present change, the most rapid of the last 10,000 years, into perspective.

We must address climate change in an aggressive and coordinated fashion, but how do we get all nations of the world to adopt the necessary approaches?

Many policymakers say that we should wait until scientific knowledge of climate change is complete. This reasoning is flawed because Earth's climate system is extremely complex, and we may never completely understand it.

For example, we often say that an increase in atmospheric CO_2 leads to climate warming. However, the increase in CO_2, like other human impacts, is not a simple cause-and-effect relationship but instead a cascade of interacting responses that ripple through the environment (**FIGURE 2.13**) We cannot begin to predict how these changes will affect humans or other organisms.

Stabilizing the climate requires a comprehensive energy plan to include phasing out fossil fuels in favor of renewable energy (such as solar and wind power), increasing energy conservation, and improving energy efficiency. Many national and local governments as well as corporations and environmentally aware individuals are

setting goals to cut carbon emissions. Other nations, however, have not recognized the urgency of the global climate problem. We need a global consensus to address climate change.

RECOMMENDATION 5: DESIGN SUSTAINABLE CITIES

At the beginning of the Industrial Revolution, in approximately 1800, only 3 percent of the world's people lived in cities, and 97 percent were rural, living on farms or in small towns. In the two centuries since then, population distribution has changed radically—toward the cities. More people live in Mexico City today than were living in all the cities of the world 200 years ago. This is a staggering difference in the way people live. Almost 50 percent of the world's population now lives in cities, and the per-

Process Diagram

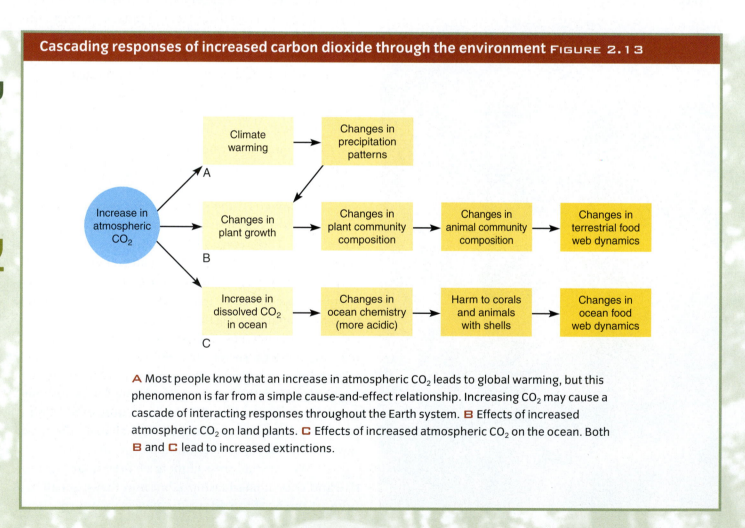

Cascading responses of increased carbon dioxide through the environment FIGURE 2.13

A Most people know that an increase in atmospheric CO_2 leads to global warming, but this phenomenon is far from a simple cause-and-effect relationship. Increasing CO_2 may cause a cascade of interacting responses throughout the Earth system. **B** Effects of increased atmospheric CO_2 on land plants. **C** Effects of increased atmospheric CO_2 on the ocean. Both **B** and **C** lead to increased extinctions.

centage continues to grow. In industrialized countries such as the United States and Canada, almost 80 percent of the people live in cities.

City planners around the world are trying a variety of approaches to make cities more livable. Many cities are developing urban transportation systems to reduce the use of cars and the problems associated with them, such as congested roads, large areas devoted to parking, and air pollution. Urban transportation ranges from mass transit subways and light rails to pedestrian and bicycle pathways.

Squatter settlement FIGURE 2.14

Manila, in the Philippines, is a city of contrasts, with gleaming modern skyscrapers and abjectly poor squatter settlements.

Investing in urban transportation in ways other than building more highways encourages commuters to use forms of transportation other than automobiles. To encourage mass transit, some cities also tax people using highways into and out of cities during business hours. When a city is built around people instead of cars—such as establishing parks and open spaces instead of highways and parking lots—urban residents gain an improved quality of life. Air pollution, including the emission of climate-warming CO_2, is substantially reduced.

Water scarcity is a major issue for many cities of the world. Some city planners think that innovative approaches must be adopted where water resources are scarce. These approaches would replace the traditional one-time water use that involves water purification before use, treatment of sewerage and industrial wastes after use, and then discharge of the treated water. For example, certain cities, such as Singapore, recycle some of their wastewater after it has been treated.

Effectively dealing with the problems in squatter settlements is an urgent need. Evicting squatters does not address the underlying problem of poverty. Instead, cities should incorporate some sort of plan for the eventual improvement of squatter settlements (FIGURE 2.14). Providing basic services—such as clean water to drink, transportation (so people can find gainful employment), and garbage pickup—would help improve the quality of life for the poorest of the poor.

CONCEPT CHECK STOP

What is the global extent of poverty?

What are two ecosystem services provided by natural resources such as forests and biological diversity?

What is food insecurity?

How is stabilizing climate related to energy use?

What are two serious problems in urban environments?

JAKARTA, INDONESIA

There is an urgent need to improve the environment and quality of life in cities, particularly in rapidly growing cities of developing countries. Consider Jakarta, Indonesia, which had a population of 16.9 million in 2006 and has continued to grow. (This number includes several nearby cities that have merged with Jakarta.)

Jakarta is plagued with many of the problems found in other rapidly growing cities in the developing world. The air in Jakarta is badly polluted with the exhaust from cars, buses, and motorbikes that transport about 2 million commuters into the city each day. Water pollution is also a critical problem. At least 95 percent of human wastes produced in the city are not cleaned up at sewage treatment plants. Instead, human sewage and garbage are dumped directly into nearby rivers. Jakarta's municipal water supply is so polluted that piped water must be boiled before people can drink it.

As groundwater has been depleted to meet the city's needs, parts of Jakarta have subsided so that many areas are increasingly flood prone, particularly during the rainy season (see photograph). Illegal squatter settlements proliferate; the poorest inhabitants build dwellings on vacant land using whatever materials they can scavenge. As in other cities around the world, in Jakarta, squatter settlements have the worst water, sewage, and solid waste problems.

Although we have painted a grim picture of Jakarta, it should be noted that improvements are slowly beginning to occur. As part of an ambitious long-term transportation upgrade, Jakarta developed a busway with dedicated bus lines that reduce commuter times and encourage people to commute by bus rather than by automobile. A monorail and subway system are also planned.

VIEW THIS IN ACTION in your WileyPLUS course

Poorly constructed houses lining a canal in Jakarta

SUMMARY

1 Human Use of the Earth

1. **Sustainable development** is economic growth that meets the needs of the present without compromising the ability of future generations to meet their own needs. Environmentally sound decisions, economically viable decisions, and socially equitable decisions interact to promote sustainable development.

2. **Sustainable consumption** is the use of goods and services that satisfy basic human needs and improve the quality of life but that also minimize the use of resources so they are available for future use.

3. **Voluntary simplicity** recognizes that individual happiness and quality of life are not necessarily linked to the accumulation of material goods.

2 Human Values and Environmental Problems

1. **Environmental ethics** is a field of applied ethics that considers the moral basis of environmental responsibility and how far this responsibility extends. Environmental ethicists consider how humans should relate to the natural environment.

2. **An environmental worldview** is a worldview that helps us make sense of how the environment works, our place in the environment, and right and wrong environmental behaviors. The **Western worldview** is an understanding of our place in the world based on human superiority and dominance over nature, the unrestricted use of natural resources, and increased economic growth to manage an expanding industrial base. The **deep ecology worldview** is an understanding of our place in the world based on harmony with nature, a spiritual respect for life, and the belief that humans and all other species have equal worth.

3 Environmental Justice

1. **Environmental justice** is the right of every citizen, regardless of age, race, gender, social class, or other factor, to adequate protection from environmental hazards. Environmental justice is a fundamental human right in an ethical society. A growing environmental justice movement has emerged at the grassroots level.

4 An Overall Plan for Sustainable Living

1. Failing to confront the problem of poverty makes it impossible to attain global sustainability. To stay within Earth's **carrying capacity**, the maximum population that can be sustained indefinitely, it will be necessary to reach a stable population and reduce excessive consumption.

2. The world's forests are being cut, burned, and seriously altered for timber and other products that the global economy requires. Also, rapid population growth and poverty are putting pressure on forests. **Biological diversity**, the number and variety of Earth's organisms, is declining at an alarming rate. Humans are part of Earth's web of life and are entirely dependent on that web for survival.

3. **Food insecurity** is the condition in which people live with chronic hunger and malnutrition. Globally, more than 800 million people lack access to the food needed for healthy, productive lives.

4. The **enhanced greenhouse effect** is the additional warming produced by increased levels of gases that absorb infrared radiation. An increase in atmospheric CO_2, mostly produced when fossil fuels are burned, leads to climate warming. To stabilize climate, we must phase out fossil fuels in favor of renewable energy, increased energy conservation, and improved energy efficiency.

5. The air in cities in the developing world is badly polluted with exhaust from motor vehicles. Illegal squatter settlements proliferate in cities; the poorest inhabitants build dwellings using whatever materials they can scavenge. Squatter settlements have the worst water, sewage, and solid waste problems.

- **sustainable development** p. 28
- **sustainable consumption** p. 29
- **environmental ethics** p. 31
- **environmental worldview** p. 32

- **Western worldview** p. 32
- **deep ecology worldview** p. 33
- **environmental justice** p. 36
- **carrying capacity** p. 39

- **biological diversity** p. 39
- **food insecurity** p. 40
- **enhanced greenhouse effect**, p. 41

CRITICAL AND CREATIVE THINKING QUESTIONS

1. State whether each of the following statements reflects the Western worldview, the deep ecology worldview, or both. Explain your answers.
   a. Species exist to be used by humans.
   b. All organisms, humans included, are interconnected and interdependent.
   c. There is a unity between humans and nature.
   d. Humans are a superior species capable of dominating other organisms.
   e. Humans should protect the environment.
   f. Nature should be used, not preserved.
   g. Economic growth will help Earth manage an expanding human population.
   h. Humans have the right to modify the environment to benefit society.
   i. All forms of life are intrinsically valuable and therefore have the right to exist.

2. How are sustainable consumption and voluntary simplicity related?

3. The graphs at right show a computer simulation by the U.S. National Climate Assessment. In (A), the level of atmospheric CO_2 is projected for the 21st century. As a result of increasing levels of CO_2 in the atmosphere, more CO_2 dissolves in ocean water, where it forms carbonic acid. The increasing acidity dissolves and weakens coral skeletons, which are composed of calcium carbonate (B). (Values in parts A and B are midrange projections.)
   a. Why could rising CO_2 levels in the atmospheric be catastrophic to corals and other shell-forming organisms?
   b. How do these graphs relate to Figure 2.13?

4. What social groups generally suffer the most from environmental pollution and degradation? What social groups generally benefit from this situation?

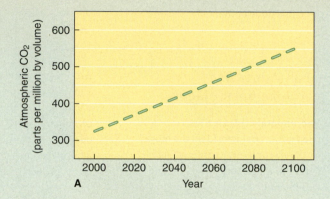

A

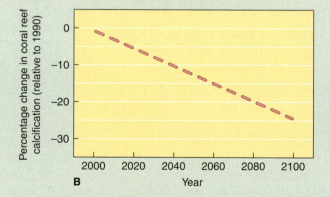

B

5. In its broadest sense, how does environmental justice relate to highly developed countries and less developed countries?

6. What is the role of education in changing personal attitudes and practices that affect the environment?

7. Write a one-page essay describing what kind of world you want to leave for your children.

8. Many conservationists today think that human health and well-being should be an important part of conservation efforts. Explain why humans and the natural world are interconnected, using tropical rain forests and indigenous people as an example.

9. Are an improved standard of living and a reduction in the level of consumption mutually exclusive? Give an example that supports your answer.

10. Write a paragraph that briefly describes your environmental worldview.

11. Development is sometimes equated with economic growth. Explain the difference between sustainable development and development as an indicator of economic growth, using the figure shown to the right.

12. How do the three factors shown in the figure interact to promote sustainable development?

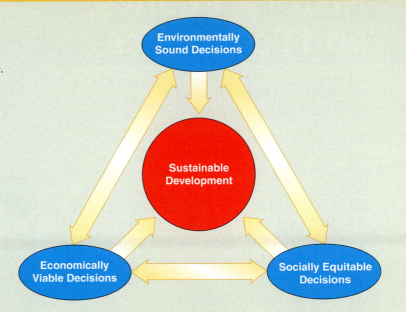

What is happening in this picture ❓

- These fishermen have seen their catches decline in recent years. Why do you think fish shortages are occurring around the world?

- How does Garrett Hardin's description of the tragedy of the commons in medieval Europe relate to the ocean's fisheries today?

- Name an additional example of a global commons other than the one shown in this photograph.

Environmental History, Politics, and Economics

OLD-GROWTH FORESTS OF THE PACIFIC NORTHWEST

During the late 1980s and 1990s, an environmental controversy began in western Oregon, Washington, and northern California that continues today. At stake were thousands of jobs and the future of large tracts of *old-growth coniferous forest* (forest that has never been logged). The northern spotted owl (see photograph), listed as a threatened species since 1990 under the *Endangered Species Act*, came to symbolize the confrontation.

In 1991 a court ordered the suspension of logging in about 1.2 million hectares (3 million acres) of federal forest in the Pacific Northwest, home to the northern spotted owl and 40 other endangered or threatened species. The timber industry stated that thousands of jobs would be lost.

The 1994 *Northwest Forest Plan*, developed under the Clinton administration, represented a compromise between environmental and timber interests. The plan protected habitat of endangered species and allowed some logging to resume on federal forests in the Pacific Northwest. Thanks to a healthy infusion of state and federal aid to the area, many timber workers were retrained for other careers.

Further legislative wrangling led to changes on both sides of the issue. Loggers gained greater access to parts of the forest that the Northwest Forest Plan had declared off-limits, but the courts ordered the federal agencies overseeing logging on federal land to complete surveys of endangered and threatened species before granting the timber industry permission to log. In 2007 the U.S. Fish and Wildlife Service released a draft recovery plan for the northern spotted owl that seeks to extend conservation efforts for the species.

In this chapter, we consider environmental science in the context that frames the story of the northern spotted owl—history, politics, and economics.

NATIONAL GEOGRAPHIC

Conservation and Preservation of Resources

Resources are any part of the natural environment used to promote the welfare of people or other species. Examples of resources include air, water, soil, forests, minerals, and wildlife. **Conservation** is the sensible and careful management of natural resources. Humans have practiced conservation of natural resources for thousands of years. More than 3000 years ago, the Phoenicians terraced hilly farmland to prevent soil erosion. More than 2000 years ago, the Greeks practiced crop rotation to maintain yields on farmlands, and the Romans practiced irrigation. Other Europeans gradually adopted and further refined these and other conservation techniques (**FIGURE 3.1 A**).

Conservation is different from preservation. Conservation involves sustainability—that is, using resources without inflicting excessive environmental damage, so that the resources are available not only for current needs but also for the needs of future generations. **Preservation** is concerned with setting aside undisturbed areas, maintaining them in a pristine state, and protecting them from human activities that might alter their "natural" state (**FIGURE 3.1 B**).

Conservation did not become a popular movement until the early 20th century. At that time, expanding industrialization, coupled with enormous growth in the human population, began to increase pressure on the world's supply of natural resources.

CONCEPT CHECK **STOP**

What is conservation?

What is preservation?

How does conservation differ from preservation?

Conservation and preservation FIGURE 3.1

A Plowing and planting fields in curves that conform to the natural contours of the land conserves soil by reducing erosion.

B The Arctic National Wildlife Reserve preserves animal populations and their habitats.

Environmental History

LEARNING OBJECTIVES

Briefly outline the environmental history of the United States.

Describe the contributions of the following people to our understanding of the environment: John James Audubon, Henry David Thoreau, George Perkins Marsh, Theodore Roosevelt, Gifford Pinchot, John Muir, Franklin Roosevelt, Aldo Leopold, Wallace Stegner, Rachel Carson, and Paul Ehrlich.

Distinguish between utilitarian conservationists and biocentric preservationists.

From the establishment of the first permanent English colony at Jamestown, Virginia, in 1607, the first two centuries of U.S. history were a time of widespread environmental destruction. Land, timber, wildlife, rich soil, clean water, and other resources were cheap and seemingly inexhaustible. The European settlers did not dream that the bountiful natural resources of North America would one day become scarce. During the 1700s and most of the 1800s, many Americans had a **frontier attitude**, a desire to conquer nature and put its resources to use in the most lucrative manner possible.

PROTECTING FORESTS

The great forests of the Northeast were cut down within a few generations, and, shortly after the Civil War in the 1860s, loggers began deforesting the Midwest at an alarming rate. Within 40 years, they had deforested an area the size of Europe, stripping Minnesota, Michigan, and Wisconsin of virgin forest. By 1897 the sawmills of Michigan had processed 160 billion board feet of white pine, leaving less than 6 billion board feet standing in the whole state.

During the 19th century, many U.S. naturalists began to voice concerns about conserving natural resources. **John James Audubon** (1785–1851) painted lifelike portraits of birds and other animals in their natural surroundings that aroused widespread public interest in the wildlife of North America (**FIGURE 3.2**). **Henry David Thoreau** (1817–1862), a prominent U.S. writer, lived for two years on the shore of Walden Pond near Concord, Massachusetts. There he observed nature and

Tangers FIGURE 3.2

This portrayal is one of 500 engravings in Audubon's classic, *The Birds of America*, completed in 1844. Shown are two male Louisiana tanagers (also called western tanagers, top) and male and female scarlet tanagers (bottom).

contemplated how people could economize and simplify their lives to live in harmony with the natural world. **George Perkins Marsh** (1801–1882) was a farmer, linguist, and diplomat at various times during his life. Today he is most remembered for his book *Man and Nature*, published in 1864, which provided one of the first discussions of humans as agents of global environmental change.

In 1875 a group of public-minded citizens formed the American Forestry Association with the intent of influencing public opinion against the wholesale destruction of America's forests. Sixteen years later, in 1891, the **Forest Reserve Act** (which was part of the General Land Law Revision Act) gave the president the authority to establish forest reserves on public (federally owned) land. Benjamin Harrison (1833–1901), Grover Cleveland (1837–1908), William McKinley (1843–1901), and Theodore Roosevelt (1858–1919) used this law to put a total of 17.4 million hectares (43 million acres) of forest, primarily in the West, out of the reach of loggers.

In 1907 angry Northwest congressmen pushed through a bill stating that national forests could no longer be created by the president but would require an act of Congress. Roosevelt signed the bill into law, but not before designating 21 new national forests that totaled 6.5 million hectares (16 million acres).

Roosevelt appointed **Gifford Pinchot** (1865–1946) the first head of the U.S. Forest Service. Both Roosevelt and Pinchot were **utilitarian conservationists** who viewed forests in terms of their usefulness to people—such as in providing jobs and renewable resources. Pinchot supported expanding the nation's forest reserves and managing them scientifically (for instance, harvesting trees only at the rate at which they regrow). Today, national forests are managed for multiple uses, from biological habitats to recreation to timber harvest to cattle grazing.

> ■ **utilitarian conservationist**
> A person who values natural resources because of their usefulness to humans but uses them sensibly and carefully.

> ■ **biocentric preservationist**
> A person who believes in protecting nature from human interference because all forms of life deserve respect and consideration.

ESTABLISHING NATIONAL PARKS AND MONUMENTS

Congress established the world's first national park in 1872 after a party of Montana explorers reported on the natural beauty of the canyon and falls of the Yellowstone River. Yellowstone National Park now includes parts of Idaho, Montana, and Wyoming. In 1890 the **Yosemite National Park Bill** established the Yosemite and Sequoia national parks in California, largely in response to the efforts of a single man, naturalist and writer **John Muir** (1838–1914) (**FIGURE 3.3**). Muir, who as a child emigrated from Scotland with his family, was a **biocentric preservationist**. Muir also founded the *Sierra Club*, a national conservation organization that is still active on a range of environmental issues.

President Theodore Roosevelt *(left)* and John Muir FIGURE 3.3

Photo was taken on Glacier Point above Yosemite Valley, California.

Some environmental battles involving the protection of national parks were lost. John Muir's Sierra Club fought with the city of San Francisco over its efforts to dam a river and form a reservoir in the beautiful Hetch Hetchy Valley, which lay within Yosemite National Park. In 1913 Congress approved the dam. The State of California is considering restoring Hetch Hetchy, at an estimated cost as high as $10 billion. Hetch Hetchy Valley before (A) and after (B) the dam was built.

In 1906 Congress passed the **Antiquities Act**, which authorized the president to set aside sites that had scientific, historic, or prehistoric importance. By 1916 there were 16 national parks and 21 national monuments, under the loose management of the U.S. Army. Today there are 58 national parks and 73 national monuments under the management of the National Park Service.

Controversy over preservation battles, such as the Hetch Hetchy Valley conflict, generated a strong sentiment that the nation should better protect its national parks (FIGURE 3.4). In 1916 Congress created the National Park Service to manage the national parks and monuments for the enjoyment of the public, "without impairment." It was this clause that gave a different outcome to another battle, fought in the 1950s between conservationists and dam builders over the construction of a dam within Dinosaur National Monument. No one could deny that to fill the canyon with 400 feet of water would "impair" it. This victory for conservation established the "use without impairment" clause as the firm backbone of legal protection afforded our national parks and monuments.

CONSERVATION IN THE MID-20TH CENTURY

During the Great Depression, the federal government financed many conservation projects to provide jobs for the unemployed. During his administration, **Franklin Roosevelt** (1882–1945) established the Civilian Conservation Corps, which employed 500,000 young men to plant trees, make paths and roads in national parks and forests, build dams to control flooding, and perform other activities that protected natural resources.

During the droughts of the 1930s, windstorms carried away much of the topsoil in parts of the Great Plains, forcing many farmers to abandon their farms and search for work elsewhere. The *American Dust Bowl* alerted the United States to the need for soil conservation, and President Roosevelt formed the Soil Conservation Service in 1935.

Aldo Leopold (1886–1948) was a wildlife biologist and environmental visionary who was extremely influential in the conservation movement of the mid- to late 20th century (**FIGURE 3.5**). His textbook *Game Management*, published in 1933, supported the passage of a 1937 act in which new taxes on sporting weapons and ammunition funded wildlife management and research. Leopold also wrote philosophically about humanity's relationship with nature and about the need to conserve wilderness areas in *A Sand County Almanac*, published in 1949. Leopold argued persuasively for a land ethic and the sacrifices that such an ethic requires.

Leopold had a profound influence on many American thinkers and writers, including **Wallace Stegner** (1909–1993), who penned his famous "Wilderness Essay" in 1962. Stegner's essay, written to a commission that was conducting a national inventory of wilderness lands, helped create support for the passage of the *Wilderness Act* of 1964. Stegner wrote:

> *Something will have gone out of us as a people if we ever let the remaining wilderness be destroyed; if we permit the last virgin forests to be turned into comic books and plastic cigarette cases; if we drive the few remaining members of the wild species into zoos or to extinction; if we pollute the last clean air and dirty the last clean streams and push our paved roads through the last of the silence, so that never again will*

Aldo Leopold FIGURE 3.5

Leopold's *A Sand County Almanac* is widely considered an environmental classic.

> *Americans be free in their own country from the noise, the exhausts, the stinks of human and automotive waste . . .*
>
> *We simply need that wild country available to us, even if we never do more than drive to its edge and look in. For it can be a means of reassuring ourselves of our sanity as creatures, a part of the geography of hope.*

During the 1960s, public concern about pollution and resource quality began to increase, in large part due to the work of marine biologist **Rachel Carson** (1907–1964). Carson wrote about interrelationships among living organisms, including humans, and the natural environment (**FIGURE 3.6**). Her most famous work, *Silent Spring*, was published in 1962. In it Carson wrote against the indiscriminate use of pesticides:

> *Pesticide sprays, dusts, and aerosols are now applied almost universally to farms, gardens, forests, and homes—nonselective chemicals that have the power to kill every insect, the "good" and the "bad," to still the song of birds and the leaping of fish in the streams, to coat the leaves with a deadly film, and to linger on in soil—all this though the intended target may be only a few weeds or insects. Can anyone believe it is possible to lay down such a barrage of poisons on the surface of the earth without making it unfit for all life? They should not be called "insecticides," but "biocides."*

Rachel Carson FIGURE 3.6

Carson's book *Silent Spring* heralded the beginning of the environmental movement.

Silent Spring heightened public awareness and concern about the dangers of uncontrolled use of DDT and other pesticides, including poisoning birds and other wildlife and contaminating human food supplies. Ultimately, the book led to restrictions on the use of certain pesticides. Around this time, the media began to increase its coverage of environmental incidents, such as hundreds of deaths in New York City from air pollution (1963), closed beaches and fish kills in Lake Erie from water pollution (1965), and detergent foam in a creek in Pennsylvania (1966).

In 1968, when the population of Earth was "only" 3.5 billion people, ecologist **Paul Ehrlich** published *The Population Bomb*. In it he described the damage occurring to Earth's life support system because it was supporting such a huge population, including the depletion of essential resources such as fertile soil, groundwater, and other living organisms. Ehrlich's book raised the public's awareness of the dangers of overpopulation and triggered debates about how to deal effectively with population issues.

THE ENVIRONMENTAL MOVEMENT

Until 1970 the voice of **environmentalists**, people concerned about the environment, was heard in the United States primarily through societies such as the Sierra Club and the National Wildlife Federation. There was no generally perceived **environmental movement** until the spring of 1970, when **Gaylord Nelson**, former senator of Wisconsin, urged Harvard graduate student **Denis Hayes** to organize the first nationally celebrated Earth Day. This event awakened U.S. environmental consciousness to population growth, overuse of resources, and pollution and degradation of the environment. On Earth Day 1970, an estimated 20 million people in the United States planted trees, cleaned roadsides and riverbanks, and marched in parades to demonstrate their support of improvements in resource conservation and environmental quality.

In the years that followed the first Earth Day, environmental awareness and the belief that individual actions could repair the damage humans were doing to Earth became a pervasive popular movement. Musicians and other celebrities popularized environmental concerns. Many of the world's religions—such as Christianity, Judaism, Islam, Hinduism, Buddhism, Taoism,

Shintoism, Confucianism, and Jainism—embraced environmental themes such as protecting endangered species and controlling global climate change.

By Earth Day 1990, the movement had spread around the world, signaling the rapid growth in environmental consciousness. An estimated 200 million people in 141 nations demonstrated to increase public awareness of the importance of individual efforts ("Think globally, act locally") (FIGURE 3.7). The theme of Earth Day 2000, "Clean Energy Now," reflected the dangers of global climate change and what individuals and communities could do: replace fossil fuel energy sources, which produce greenhouse gases, with solar electricity, wind power, and the like. However, by 2000 many environmental activists had begun to think that the individual actions Earth Day espouses, while collectively important, are not as important as pressuring governments and large corporations to make environmentally friendly decisions. In 2007, global environmental concern was expressed in the efforts of Live Earth, a series of concerts held worldwide to launch programs to combat climate change. FIGURE 3.8 shows a timeline of selected environmental events since Earth Day 1970.

CONCEPT CHECK STOP

What role did each of the following have in U.S. environmental history: protecting forests; establishing and protecting national parks and monuments; conservation in the mid-20th century; and the environmental movement of the late-20th century?

What was the environmental contribution of Rachel Carson?

How did Aldo Leopold influence the conservation movement of the mid- to late-20th century?

Timeline of selected environmental events, from 1970 to the present FIGURE 3.8

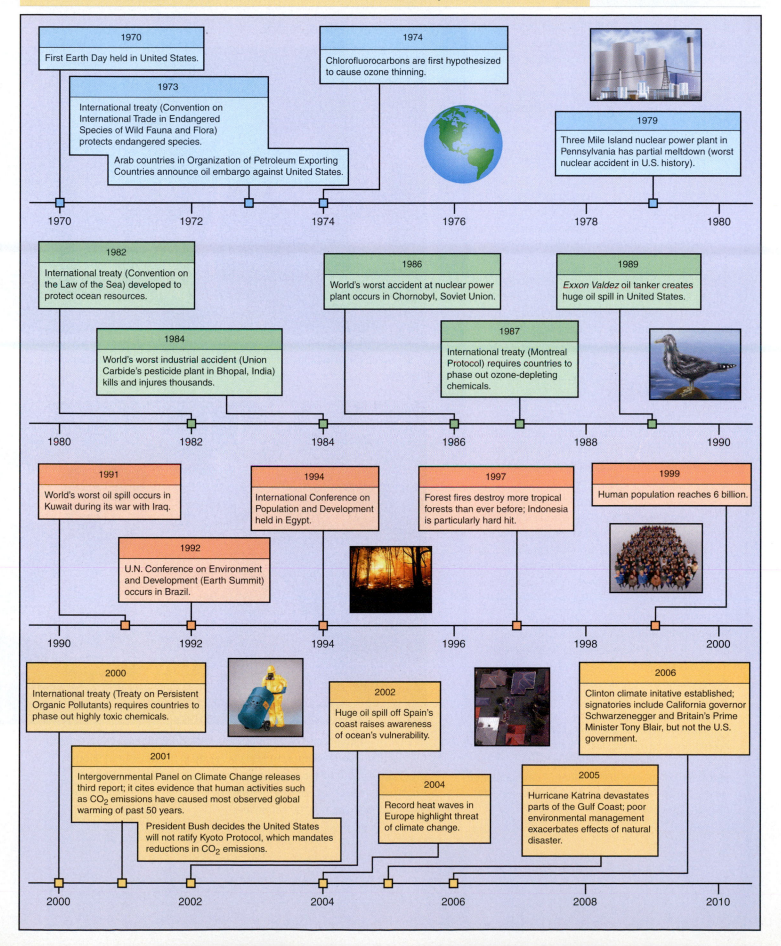

1970
First Earth Day held in United States.

1973
International treaty (Convention on International Trade in Endangered Species of Wild Fauna and Flora) protects endangered species.

Arab countries in Organization of Petroleum Exporting Countries announce oil embargo against United States.

1974
Chlorofluorocarbons are first hypothesized to cause ozone thinning.

1979
Three Mile Island nuclear power plant in Pennsylvania has partial meltdown (worst nuclear accident in U.S. history).

1970   1972   1974   1976   1978   1980

1982
International treaty (Convention on the Law of the Sea) developed to protect ocean resources.

1984
World's worst industrial accident (Union Carbide's pesticide plant in Bhopal, India) kills and injures thousands.

1986
World's worst accident at nuclear power plant occurs in Chornobyl, Soviet Union.

1987
International treaty (Montreal Protocol) requires countries to phase out ozone-depleting chemicals.

1989
Exxon Valdez oil tanker creates huge oil spill in United States.

1980   1982   1984   1986   1988   1990

1991
World's worst oil spill occurs in Kuwait during its war with Iraq.

1992
U.N. Conference on Environment and Development (Earth Summit) occurs in Brazil.

1994
International Conference on Population and Development held in Egypt.

1997
Forest fires destroy more tropical forests than ever before; Indonesia is particularly hard hit.

1999
Human population reaches 6 billion.

1990   1992   1994   1996   1998   2000

2000
International treaty (Treaty on Persistent Organic Pollutants) requires countries to phase out highly toxic chemicals.

2001
Intergovernmental Panel on Climate Change releases third report; it cites evidence that human activities such as CO_2 emissions have caused most observed global warming of past 50 years.

President Bush decides the United States will not ratify Kyoto Protocol, which mandates reductions in CO_2 emissions.

2002
Huge oil spill off Spain's coast raises awareness of ocean's vulnerability.

2004
Record heat waves in Europe highlight threat of climate change.

2005
Hurricane Katrina devastates parts of the Gulf Coast; poor environmental management exacerbates effects of natural disaster.

2006
Clinton climate initative established; signatories include California governor Schwarzenegger and Britain's Prime Minister Tony Blair, but not the U.S. government.

2000   2002   2004   2006   2008   2010

Because responses to environmental problems depend on the public's awareness and understanding of the issues and the underlying scientific concepts involved, environmental education is critical to appropriate decision making. The emphasis on environmental education has grown dramatically over the years:

- Three international treaties supporting environmental education went into effect between 1975 and 1990.

- In 1990, 22 university presidents from 13 nations issued a declaration of their commitment to environmental education and research at their institutions. More than 300 university presidents from at least 40 countries have since followed suit.

- Prepared by coursework at their schools, more than 50,000 U.S. high school students in 2007 took the College Board Advanced Placement exam in Environmental Science, a test accepted by approximately 1700 colleges.

- More than 30 states require some form of environmental education in primary and secondary schools.

- The National Environmental Education Act of 1990 requires the Environmental Protection Agency to provide leadership in promoting environmental education and increasing public awareness and knowledge of environmental issues.

- The U.N. Decade of Education for Sustainable Development (2005–2014) is dedicated to improving basic education, including public understanding about environmental sustainability. Programs focus on major themes, such as water, climate change, biodiversity, and disaster prevention.

The North American Association for Environmental Education has issued guidelines for educators to help them select materials such as textbooks and films that are based on sound scientific evidence and that present a balanced perspective on environmental problems. Fairness and accuracy are emphasized in these guidelines.

However, a backlash against environmental education occurred beginning in the late 1990s. Some conservative research groups criticized what they perceived as a biased presentation of environmental issues, particularly the promotion of environmental activism, in schools.

Environmental education

A Schoolchildren in China participate in an environmental education exercise.

B Environmentalist and primatologist Jane Goodall meets with Connecticut middle school students involved in the Roots and Shoots program, a youth-based environmental action organization that Goodall started. The program includes tens of thousands of members and chapters in nearly 100 countries.

Environmental Legislation

LEARNING OBJECTIVES

Explain why the National Environmental Policy Act is the cornerstone of U.S. environmental law.

Describe how environmental impact statements provide powerful protection of the environment.

Define *full cost accounting*.

Well-publicized ecological disasters, such as the 1969 oil spill off the coast of Santa Barbara, California, and overwhelming public support for the Earth Day movement, resulted in the January 1970 signing of the **National Environmental Policy Act (NEPA)**. The Environmental Protection Agency (EPA) was created in July of the same year. A key provision of NEPA states that the federal government must consider the environmental impact of a proposed federal action, such as financing highway or dam construction, when making decisions about that action. NEPA provides the basis for developing detailed **environmental impact statements (EISs)** to accompany every federal recommendation or proposal for legislation. An EIS is a document that describes the nature of the proposal, why it is needed, short- and long-term environmental impacts of the proposal, and possible alternatives to the proposed action that would create fewer adverse effects. NEPA also requires solicitation of public comments when preparing an EIS, which generally provides a broader perspective on the proposal and its likely effects.

NEPA established the **Council on Environmental Quality** to monitor the required EISs and report directly to the president. Because this council had no enforcement powers, NEPA was originally considered innocuous, more a statement of good intentions than a regulatory policy. During the next few years, however, environmental activists took people, corporations, and the federal government to court to challenge their EISs or use them to block proposed development. The courts decreed that EISs had to thoroughly analyze the environmental consequences of anticipated projects on soil, water, and endangered species and that EISs be made available to the public (**FIGURE 3.9**). These rulings put

Environmental impact statements FIGURE 3.9

Detailed environmental impact statements help federal agencies consider the environmental impacts of proposed actives. When the environmental impacts are judged too severe, alternative actions are considered.

sharp teeth into NEPA—particularly the provision for public scrutiny, which places intense pressure on federal agencies to respect EIS findings.

NEPA revolutionized environmental protection in the United States. Federal agencies manage federal highway construction, flood and erosion control, military projects, and many other public works. They oversee nearly one-third of the land in the United States. Federal holdings include fossil fuel and mineral reserves, millions of hectares of public grazing land, and public forests. Since 1970 very little has been done to any of them without some sort of environmental review. NEPA has also influenced environmental legislation in at least 36 states and in other countries, including Canada, Australia, France, New Zealand, and Sweden.

Although almost everyone agrees that NEPA has successfully reduced adverse environmental impacts of federal activities and projects, it has its critics. Some environmentalists complain that EISs are sometimes incomplete or that reports are ignored when decisions are made. Other critics think the EISs delay important projects ("paralysis by analysis") because the documents are too involved, take too long to prepare, and are often the targets of lawsuits.

ENVIRONMENTAL REGULATIONS

When an environmental problem becomes widely recognized, the process of environmental regulation begins with a U.S. congressperson drafting legislation. Ideally, before the legislation is drafted, the trade-offs for several proposed alternative actions are evaluated. This process, known as **full cost accounting**, is a valuable economic tool in environmental decision making.

After the legislation is passed and the president signs it, it usually goes to the EPA, which was created to translate the law's language into regulations that specify, for example, allowable levels of pollution. Before the regulations officially become law, several rounds of public comments allow affected parties to present their views; the EPA is required to respond to all of

full cost accounting
The process of evaluating and presenting to decision makers the relative benefits and costs of various alternatives.

these comments. Then the Office of Management and Budget reviews the new regulations. Implementation and enforcement of the new law often fall to state governments, which must send the EPA details for achieving the goals of the new regulations.

ACCOMPLISHMENTS OF ENVIRONMENTAL LEGISLATION

During the period since Earth Day 1970, Congress has passed almost 40 major environmental laws that address a wide range of issues, such as endangered species, clean water, clean air, energy conservation, hazardous wastes, and pesticides. This tough interlocking mesh of laws greatly increased federal regulation of pollution to improve environmental quality.

Despite imperfections, environmental legislation has had overall positive effects. Since 1970,

- Fifteen national parks have been established (**FIGURE 3.10**), and the National Wilderness Preservation System now totals more than 43 million hectares (106 million acres).

Joshua Tree National Park, California
FIGURE 3.10

Formerly a national monument, Joshua Tree was declared a national park in 1994.

- Millions of hectares of farmland particularly vulnerable to erosion have been withdrawn from production, reducing soil erosion by more than 60 percent.

- Many endangered species are recovering, and the American alligator, California gray whale, and bald eagle have recovered enough to be removed from the endangered species list. (However, dozens of other species, such as the manatee and Kemp's ridley sea turtle, have suffered further declines or extinction since 1970.)

Although we still have a long way to go, pollution control efforts through legislation have been particularly successful. According to the EPA's "Draft Report on the Environment 2003":

- Emissions of six important air pollutants have dropped by more than 25 percent.

- Since 1990, levels of wet sulfate, a major component of acid rain, have dropped by 20 to 30 percent.

- Releases of toxic chemicals into water and air from industrial sources have declined by 48 percent since 1988.

- Fewer rivers and streams are in violation of water quality standards. However, fish-consumption advisories related to specific toxins such as mercury or polychlorinated biphenyls (PCBs) have increased, possibly because more consistent monitoring has led to more accurate measurements of these pollutants.

- In 2002, 94 percent of the U.S. population received water from community water systems that met health-based drinking-water standards, up from 79 percent in 1993 (FIGURE 3.11).

- Although 10,753 contaminated sites are currently listed on the Superfund inventory, more than 1000 have been cleaned up and removed from the inventory since 1980.

In the 1960s and 1970s, pollution was often obvious—witness the Cuyahoga River in Cleveland, Ohio, which burst into flames from the oily pollutants on

Water treatment plant FIGURE 3.11

The water supply for a town or city is treated before use so it is safe to drink. Photographed in Miami, Florida.

its surface several times, including a highly publicized burn in 1969. Legislators, the media, and the public typically perceive things like burning rivers as serious threats that require immediate attention without regard to the cost. Now that the most obvious pollution problems in the United States are largely addressed, more and more people look at environmental cleanup in terms of benefit versus cost. Thus, economics is increasingly important in environmental legislation and policymaking.

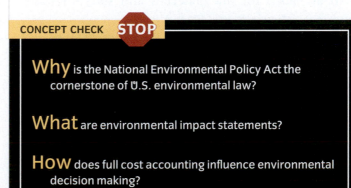

CONCEPT CHECK STOP

Why is the National Environmental Policy Act the cornerstone of U.S. environmental law?

What are environmental impact statements?

How does full cost accounting influence environmental decision making?

Environmental Economics

Economics is the study of how people use their limited resources to try to satisfy their unlimited wants. Economists try to understand the consequences of the ways in which people, businesses, and governments allocate their resources. Seen through an economist's eyes, the world is one large marketplace, where resources are allocated to a variety of uses, and where goods—a car, a pair of shoes, a hog—and services—a haircut, a museum tour, an education—are consumed and paid for. In a free market, supply and demand determine the price of a good. If something in great demand is in short supply, its price will be high. High prices encourage suppliers to produce more of a good or service, as long as the selling price is higher than the cost of producing the good or service. This interaction of demand, supply, price, and cost underlies much of what happens in the U.S. economy, from the price of a hamburger to the cycles of economic expansion (increase in economic activity) and recession (slowdown in economic activity).

Economies depend on the natural environment as *sources* for raw materials and *sinks* for waste products (**FIGURE 3.12**). Both sources and sinks contribute to **natural capital**. According to economists, the environment provides natural capital for human production and consumption. Resource degradation and pollution represent the overuse of natural capital. *Resource degradation* is the overuse of sources,

> **natural capital**
> Earth's resources and processes that sustain living organisms, including humans; includes minerals, forests, soils, groundwater, clean air, wildlife, and fisheries.

and *pollution* is the overuse of sinks; both threaten our long-term economic future.

NATIONAL INCOME ACCOUNTS AND THE ENVIRONMENT

Much of our economic well-being flows from natural assets—such as land, rivers, the ocean, oil, timber, and the air we breathe—rather than human-made assets.

Ideally, for the purposes of economic and environmental planning, the **national income accounts** should include the use and misuse of natural resources and the environment. Two measures used in national income accounting are *gross domestic product* (*GDP*) and *net domestic product* (*NDP*). Both GDP and NDP provide estimates of national economic performance that are used to make important policy decisions.

> **national income accounts**
> A measure of the total income of a nation's goods and services for a given year.

Unfortunately, current national income accounting practices provide an incomplete or inaccurate measure of income because they do not incorporate environmental factors. Two important conceptual problems exist with the way the national income accounts currently handle the economic use of natural resources and the environment: natural resource depletion and the costs and benefits of pollution control. Better accounting for environmental quality—both depletion of natural capital and pollution—would help address whether for any given activity the benefits (both economic and environmental) exceed the costs.

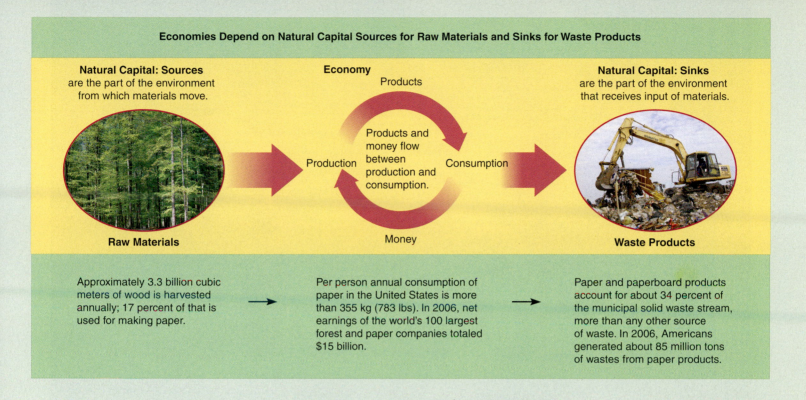

Economies Depend on Natural Capital Sources for Raw Materials and Sinks for Waste Products

Natural Capital: Sources are the part of the environment from which materials move.

Economy

Products

Products and money flow between production and consumption.

Production

Consumption

Money

Natural Capital: Sinks are the part of the environment that receives input of materials.

Raw Materials

Waste Products

Approximately 3.3 billion cubic meters of wood is harvested annually; 17 percent of that is used for making paper.

Per person annual consumption of paper in the United States is more than 355 kg (783 lbs). In 2006, net earnings of the world's 100 largest forest and paper companies totaled $15 billion.

Paper and paperboard products account for about 34 percent of the municipal solid waste stream, more than any other source of waste. In 2006, Americans generated about 85 million tons of wastes from paper products.

Natural resource depletion

If a manufacturing firm produces some product (output) but in the process wears out a portion of its plant and equipment, the firm's output is counted as part of GDP, but the depreciation of capital is subtracted in the calculation of NDP. Thus NDP is a measure of the net production of the economy, after a deduction for used-up capital. In contrast, when an oil company drains oil from an underground field, the value of the oil produced is counted as part of the nation's GDP, but no offsetting deduction to NDP is made to account for the fact that nonrenewable resources were used up (**FIGURE 3.13**).

In principle, the draining of the oil field is a type of depreciation, and the oil company's net product should be accordingly reduced. The same point applies to any other natural resource that is depleted in the process of production. Natural capital is a very large part of a country's economic wealth, and we should treat it the same as human-made capital.

GRAND BANKS

Global Locator

The Hibernia oil platform on the Grand Banks in the Atlantic Ocean FIGURE 3.13

The oil field being drained under the seafloor is part of the U.S. GDP. The fact that it will be drained dry someday is not taken into account.

Hazardous materials workers in protective gear remove toxic waste from a Texas beach. These cleanup costs should be added to the GDP accounts because the cleanup improves the environment.

The costs and benefits of pollution control

Imagine that a company has the following choices: It can produce $100 million worth of output and, at the same time, dump its wastes, polluting the local river. Alternatively, if the company uses 10 percent of its workers to properly dispose of its wastes, it avoids polluting but gets only $90 million of output. Under current national income accounting rules, if the firm chooses to pollute rather than not to pollute, it will make a larger contribution to GDP ($100 million rather than $90 million) because the national income accounts attach no explicit value to a clean river. In an ideal accounting system, the economic cost of environmental degradation is subtracted in the calculation of a firm's contribu-

tion to GDP, and activities that improve the environment—because they provide real economic benefits—are added to GDP (FIGURE 3.14). To summarize, we should subtract estimates of environmental damage from GDP.

Incorporating resource depletion and pollution into national income accounting is important because GDP and related statistics are used continually in policy analyses. An increasing number of economists, government planners, and scientists support replacing GDP and NDP with a more comprehensive measure of national income accounting that includes estimates of both depletion of natural capital and the environmental cost of economic activities.

AN ECONOMIST'S VIEW OF POLLUTION

An important aspect of the operation of a free-market system is that the person consuming a product should pay for all the cost of producing it. However, production or consumption of a product often has an **external cost**.

A product's market price does not usually reflect an external cost—that is, the buyer or seller doesn't pay for the external cost. As a result, the market system generally does not operate in the most efficient way.

> **external cost**
>
> A harmful environmental or social cost that is borne by people not directly involved in selling or buying a product.

Consider the following example of an external cost. If an industry makes a product and, in so doing, also releases a pollutant into the environment, the product is bought at a price that reflects the cost of making it but not the cost of the pollutant's damage to the environment. This damage is the external cost of the product. (One common external cost of many products is air pollution released when fossil fuels are burned to transport manufacturing components or finished goods.)

Because this damage is not included in the product's price and because the consumer may not know that the pollution exists or that it harms the environment, the cost of the pollution has no impact on the consumer's decision to buy the product. As a result, consumers of the product may buy more of it than they would if its true cost, including the cost of pollution, were known or reflected in the selling price.

The failure to add the price of environmental damage to the cost of products generates a market force that increases pollution. From the perspective of economics, then, one of the causes of the world's pollution problem is the failure to consider negative external costs in the pricing of goods.

We now examine industrial pollution from an economist's viewpoint, as a policymaking failure. Keep in mind, however, that lessons about the economics of industrial pollution also apply to other environmental issues (such as resource degradation) where harm to the environment is a consequence of economic activity.

How much pollution is acceptable? To assign a proper price to pollution, economists first try to answer the basic question "How much pollution should we allow in our environment?" Imagine two environmental extremes: a wilderness in which no pollution is produced but neither are goods, and a "sewer" that is completely polluted from excess production of goods. In our world, a move toward a better environment almost always entails a cost in terms of goods.

How do we, as individuals, as a country, and as part of the larger international community, decide where we want to be between the two extremes of a wilderness and a sewer? Economists analyze the marginal costs of environmental quality and of other goods to answer such questions. A **marginal cost** is the additional cost associated with one more unit of something. Two examples of marginal costs associated with pollution are the effects of pollution on human health and on organisms in the natural environment.

The trade-off between protecting environmental quality and producing more goods involves balancing marginal costs of two kinds: (1) the cost, in terms of environmental damage, of more pollution (the marginal cost of pollution) and (2) the cost, in terms of giving up goods, of eliminating pollution (the marginal cost of pollution abatement).

Determining the **marginal cost of pollution** involves assessing the risks associated with the pollution—for example, damage to health, property, or agriculture. (See Chapter 4 for a discussion of risk assessment.)

> **marginal cost of pollution**
>
> The added cost of an additional unit of pollution.

Let's consider a simple example involving the marginal cost of sulfur dioxide, a type of air pollution produced during the combustion of fuels containing sulfur. Sulfur dioxide is removed from the atmosphere as acid rain, which causes damage to the environment, particularly aquatic ecosystems. Economists add up the harm of each additional unit of pollution—in this example, each ton of sulfur dioxide added to the atmosphere. As the total amount of pollution increases, the harm of each additional unit usually also increases, and as a result, the

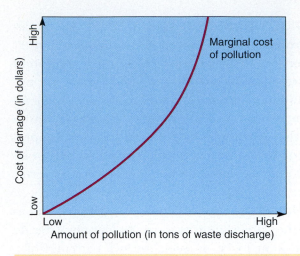

Marginal cost of pollution FIGURE 3.15

At low pollution levels, the environment may absorb the damage, so that the marginal cost of one added unit of pollution is near zero. As the level of pollution rises, the cost in terms of human health and a damaged environment increases sharply. At very high levels of pollution, the cost soars.

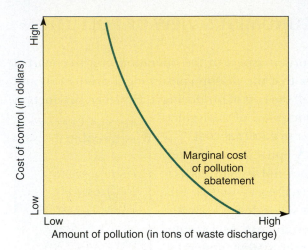

Marginal cost of pollution abatement FIGURE 3.16

At high pollution levels, the marginal cost of eliminating one unit of pollution is low. As more and more pollution is eliminated from the environment, the cost of removing each additional (marginal) unit of pollution increases.

> **marginal cost of pollution abatement** The added cost of reducing one unit of a given type of pollution.

> **cost–benefit diagram** A diagram that helps policymakers make decisions about costs of a particular action and benefits that would occur if that action were implemented.

curve showing the marginal cost of pollution slopes upward, as in FIGURE 3.15.

The **marginal cost of pollution abatement** tends to rise as the level of pollution declines, as shown in FIGURE 3.16. It is relatively inexpensive to reduce automobile exhaust emissions by half, but costly devices are required to reduce the remaining emissions by half again. For this reason, the curve showing the marginal cost of pollution abatement slopes downward.

In FIGURE 3.17, the two marginal-cost curves from Figures 3.15 and 3.16 are plotted together on one graph, called a **cost-benefit diagram**. Economists use this diagram to identify the point at which the marginal cost of pollution equals the marginal cost of abatement—that is, the point where the two curves intersect. As far as economics is concerned, this point represents an **optimum amount of pollution**. At this optimum, the cost to society of having less pollution is offset by the benefits to society of the activity creating the pollution.

> **optimum amount of pollution** The amount of pollution that is economically most desirable.

There are two major flaws in the economist's concept of optimum pollution. First, it is difficult to determine the true cost of environmental damage caused by pollution. Usually, there are many polluters and many affected individuals. Second, when economists add up pollution costs, they do not take into account the possible disruption or destruction of the environment. The web of relationships within the environment is extremely intricate and may be more vulnerable to pollution damage than is initially obvious, sometimes with disastrous results. This is truly a case where the whole is much greater than the sum of its parts, and it is inappropriate for economists to simply add up the costs of lost elements in a polluted environment.

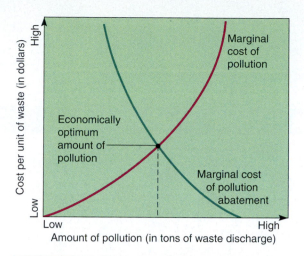

Cost-benefit diagram FIGURE 3.17

Economists identify the optimum amount of pollution as the amount at which the marginal cost of pollution equals the marginal cost of pollution abatement (the point at which the two curves intersect). If more pollution than the optimum is allowed, the social cost is unacceptably high. If less than the optimum amount of pollution is allowed, the pollution abatement cost is unacceptably high.

ECONOMIC STRATEGIES FOR POLLUTION CONTROL

Command and control regulations and incentive-based regulations are two ways that governments control pollution (**FIGURE 3.18**). To date, most pollution control efforts in the United States have involved **command and control regulation**. Sometimes command and control

Pollution control FIGURE 3.18

Command and control regulations restrict the emission of pollutants, often by requiring the use of a specific type of pollution abatement technology. Incentive-based regulations set emission targets and let the industry decide how to meet them.

Economic Strategies for Pollution Control	
Command and Control Regulations	Incentive-Based Regulations

laws require use of a specific pollution control method, such as the use of catalytic converters in cars to decrease polluting exhaust emissions. In other cases, a quantitative goal is set. The Clean Air Act Amendments of 1990 established a goal of a 60 percent reduction in nitrogen oxide emissions in passenger cars by the year 2003. Usually, all polluters must comply with the same rules and regulations regardless of their particular circumstances.

Some economists criticize command and control regulation for being more costly than necessary. They think command and control regulation sets environmental pollution levels much lower than the economically optimum level of pollution. Most economists, whether progressive or conservative, prefer **incentive-based regulation** over command and control regulation. Incentive-based regulation, such as environmental taxes and tradable permits, is a market-oriented strategy. It seeks to use the economic forces of a free market to alleviate the pollution problem—that is, it depends on market incentives to reduce pollution and minimize the cost of control.

command and control regulation Pollution control laws that work by setting limits on levels of pollution.

incentive-based regulation Pollution control laws that work by establishing emission targets and providing industries with incentives to reduce emissions.

CONCEPT CHECK STOP

What is natural capital? How is economics related to natural capital?

Why are national income accounts incomplete estimates of total national economic performance?

How do command and control regulation and incentive-based regulation differ regarding pollution control?

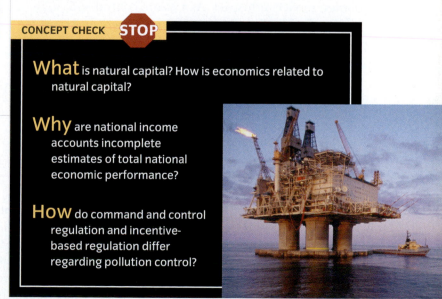

ENVIRONMENTAL PROBLEMS IN CENTRAL AND EASTERN EUROPE

The fall of the Soviet Union and communist governments in central and eastern Europe during the late 1980s revealed a grim legacy of environmental destruction. Chemicals leaking out of dump sites contaminated agricultural soil and water. Buildings and statues eroded, and entire forests died because of air pollution and acid rain. One of the most polluted areas in the world is the "Black Triangle," which consists of the bordering regions of eastern Germany, the northern Czech Republic, and southwest Poland.

Massive pollution affects human health in the region. Many people suffer from asthma, emphysema, chronic bronchitis, and other respiratory diseases. Levels of cancer, miscarriages, and birth defects are extremely high. Life expectancies are still lower there than in other industrialized nations. In 2007, the average eastern European lived to age 69—11 years less than the average western European.

The economic assumption behind communism was one of high production and economic self-sufficiency. Governments supported heavy industry—power plants, chemicals, metallurgy, and large machinery—at the expense of more environmentally benign service industries. As a result, central and eastern Europe became over-industrialized, and most of the region's plants lacked the pollution abatement equipment now required in factories in most industrialized countries.

Experts predict that eliminating the pollution legacy of communism will take decades. How much will it cost? The numbers are staggering. Improving the

Birth defects in Moscow
Since 1973, at least 90 children with terminal-limb defects were born in two neighborhoods of Moscow. Medical researchers have not investigated limb defects thoroughly and do not know if their cause is environmental, genetic, or both.

MOSCOW, RUSSIA

Global Locator

environment in eastern Germany alone will cost up to an estimated $300 billion.

While switching from communism to democracy with a free-market economy, current central and eastern European governments also face the responsibility of improving the environment. Hungary, Poland, and the Czech Republic have moved relatively successfully toward a market economy and have generated enough money to invest in environmental cleanup.

Economic recovery is slow in other countries, such as Bulgaria, Romania, and Russia; severe budgetary problems there have forced the environment to take a back seat to political and economic reform.

VIEW THIS IN ACTION
in your WileyPLUS course

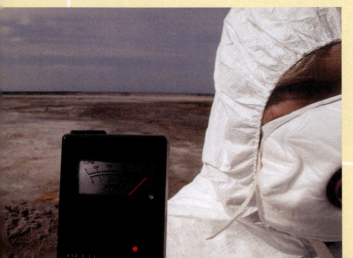

Pollution problems in former communist countries
A former uranium processing plant produced radioactive waste that damaged this land in the Estonian town of Sillamae. Note the exceedingly high reading on the Geiger counter. Thousands of polluted sites exist in former communist countries, the result of rapid expansion of industrialization without regard for the environment. Cleaning up these contaminated sites will take decades.

SUMMARY

1 Conservation and Preservation of Resources

1. **Conservation** is the sensible and careful management of natural **resources**, such as air, water, soil, forests, minerals, and wildlife. **Preservation** is concerned with setting aside undisturbed areas, maintaining them in a pristine state, and protecting them from human activities.

2 Environmental History

1. The first two centuries of U.S. history were a time of widespread environmental destruction. During the 1700s and early 1800s, most Americans had a desire to conquer and exploit nature as quickly as possible. During the 19th century, many U.S. naturalists became concerned about conserving natural resources. The earliest conservation legislation revolved around protecting land—forests, parks, and monuments. By the late 20th century, environmental awareness had become a pervasive popular movement.

2. **John James Audubon**'s art aroused widespread interest in the wildlife of North America. **Henry David Thoreau** wrote about living in harmony with the natural world. **George Perkins Marsh** wrote about humans as agents of global environmental change. **Theodore Roosevelt** appointed **Gifford Pinchot** as the first head of the U.S. Forest Service. Pinchot supported expanding the nation's forest reserves and managing forests scientifically. The Yosemite and Sequoia national parks were established largely in response to the efforts of naturalist **John Muir**. **Franklin Roosevelt** established the Civilian Conservation Corps and the Soil Conservation Service. In *A Sand County Almanac*, **Aldo Leopold** wrote about humanity's relationship with nature. **Wallace Stegner** helped create support for the passage of the Wilderness Act of 1964. **Rachel Carson** published *Silent Spring*, alerting the public about the dangers of uncontrolled pesticide use. **Paul Ehrlich** published *The Population Bomb*, which raised the public's awareness of the dangers of overpopulation.

3. A **utilitarian conservationist** is a person who values natural resources because of their usefulness to humans but uses them sensibly and carefully. A **biocentric preservationist** is a person who believes in protecting nature because all forms of life deserve respect and consideration.

3 Environmental Legislation

1. Since 1970 the federal government has addressed many environmental problems. The **National Environmental Policy Act (NEPA)** was passed in 1970. NEPA established the **Council on Environmental Quality** to monitor required **environmental impact statements (EISs)** and report directly to the president.

2. By requiring EISs that are open to public scrutiny, NEPA initiated serious environmental protection in the United States. NEPA allows citizen suits, in which private citizens take violators, whether they are private industries or government-owned facilities, to court for noncompliance.

3. **Full cost accounting** is the process of evaluating and presenting to decision makers the relative benefits and costs of various alternatives.

4 Environmental Economics

1. **Economics** is the study of how people use their limited resources to try to satisfy their unlimited wants. Economies depend on the natural environment as sources for raw materials and sinks for waste products. Both sources and sinks contribute to **natural capital**, which is Earth's resources and processes that sustain living organisms, including humans. Natural capital includes minerals, forests, soils, groundwater, clean air, wildlife, and fisheries.

2. **National income accounts** are a measure of the total income of a nation's goods and services for a given year. An **external cost** is a harmful environmental or social cost that is borne by people not directly involved in buying or selling a product. National income accounts are incomplete estimates of national economic performance because they do not include both natural resource depletion and the environmental costs of economic activities. Many economists, government planners, and scientists support more comprehensive income accounting that includes these estimates.

3. From an economic point of view, the appropriate amount of pollution is a trade-off between harm to the environment and inhibition of development. The **marginal cost of pollution** is the added cost of an additional unit of pollution. The marginal cost of pollution abatement is the added cost of reducing one unit of a given type of pollution. Economists think the use of resources for pollution abatement should increase only until the cost of abatement equals the cost of the pollution damage. This results in the **optimum amount of pollution**—the amount of pollution that is economically most desirable.

4. To control pollution, government often uses **command and control regulations**, which are pollution control laws that work by setting limits on levels of pollution. **Incentive-based regulations** are pollution control laws that work by establishing emission targets and providing industries with incentives to reduce emissions.

KEY TERMS

- **utilitarian conservationist** p. 52
- **biocentric preservationist** p. 52
- **full cost accounting** p. 60
- **natural capital** p. 62
- **national income accounts** p. 62
- **external cost** p. 65
- **marginal cost of pollution** p. 65
- **marginal cost of pollution abatement** p. 66
- **cost–benefit diagram** p. 66
- **optimum amount of pollution** p. 66
- **command and control regulation** p. 67
- **incentive-based regulation** p. 67

CRITICAL AND CREATIVE THINKING QUESTIONS

1. The National Environmental Policy Act (NEPA) is sometimes called the "Magna Carta of environmental law." What is meant by such a comparison?

2. If you were a member of Congress, what legislation would you introduce to deal with each of the following problems?
   - Poisons from a major sanitary landfill are polluting your state's groundwater.
   - Acid rain from a coal-burning power plant in a nearby state is harming the trees in your state. Loggers and foresters are upset.
   - There is a high incidence of cancer in the area of your state where heavy industry is concentrated.

3. How would an economist approach each of the problems listed in question 2? How would an environmentalist?

4. Draw a diagram showing how the following are related: natural environment, sources, sinks, economy, production, consumption, raw materials, and waste products.

5. What is the message of this cartoon? What environmental issue probably inspired it, and when was it likely published?

6. Discuss the events that led to the Northwest Forest Plan of 1994.

7. Describe the extent of environmental destruction in formerly communist countries, and compare that to the most urgent environmental issues in your area.

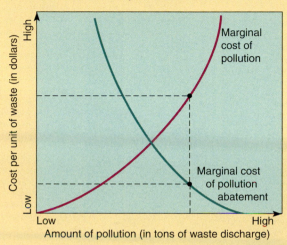

8. In the graph shown above, is the amount of pollution indicated by the vertical dashed line more or less than the economically optimum amount of pollution? Explain your answer.

9. If you were an economist examining the previous graph, what would you recommend, increasing or decreasing pollution abatement measures? Why?

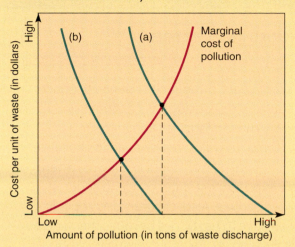

10. The graph above shows two curves, labeled a and b, that represent marginal cost of pollution abatement. In this hypothetical situation, technological innovations were developed between 2003 and 2006 that lowered the abatement cost. Which curve corresponds to 2003 and which to 2006? Explain your answer.

What is happening in this picture ?

- This photo was taken in Hong Kong in 1990. What event is taking place?

- Note the ages of the individuals in this photo. How do their ages contribute to growing environmental awareness?

- Do you think a major international celebration of Earth Day will be held in 2010? Why or why not?

Risk Analysis and Environmental Health Hazards

PESTICIDES AND CHILDREN

VIEW THIS IN ACTION in your WileyPLUS course

In recent years, attention to the health effects of household pesticides on children has increased. These pesticides appear to be a greater threat to children than to adults because children tend to play on floors and lawns, where they are exposed to greater concentrations of pesticide residues. Also, children are probably more sensitive to pesticides because their bodies are still developing. Several preliminary studies suggest that exposure to household pesticides may cause brain cancer and leukemia in children, but scientists must do more research before they can reach any firm conclusions.

Research supports the hypothesis that exposure to pesticides may affect the development of intelligence and motor skills in young children. One study, published in *Environmental Health Perspectives* in 1998, compared two groups of rural Yaqui Indian preschoolers in Mexico. These two nearly identical groups differed mainly in their exposure to pesticides: One group lived in a farming community where pesticides were used frequently (see photograph), and the other lived in an area where pesticides were rarely used. When asked to draw a person, most of the 17 children from the low-pesticide area drew recognizable stick figures (see part a of inset), whereas most of the 34 children from the high-pesticide area drew meaningless lines and circles (see part b). Additional tests of simple mental and physical skills revealed similar striking differences between the two groups of children.

NATIONAL GEOGRAPHIC

Drawings of a person
(by 4-year-olds)

Foothills   Valley

(a) 54 mo. female   (b) 54 mo. female

A Perspective on Risks

Threats to our health, particularly from toxic chemicals in the environment, make big news. Many of these stories are more sensational than factual. In fact, human health is generally better today than at any previous time in our history, and our life expectancy continues to increase rather than decline. This does not mean that you should ignore chemicals that humans introduce into the environment. Nor does it mean you should discount the stories that the news media sometimes sensationalize. These stories serve an important role in getting the regulatory wheels of the government moving to protect us as much as possible from the dangers of our technological and industrialized world.

Risk is inherent in all our actions and in everything in our environment. All of us take risks every day of our lives. Walking on stairs involves a small risk, but a risk nonetheless because sometimes people die from falls on stairs. Using household appliances is slightly risky because sometimes people die from electrocution when they operate appliances with faulty wiring or use appliances in an unsafe manner. Driving or riding in a car or flying in a jet has risks that are easier for most of us to recognize. Yet few of us hesitate to get in a car or board a plane because of the associated risk. It is important to have an adequate understanding of the nature and size of risks before deciding what actions are appropriate to avoid them (**FIGURE 4.1**).

> **risk** The probability of harm (such as injury, disease, death, or environmental damage) occurring under certain circumstances.

Lifetime probability of death by selected causes FIGURE 4.1

These 2003 data from the National Safety Council are for U.S. residents.

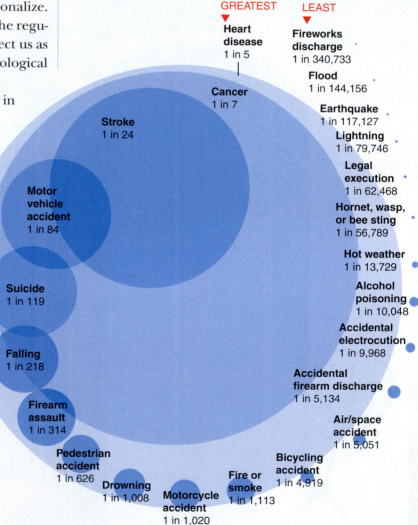

GREATEST ▼   LEAST ▼

Heart disease 1 in 5

Fireworks discharge 1 in 340,733

Flood 1 in 144,156

Earthquake 1 in 117,127

Lightning 1 in 79,746

Legal execution 1 in 62,468

Hornet, wasp, or bee sting 1 in 56,789

Hot weather 1 in 13,729

Alcohol poisoning 1 in 10,048

Accidental electrocution 1 in 9,968

Accidental firearm discharge 1 in 5,134

Air/space accident 1 in 5,051

Cancer 1 in 7

Stroke 1 in 24

Motor vehicle accident 1 in 84

Suicide 1 in 119

Falling 1 in 218

Firearm assault 1 in 314

Pedestrian accident 1 in 626

Drowning 1 in 1,008

Motorcycle accident 1 in 1,020

Fire or smoke 1 in 1,113

Bicycling accident 1 in 4,919

The four steps of risk assessment for adverse health effects FIGURE 4.2

1 Hazard identification
Does exposure to substance cause increased likelihood of adverse health effect such as cancer or birth defects?

2 Dose–response assessment
What is the relationship between amount of exposure (dose) and seriousness of adverse health effect? A person exposed to a low dose may have no symptoms, whereas exposure to a high dose may result in illness.

3 Exposure assessment
How much, how often, and how long are humans exposed to substance in question? Where humans live relative to emissions is also considered.

4 Risk characterization
What is probability of individual or population having adverse health effect? Risk characterization evaluates data from dose–reponse assessment and exposure assessment (steps 2 and 3). Risk characterization indicates that Mexican-Americans, many of whom are agricultural workers, are more vulnerable to pesticide exposure than other groups (see graph).

Agricultural workers have a greater than average exposure to chemicals such as pesticides.

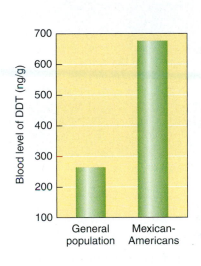

Each of us uses our intuition, habit, and experience to make many decisions regarding risk every day. However, environmental and health risks often affect many individuals, and the best choices cannot always be made on an intuitive level. **Risk management** is the process of identifying, assessing, and reducing risks. The four steps involved in **risk assessment** for adverse health effects are summarized in **FIGURE 4.2**. Once a risk assessment is performed, its results are combined with relevant political, social, and economic considerations to deter-

mine whether we should reduce or eliminate a particular risk and, if so, what we should do. This evaluation includes the development and implementation of laws to regulate hazardous substances.

Risk assessment helps estimate the probability that an event will occur and lets us set priorities and manage risks in an appropriate way. As an example, consider a person who smokes a pack of cigarettes a day and drinks well water containing traces of the cancer-causing chemical trichloroethylene (in acceptable amounts,

risk assessment
The use of statistical methods to quantify risks so they can be compared and contrasted.

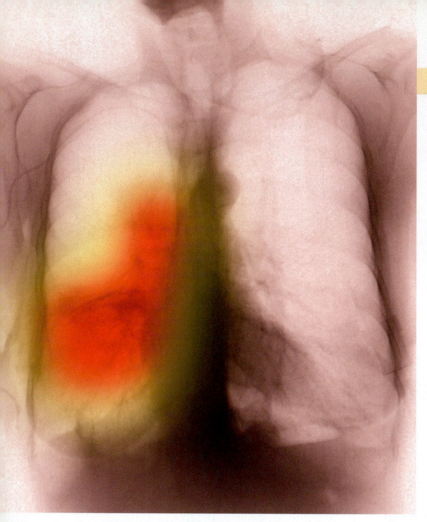

Lung cancer FIGURE 4.3

Cancer was diagnosed in the right lung of this 73-year-old woman (shown by the red areas) after years of heavy smoking, a high-risk-behavior. Fortunately, some data suggest that smoking is on the decline among younger people.

Percentage of U.S. students (grades 9–12) who smoked cigarettes on 20 or more of the past 30 days TABLE 4.1			
Year	Total	Females	Males
2005	9.3	9.3	9.3
2003	9.7	9.7	9.6
2001	13.8	12.9	14.9
1999	16.8	15.6	17.9
1997	16.7	15.7	17.6
1995	16.1	15.9	16.3
1993	13.8	13.5	14.0
1991	12.7	12.4	13.0

as established by the Environmental Protection Agency [EPA]). Without a knowledge of risk assessment, this person might buy bottled water in an attempt to reduce his or her chances of getting cancer. Based on risk assessment calculations, the annual risk from smoking is 0.00059, or 5.9×10^{-4}, whereas the annual risk from drinking water with EPA-accepted levels of trichloroethylene is 0.000000002, or 2.0×10^{-9}. This means that this person is almost *300,000 times* more likely to get cancer from smoking than to get it from ingesting such low levels of trichloroethylene (FIGURE 4.3). Knowing this, the person in our example would, we hope, stop smoking.

One of the most perplexing dilemmas of risk assessment is that people often ignore substantial risks but get extremely upset about minor risks. The average life expectancy of smokers is more than eight years shorter than that of nonsmokers, and almost one-third of all smokers die from diseases that the habit causes or exacerbates. Yet many people get more upset over a one-in-a-

million chance of getting cancer from pesticide residues on food than they do over the relationship between smoking and cancer. Perhaps part of the reason for this attitude is that behaviors such as diet, smoking, and exercise are parts of our lives that we can change if we choose to (TABLE 4.1). Risks over which most of us have no control, such as pesticide residues or nuclear wastes, tend to evoke more fearful responses.

CONCEPT CHECK STOP

What are risk and risk assessment?

What are the four steps of risk assessment?

Environmental Health Hazards

LEARNING OBJECTIVES

Define *toxicology* and distinguish between acute toxicity and chronic toxicity.

Explain why public water supplies are monitored for fecal coliform bacteria despite the fact that most strains of *E. coli* do not cause disease.

Describe the link between environmental changes and emerging diseases, such as avian influenza.

The human body is exposed to many kinds of chemicals in the environment. Both natural and synthetic chemicals are in the air we breathe, the water we drink, and the food we eat. All chemicals, even "safe" chemicals such as sodium chloride (table salt), are toxic if exposure is high enough. For example, a 1-year-old child will die from ingesting about 2 tablespoons of table salt; table salt is also harmful to people with heart or kidney disease. Chemicals with adverse effects are known as **toxicants**.

Toxicology involves (a) studying the effects of toxicants on living organisms, (b) studying the mechanisms that cause toxicity, and (c) developing ways to prevent or minimize adverse effects. (Developing appropriate handling or exposure guidelines for specific toxicants is one of these ways.)

The effects of toxicants following exposure can be immediate (*acute toxicity*) or prolonged (*chronic toxicity*). Symptoms of **acute toxicity** range from dizziness and nausea to death. Acute toxicity occurs immediately to several days following a single exposure. In comparison, **chronic toxicity** generally produces damage to vital organs, such as the kidneys or liver, following long-term, low-level exposure to a toxicant. Human diseases that are the result of chronic toxicity are *noninfectious*—that is, they are not transmitted from one human host to another. Toxicologists know far less about chronic toxicity than they do about acute toxicity, partly because the symptoms of chronic toxicity often mimic those of other chronic diseases associated with risky

lifestyle patterns, poor nutrition, and aging. Also, it is difficult to isolate a causative agent from among the multiple toxicants we are routinely exposed to.

DISEASE-CAUSING AGENTS IN THE ENVIRONMENT

Disease-causing agents are *infectious* organisms, such as bacteria, viruses, protozoa, and parasitic worms that cause diseases. Typhoid, cholera, bacterial dysentery, polio, and infectious hepatitis are some of the most common bacterial or viral diseases that are transmissible through contaminated food and water. Diseases such as these are considered environmental health hazards. Other human diseases, such as acquired immunodeficiency syndrome (AIDS), are not transmissible through the environment and aren't discussed here.

The vulnerability of our public water supplies to waterborne disease-causing agents was dramatically demonstrated in 2000, when the first waterborne outbreak in North America of a deadly strain of *Escherichia coli* occurred in Ontario, Canada. Several people were killed, and several thousand became sick. Prior to this outbreak, this deadly *E. coli* strain had been transmitted almost exclusively through contaminated food.

The largest outbreak of a waterborne disease ever recorded in the United States occurred in 1993, when a microorganism (*Cryptosporidium*) contaminated the water supply in the greater Milwaukee area. About 370,000

toxicology
The study of toxicants, chemicals with adverse effects on health.

acute toxicity
Adverse effects that occur within a short period after high-level exposure to a toxicant.

chronic toxicity
Adverse effects that occur after a long period of low-level exposure to a toxicant.

Some human diseases transmitted by polluted water TABLE 4.2

Disease	Type of organism	Symptoms
Cholera	Bacterium	Severe diarrhea, vomiting; fluid loss of as much as 20 quarts per day causes cramps and collapse
Dysentery	Bacterium	Infection of the colon causes painful diarrhea with mucus and blood in the stools; abdominal pain
Enteritis	Bacterium	Inflammation of the small intestine causes general discomfort, loss of appetite, abdominal cramps, and diarrhea
Typhoid	Bacterium	Early symptoms include headache, loss of energy, fever; later, a pink rash appears, along with (sometimes) hemorrhaging in the intestines
Infectious hepatitis	Virus	Inflammation of liver causes jaundice, fever, headache, nausea, vomiting, severe loss of appetite, muscle aches, and general discomfort
Poliomyelitis	Virus	Early symptoms include sore throat, fever, diarrhea, and aching in limbs and back; when infection spreads to spinal cord, paralysis and atrophy of muscles occur
Cryptosporidiosis	Protozoon	Diarrhea and cramps that last up to 22 days
Amoebic dysentery	Protozoon	Infection of the colon causes painful diarrhea with mucus and blood in the stools; abdominal pain
Schistosomiasis	Fluke	Tropical disorder of the liver and bladder causes blood in urine, diarrhea, weakness, lack of energy, repeated attacks of abdominal pain
Ancylostomiasis	Hookworm	Severe anemia, sometimes symptoms of bronchitis

people developed diarrhea. These and similar outbreaks raise concerns about the safety of our drinking water. Because sewage-contaminated water is an environmental threat to public health, periodic tests are made for the presence of sewage in our drinking water supplies. The best indicator of sewage-contaminated water is the presence of the common intestinal bacterium *E. coli* because it doesn't appear in the environment except from human and animal feces. Tests such as those for *E. coli* are used to indicate the possible presence of various disease-causing agents (TABLE 4.2). Although most strains of coliform bacteria found in sewage do not cause disease, testing for these bacteria is a reliable way to indicate the likely presence of **pathogens** in water.

> **pathogen**
> An agent (usually a microorganism) that causes disease.

To test for the presence of *E. coli* in water, the **fecal coliform test** is performed (FIGURE 4.4). A small sample of water is passed through a filter to trap the bacteria, which are then transferred to a petri dish that contains nutrients. After an incubation period, the number of greenish colonies present indicates the number of *E. coli*. Safe drinking water should contain no more than one coliform bacterium per 100 mL of water (about ½ cup); safe swimming water should have no more than 200 per 100 mL of water; and general recreational water (for boating) should have no more than 2,000 per 100 mL. In contrast, raw sewage may contain several million coliform bacteria per 100 mL of water. Water pollution and purification are discussed further in Chapter 10.

ENVIRONMENTAL CHANGES AND EMERGING DISEASES

Human health has improved significantly over the past several decades, but environmental factors remain a significant cause of human disease in many areas of the world. **Epidemiologists**, scientists who investigate the out-

Fecal coliform test FIGURE 4.4

This test indicates the likely presence of disease-causing agents in water. A water sample is first passed through a filtering apparatus. **A** The filter disk is then placed on a medium that supports coliform bacteria for 24 hours. **B** After incubation, the number of bacterial colonies is counted. Each colony of *Escherichia coli* arose from a single coliform bacterium in the original water sample.

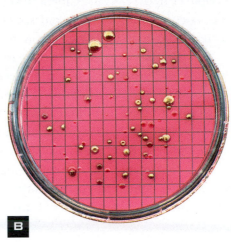

breaks of both infectious and noninfectious diseases in a population, have established links between human health and human activities that alter the environment. The World Health Organization released a 1997 report which concluded that about 25 percent of disease and injury worldwide is related to human activities that cause environmental changes. The environmental component of human health is sometimes direct and obvious, as when people drink unsanitary water and contract dysentery, a waterborne disease that causes diarrhea. Diarrhea causes 4 million deaths worldwide each year, mostly in children.

The health effects of many human activities are complex and often indirect. The disruption of natural environments may give disease-causing agents an opportunity to thrive. Development activities such as cutting down forests, building dams, and expanding agriculture may bring more humans into contact with new or previously rare disease-causing agents. Such projects may increase the population and distribution of disease-carrying organisms such as mosquitoes, thereby increasing the spread of disease (FIGURE 4.5). Social factors may also contribute to disease epidemics. Highly

Road clearing in the Amazon rain forest FIGURE 4.5

The drainage ditches that will be added to each side of the road wil hold standing water where mosquito larvae thrive.

Global Locator

concentrated urban populations promote the rapid spread of infectious organisms among large numbers of people (**FIGURE 4.6**). Global travel also has the potential to contribute to the rapid spread of disease as infected individuals move easily from one place to another.

Consider malaria, a disease that mosquitoes transmit to humans. Each year, between 300 million and 500 million people worldwide contract malaria, and it causes more than 1 million deaths. About 60 different species of *Anopheles* mosquito transmit the parasites that cause malaria. Each mosquito species thrives in its own unique combination of environmental conditions (such as elevation, amount of precipitation, temperature, relative humidity, and availability of surface water).

Another recent concern is the possibility of a pandemic of avian influenza, or bird flu. A **pandemic** is a disease that reaches nearly every part of the world and has the potential to infect almost every person. Avian influenza is a strain of influenza virus that is common in birds (**FIGURE 4.7**). It tends to be difficult for humans to contract because it is usually transferred from bird to human but not from human to human. It is extremely potent once contracted, however, and has a high fatality rate. A major concern for scientists is the possible evolution of a strain easily transferred from human to human. Such a strain, which might involve a single mutation in the viral genes, could kill millions of people in a single year.

Crowds on a street in Hong Kong, China FIGURE 4.6

The development of cities, and the concentration of people living in them, permit the rapid spread of infectious disease-causing agents.

Deaths of wild swans provided early evidence that avian influenza had migrated from Asia to Europe.

Understanding and controlling an avian influenza pandemic requires study of the environment that allows the virus to survive and travel, as well as cooperation among many governments and individuals. The virus often originates in areas that have dense populations of domestic birds, especially chickens, raised in small cages. In the past several years, large numbers of domestic poultry have been killed and burned to prevent or stop disease outbreaks. Avian flu is now endemic in domestic and wild birds in Asia, Europe, and Northern Africa, and it may have reached the United States by the time this book is published.

CONCEPT CHECK **STOP**

What is the difference between acute and chronic toxicity?

How is the incidence of malaria related to human activities that alter the environment?

Why is the fecal coliform test performed on public drinking water supplies?

Movement and Fate of Toxicants

LEARNING OBJECTIVES

Distinguish among persistence, bioaccumulation, and biological magnification of toxicants.

Discuss the mobility of persistent toxicants in the environment.

Describe the purpose of the Stockholm Convention on Persistent Organic Pollutants.

S ome chemically stable toxicants are particularly dangerous because they resist degradation and readily move around in the environment. These include certain pesticides, radioactive isotopes, heavy metals such as mercury, flame retardants such as PBDEs (polybrominated diphenyl ethers), and industrial chemicals such as PCBs (polychlorinated biphenyls).

The effects of the pesticide **DDT (dichlorodiphenyltrichloroethane)** on many bird species demonstrate the problem. Falcons, pelicans, bald eagles, and many other birds are sensitive to traces of DDT in their tissues. Substantial evidence indicates that DDT causes these birds to lay eggs with thin, fragile shells that usually break during incubation, causing the chicks' deaths. After 1972, the year DDT was banned in the United States, the reproductive success of many birds began to slowly improve.

The impact of DDT on birds is the result of (1) its persistence, (2) bioaccumulation, and (3) biological magnification. **Persistence** means that the substance is extremely stable and may take many years to break down into a less toxic form. When an organism can't

Effect of DDT on birds FIGURE 4.8

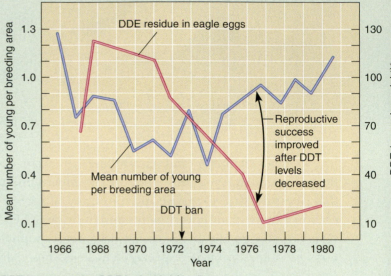

*DDT is converted to DDE in the birds' bodies

A A bald eagle feeds its chick.

B A comparison of the number of successful bald eagle offspring with the level of DDT residues in their eggs.

Biological magnification of DDT in a Long Island salt marsh FIGURE 4.9

A A ring-billed gull stands at the edge of a lake.

B Note how the level of DDT, expressed as parts per million, ▶ increased in the tissues of various organisms as DDT moved through the food chain from producers to consumers (bottom to top of figure). The ring-billed gull at the top of the food chain had approximately 1.5 million times more DDT in its tissues than the water contained.

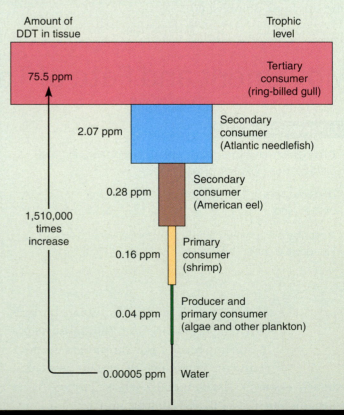

Amount of DDT in tissue | Trophic level

75.5 ppm — Tertiary consumer (ring-billed gull)

2.07 ppm — Secondary consumer (Atlantic needlefish)

0.28 ppm — Secondary consumer (American eel)

0.16 ppm — Primary consumer (shrimp)

0.04 ppm — Producer and primary consumer (algae and other plankton)

1,510,000 times increase

0.00005 ppm — Water

metabolize (break down) or excrete a toxicant, it is simply stored, usually in fatty tissues. Over time, the organism may accumulate high concentrations of the toxicant. The buildup of a persistent toxicant in an organism is **bioaccumulation** (**FIGURE 4.8**).

Organisms at the top of the food chain tend to store greater concentrations of bioaccumulated toxicants in their bodies than those lower on the food chain. As an example of **biological magnification**, consider a food chain studied in a Long Island salt marsh that was sprayed with DDT over several years for mosquito control: algae and plankton → shrimp → American eel →

biological magnification
The increase in toxicant concentrations as a toxicant passes through successive levels of the food chain.

Atlantic needlefish → ring-billed gull (**FIGURE 4.9**). All top carnivores, from fishes to humans, are at risk of health problems from biological magnification. Scientists therefore test pesticides to ensure that they do not persist and accumulate in the environment.

MOBILITY IN THE ENVIRONMENT

Persistent toxicants tend to move through the soil, water, and air, sometimes long distances. For example, pesticides applied to agricultural lands may be washed into rivers and streams by rain, harming aquatic life (**FIGURE 4.10**). If the pesticide level in their aquatic ecosystem is high enough, plants and animals may die. At lower pesticide levels, aquatic life may still suffer from symptoms of chronic toxicity such as bone degeneration in fishes. These symptoms may, for example, decrease fishes' competitiveness or increase their chances of being eaten by predators.

Mobility of persistent toxicants is also a risk for humans. In 1994 the Environmental Working Group, a private organization, analyzed five common herbicides (weed-killing chemicals) found in drinking water. It concluded that 3.5 million people in the Midwest face a slightly elevated cancer risk because of exposure to the herbicides. The EPA has since mandated a reduction in use of the five herbicides.

Mobility of pesticides in the environment FIGURE 4.10

The intended pathway of pesticides in the environment is shown in the green band across the top of the figure, and the actual pathways are shown across the bottom.

VIEW THIS IN ACTION
in your WileyPLUS course

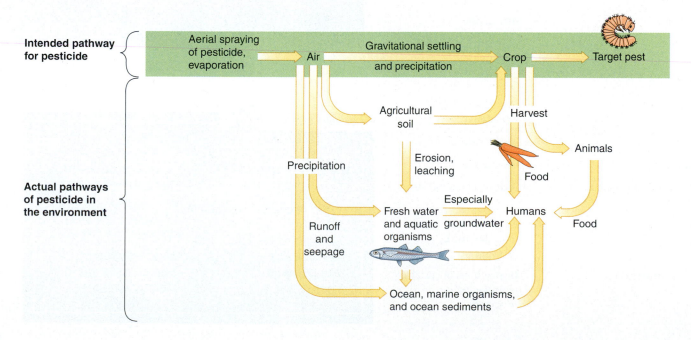

THE GLOBAL BAN OF PERSISTENT ORGANIC POLLUTANTS

The **Stockholm Convention on Persistent Organic Pollutants**, which was adopted in 2001, is an important U.N. treaty that seeks to protect human health and the environment from the 12 most toxic **persistent organic pollutants**, or **POPs**, on Earth (**TABLE 4.3**). Some POPs disrupt the endocrine system (discussed later in this chapter), cause cancer, or adversely affect the developmental processes of organisms.

persistent organic pollutants (POPs) Persistent toxicants that bioaccumulate in organisms and travel through air and water to contaminate sites far from their source.

The Stockholm Convention requires countries to develop plans to eliminate the production and use of intentionally produced POPs. A notable exception to this requirement is that DDT is still produced and used to control malaria-carrying mosquitoes in countries where no affordable alternatives exist (**FIGURE 4.11**).

Mosquito net FIGURE 4.11

This net was treated with DDT to kill malaria-transmitting mosquitoes. Photographed in Kenya.

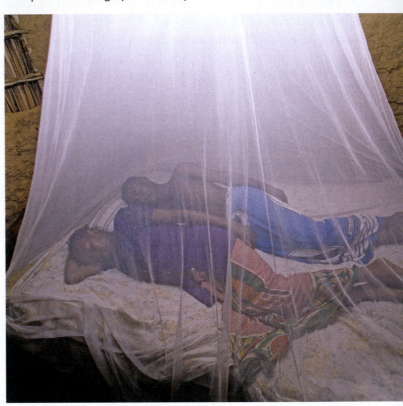

Persistent organic pollutants: The "dirty dozen" TABLE 4.3	
Persistent organic pollutant	**Use**
Aldrin	Insecticide
Chlordane	Insecticide
DDT (dichlorodiphenyl-trichloroethane)	Insecticide
Dieldrin	Insecticide
Endrin	Rodenticide and insecticide
Heptachlor	Fungicide
Hexachlorobenzene	Insecticide; fire retardant
Mirex™	Insecticide
Toxaphene™	Insecticide
PCBs (polychlorinated biphenyls)	Industrial chemicals
Dioxins	By-products of certain manufacturing processes
Furans (dibenzofurans)	By-products of certain manufacturing processes

CONCEPT CHECK STOP

What is meant by a persistent toxicant?

What problems are associated with bioaccumulation and biological magnification?

What are POPs, and why has the international community banned them?

What is the Stockholm Convention on Persistent Organic Pollutants?

How We Determine the Health Effects of Pollutants

LEARNING OBJECTIVES

Describe how a dose–response curve is used to determine the health effects of environmental pollutants.

Describe the most common method of determining whether a chemical causes cancer.

Distinguish among additive, synergistic, and antagonistic interactions in chemical mixtures.

Explain why children are particularly susceptible to toxicants.

We measure toxicity by the dose at which adverse effects are produced. A **dose** of a toxicant is the amount that enters the body of an exposed organism. The **response** is the type and amount of damage that exposure to a particular dose causes. A dose may cause death (*lethal dose*) or cause harm but not death (*sub-lethal dose*). Lethal doses, which are usually expressed in milligrams of toxicant per kilogram of body weight, vary depending on the organism's age, sex, health, and metabolism, and on how the dose was administered (all at once or over a period of time). The lethal doses, for humans, of many toxicants are known through records of homicides and accidental poisonings.

One way to determine acute toxicity is to administer different-sized doses to populations of laboratory animals, measure the responses, and use these data to predict the chemical effects on humans (**FIGURE 4.12**).

LD_{50} values for selected chemicals TABLE 4.4	
Chemical	**LD_{50} (mg/kg)***
Aspirin	1,750.0
Ethanol	1,000.0
Morphine	500.0
Caffeine	200.0
Heroin	150.0
Lead	20.0
Cocaine	17.5
Sodium cyanide	10.0
Nicotine	2.0
Strychnine	0.8

*Administered orally to rats.

Laboratory rat FIGURE 4.12

The results of chemicals administered orally to laboratory animals are extrapolated to humans.

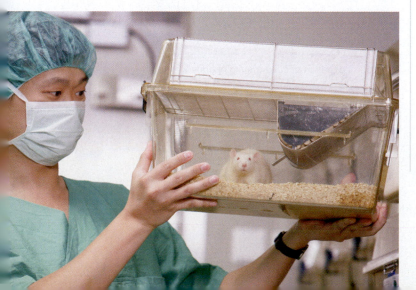

The dose that is lethal to 50 percent of a population of test animals is called the **lethal dose–50 percent**, or **LD_{50}**. It is usually reported in milligrams of chemical toxicant per kilogram of body weight. An inverse relationship exists between the size of the LD_{50} and the acute toxicity of a chemical: The smaller the LD_{50}, the more toxic the chemical, and conversely, the greater the LD_{50}, the less toxic the chemical (**TABLE 4.4**). The LD_{50} is determined for new synthetic chemicals—thousands are produced each year—as a way of estimating their toxic potential. It is generally assumed that a chemical with a low LD_{50} for several species of test animals is also very toxic in humans.

The **effective dose–50 percent**, or ED_{50}, measures a wide range of biological responses, such as stunted development in the offspring of a pregnant animal, reduced enzyme activity, or onset of hair loss. The ED_{50} is the dose that causes 50 percent of a population to exhibit whatever response is under study.

To develop a **dose–response curve**, scientists first test the effects of high doses and then work their way down to a **threshold** level, the maximum dose that has no measurable effect (or, alternatively, the minimum dose that produces a measurable effect) (**FIGURE 4.13**). Scientists assume that doses lower than the threshold level are safe.

A growing body of evidence, however, suggests that for certain toxicants there is no safe dose. A threshold does not exist for these chemicals, and even the smallest amount causes a measurable response.

dose–response curve In toxicology, a graph that shows the effects of different doses on a population of test organisms.

CANCER-CAUSING SUBSTANCES

Because cancer is so feared, for many years it was the only disease evaluated in chemical risk assessment. Environmental contaminants are linked to many serious health concerns, including other diseases, birth defects, damage to the immune system, reproductive problems, and damage to the nervous system or other body systems. We focus here on risk assessment as it relates to cancer, but non-cancer hazards are assessed in similar ways.

The most common method of determining whether a chemical causes cancer is to expose laboratory animals, such as rats, to extremely large doses of it and see whether they develop cancer. This method is indirect and uncertain, however. For one thing, although humans and rats are both mammals, they are different organisms and

---

Dose–response curves FIGURE 4.13

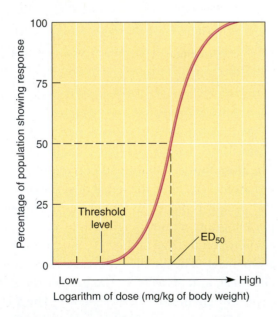

A This hypothetical dose–response curve demonstrates two assumptions: first, that the biological response increases as the dose is increased; second, that harmful responses occur only above a certain threshold level.

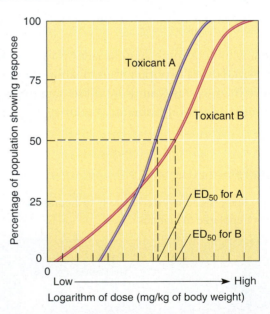

B Dose–response curves for two hypothetical toxicants, A and B. In this example, A has a lower effective dose–50 percent (ED_{50}) than B. However, at lower doses, B is more toxic than A.

may respond differently to exposure to the same chemical. (Even rats and mice often respond differently to the same toxicant.)

Another problem is that lab rats are exposed to massive doses of the suspected **carcinogen** relative to their body size, whereas humans are usually exposed to much lower amounts. Researchers must use large doses to cause cancer in a small group of laboratory animals within a reasonable amount of time. Otherwise, such tests would take years, require thousands of test animals, and be prohibitively expensive to produce enough data to have statistically significant results.

Risk assessment assumes that you can extrapolate (work backward) from the huge doses of chemicals and the high rates of cancer they cause in rats to determine the expected rates of cancer in humans exposed to lower amounts of the same chemicals. However, little evidence exists to indicate that extrapolating backward is scientifically sound. Even if you are reasonably sure that exposure to high doses of a chemical causes the same effects for the same reasons in both rats and humans, you cannot assume that these same mechanisms work at low doses in humans. The body metabolizes small and large doses of a chemical in different ways. For example, enzymes in the liver may break down carcinogens in small quantities, but an excessive amount of carcinogen might overwhelm the liver enzymes.

In short, extrapolating from one species to another and from high doses to low doses is uncertain and may overestimate or underestimate a toxicant's danger. However, animal carcinogen studies provide valuable information: A toxicant that does not cause cancer in laboratory animals at high doses is not likely to cause cancer in humans at levels found in the environment or in occupational settings.

Scientists do not currently have a reliable way to determine whether exposure to small amounts of a substance causes cancer in humans. However, the EPA is planning to change how toxic chemicals are evaluated and regulated. Toxicologists are developing methods to provide direct evidence of the risk involved in exposure to low doses of cancer-causing chemicals (**FIGURE 4.14**).

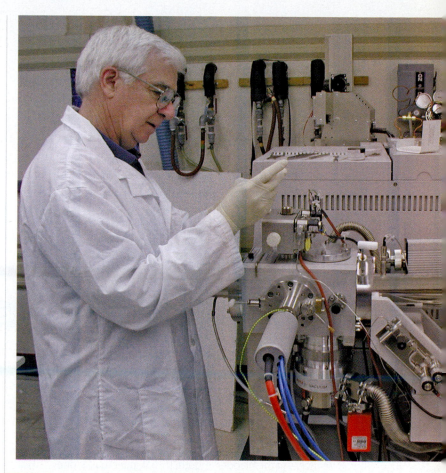

Measuring low doses of a toxicant (dioxin) in human blood serum FIGURE 4.14

Scientists are developing increasingly sophisticated methods of biomonitoring to analyze human tissues and fluids. Photographed at the Centers for Disease Control and Prevention's Environmental Health Laboratory.

Epidemiological evidence, including studies of human groups accidentally exposed to high levels of suspected carcinogens, is also used to determine whether chemicals are carcinogenic. For example, in 1989 epidemiologists in Germany established a direct link between cancer and a group of persistent organic pollutants called *dioxins* (see Table 4.3 on page 84). They observed the incidence of cancer in workers exposed to high concentrations of dioxins during an accident at a chemical plant in 1953 and found unexpectedly high levels of cancers in their digestive and respiratory tracts.

RISK ASSESSMENT OF CHEMICAL MIXTURES

Humans are frequently exposed to various combinations of chemical compounds, in the air we breathe, the food we eat, and the water we drink. For example, cigarette smoke contains a mixture of chemicals, as does automobile exhaust. Cigarette smoke is a mixture of air pollutants that includes hydrocarbons, carbon dioxide, carbon monoxide, particulate matter, cyanide, and a small amount of radioactive materials that come from the fertilizer used to grow the tobacco plants.

The vast majority of toxicology studies have been performed on single chemicals rather than chemical mixtures, and for good reason. Mixtures of chemicals interact in a variety of ways, increasing the level of complexity in risk assessment, a field already complicated by many uncertainties. Moreover, there are simply too many chemical mixtures to evaluate.

Chemical mixtures interact by **additivity**, **synergy**, or **antagonism**. When a chemical mixture is *additive*, the effect is exactly what you would expect, given the individual effects of each component of the mixture. If a chemical with a toxicity level of 1 is mixed with a differ-

---

EnviroDiscovery   SMOKING: A SIGNIFICANT RISK

Tobacco use is the single largest cause of preventable death. Smoking causes serious diseases such as lung cancer, emphysema, and heart disease and is responsible for the premature deaths of nearly half a million people in the United States each year. Cigarette smoking annually causes about 120,000 of the 140,000 deaths from lung cancer in the United States. Smoking also contributes to heart attacks and strokes and to cancers of the bladder, mouth, throat, pancreas, kidney, stomach, voice box, and esophagus.

Passive smoking, which is nonsmokers' chronic breathing of smoke from cigarette smokers, also increases the risk of cancer. Passive smokers suffer more respiratory infections, allergies, and other chronic respiratory diseases than other nonsmokers. Passive smoking is particularly harmful to infants and young children, pregnant women, the elderly, and people with chronic lung disease.

There is good news and bad news about smoking. The good news is that fewer people in highly developed nations such as the United States are smoking. The bad news is that more and more people are taking up the habit in China, Brazil, Pakistan, and other developing nations. Tobacco companies in the United States promote smoking abroad, and a substantial portion of the U.S. tobacco crop is exported. The World Health Organization (WHO) estimates that worldwide, 5 million people die each year of smoking-related causes. In an attempt to establish a global ban on tobacco advertising, WHO developed the Framework Convention on Tobacco Control. The treaty went into effect for signatory nations in 2005 but had not been ratified by the United States as of mid-2008.

A sugarcane worker smokes during a break in his long, hard day

U.S. tobacco companies export the cigarette habit abroad to compensate for the reduced consumption in the United States. Photographed in the Philippines.

---

ent chemical, also with a toxicity level of 1, the combined effect of exposure to the mixture is 2. A chemical mixture that is *synergistic* has a greater combined effect than expected; two chemicals, each with a toxicity level of 1, might have a combined toxicity of 3. An *antagonistic* interaction in a chemical mixture results in a smaller combined effect than expected; for example, two chemicals, each with a toxicity level of 1, might have a combined effect of 1.3.

If toxicological studies of chemical mixtures are lacking, how is risk assessment for chemical mixtures assigned? Toxicologists use *additivity* to assign risk to mixtures—that is, they add the known effects of each compound in the mixture. Such an approach sometimes overestimates or underestimates the actual risk involved, but it is the best method currently available. The alternative—waiting for years or decades until numerous studies have been designed, funded, and completed—is unreasonable.

CHILDREN AND CHEMICAL EXPOSURE

Because they weigh less than adults, children are more susceptible to the effects of chemicals. Consider a toxicant with an LD_{50} of 100 mg/kg. A potentially lethal dose for a child who weighs 11.3 kg (25 lb) is $100 \times 11.3 = 1130$ mg, which is equal to a scant ¼ teaspoon if the chemical is a liquid. In comparison, the potentially lethal dose for an adult who weighs 68 kg (150 lb) is 6,800 mg, or about 2 teaspoons. This exercise demonstrates that we must protect children from exposure to environmental chemicals because lethal doses are smaller for children than for adults.

Children and pollution
Consider the toxicants in air pollution. Air pollution is a greater health threat to children than it is to adults (**FIGURE 4.15**). The lungs continue to develop throughout childhood, and air pollution restricts lung development. In addition, a child has a higher metabolic rate than an adult and therefore needs more oxygen. To obtain this oxygen, a child breathes more air—about two times as much air per pound of body weight as an adult. This means that a child also breathes more air pollutants into the lungs. A 1990 study in which autopsies were performed on 100

Los Angeles children who died for reasons unrelated to respiratory problems found that more than 80 percent had subclinical lung damage, which is lung disease in its early stages, before clinical symptoms appear. (Los Angeles has some of the worst air quality in the world.)

Air pollution and respiratory disease in children FIGURE 4.15

A Honduran mother gives oxygen to her baby, who suffers from environmentally linked respiratory disease. Farmers nearby burn land to prepare for the planting season; the resulting smoke triggers breathing problems, mostly in children and the elderly.

CONCEPT CHECK STOP

What is a dose–response curve?

What are the three ways that chemical mixtures interact?

What is one way that scientists determine whether a chemical causes cancer? What are two problems with this method?

The Precautionary Principle

LEARNING OBJECTIVE

Discuss the precautionary principle as it relates to the introduction of new technologies or products.

You've probably heard the expression "An ounce of prevention is worth a pound of cure." This statement is the heart of the **precautionary principle** that many politicians and environmental activists advocate. According to the precautionary principle, we should not introduce new technology or chemicals until it is demonstrated that (a) the risks are small and (b) the benefits outweigh the risks. The precautionary principle puts the burden of proof on the developers of the new technology or substance. It asks them to prove that a

precautionary principle A practice that involves making decisions about adopting a new technology or chemical product by assigning the burden of proof of its safety to its developers.

product is safe beyond a reasonable doubt instead of society discovering its harmful effects after it has been introduced.

The precautionary principle is also applied to existing technologies when new evidence suggests that they are more dangerous than originally thought. For example, when observations and experiments suggested that lead added to gasoline as an anti-knock ingredient was contaminating soil, particularly in inner cities near major highways, the precautionary principle led to the phase-out of leaded gasoline (**FIGURE 4.16**).

To many people the precautionary principle is just common sense, given that science and risk assessment often cannot provide definitive answers to policymakers dealing with environmental and public health problems. The developers of a new technology or substance must prove that it is

Children in the South Bronx of New York City FIGURE 4.16

In addition to air pollution, these children are probably exposed to lead in the soil (from leaded gasoline) and in paint from old buildings. Children with even low levels of lead in their blood may suffer from partial hearing loss, hyperactivity, attention deficit, lowered IQ, and learning disabilities.

Precautionary principle or economic protectionism? FIGURE 4.17

Is Europe's ban of U.S. beef the result of its concern over the safety of hormone-treated beef or an excuse to support the European beef industry? Photographed on a range in Wyoming.

safe instead of society proving that it is harmful after it has already been introduced. However, the precautionary principle does not require that developers provide absolute proof that their product is safe; such proof would be impossible to provide.

Certain laws and decisions in many European Union nations have incorporated the precautionary principle, and some laws in the United States have a precautionary tone. In 2000, Christine Todd Whitman, then governor of New Jersey, said in a speech to the National Academy of Sciences,

> *Policy makers need to take a precautionary approach to environmental protection . . . We must acknowledge that uncertainty is inherent in managing natural resources, recognize it is usually easier to prevent environmental damage than to repair it later, and shift the burden of proof away from those advocating protection toward those proposing an action that may be harmful.*

The precautionary principle has generated much controversy. Some scientists fear that the precau-

tionary principle challenges the role of science and endorses making decisions without the input of science. Some critics contend that its imprecise definition reduces trade and limits technological innovations. For example, several European countries made precautionary decisions to ban beef from the United States and Canada because these countries use growth hormones to make the cattle grow faster. Europeans contend that the growth hormone might harm humans eating the beef, but the ban, in effect since 1989, is widely viewed as protecting the European beef industry (**FIGURE 4.17**). Another international controversy in which the precautionary principle is involved is the cultivation of genetically modified foods (discussed further in Chapter 14).

CONCEPT CHECK STOP

What is the precautionary principle?

What are two criticisms of the precautionary principle?

ENDOCRINE DISRUPTERS

Mounting evidence suggests that dozens of widely used industrial and agricultural chemicals are **endocrine disrupters**, which interfere with the normal actions of the endocrine system (the body's hormones) in humans and animals. These chemicals include chlorine-containing industrial compounds known as PCBs and dioxins, the heavy metals lead and mercury, pesticides such as DDT, and certain plastics and plastic additives.

Hormones are chemical messengers that organisms produce to regulate their growth, reproduction, and other important biological functions. Some endocrine disrupters mimic the *estrogens*, a class of female sex hormones. Other endocrine disrupters mimic *androgens* (male hormones such as testosterone) or *thyroid hormones*. Like hormones, endocrine disrupters are active at very low concentrations and therefore may cause significant health effects at relatively low doses.

Many endocrine disrupters appear to alter the reproductive development of various animal species.

A chemical spill in 1980 contaminated Lake Apopka, Florida's third largest lake, with DDT and other agricultural chemicals that have known estrogenic properties. In the years following the spill, male alligators had low levels of testosterone and elevated levels of estrogen. The mortality rate for eggs in this lake was extremely high, which reduced the alligator population for many years (see photo at left).

Humans may also be at risk from endocrine disrupters. The number of reproductive disorders, infertility cases, and hormonally related cancers (such as testicular cancer and breast cancer) appears to be increasing. However, we cannot make definite connections between environmental endocrine disrupters and human health problems at this time because of the limited number of human studies. Complicating such assessments is the fact that humans are also exposed to *natural*, hormone-mimicking substances in the plants we eat. For example, soy-based foods such as bean curd and soymilk contain natural estrogens.

Congress amended the *Food Quality Protection Act* and the *Safe Drinking Water Act* in 1996 to require the Environmental Protection Agency to develop a plan and establish priorities to test thousands of chemicals for their potential to disrupt endocrine systems. Chemicals testing positive are tested further to determine what specific damage, if any, they cause to reproduction and other biological functions. These tests, which may take decades to complete, should reveal the level of human and animal exposure to endocrine disrupters and the effects of this exposure.

Lake Apopka alligators

A young American alligator hatches from eggs that University of Florida researchers took from Lake Apopka, Florida. Many of the young alligators that hatch have abnormalities in their reproductive systems. This young alligator may not leave any offspring.

SUMMARY

1 A Perspective on Risks

1. A **risk** is the probability of harm (such as injury, disease, death, or environmental damage) occurring under certain circumstances. **Risk assessment** is the use of statistical methods to quantify the risks of an action so they can be compared and contrasted with other risks.

2. The four steps of risk assessment are hazard identification (Does exposure cause an increased risk of an adverse health effect?), dose–response assessment (What is the relationship between the dose and the extent of the adverse health effect?), exposure assessment (How often and how long are humans exposed to the substance?), and risk characterization (What is the probability of an adverse effect?).

2 Environmental Health Hazards

1. **Toxicology** is the study of **toxicants**, chemicals that have adverse effects on health. **Acute toxicity** is adverse effects that occur within a short period after high-level exposure to a toxicant. **Chronic toxicity** is adverse effects that occur after a long period of low-level exposure to a toxicant.

2. A **pathogen** is an agent (usually a microorganism) that causes disease. Although most strains of coliform bacteria do not cause disease, the **fecal coliform test** is a reliable way to indicate the likely presence of pathogens in water.

3. About 25 percent of disease and injury worldwide is related to human-caused environmental changes. The environmental component of human health is sometimes direct, as when people drink unsanitary water and contract a waterborne disease. The health effects of other human activities are complex and indirect, as when the disruption of natural environments gives disease-causing agents an opportunity to break out of isolation.

3 Movement and Fate of Toxicants

1. Some toxicants exhibit **persistence**—they are extremely stable in the environment and may take many years to break down into less toxic forms. **Bioaccumulation** is the buildup of a persistent toxicant in an organism's body. **Biological magnification** is the increase in toxicant concentration as a toxicant passes through successive levels of the food chain.

2. Persistent toxicants do not stay where they are applied but tend to move through the soil, water, and air, sometimes long distances. For example, pesticides applied to agricultural lands may wash into rivers and streams, harming fishes.

3. The Stockholm Convention on Persistent Organic Pollutants requires countries to eliminate the production and use of the 12 worst **persistent organic pollutants (POPs)**. POPs are a group of persistent toxicants that bioaccumulate in organisms and travel thousands of kilometers through air and water, contaminating sites far removed from their source.

4 How We Determine the Health Effects of Pollutants

1. A **dose–response curve** is a graph that shows the effect of different doses on a population of test organisms. Scientists test the effects of high doses and work their way down to a **threshold** level, the maximum dose that has no measurable effect. It is assumed that doses lower than the threshold level will not have an effect on the organism and are safe.

2. A **carcinogen** is any substance (for example, chemical, radiation, virus) that causes cancer. The most common method of determining whether a chemical is carcinogenic is to expose laboratory animals such as rats to extremely large doses of that chemical and see if they develop cancer. It is assumed that we can extrapolate from the huge doses of chemicals and the high rates of cancer they cause in rats to determine the rates of cancer expected in humans exposed to lower amounts of the same chemicals.

3. When a chemical mixture is **additive**, the effect is exactly what you would expect, given the individual effects of each component of the mixture. A chemical mixture that is **synergistic** has a greater combined effect than expected. An **antagonistic** interaction in a chemical mixture results in a smaller combined effect than expected.

4. Because they weigh less than adults, children are more susceptible to chemicals; the potentially lethal dose for a child is considerably less than the potentially lethal dose for an adult. Air pollution restricts a child's lung development. Also, a child has a higher metabolic rate than an adult and therefore breathes more air and more air pollutants into the lungs.

5 The Precautionary Principle

1. The **precautionary principle** argues for making decisions about adopting a new technology or chemical product by assigning the burden of proof of its safety to the developers of that technology or product.

KEY TERMS

- **risk** p. 74
- **risk assessment** p. 75
- **toxicology** p. 77
- **acute toxicity** p. 77
- **chronic toxicity** p. 77
- **pathogen** p. 78
- **biological magnification** p. 83
- **persistant organic pollutants (POPs)** p. 84
- **dose–response curve** p. 86
- **carcinogen** p. 87
- **precautionary principle** p. 90

CRITICAL AND CREATIVE THINKING QUESTIONS

1. Which risk—an extremely small amount of a cancer-causing chemical in drinking water or smoking cigarettes—tends to generate the greatest public concern? Explain why this view is counterproductive.

2. Should public policymakers be more concerned with public risk perception or with actual risks? Explain your answer.

3. Describe the four steps of risk assessment for adverse health effects.

4. What is the Stockholm Convention on Persistent Organic Pollutants?

5. Is the absence of scientific certainty about the health effects of an environmental pollutant synonymous with the absence of risk? Explain your answer.

6. Would you support the United States adopting the precautionary policy in all of its legislation? Why or why not?

7. Is risk assessment important in toxicology? Explain.

8. Describe how a persistent pesticide might move around in the environment.

9. Why is air pollution a greater threat to children than it is to adults?

10. Distinguish among persistence, bioaccumulation, and biological magnification.

11. How do acute and chronic toxicity differ?

12–14. The figure to the right shows the organisms sampled in the Long Island salt marsh study of DDT (also see Figure 4.9 on page 82).

12. If DDT is sprayed on land to control insects, how does it get into the bodies of aquatic species?

13. Why does the Atlantic needlefish (5) contain more DDT in its body than an American eel (4)?

14. How does high concentration of DDT cause reproductive failure in birds at the top of the food chain?

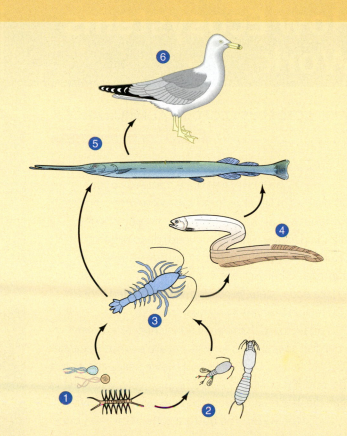

What is happening in this picture ?

■ Why do we use animal testing to determine whether a new pesticide causes cancer?

■ This mouse developed cancer after exposure to high levels of a toxicant. What uncertainties are associated with extrapolating this result to low levels of exposure of that toxicant in humans?

■ Chemical manufacturers have sometimes paid human subjects to be exposed to low doses of new chemicals, but the EPA currently has a moratorium on such testing. Suggest a possible reason for the moratorium. Do you consider such testing ethical? Why or why not?

Critical and Creative Thinking Questions   95

LAKE VICTORIA'S ECOLOGICAL IMBALANCE

VIEW THIS IN ACTION
in your WileyPLUS course

Africa's Lake Victoria, the world's second largest freshwater lake, was once home to about 400 species of small, colorful fishes known as cichlids (pronounced "SICK'-lids"). Some of these cichlids grazed on algae; others consumed dead organic material at the lake bottom; still others ate insects, shrimp, or other cichlids. They thrived throughout the lake ecosystem, providing much-needed protein to 30 million humans living near the lake.

Today, more than half of the cichlids and other native fish species in Lake Victoria are extinct. Because most of the algae-eating cichlids have disappeared, the algal population has increased explosively. When these algae die, their decomposition uses up the dissolved oxygen in the water. The bottom zone of the lake, once filled with cichlids, is empty because it contains too little dissolved oxygen. Any fishes venturing into the oxygen-free zone suffocate. Local fishermen, who once caught and ate hundreds of types of fishes, now catch only a few types.

How did this happen? In the early 1960s, the Nile perch (see inset) was introduced into the lake to stimulate the local economy. For about 20 years, as its population slowly increased, the Nile perch didn't have an appreciable effect on the lake. But in 1980, fishermen noticed increasing catches of Nile perch and decreasing harvests of native fishes. By 1985, most of the catch was Nile perch, which was increasing in number because it had an abundant food supply—the cichlids.

In 1988 the water hyacinth, a South American plant, invaded Lake Victoria, adding to the ecological havoc (see large photo). Efforts to release hyacinth-eating weevils into the lake, supplemented by chemical and mechanical treatment, eventually provided effective control of the weed.

Ecological imbalances such as that in Lake Victoria can occur with disruptions large and small, from the clearing of a small forest to the pollution of a continent's largest river.

What Is Ecology?

I n the 19th century the German biologist Ernst Haeckel first developed the concept of ecology and devised its name—*eco* from the Greek word for "house" and *logy* from the Greek word for "study." Thus, **ecology** literally means "the study of one's house." The environment—one's house—consists of two parts: the *biotic* (living) environment, which includes all organisms, and the *abiotic* (nonliving, or physical) environment, which includes such physical factors as living space, temperature, sunlight, soil, wind, and precipitation.

ecology The study of the interactions among organisms and between organisms and their abiotic environment.

The focus of ecology is local or global, specific or generalized, depending on what questions the scientist is trying to answer. One ecologist might determine the temperature or light requirements of a single oak, another might study all the organisms that live in a forest where the oak is found, and another might examine how nutrients flow between the forest and surrounding communities.

Ecology is the broadest field within the biological sciences. It is linked to every other biological discipline and to other fields as well. Geology and earth science are extremely important to ecology, especially when ecologists examine the physical environment of Earth. Chemistry and physics are also important. Humans are biological organisms, and all our activities have a bearing on ecology. Even economics and politics have profound ecological implications, as discussed in Chapter 3.

Ecologists are most interested in the levels of biological organization including and above the level of the individual organism. Individuals of the same species occur in **populations**. (A *species* is a group of similar organisms whose members freely interbreed with one another in the wild to produce fertile offspring; members of one species generally don't interbreed with other species of organisms.) A population ecologist might study a population of walruses or a population of marsh grass.

population A group of organisms of the same species that live together in the same area at the same time.

Populations are organized into **communities**. The number and kinds of species that live within a community, along with their relationships with one another, characterize the community. A community ecologist might study how organisms interact with one another—including feeding relationships (who eats whom)—in a coral reef community or in an alpine meadow community (**FIGURE 5.1**).

Ecosystem is a more inclusive term than community. An ecosystem includes all the biotic interactions of a community as well as the interactions between organisms and their abiotic environment. In an ecosystem, all the biological, physical, and chemical components of an area form an extremely complicated interacting network of energy flow and materials cycling. An ecosystem ecologist might examine how energy, nutrient composition, or water affects the organisms living in a desert community or a coastal bay ecosystem.

community
A natural association that consists of all the populations of different species that live and interact together within an area at the same time.

ecosystem
A community and its physical environment.

The ultimate goal of ecosystem ecologists is to understand how ecosystems function. They study how ecosystem processes collectively regulate the global cycles of water, carbon, nitrogen, and oxygen that are essential to the survival of humans and all other organisms. As humans increasingly alter ecosystems for their own

An alpine meadow community FIGURE 5.1

Alpine flowers put on a colorful show during their brief blooming period. Photographed in Berchtesgaden National Park, Germany.
(*Inset*) The golden-mantled ground squirrel, which is larger than a chipmunk, is a burrowing mammal common in alpine meadows in western North America.

uses, the natural functioning of ecosystems is changed, and we must learn whether these changes will affect the *sustainability* of our life-support system.

    Landscape ecology is a subdiscipline in ecology that studies the connections among ecosystems. Consider, for example, a **landscape** consisting of a forest ecosystem and a pond ecosystem located adjacent to the forest. One possible connection between these two ecosystems is the great blue

landscape A region that includes several interacting ecosystems.

biosphere The layer of Earth that contains all living organisms.

heron, which eats fish, frogs, insects, crustaceans, and snakes along the shallow water of the pond but often builds nests and raises its young in the secluded treetops of the nearby forest (FIGURE 5.2 next page). Landscapes, then, are based on larger land areas that include several ecosystems.

    The organisms of the **biosphere**— Earth's communities, ecosystems, and landscapes—depend on one another and on the other realms of Earth's physical

A A great blue heron searches for food in shallow water.

B Herons usually nest in trees adjacent to the pond, lake, or other wetland ecosystem where they feed.

environment: the atmosphere, hydrosphere, and lithosphere. The *atmosphere* is the gaseous envelope surrounding Earth; the *hydrosphere* is Earth's supply of water—liquid and frozen, fresh and salty; and the *lithosphere* is the soil and rock of Earth's crust. Ecologists who study the biosphere examine the global interrelationships among Earth's atmosphere, land, water, and organisms.

The biosphere teems with life. Where do these organisms get the energy to live? And how do they harness this energy? We now examine the importance of en-

ergy to organisms, which survive only as long as the environment continuously supplies them with energy. We will revisit energy as it relates to human endeavors in many chapters throughout this text.

CONCEPT CHECK STOP

What is the difference between a community and an ecosystem?

What is the difference between an ecosystem and a landscape?

The Flow of Energy Through Ecosystems

Energy is the capacity or ability to do work. Organisms require energy to grow, move, reproduce, and maintain and repair damaged tissues. Energy exists as stored energy—called **potential energy**—and as **kinetic energy**, the energy of motion. We can think of potential energy as an arrow on a drawn bow (**FIGURE 5.3**). When the string is released, this potential energy is converted to kinetic energy as the motion of the bow propels the arrow. Similarly, the grass a bison eats has chemical potential energy, some of which is converted to kinetic energy and heat as the bison runs across the prairie. Thus, energy changes from one form to another.

THE FIRST AND SECOND LAWS OF THERMODYNAMICS

Thermodynamics is the study of energy and its transformations. Two laws about energy apply to all things in the universe: the first and second laws of thermodynamics. According to the **first law of thermodynamics**, an organism may absorb energy from its surroundings, or it may give up some energy into its

> **first law of thermodynamics**
> A physical law which states that energy cannot be created or destroyed, although it can change from one form to another.

Potential and kinetic energy FIGURE 5.3

Potential energy is stored in the drawn bow (**A**) and is converted to kinetic energy (**B**) as the arrow speeds toward its target. Photographed in Athens, Greece, during the 2004 Summer Olympics.

The sun powers photosynthesis, producing chemical energy stored in the leaves and seeds of this umbrella tree. Photographed in Hanging Rock State Park, North Carolina.

surroundings, but the total energy content of the organism and its surroundings is always the same. An organism can't create the energy it requires to live. Instead, it must

> **photosynthesis**
> The biological process that captures light energy and transforms it into the chemical energy of organic molecules, which are manufactured from carbon dioxide and water.

capture energy from the environment to use for biological work, a process that involves the transformation of energy from one form to another. In **photosynthesis**, for example, plants absorb the radiant energy of the sun and convert it into the chemical energy contained in the bonds of sugar molecules (**FIGURE 5.4**). Later, an animal that eats the plant may transform some of the chemical energy into the mechanical energy of muscle contraction, enabling the animal to walk, run, slither, fly, or swim.

As each energy transformation occurs, some of the energy is changed to heat energy that is released into the cooler surroundings. No organism can ever use this energy again for biological work; it is "lost" from the biological point of view. However, it isn't gone from a ther-

modynamic point of view because it still exists in the surrounding physical environment. The use of food to enable you to walk or run doesn't destroy the chemical energy that was once present in the food molecules. After you have performed the task of walking or running, the energy still exists in your surroundings as heat energy.

According to the **second law of thermodynamics**, the amount of usable energy available to do work in the universe decreases over time. The second law of thermodynamics is consistent with the first law—that is, the total amount of energy in the universe isn't decreasing with time. However, the total amount of energy in the universe available to do biological work is decreasing over time.

Less usable energy is more diffuse, or disorganized, than more usable energy.

> **second law of thermodynamics**
> A physical law which states that when energy is converted from one form to another, some of it is degraded into heat, a less usable form that disperses into the environment.

Entropy is a measure of this disorder or randomness. Organized, usable energy has low entropy, whereas disorganized energy such as heat has high entropy. Another way to explain the second law of thermodynamics is that entropy, or disorder, in a system tends to increase over time. As a result of the second law of thermodynamics, no process that requires an energy conversion is ever 100 percent efficient because much of the energy is dispersed as heat, resulting in an increase in entropy. For example, an automobile engine, which converts the chemical energy of gasoline to mechanical energy, is between 20 and 30 percent efficient: Only 20 to 30 percent of the original energy stored in the chemical bonds of the gasoline molecules is actually transformed into mechanical energy, or work.

PRODUCERS, CONSUMERS, AND DECOMPOSERS

The organisms of an ecosystem are divided into three categories based on how they obtain nourishment: producers, consumers, and decomposers. Virtually all ecosystems contain representatives of all three groups, which interact extensively with one another, both directly and indirectly.

Plants and other photosynthetic organisms are **producers** and manufacture large organic molecules from simple inorganic substances, generally carbon dioxide and water, usually using the energy of sunlight. Producers are potential food resources for other organisms because they incorporate the chemicals they manufacture into their own bodies. Plants are the most significant producers on land, and algae and certain types of bacteria are important producers in aquatic environments.

Animals are **consumers**—they consume other organisms as a source of food energy and bodybuilding materials. Consumers that eat producers are *primary consumers*, or *herbivores*. Grasshoppers, deer, moose, and rabbits are examples of primary consumers (**FIGURE 5.5A**). *Secondary consumers* eat primary consumers, whereas *tertiary consumers* eat secondary consumers. Both secondary and tertiary consumers are *carnivores* that eat other animals. Lions, spiders, and lizards are examples of carnivores (**FIGURE 5.5B**). Other consumers, called *omnivores*, eat a variety of organisms. Bears, pigs, and humans are examples of omnivores.

Some consumers, called detritus feeders, consume **detritus**, organic matter that includes animal carcasses, leaf litter, and feces (**FIGURE 5.5C**). Detritus feeders, such as snails, crabs, clams, and worms, are abundant in aquatic environments. Earthworms, termites, beetles, snails, and millipedes are terrestrial (land-dwelling) detritus feeders. Detritus feeders work together with microbial decomposers to destroy dead organisms and waste products.

Bacteria and fungi are important examples of **decomposers**, organisms that break down dead organisms and waste products (**FIGURE 5.5D**). Decomposers release simple inorganic molecules, such as carbon dioxide and mineral salts, which producers can then reuse.

Consumers and decomposers FIGURE 5.5

A The moose is an herbivore, or a primary consumer. The chemical energy stored in grasses transfers to the moose cow as it eats.

C A crab forages for detritus along stones near a stream. Photographed in the Philippines.

D These mushrooms are growing on a dead beech tree in Ostmuritz/Serrahn National Park, Germany. The mushrooms you see are reproductive structures; the invisible branching, threadlike body of the mushroom grows in the tree trunk, decomposing dead organic material.

B A Madagascar day gecko (a tertiary consumer) feeds on a spider (a secondary consumer). Both the gecko and the spider are carnivores.

THE PATH OF ENERGY FLOW IN ECOSYSTEMS

In an ecosystem, **energy flow** occurs in **food chains**, in which energy from food passes from one organism to the next in a sequence. When a food chain is diagrammed, it consists of a series of arrows, each of which points from the species that is consumed to the species that uses it as food (**FIGURE 5.6**). Each level, or "link," in a food chain is a **trophic level** (the Greek *tropho* means "nourishment"). An organism is assigned a trophic level based on the number of energy transfer steps to that level.

Producers form the first trophic level, primary consumers form the second trophic level, secondary consumers form the third trophic level, and so on. At every step in a food chain are decomposers, which respire organic molecules in the carcasses and body wastes of all members of the food chain.

Simple food chains rarely occur in nature because few organisms eat just one kind of organism. The plants, for example, are eaten by a variety of insects, birds, and mammals; and most of these herbivores are consumed by several different predators. Thus, the flow of energy through an ecosystem typically takes place in accordance with a range of food choices for each organism involved. In an ecosystem of average complexity, numerous alternative pathways are possible. An owl eating a rabbit is a different energy pathway than an owl eating a snake. A **food web**, a complex of interconnected food chains in an ecosystem, is a more realistic model of the flow of energy and materials through ecosystems (**FIGURE 5.7**). A food web

> **energy flow**
> The passage of energy in a one-way direction through an ecosystem.

Process Diagram

Energy flow through a food chain FIGURE 5.6

Much of the energy acquired by a given level of a food chain is used and escapes into the surrounding environment as heat. This energy, as the second law of thermodynamics stipulates, is unavailable to the next level of the food chain.

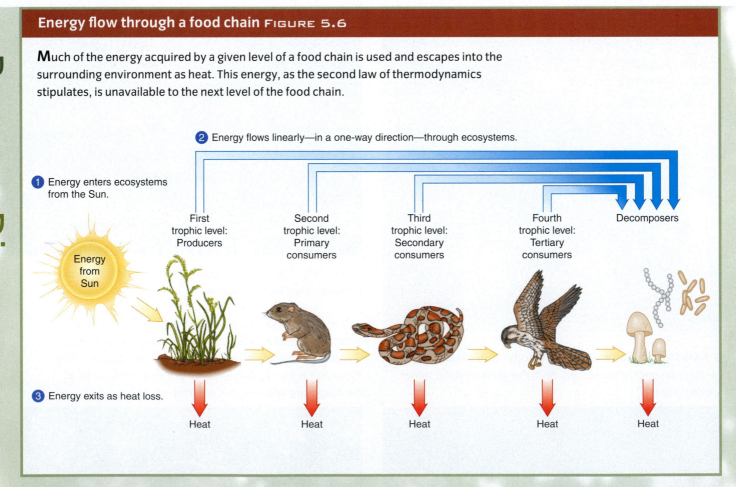

2 Energy flows linearly—in a one-way direction—through ecosystems.

1 Energy enters ecosystems from the Sun.

Energy from Sun

3 Energy exits as heat loss.

First trophic level: Producers

Second trophic level: Primary consumers

Third trophic level: Secondary consumers

Fourth trophic level: Tertiary consumers

Decomposers

Heat   Heat   Heat   Heat   Heat

Food web at the edge of an eastern U.S. deciduous forest FIGURE 5.7

This food web is greatly simplified compared to what actually happens in nature. Many species aren't included, and numerous links in the web aren't shown.

1 Pitch pine	5 Eastern chipmunk	9 Red-tailed hawk	13 American robin	16 Bacteria and fungi	20 Spiders
2 White oak	6 Eastern cottontail	10 Eastern bluebird	14 Red-headed woodpecker	17 Worms and ants	21 Insect larvae
3 Barred owl	7 Red fox	11 Red-winged blackbird		18 Moths	22 Insects
4 Gray squirrel	8 White-tailed deer	12 Blackberry	15 Red clover	19 Deer mouse	23 Fungi

helps us visualize feeding relationships that indicate how a community is organized.

The most important thing to remember about energy flow in ecosystems is that it is *linear*, or *one-way*. Energy moves along a food chain or food web from one organism to the next as long as it isn't used for biological work. Once an organism uses energy, it is lost as heat (recall the second law of thermodynamics) and is unavailable for any other organism in the ecosystem.

Organisms at each step of a food chain use a large amount of the potential energy available to them. Because this energy is ultimately lost into the environment as heat, the number of steps in any food chain is limited. The longer the food chain, the less energy is available for organisms at the higher trophic levels.

The Cycling of Matter in Ecosystems

LEARNING OBJECTIVE

Diagram and explain the carbon, hydrologic, nitrogen, sulfur, and phosphorus cycles.

In contrast to energy flow, matter, the material of which organisms are composed, moves in numerous cycles from one part of an ecosystem to another—from one organism to another and from living organisms to the abiotic environment and back again. We call these cycles of matter **biogeochemical cycles** because they involve biological, geological, and chemical interactions. Five different biogeochemical cycles of matter—carbon, hydrologic, nitrogen, sulfur, and phosphorus—are representative of all biogeochemical cycles. These five cycles are particularly important to organisms, for these materials make up the chemical compounds of cells.

THE CARBON CYCLE

Proteins, carbohydrates, and other molecules that are essential to living organisms contain carbon, so organisms must have carbon available to them. Carbon makes up approximately 0.04 percent of the atmosphere as a gas, carbon dioxide (CO_2). It is present in the ocean in several chemical forms, such as carbonate (CO_3^{2-}) and bicarbonate (HCO_3^-) and in rocks such as limestone. The global movement of carbon between the abiotic environment, including the atmosphere and ocean, and organisms is the *carbon cycle* (**FIGURE 5.8**).

Photosynthesis removes carbon (as CO_2) from the abiotic environment and fixes it into the biological compounds of producers. Carbon dioxide is returned to the atmosphere when living organisms—producers, consumers, and decomposers—respire. During respiration, sugar is broken down to carbon dioxide, and energy is released. A similar carbon cycle occurs in aquatic ecosystems, involving carbon dioxide dissolved in the water.

The carbon cycle FIGURE 5.8

The movement of carbon between the abiotic environment (the atmosphere and ocean) and living organisms is known as the carbon cycle. Because proteins, carbohydrates, and other molecules contain carbon, the process is essential to life.

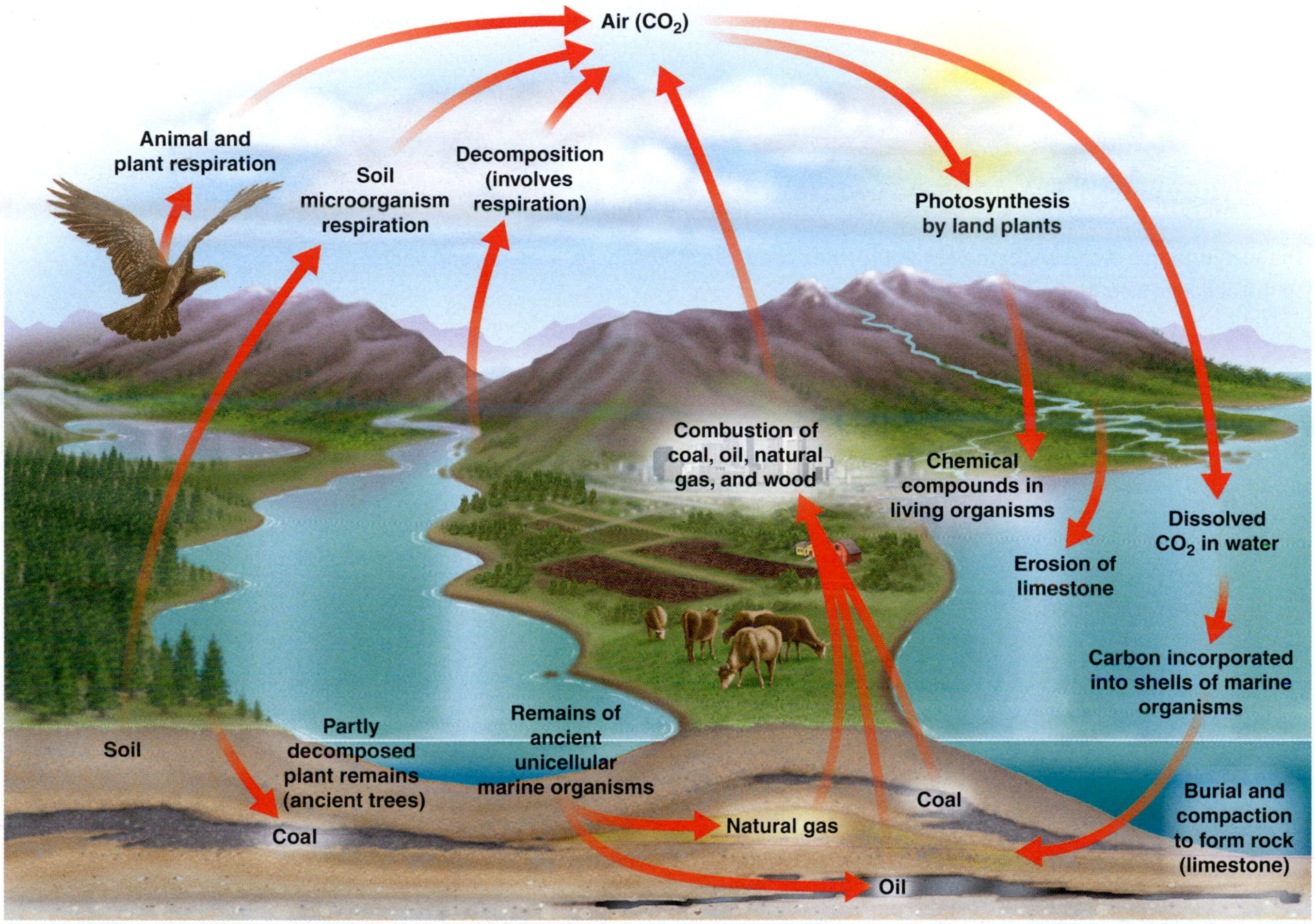

Air (CO_2)

Animal and plant respiration

Soil microorganism respiration

Decomposition (involves respiration)

Photosynthesis by land plants

Combustion of coal, oil, natural gas, and wood

Chemical compounds in living organisms

Dissolved CO_2 in water

Erosion of limestone

Carbon incorporated into shells of marine organisms

Soil

Partly decomposed plant remains (ancient trees)

Remains of ancient unicellular marine organisms

Coal

Natural gas

Coal

Oil

Burial and compaction to form rock (limestone)

Sometimes the carbon in biological molecules isn't recycled back to the abiotic environment for quite a while. For example, a large amount of carbon is stored in the wood of trees, where it may stay for several hundred years or even longer. Coal, oil, and natural gas, called *fossil fuels* because they formed from the remains of ancient organisms, are vast deposits of carbon compounds—the end products of photosynthesis that occurred millions of years ago. In *combustion*, organic molecules in wood,

coal, oil, and natural gas are burned, with accompanying releases of heat, light, and carbon dioxide.

The thick deposits of shells of marine organisms contain carbon. These shells settle to the ocean floor and are eventually cemented together to form the sedimentary rock limestone. The crust is dynamically active, and over millions of years, sedimentary rock on the bottom of the seafloor may lift to form land surfaces. The summit of Mt. Everest, for example, is composed of sedimentary rock.

THE HYDROLOGIC CYCLE

You are probably most familiar with the *hydrologic cycle*. Water continuously circulates from the ocean to the atmosphere to the land and back to the ocean. It provides a renewable supply of purified water for terrestrial organisms. This cycle results in a balance among water in the ocean, on the land, and in the atmosphere (**Figure 5.9**).

Water may evaporate from land and reenter the atmosphere directly. Alternatively, it may flow in rivers and streams to coastal *estuaries*, bodies of water with access to both the open ocean and fresh water. The movement of water from land to rivers, lakes, wetlands, and ultimately the ocean is *runoff*, and the area of land where runoff drains is a *watershed*. Regardless of its physical form—solid, liquid, or vapor—or location, every molecule of water eventually moves through the hydrologic cycle.

Process Diagram

The hydrologic cycle FIGURE 5.9

In the hydrologic cycle, water moves among the ocean, the atmosphere, the land, and back to the ocean in a continuous process that supports life.

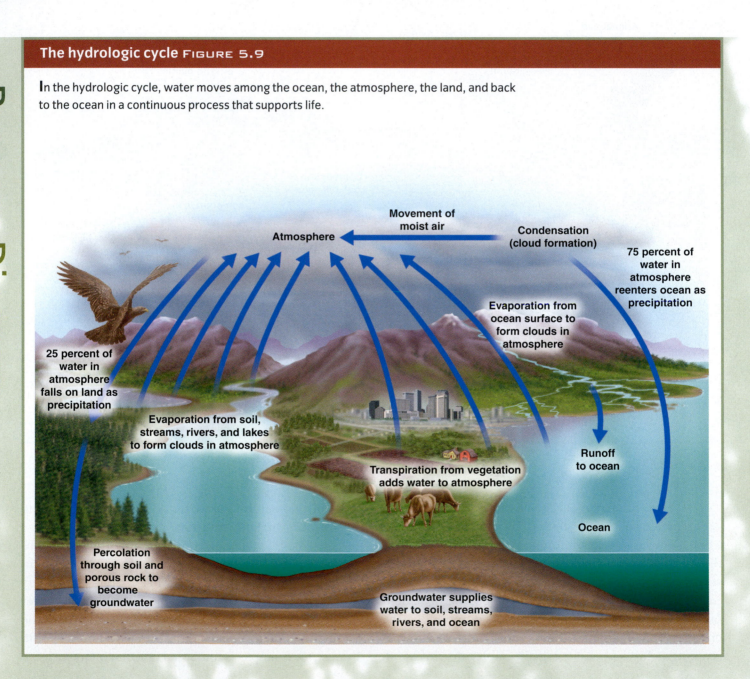

Movement of moist air

Atmosphere

Condensation (cloud formation)

75 percent of water in atmosphere reenters ocean as precipitation

Evaporation from ocean surface to form clouds in atmosphere

25 percent of water in atmosphere falls on land as precipitation

Evaporation from soil, streams, rivers, and lakes to form clouds in atmosphere

Transpiration from vegetation adds water to atmosphere

Runoff to ocean

Ocean

Percolation through soil and porous rock to become groundwater

Groundwater supplies water to soil, streams, rivers, and ocean

THE NITROGEN CYCLE

Nitrogen is an essential part of biological molecules such as proteins and nucleic acids (for example, DNA). At first glance, a shortage of nitrogen for organisms appears impossible. The atmosphere is 78 percent nitrogen gas (N_2). But atmospheric nitrogen is so stable that it does not readily combine with other elements. Atmospheric nitrogen must first break apart before the nitrogen atoms combine with other elements to form proteins

and nucleic acids. There are five steps in the *nitrogen cycle*, in which nitrogen cycles between the abiotic environment and organisms: nitrogen fixation, nitrification, assimilation, ammonification, and denitrification (**FIGURE 5.10**).

Nitrogen-fixing bacteria carry out *nitrogen fixation* in soil and aquatic environments. Nitrogen-fixing bacteria split atmospheric nitrogen and combine the resulting nitrogen atoms with hydrogen. Some nitrogen-fixing bacteria, *Rhizobium*, live inside swellings, or nodules, on the

Process Diagram

The nitrogen cycle FIGURE 5.10

The movement of nitrogen between the abiotic environment (primarily the atmosphere) and living organisms is known as the nitrogen cycle. The five steps of the nitrogen cycle are nitrogen fixation, nitrification, assimilation, ammonification, and denitrification.

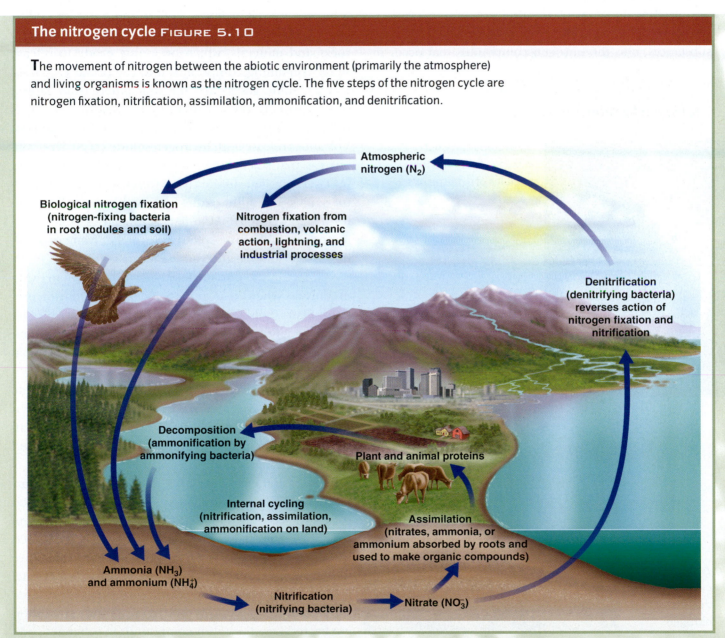

Atmospheric nitrogen (N_2)

Biological nitrogen fixation (nitrogen-fixing bacteria in root nodules and soil)

Nitrogen fixation from combustion, volcanic action, lightning, and industrial processes

Denitrification (denitrifying bacteria) reverses action of nitrogen fixation and nitrification

Decomposition (ammonification by ammonifying bacteria)

Plant and animal proteins

Internal cycling (nitrification, assimilation, ammonification on land)

Assimilation (nitrates, ammonia, or ammonium absorbed by roots and used to make organic compounds)

Ammonia (NH_3) and ammonium (NH_4^+)

Nitrification (nitrifying bacteria)

Nitrate (NO_3^-)

roots of legumes such as beans or peas and some woody plants (**Figure 5.11 A**). In aquatic environments, photosynthetic bacteria called *cyanobacteria* perform most of the nitrogen fixation (**Figure 5.11 B**).

During *nitrification*, soil bacteria convert ammonia to nitrate. The process of nitrification furnishes these bacteria, called nitrifying bacteria, with energy. In *assimilation*, plants absorb ammonia or nitrate through their roots and convert the nitrogen into plant compounds such as proteins. Animals assimilate nitrogen when they consume plants or other animals and convert the proteins into animal proteins.

Ammonification occurs when organisms produce nitrogen-containing waste products such as urine. These substances, plus the nitrogen compounds that occur in dead organisms, are decomposed, releasing the nitrogen into the abiotic environment as ammonia. The bacteria that perform this process are called ammonifying bacteria. Other bacteria perform *denitrification*, in which nitrate is converted back to nitrogen gas. Denitrifying bacteria typically live and grow where there is little or no free oxygen. For example, they are found deep in the soil near the water table, an environment that is nearly oxygen-free.

THE SULFUR CYCLE

Scientists are still piecing together how the global sulfur cycle works. Most sulfur is underground in sedimentary rocks and minerals, which over time erode to release sulfur-containing compounds into the ocean (**Figure 5.12**). Sulfur gases enter the atmosphere from natural sources in both the ocean and land. Sea spray delivers sulfates into the air, as do forest fires and dust storms. Volcanoes release both hydrogen sulfide (H_2S), a poisonous gas that smells like rotten eggs, and sulfur oxides (SO_x). Hydrogen sulfide reacts with oxygen to form sulfur oxides, and sulfur oxides react with water to form sulfuric acid (H_2SO_4). Although sulfur gases comprise a minor part of the atmosphere, the total movement of sulfur to and from the atmosphere is substantial.

A tiny fraction of global sulfur is present in living organisms, where it is an essential component of proteins. Plant roots absorb sulfate and incorporate the sulfur into plant proteins. Animals assimilate sulfur when they consume plant proteins and convert them to animal proteins. In the ocean, certain marine algae release a compound that bacteria convert to dimethyl sulfide (DMS). DMS is released into the atmosphere, where it helps condense water into droplets in clouds and may affect weather and climate. Atmospheric DMS is converted to sulfate, most of which is deposited into the ocean.

As in the nitrogen cycle, bacteria drive the sulfur cycle. In freshwater wetlands, tidal flats, and flooded soils, which are oxygen-deficient, certain bacteria convert sulfates to hydrogen sulfide gas, which is released into the atmosphere, or to metallic sulfides, which are deposited as rock. In the absence of oxygen, other bacteria perform a type of photosynthesis that uses hydrogen sulfide instead of water. Where oxygen is present, different bacteria oxidize sulfur compounds to sulfates.

Nitrogen fixation Figure 5.11

A Bacteria carry out nitrogen fixation in the nodules of a pea plant's roots.

B *Nostoc* is a cyanobacterium that fixes nitrogen.

The sulfur cycle FIGURE 5.12

The largest sources of sulfur on Earth are sedimentary rock and the ocean. In the sulfur cycle, sulfur compounds are incorporated into organisms and move among them, the atmosphere, the ocean, and land.

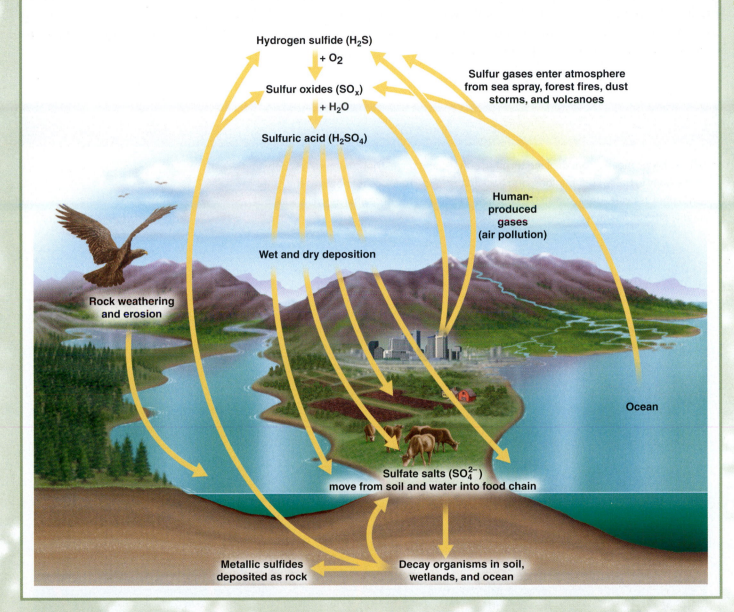

Hydrogen sulfide (H_2S)

+ O_2

Sulfur oxides (SO_x)

+ H_2O

Sulfuric acid (H_2SO_4)

Sulfur gases enter atmosphere from sea spray, forest fires, dust storms, and volcanoes

Human-produced gases (air pollution)

Wet and dry deposition

Rock weathering and erosion

Ocean

Sulfate salts (SO_4^{2-}) move from soil and water into food chain

Metallic sulfides deposited as rock

Decay organisms in soil, wetlands, and ocean

THE PHOSPHORUS CYCLE

Unlike the biogeochemical cycles just discussed, the phosphorus cycle doesn't have an atmospheric component. Phosphorus cycles from the land into living organisms, then from one organism to another, and finally back to the land (**FIGURE 5.13**). The erosion of phosphorus-containing minerals releases phosphorus into the soil, where plant roots absorb it in the form of inorganic phosphates. Phosphates are used in biological molecules such as nucleic acids and ATP, a compound that is important in energy transfer reactions in cells. Like carbon and nitrogen, phosphorus moves through the food web as one organism consumes another.

Phosphorus cycles through aquatic communities in much the same way it does through terrestrial communities. Dissolved phosphorus enters aquatic communities as algae and plants absorb and assimilate it; plankton and larger organisms obtain phosphorus when they consume the algae and plants. A variety of fishes and molluscs eat plankton in turn. Ultimately, decomposers release inorganic phosphorus into the water, where it is available for aquatic producers to use again.

Process Diagram

The phosphorus cycle FIGURE 5.13

Phosphorus moves from the land through aquatic and terrestrial communities, between organisms in these communities, and back to the land in a process known as the phosphorus cycle. Unlike other biogeochemical cycles, the phosphorus cycle does not involve the atmosphere.

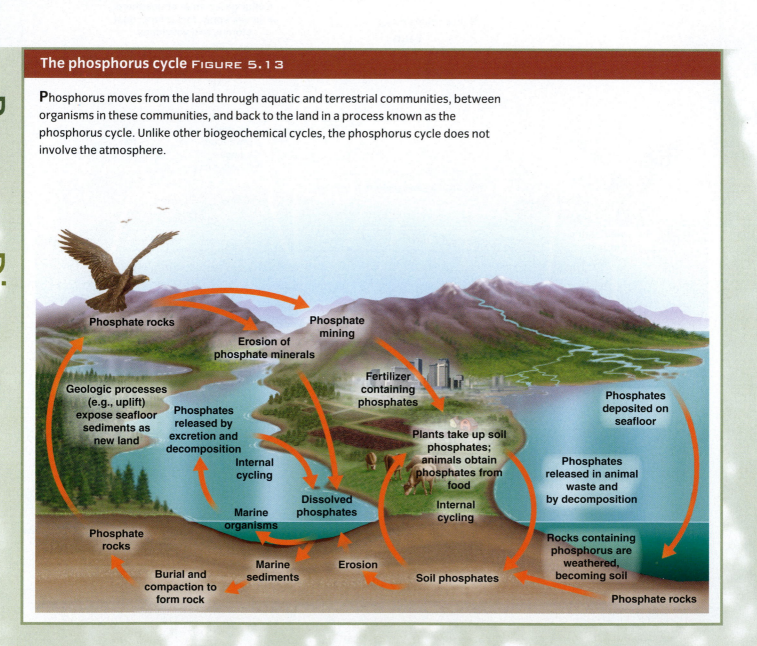

Phosphate can be lost from biological cycles. Some phosphate is carried from the land by streams and rivers to the ocean, where it can be deposited on the seafloor and remain for millions of years.

A small portion of the phosphate in the aquatic food web finds its way back to the land. A few fishes and aquatic invertebrates are eaten by seabirds, which may defecate on land where they roost. The manure of sea birds, called guano, contains large amounts of phosphate and nitrate. Once on land, these minerals may be absorbed by the roots of plants. The phosphate contained in guano may enter terrestrial food webs in this way, although the amounts involved are small.

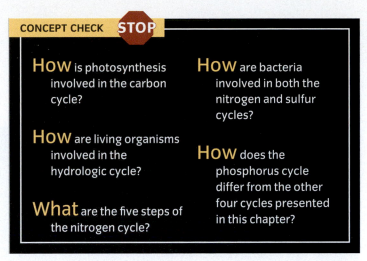

CONCEPT CHECK STOP

How is photosynthesis involved in the carbon cycle?

How are living organisms involved in the hydrologic cycle?

What are the five steps of the nitrogen cycle?

How are bacteria involved in both the nitrogen and sulfur cycles?

How does the phosphorus cycle differ from the other four cycles presented in this chapter?

Ecological Niches

LEARNING OBJECTIVES

Describe the factors that contribute to an organism's ecological niche.

Explain the concept of resource partitioning.

You have seen that a diverse assortment of organisms inhabits each community and that these organisms obtain nourishment in a variety of ways. You have also considered energy flow and biogeochemical cycles. Now let's examine the way of life of a given species in its ecosystem. An ecological description of a species typically includes whether it is a producer, consumer, or decomposer. However, we need other details to provide a complete picture.

Every organism is thought to have its own role, or **ecological niche**, within the structure and function of an ecosystem. The ecological niche describes a species' place and function within a complex system of biotic and abiotic factors.

An ecological niche is difficult to define precisely because it takes into account all aspects of the organism's existence—all physical, chemical, and biological factors the organism needs to survive, remain healthy, and reproduce. Among other things, the niche includes the local environment in which an organism lives—its **habitat**. An organism's niche also encompasses what it eats, what organisms eat it, what organisms it competes with, and how the abiotic components of its environment, such as light, temperature, and moisture, interact with and influence it. A complete description of an organism's ecological niche involves numerous dimensions to explain when, where, and how an organism makes its living.

An organism's ecological niche may be much broader potentially than it actually is in nature. Put differently, an organism is potentially capable of using much more of its environment's resources or of living in a wider assortment of habitats than it actually does. The potential, idealized ecological niche of an organism is its **fundamental niche**, but various factors, such as competition with other species, usually exclude it from part of its fundamental niche. The lifestyle an organism actually

ecological niche

The totality of an organism's adaptations, its use of resources, and the lifestyle to which it is fitted.

The wildebeest's realized niche FIGURE 5.14

Wildebeests live in herds on the open plains of East Africa and graze on the grasses. When the dry season begins, wildebeests living in East Africa's Serengeti Plain migrate more than 1,200 km (746 mi) in search of water and grass. Females usually bear one calf at a time. Adult wildebeests are swift runners. Lions and crocodiles prey on adult wildebeests. Hyenas, lions, cheetahs, leopards, and wild dogs prey on the calves.

pursues and the resources it actually uses make up its **realized niche** (**FIGURE 5.14**).

When two species are similar, their ecological niches may appear to overlap. However, many ecologists think no two species indefinitely occupy the same niche in the same community. Resource partitioning is one way some species avoid or at least reduce niche overlap. **Resource partitioning** is the reduction in competition for environmental resources such as food among coexisting species as a result of the niche of each species differing from the niches of others in one or more ways. Evidence of resource partitioning in animals is well documented and includes studies in tropical forests of Central and South America that demonstrate little overlap in the diets of fruit-eating birds, primates, and bats that coexist in the same habitat. Although fruits are the primary food for several hundred bird, primate, and bat species, the wide variety of fruits available has allowed fruit eaters to specialize, thereby reducing competition. Resource partitioning also includes timing of feeding, location of feeding, nest sites, and other aspects of an organism's ecological niche (see What a Scientist Sees).

CONCEPT CHECK STOP

What are three aspects of the wildebeest's realized niche?

What is resource partitioning?

Yellow-rumped warbler   Bay-breasted warbler   Cape May warbler   Black-throated green warbler   Blackburnian warbler

Resource Partitioning

Robert MacArthur's study of five American warbler species is a classic example of resource partitioning. Although it initially appeared that the niches of the species were nearly identical, MacArthur determined that individuals of each species spend most of their feeding time in different portions of spruces and other conifer trees. They also move in different directions through the canopy, consume different combinations of insects, and nest at slightly different times. The photo shows a male black-throated green warbler in a spruce tree.

Interactions Among Organisms

LEARNING OBJECTIVES

Distinguish among mutualism, commensalism, and parasitism.

Define *predation* and describe predator-prey relationships.

Define *competition* and distinguish between intraspecific and interspecific competition.

Discuss an example of a keystone species.

No organism exists independently of other organisms. The producers, consumers, and decomposers of an ecosystem interact with one another in a variety of ways, and each forms associations with other organisms. Three main types of interactions occur among species in an ecosystem: symbiosis, predation, and competition.

symbiosis
An intimate relationship or association between members of two or more species; includes mutualism, commensalism, and parasitism.

SYMBIOSIS

In **symbiosis**, one species usually lives in or on another species. The partners in a symbiotic relationship may benefit, be unaffected, or be harmed by the relationship.

Symbiosis is the result of **coevolution**, the interdependent evolution of two interacting species. Flowering plants and their animal pollinators are an excellent example of coevolution. Bees, beetles, hummingbirds, bats, and other animals transport pollen from one plant to another. During the millions of years over which these associations developed, flowering plants evolved several ways to attract animal pollinators. One of the rewards for the pollinator is food—nectar (a sugary solution) and pollen. Plants possess a variety of ways to get the pollinator's attention, most involving showy petals and scents.

While plants were acquiring specialized features to attract pollinators, animals coevolved specialized body parts and behaviors to aid pollination and obtain nectar and pollen as a reward. Coevolution is responsible for the hairy bodies of bumblebees, which catch and hold the sticky pollen for transport from one flower to another. Coevolution is also responsible for the long, curved beaks of certain Hawaiian birds that insert their beaks into tubular flowers to obtain nectar (**FIGURE 5.15**).

The thousands, or even millions, of symbiotic associations that result from coevolution fall into three categories: mutualism, commensalism, and parasitism. One example of **mutualism**, an association in which both organisms benefit, is the interaction between acacia ants and the bull's horn acacia plant (**FIGURE 5.16**). The ants make hollow nests out of thorns at the base of the plant's leaves and gain special nutrients from the leaf tips. In return, the ants effectively protect the plant from invertebrate and vertebrate herbivores and clear away competing plants. Both ant and acacia depend on this association for survival.

Commensalism is a symbiotic relationship in which one species benefits and the other is neither harmed nor helped. One example of commensalism is the relationship between a tropical tree and its *epiphytes*, smaller plants such as mosses, orchids, and ferns that live attached to the bark of the tree's branches (**FIGURE 5.17**). An epiphyte anchors itself to a tree but typically doesn't obtain nutrients or water directly from the tree. Its location on the tree enables it to obtain adequate light, water (as rainfall dripping down the branches), and required nutrient minerals (which rainfall washes out of the tree's leaves). The epiphyte benefits from the association, whereas the tree is apparently unaffected.

Parasitism is a symbiotic relationship in which one species (the *parasite*) benefits at the expense of the other (the *host*). Parasitism is a successful lifestyle; more than 100 parasites live in or on the human species (**FIGURE 5.18**). A parasite, usually much smaller

Coevolution FIGURE 5.15

This Hawaiian honeycreeper uses its gracefully curved bill to sip nectar from the long, tubular flowers of the lobelia.

Mutualism FIGURE 5.16

Most common in Central America, the acacia ant gains shelter and nutrients from the acacia plant, in turn protecting the plant from predators. Photographed in Costa Rica.

Commensalism FIGURE 5.17

Epiphytes are small plants that attach to the branches and trunks of larger trees. Photographed in the Fiji Islands.

Parasitism FIGURE 5.18

Close-up of body lice feeding on a human arm. Each louse is about 3 mm (0.12 in) long.

NATIONAL GEOGRAPHIC

than its host, obtains nourishment from its host, but although a parasite may weaken its host, it rarely kills it quickly. (A parasite would have a difficult life if it kept killing off its hosts!) Some parasites, such as ticks, live outside the host's body; other parasites, such as tapeworms, live within the host.

TABLE 5.1 summarizes the three categories of symbiosis.

Categories of symbiosis TABLE 5.1			
	Organism 1	*Organism 2*	*Characteristic of relationship*
Mutualism	Benefits	Benefits	Each organism depends on the other
Commensalism	Benefits	Not affected	Only one organism depends on the other
Parasitism	Benefits	Harmed	Host harmed, rarely killed; host usually much larger than parasite

EnviroDiscovery   BEE COLONIES UNDER THREAT

Since late 2006, many U.S. beekeepers have experienced major losses in their bee colonies, 30 to 90 percent of total individuals. The cause of these sudden declines, now known as colony collapse disorder (CCD), remains unknown. Researchers are investigating three major potential causes of CCD:

- The negative effects of pesticides;

- Damage caused by pathogens or parasites, such as *Varroa* and tracheal mites;

- Deaths resulting from viruses, including the Israeli acute-paralysis virus (IAPV), which can be spread by mites.

Bees in the United States are necessary for the pollination of many important crops, particularly nuts, fruits, vegetables, and berries. While the number of bee colonies being maintained by keepers has dropped by half during the past 50 years, the demand on bees to pollinate crop species has increased. Some agricultural researchers believe that the added stress placed on bee colonies as they are transported to carry out pollination has increased the susceptibility of these colonies to health threats. Individuals in stressed colonies appear to have compromised immune systems that likely hasten their decline when they are faced with health threats such as mites or viruses, subsequently leading to colony collapse.

Bees that have pollinated a blueberry crop in Maine gather along hives that will soon be moved to another agricultural site.

PREDATION

Predation includes both animals eating other animals (for example, herbivore–carnivore interactions) and animals eating plants (producer–herbivore interactions). Predation has resulted in an "arms race," with the coevolution of predator strategies—more efficient ways to catch prey—and prey strategies—better ways to escape the predator. An efficient predator exerts a strong selective force on its prey, and over time the prey species may evolve some sort of countermeasure that reduces the probability of its being captured. The countermeasure that the prey acquires in turn may act as a strong selective force on the predator.

The cheetah is the world's fastest animal and can sprint at 110 km (68 mi) per hour for short distances (**FIGURE 5.19**). Orcas (formerly known as killer

predation
The consumption of one species (the prey) by another (the predator).

whales), hunt in packs and often herd salmon or tuna into a cove so that they are easier to catch. Any trait that increases hunting efficiency, such as the speed of a cheetah or the intelligence of orcas, favors predators that pursue their prey. Ambush is another effective way to catch prey. The goldenrod spider is the same color as the white or yellow flowers in which it hides. This camouflage prevents unwary insects that visit the flower for nectar from noticing the spider until it is too late.

Many potential animal prey, such as woodchucks, run to their underground burrows to escape predators. Others have mechanical defenses, such as the barbed quills of a porcupine and the shell of a pond turtle. Some animals live in groups—a herd of antelope, colony of honeybees, school of anchovies, or flock of pigeons. This social behavior decreases the likelihood of a predator catching an individual unaware because the

Predation FIGURE 5.19

A cheetah is in a full sprint as it pursues possible prey. Photographed in Masai Mara National Park, Kenya.

Global Locator

A Two adult meerkats stand watch over a young meerkat pup. If one of the sentries spies a predator such as an eagle or a hawk, they will alert the other meerkats, and all will scramble into their burrows. Photographed in the Kalahari Desert, South Africa.

B Predators often overlook caterpillars that closely resemble small branches. Can you find the caterpillar? (*Hint:* The caterpillar is mimicking an adjacent twig.)

group has so many eyes, ears, and noses watching, listening, and smelling for predators (**FIGURE 5.20A**).

Chemical defenses are common among animal prey. The South American poison arrow frog has poison glands in its skin and bright warning colors that experienced predators avoid. Some animals blend into their surroundings and so hide from predators. Certain caterpillars resemble twigs so closely you would not guess that they are animals until they move (**FIGURE 5.20B**).

Plants possess adaptations that protect them from being eaten. The presence of spines, thorns, tough leathery leaves, or even thick wax on leaves discourages foraging herbivores from grazing. Other plants produce an array of protective chemicals that are unpalatable or even toxic to herbivores. The nicotine found in tobacco is so effective at killing insects that it is an ingredient in many commercial insecticides.

COMPETITION

Competition occurs when two or more individuals attempt to use an essential common resource such as food, water, shelter, living space, or sunlight. Resources are often in limited supply in the environment, and their use by one individual decreases the amount available to others. If a tree in a dense forest grows taller than surrounding trees, it absorbs more of the incoming sunlight. Less sunlight is available for nearby trees that the taller tree shades. Competition occurs among individuals within a population (*intraspecific competition*) and between different species (*interspecific competition*).

Competition isn't always a straightforward, direct interaction. Consider a variety of flowering plants that live in a young pine forest and compete with conifers for such resources as soil moisture and soil nutrient minerals. Their relationship is more involved than simple competition. The flowers produce nectar that some insect species consume; these insects also prey on needle-eating insects, reducing the number of insects feeding on pines. It is therefore difficult to assess the overall effect of flowering plants on pines. If the

competition
The interaction among organisms that vie for the same resources in an ecosystem (such as food or living space).

flowering plants were removed from the community, would the pines grow faster because they were no longer competing for necessary resources? Or would the increased presence of needle-eating insects (caused by fewer omnivorous insects) inhibit pine growth?

Short-term experiments in which one competing plant species is removed from a forest community have in several instances demonstrated improved growth for the remaining species. However, few studies have tested the long-term effects on forest species of removing one competing species. These long-term effects may be subtle, indirect, and difficult to assess. They may reduce or negate the negative effects of competition for resources.

KEYSTONE SPECIES

Certain species are more crucial to the maintenance of their ecosystem than others. Such **keystone species** are vital in determining an ecosystem's species composition and how the ecosystem functions. The fact that other species depend on or are greatly affected by the keystone species is revealed when the keystone species is removed. Keystone species are usually not the most abundant species in the ecosystem. Although present in relatively small numbers, keystone species exert a profound influence on the entire ecosystem because they often affect the available amount of food, water, or some other resource.

Identifying and protecting keystone species are crucial goals of conservation biologists because if a keystone species disappears from an ecosystem, other organisms may become more common or more rare, or they may even disappear. One example of a keystone species is a top predator such as the gray wolf (FIGURE 5.21). Where wolves were hunted to extinction, the populations of deer, elk, and other herbivores increased explosively. As these herbivores overgrazed the vegetation, plant species that couldn't tolerate such grazing pressure disappeared. Smaller animals such as insects were lost from the ecosystem because the plants they depended on for food had become less abundant. Thus, the disappearance of the wolf resulted in the ecosystem having considerably less biological diversity.

Some scientists think we should abandon the concept of keystone species because it is problematic.

Keystone species FIGURE 5.21

A The gray wolf is considered a keystone species in its ecosystem.

B Deer populations may grow exponentially when wolves, one of their key predators, disappear from an ecosystem.

For one thing, most of the information about keystone species is anecdotal. Scientists have performed few long-term studies to identify keystone species and to determine the nature and magnitude of their effects on the ecosystems they inhabit.

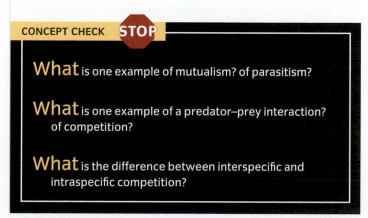

CONCEPT CHECK STOP

What is one example of mutualism? of parasitism?

What is one example of a predator–prey interaction? of competition?

What is the difference between interspecific and intraspecific competition?

GLOBAL CLIMATE CHANGE: IS THERE AN IMBALANCE IN THE CARBON CYCLE?

Hybrid vehicles such as this model of the Ford Escape allow consumers to lower their gasoline consumption and thus reduce their CO₂ emissions.

During the past two centuries, the level of carbon dioxide (CO₂) in the atmosphere has increased dramatically due to the burning of fossil fuels such as coal, oil, and natural gas, as well as the clearing and burning of forests. Environmental scientists are increasingly concerned that the rising levels of CO₂ may change Earth's climate. CO₂ levels rose from 315 parts per million (ppm) in 1958 to 382 ppm in 2006—a 21 percent increase. CO₂ in the atmosphere allows solar radiation to pass through but does not allow heat to radiate into space. Instead, the heat is radiated back to Earth's surface. As the CO₂ accumulates, it may trap enough heat to warm the planet.

Observations confirm that our climate is changing rapidly, particularly in the past few decades. The 1990s and early 2000s saw some of the warmest years since scientists started keeping records in 1820. The 10 warmest years have occurred since 1990. Earth's temperature in 2005 was the warmest in the weather records; the second warmest temperature on record occurred in 1998, and the third warmest in 2002.

Environmental scientists estimate that if trends don't change, Earth's mean temperature could rise 1.8° to 4.0°C (3.2° to 7.2°F) by the end of the 21st century. This temperature increase would make the atmosphere warmer than tree ring and glacial ice analyses estimate that it has been for the past 1,000 years. This warming could produce major shifts in patterns of rainfall and initiate melting of the West Antarctic and Southern Greenland ice sheets, as did the last warm period 120,000 years ago. Such melting would cause the ocean level to rise, which is an alarming scenario, as it might put many major cities at least partly under water.

An increasing number of environmental scientists are convinced that human-induced global climate change is under way. The U.S. National Academy of Science has stated that global climate change may be the most pressing international issue of the 21st century. In 2007, the fourth report of the U.N. Intergovernmental Panel on Climate Change (IPCC), which represents the consensus of hundreds of climate scientists worldwide, declared that the human-produced increase in greenhouse gas concentrations is likely responsible for most of the observed warming in the past 50 years.

An international climate change conference was held in Kyoto, Japan, in 1997. Although many participating countries had conflicting positions, an initial treaty known as the Kyoto Protocol was formulated. This treaty stipulated that highly developed countries must cut their emissions of CO₂ and other gases that cause warming by an average of 5.2 percent by 2012. Unless many countries ratify and observe the treaty, CO₂ emissions are expected to continue to increase. (The United States signed the protocol in 1998, but the Bush administration withdrew the United States from that commitment.)

Energy experts at the U.S. Department of Energy say that energy conservation and efficiency measures could accomplish much of the U.S. emissions reductions required if we were to observe the Kyoto Protocol. We could save the equivalent of 60 million metric tons of CO₂ emissions each year if automotive vehicles were designed to get better gas mileage (see photo). Installing high-efficiency wind turbines to generate electricity in states such as Texas and California could save as much as 20 million metric tons of CO₂ emissions per year. We say more about global climate change in Chapter 9.

SUMMARY

1 What Is Ecology?

1. **Ecology** is the study of the interaction among organisms and between organisms and their abiotic environment.

2. A **population** is a group of organisms of the same species that live together in the same area at the same time.

A **community** is a natural association that consists of all the populations of different species that live and interact together within an area at the same time. An **ecosystem** is a community and its physical environment. A **landscape** is a region that includes several interacting ecosystems. The **biosphere** is the layer of Earth that contains all living organisms.

2 The Flow of Energy Through Ecosystems

1. **Energy** is the capacity or ability to do work. According to the **first law of thermodynamics**, energy can't be created or destroyed, although it can change from one form to another. As a result of the **second law of thermodynamics**, when energy is converted from one form to another, some of it is degraded into heat, a less usable form that disperses into the environment.

2. A **producer** manufactures large organic molecules from simple inorganic substances. A **consumer** can't make its own food and uses the bodies of other organisms as a source of energy and bodybuilding materials. **Decomposers** are microorganisms that break down dead organic material and use the decomposition products to supply themselves with energy.

3. **Energy flow** is the passage of energy in a one-way direction through an ecosystem, from producers to consumers to decomposers.

3 The Cycling of Matter in Ecosystems

1. **Biogeochemical cycles** are the processes by which matter cycles from the living world to the nonliving, physical environment and back again. Carbon dioxide is the important gas of the carbon cycle; carbon enters the living world through photosynthesis and returns to the abiotic environment when organisms respire. The hydrologic cycle continuously renews the supply of water and involves an exchange of water among the land, the atmosphere, and organisms. There are five steps in the nitrogen cycle: nitrogen fixation, nitrification, ammonification, assimilation, and denitrification. In the sulfur cycle, sulfur compounds whose natural sources are the ocean and rock are incorporated by organisms into proteins and move between organisms, the atmosphere, the ocean, and land. The phosphorus cycle has no biologically important gaseous compounds; phosphorus erodes from rock and is absorbed by plant roots.

4 Ecological Niches

1. An **ecological niche** is the totality of an organism's adaptations, its use of resources, and the lifestyle to which it fits. An organism's ecological niche includes its **habitat**, its distinctive lifestyle, and its role in the community.

2. **Resource partitioning** is the reduction in competition for environmental resources, such as food, that occurs among coexisting species as a result of the niche of each species differing from the niches of other species in one or more ways.

5 Interactions Among Organisms

1. **Symbiosis** is an intimate relationship or association between members of two or more species. **Mutualism** is a symbiotic relationship in which both species benefit. **Commensalism** is a symbiotic relationship in which one species benefits and the other species is neither harmed nor helped. **Parasitism** is a symbiotic relationship in which one species (the parasite) benefits at the expense of the other (the host).

2. **Predation** is the consumption of one species (the prey) by another (the predator). During **coevolution** between predator and prey, the predator evolves more efficient ways to catch prey (such as pursuit and ambush), and the prey evolves better ways to escape the predator (such as flight, association in groups, and camouflage).

3. **Competition** is the interaction among organisms that vie for the same resources in an ecosystem (such as food or living space). Competition occurs among individuals within a population (intraspecific competition) and between species (interspecific competition).

4. A **keystone species** is crucial in determining the nature and structure of the entire ecosystem in which it lives. Though present in relatively small numbers, keystone species have disproportionate effects on ecosystems.

KEY TERMS

- **ecology** p. 98
- **population** p. 98
- **community** p. 98
- **ecosystem** p. 98
- **landscape** p. 99

- **biosphere** p. 99
- **first law of thermodynamics** p. 101
- **photosynthesis** p. 102
- **second law of thermodynamics** p. 102
- **energy flow** p. 104

- **ecological niche** p. 113
- **symbiosis** p. 116
- **predation** p. 119
- **competition** p. 120

CRITICAL AND CREATIVE THINKING QUESTIONS

1. To function, ecosystems require an input of energy. Where does this energy come from?

2. After an organism uses energy, what happens to it?

3. What is a biogeochemical cycle? Why is the cycling of matter essential to the continuance of life?

4. Describe how organisms participate in each of these biogeochemical cycles: carbon, nitrogen, sulfur, and phosphorus.

5. How are food chains important in biogeochemical cycles?

6. How are the many cichlid species in Lake Victoria an example of resource partitioning?

7. In both parasitism and predation, one organism benefits at the expense of another. What is the difference between the two relationships?

8. Some biologists think protecting keystone species would help preserve biological diversity in an ecosystem. Do you agree? Explain your answer.

9. How does the role of humans in the carbon cycle influence global climate change? How might your role in the carbon cycle compare to that of a young person on a remote South American farm that uses animal labor rather than machines?

10. Ecologists investigating interactions of two species at a study site first counted individuals of Species A and then removed all Species B individuals. Six months later, the ecologists again counted individuals of Species A. Viewing their results as graphed below, what is the likely ecological interaction between Species A and Species B? Explain your answer.

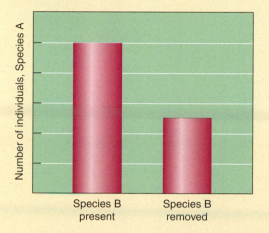

This figure shows the components of a simple food chain. Use it to answer questions 11–13.

11. Identify the producers, consumers, and decomposers in the food chain. How many trophic levels are represented?

12. Describe or indicate the flow of food and energy within this system.

13. Which forms of energy are present within this chain?

What is happening in this picture ?

■ This caterpillar has inflated its thorax to make its body look like the head of a snake. Suggest a possible reason the caterpillar does this.

■ Note the two large spots. What do they resemble? Why would this animal have such conspicuous spots?

■ If a hungry bird saw this caterpillar, do you think it would have second thoughts before eating it? Why or why not?

Ecosystems and Evolution

THE FLORIDA EVERGLADES

The Everglades, in the southernmost part of Florida, is a vast expanse of predominantly saw-grass wetlands dotted with small islands of trees. It is a haven for wildlife, including alligators (*see photograph*), snakes, panthers, otters, raccoons, and thousands of birds—great blue herons, snowy egrets, great white herons, roseate spoonbills, and osprey, to name just a few. At one time, this "river of grass" drifted south in a slow-moving sheet of water from Lake Okeechobee to Florida Bay.

The Everglades today is about half its original size of 1.6 million hectares (4 million acres) and suffers from many serious environmental problems. Wading bird populations have dropped 93 percent since 1930, and the area is now home to 50 endangered or threatened species. Changes in the Everglades illustrate how misguided human activities can cause more harm than good.

More than 70 years of engineering projects—deemed necessary to protect the human population from storm-related flooding—have reduced the quantity of water flowing into the Everglades, restricting the natural recharging process there. The water that does enter is polluted by fertilizers and pesticides used in agriculture. Flood-control measures diverted water from the wetlands, creating dry spaces that were then converted to agricultural or residential use. These developments caused habitat fragmentation, contributing to the declining wildlife populations in the Everglades.

Florida's unique river of grass will never return completely to its original condition because there are now too many sugar plantations and too many cities in the region. However, state and federal governments are working on a massive restoration project to undo some of the damage. The plan incorporates controls on agricultural runoff, the conversion of some agricultural land to marshes, and a massive project to reengineer the area's entire system of canals, levees, and pumps so that a more natural flow of water will be restored to the Everglades. Restoration efforts will take more than 20 years and cost $8 billion. All who appreciate the Everglades' uniqueness hope that its wildlife will rebound.

NATIONAL GEOGRAPHIC

Earth's Major Biomes

Earth has many different climates, which are based primarily on temperature and precipitation differences. Characteristic organisms have adapted to each climate. Because it is so large in area, a **biome** encompasses many interacting ecosystems (**FIGURE 6.1**). In terrestrial ecology, a biome is considered the next level of ecological organization above community, ecosystem, and landscape.

Near the poles, temperature is generally the overriding climate factor defining a biome, whereas in temperate and tropical regions, precipitation is more significant than temperature, as shown in **FIGURE 6.2**. Other *abiotic* factors to which certain biomes are sensitive include extreme temperatures as well as rapid temperature changes, fires, floods, droughts, and strong winds. Elevation also affects biomes: Changes in vegetation with increasing elevation resemble the changes in vegetation observed in going from warmer to colder climates. These differences across biomes can be further defined by types of vegetation present and land use patterns (**FIGURE 6.3** on pages 130 and 131).

> **biome** A large, relatively distinct terrestrial region with similar climate, soil, plants, and animals regardless of where it occurs in the world.

The world's terrestrial biomes FIGURE 6.1

Although sharp boundaries are shown in this highly simplified map, biomes actually grade together at their boundaries. Use the legend below to identify the locations of the different biomes.

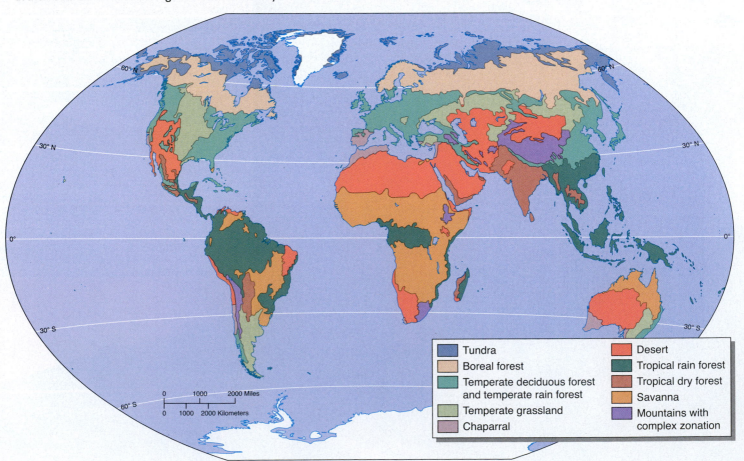

Legend:
- Tundra
- Boreal forest
- Temperate deciduous forest and temperate rain forest
- Temperate grassland
- Chaparral
- Desert
- Tropical rain forest
- Tropical dry forest
- Savanna
- Mountains with complex zonation

Two climate factors, temperature and precipitation, have a predominant effect on biome distribution.

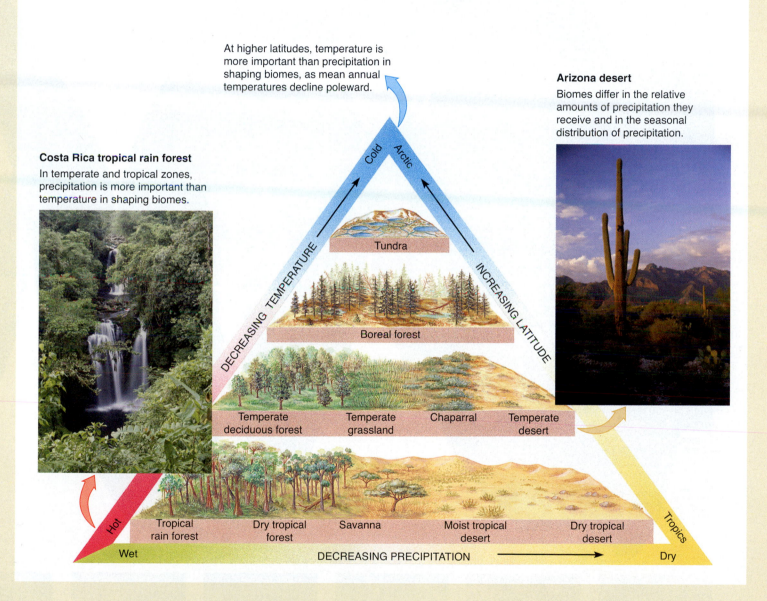

At higher latitudes, temperature is more important than precipitation in shaping biomes, as mean annual temperatures decline poleward.

Arizona desert
Biomes differ in the relative amounts of precipitation they receive and in the seasonal distribution of precipitation.

Costa Rica tropical rain forest
In temperate and tropical zones, precipitation is more important than temperature in shaping biomes.

DECREASING TEMPERATURE

INCREASING LATITUDE

Cold Arctic

Tundra

Boreal forest

Temperate deciduous forest

Temperate grassland

Chaparral

Temperate desert

Tropical rain forest

Dry tropical forest

Savanna

Moist tropical desert

Dry tropical desert

Hot

Wet

DECREASING PRECIPITATION

Dry

Tropics

Similar vegetation types can occur at many different locations.

GLOBAL LAND COVER COMPOSITION
Three characteristics underlie these categories: life-form (woody, herbaceous, or bare); leaf type (needle or broad); and leaf duration (evergreen or deciduous).

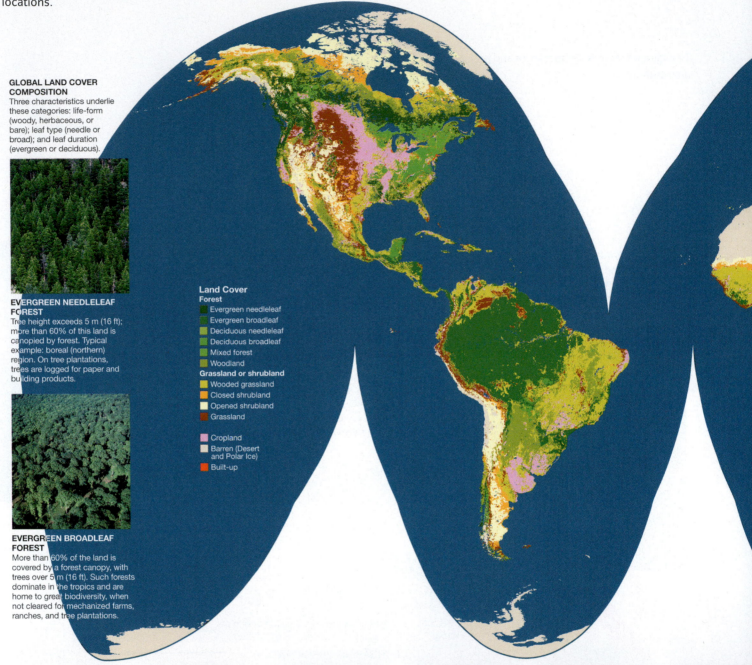

Land Cover
Forest
- Evergreen needleleaf
- Evergreen broadleaf
- Deciduous needleleaf
- Deciduous broadleaf
- Mixed forest
- Woodland

Grassland or shrubland
- Wooded grassland
- Closed shrubland
- Opened shrubland
- Grassland

- Cropland
- Barren (Desert and Polar Ice)
- Built-up

EVERGREEN NEEDLELEAF FOREST
Tree height exceeds 5 m (16 ft); more than 60% of this land is canopied by forest. Typical example: boreal (northern) region. On tree plantations, trees are logged for paper and building products.

EVERGREEN BROADLEAF FOREST
More than 60% of the land is covered by a forest canopy, with trees over 5 m (16 ft). Such forests dominate in the tropics and are home to great biodiversity, when not cleared for mechanized farms, ranches, and tree plantations.

DECIDUOUS NEEDLE-LEAF FOREST
A forest canopy covers more than 60% of the land; tree height exceeds 5 m (16 ft). This class is dominant only in Siberia, taking the form of larch forests.

DECIDUOUS BROADLEAF FOREST
More than 60% of the land is covered by a forest canopy; tree height exceeds 5 m (16 ft). In temperate regions, much of this forest has been converted to cropland.

MIXED FOREST
Both needle and deciduous types of trees appear. Mixed forest is largely found between temperate decidu-ous and boreal evergreen forests.

WOODLAND
Land has herbaceous or woody understory; trees exceed 5 m (16 ft) and may be deciduous or evergreen. Highly degraded in long-settled human environments, such as West Africa.

WOODED GRASSLAND
Woody or herbaceous understories are punctuated by trees. Examples are African savanna as well as open boreal border land between trees and tundra.

CLOSED SHRUBLAND
Found where prolonged or dry seasons limit plant growth. This cover is dominated by bushes shrubs not exceeding (16 ft). Tree canopy is than 10%.

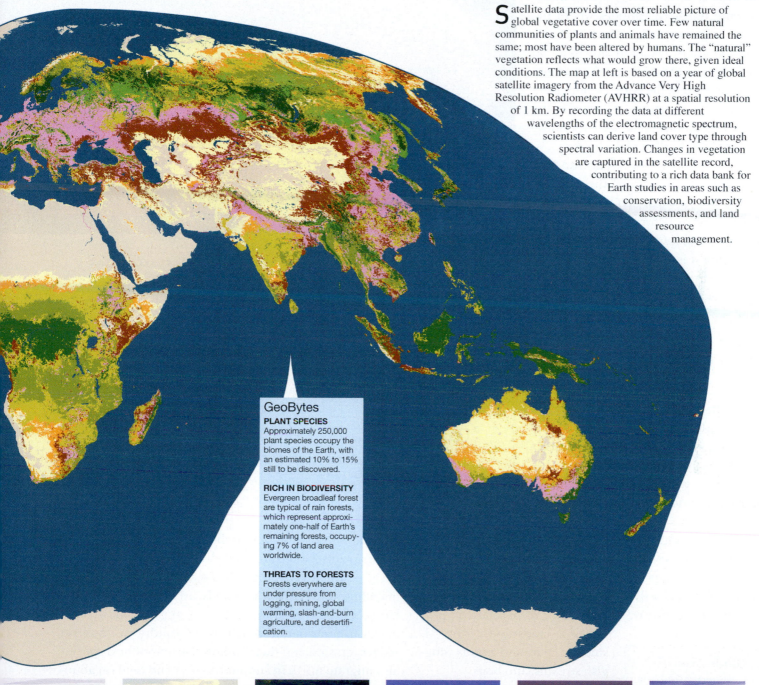

Satellite data provide the most reliable picture of global vegetative cover over time. Few natural communities of plants and animals have remained the same; most have been altered by humans. The "natural" vegetation reflects what would grow there, given ideal conditions. The map at left is based on a year of global satellite imagery from the Advance Very High Resolution Radiometer (AVHRR) at a spatial resolution of 1 km. By recording the data at different wavelengths of the electromagnetic spectrum, scientists can derive land cover type through spectral variation. Changes in vegetation are captured in the satellite record, contributing to a rich data bank for Earth studies in areas such as conservation, biodiversity assessments, and land resource management.

GeoBytes

PLANT SPECIES
Approximately 250,000 plant species occupy the biomes of the Earth, with an estimated 10% to 15% still to be discovered.

RICH IN BIODIVERSITY
Evergreen broadleaf forest are typical of rain forests, which represent approximately one-half of Earth's remaining forests, occupying 7% of land area worldwide.

THREATS TO FORESTS
Forests everywhere are under pressure from logging, mining, global warming, slash-and-burn agriculture, and desertification.

EN SHRUBLAND
ubs are dominant, not eeding 2 m (6.5 ft) in ght. They can be rgreen or deciduous. type occurs in semiarid everely cold areas.

GRASSLAND
Occurring in a wide range of habitats, this landscape has continuous herbaceous cover. The American Plains and central Russia are the premier examples.

CROPLAND
Crop-producing fields constitute over 80% of the land. Temperate regions are home to large areas of mechanized farming; in the developing world, plots are small.

BARREN (DESERT)
The land never has more than 10% vegetated cover. True deserts, such as the Sahara, as well as areas succumbing to desertification are examples.

BUILT-UP
This land cover type was mapped using digital population data from around the world. It represents the most densely inhabited areas.

BARREN (POLAR ICE)
Permanent snow cover characterizes this class, the greatest examples of which are in the polar regions, as well as on high elevations in Alaska and the Himalayas.

TUNDRA

Tundra (or *arctic tundra*) occurs in the extreme northern latitudes where the snow melts seasonally (**FIGURE 6.4**). The Southern Hemisphere has no equivalent of the arctic tundra because it has no land in the corresponding latitudes. A similar ecosystem located in the higher elevations of mountains, above the tree line, is called *alpine tundra*.

Although the arctic tundra's growing season is short, the days are long. Above the Arctic Circle, the sun does not set at all for many days in midsummer, although the amount of light at midnight is one-tenth that at noon. There is little precipitation, and most of the yearly 10 to 25 cm (4 to 10 in) of rain or snow falls during summer months.

Most tundra soils formed when glaciers began retreating after the last Ice Age, about 17,000 years ago. These soils are usually nutrient poor and have little *detritus* such as dead leaves and stems, animal droppings, or remains of organisms. Although the tundra's surface soil thaws during summer, beneath it lies a layer of *permafrost*, permanently frozen ground that varies in depth and thickness. Permafrost impedes drainage, so the thawed upper zone of soil is usually waterlogged during summer. Limited precipitation, combined with low temperatures, flat topography (or surface features), and the layer of permafrost, produces a landscape of broad, shallow lakes and ponds, sluggish streams, and bogs.

Tundra supports relatively few species compared to other biomes, but the species that do occur there often exist in great numbers. Mosses, lichens, grasses, and grasslike sedges are the dominant plants. Stunted trees and shrubs grow only in sheltered locations. As a rule, tundra plants seldom grow taller than 30 cm (12 in).

Animals adapted to live year-round in the tundra include lemmings, voles, weasels, arctic foxes, snowshoe hares, ptarmigan, snowy owls, and musk oxen. In summer, caribou migrate north to the tundra to graze on sedges, grasses, and dwarf willow. Dozens of bird species also migrate north in summer to nest and feed on abundant insects. Mosquitoes, blackflies, and deerflies survive winter as eggs or pupae and appear in great numbers during summer weeks.

Tundra recovers slowly from even small disturbances. Oil and natural gas exploration and military use have caused damage to tundra likely to persist for hundreds of years (see the Chapter 17 case study).

> **tundra** The treeless biome in the far north that consists of boggy plains covered by lichens and mosses; it has harsh, cold winters and extremely short summers.

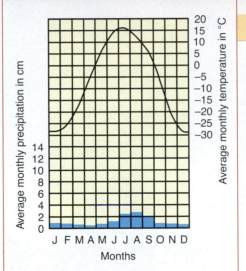

Arctic tundra FIGURE 6.4

Because of the tundra's short growing season and permafrost, only small, hardy plants grow in the northernmost biome that encircles the Arctic Ocean. Photographed in the Yukon Territory, Canada. The Ogilvie Mountains are in the background. Climate graph shows monthly temperatures and precipitation for Fort Yukon, Alaska.

BOREAL FOREST

Just south of the tundra is the **boreal forest**, or northern coniferous forest, which stretches across North America and Eurasia (**FIGURE 6.5**). There is no biome comparable to the boreal forest in the Southern Hemisphere. Winters are extremely cold and severe, although not as harsh as those in the tundra. Boreal forest receives little precipitation, perhaps 50 cm (20 in) per year, and its soil is typically acidic and mineral poor, with a thick surface layer of partly decomposed pine and spruce needles. Where permafrost occurs, it is found deep under the surface. Boreal forest has numerous ponds and lakes dug by ice sheets during the last Ice Age.

Black and white spruces, balsam fir, eastern larch, and other conifers (cone-bearing evergreens) dominate the boreal forest. Conifers have many drought-resistant adaptations, such as needle-like leaves whose minimal surface area prevents water loss by evaporation. Such an adaptation helps conifers withstand the drought of the northern winter, when roots cannot absorb water through the frozen ground. Being evergreen, conifers resume photosynthesis as soon as warmer temperatures return.

The animal life of the boreal forest consists of some larger species such as caribou, which migrate from the tundra for winter; wolves; brown and black bears; and moose. However, most boreal mammals are medium sized to small, including rodents, rabbits, and smaller predators such as lynx, sable, and mink. Birds are abundant in the summer but migrate to warmer climates for winter. Insects are plentiful.

Currently, boreal forest is the world's top source of industrial wood and wood fiber. Extensive logging, gas and oil exploration, and farming have contributed to loss of boreal forest.

boreal forest
A region of coniferous forest (such as pine, spruce, and fir) in the Northern Hemisphere; located just south of the tundra. Also called taiga.

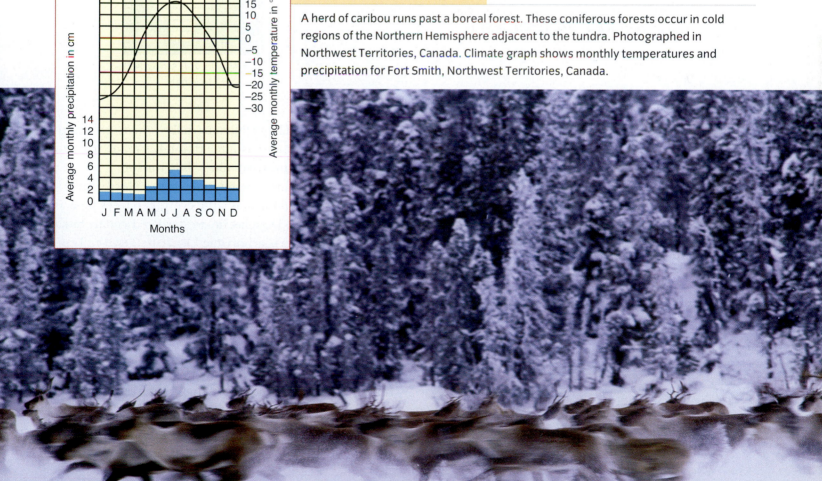

Boreal forest FIGURE 6.5

A herd of caribou runs past a boreal forest. These coniferous forests occur in cold regions of the Northern Hemisphere adjacent to the tundra. Photographed in Northwest Territories, Canada. Climate graph shows monthly temperatures and precipitation for Fort Smith, Northwest Territories, Canada.

Temperate rain forest FIGURE 6.6

This temperate biome has large amounts of precipitation. Photographed in Olympic National Park in Washington State. Climate graph shows monthly temperatures and precipitation for Estacada, Oregon.

TEMPERATE RAIN FOREST

A coniferous **temperate rain forest** occurs on the northwest coast of North America. Similar vegetation exists in southeastern Australia and in southern South America. Annual precipitation in this biome is high, from 200 to 380 cm (80 to 152 in), and condensation of water from dense coastal fogs augments the precipitation. The proximity of temperate rain forest to the coastline moderates its temperature so that the seasonal fluctuation is narrow; winters are mild, and summers are cool. Temperate rain forest has relatively nutrient-poor soil, though its organic content may be high. Cool temperatures slow the activity of bacterial and fungal decomposers.

> **temperate rain forest** A coniferous biome with cool weather, dense fog, and high precipitation.

Thus, needles and large fallen branches and trunks accumulate on the ground as litter that takes many years to decay and release nutrient minerals to the soil.

The dominant vegetation in the North American temperate rain forest is large evergreen trees such as western hemlock, Douglas fir, western red cedar, Sitka spruce, and western arborvitae (**FIGURE 6.6**). Temperate rain forests are rich in epiphytes, smaller plants that grow on the trunks and branches of large trees. Epiphytes in this biome are mainly mosses, club mosses, lichens, and ferns, all of which also carpet the ground. Squirrels, wood rats, mule deer, elk, numerous bird species, and several species of amphibians and reptiles are common temperate rainforest animals. Issues surrounding logging of old-growth temperate rain forests of the Pacific Northwest were presented in Chapter 3 and are discussed in the Chapter 13 case study.

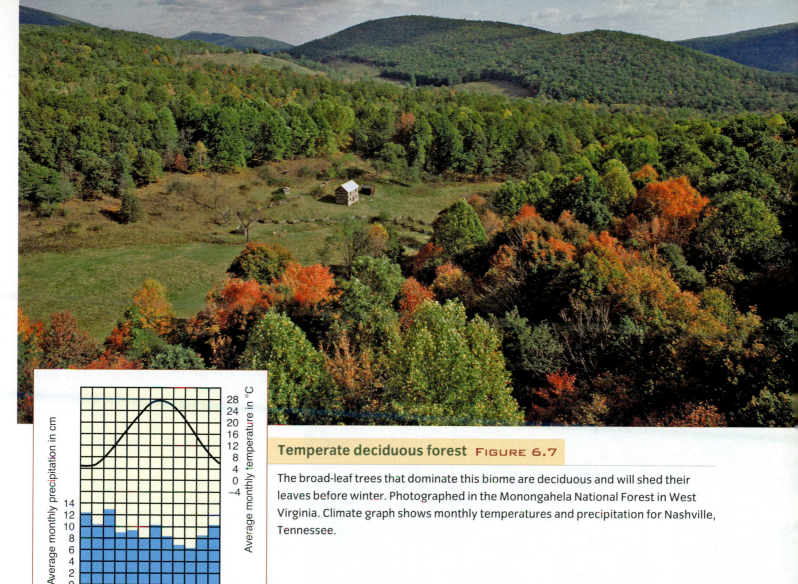

Temperate deciduous forest FIGURE 6.7

The broad-leaf trees that dominate this biome are deciduous and will shed their leaves before winter. Photographed in the Monongahela National Forest in West Virginia. Climate graph shows monthly temperatures and precipitation for Nashville, Tennessee.

TEMPERATE DECIDUOUS FOREST

Hot summers and cold winters characterize the **temperate deciduous forest**. Typically, the soil of a temperate deciduous forest consists of a topsoil rich in organic material and a deep, clay-rich lower layer. Broad-leafed hardwood trees that lose their leaves seasonally, such as oak, hickory, and beech, dominate the temperate deciduous forests of the northeastern and mideastern United States (**FIGURE 6.7**). In the southern areas of the temperate deciduous forest, the number of broad-leafed evergreen trees, such as magnolia, increases. The trees of the temperate deciduous forest form a dense canopy that overlies saplings and shrubs.

Temperate deciduous forests originally contained a variety of large mammals, such as puma, wolves, and bison, which are now absent. Other more common animals include deer, bears, and many small mammals and birds.

In Europe and North America, logging and land clearing for farms, tree plantations, and cities destroyed much of the original temperate deciduous forest. Where it has regenerated, temperate deciduous forest is often in a seminatural state that humans have modified for recreation, livestock foraging, timber harvest, and other uses. Many forest organisms have successfully reestablished themselves in these returning forests.

> **temperate deciduous forest**
> A forest biome that occurs in temperate areas where annual precipitation ranges from about 75 cm to 126 cm (30 to 50 in).

TROPICAL RAIN FOREST

Tropical rain forest occurs where temperatures are warm throughout the year and precipitation occurs almost daily. The annual precipitation in a tropical rain forest is typically between 200 and 450 cm (80 to 180 in). Tropical rain forest commonly occurs in areas with ancient, highly weathered, mineral-poor soil. Little organic matter accumulates in such soils; because temperatures are high year-round, bacteria, fungi, and detritus-feeding ants and termites decompose organic litter quite rapidly. Roots quickly absorb nutrient minerals from the decomposing material. Tropical rain forests are found in Central and South America, Africa, and Southeast Asia.

Of all the biomes, the tropical rain forest is unexcelled in species richness and variety (**FIGURE 6.8**). No single species dominates this biome. The trees are typically evergreen flowering plants. A fully developed tropical rain forest has at least three distinct stories, or layers, of vegetation. The topmost story, or emergent layer, consists of the crowns of very tall trees, some 50 m (164 ft) or more in height, which are exposed to direct sunlight. The middle story, or canopy, which reaches a height of 30 to 40 m (100 to 130 ft), forms a continuous

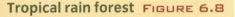

tropical rain forest A lush, species-rich forest biome that occurs where the climate is warm and moist throughout the year.

Tropical rain forest FIGURE 6.8

A broad view of tropical rainforest vegetation along a riverbank in Borneo, Southeast Asia. Except at riverbanks, tropical rain forests have a closed canopy that admits little light to the rainforest floor. Climate graph shows monthly temperatures and precipitation for Belem, Brazil.

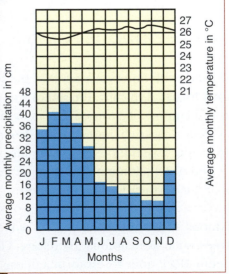

layer of leaves that lets in very little sunlight to support the smaller plants in the sparse understory. Only 2 to 3 percent of the light bathing the forest canopy reaches the forest understory. The vegetation of tropical rain forests is not dense at ground level except near stream banks or where a fallen tree has opened the canopy. Tropical rainforest trees support extensive communities of epiphytic plants such as ferns, mosses, orchids, and bromeliads.

Not counting bacteria and other soil-dwelling organisms, about 90 percent of tropical rainforest organisms are adapted to live in the canopy. Rain forests shelter the most abundant and varied insects, reptiles, and amphibians on Earth. The birds, often brilliantly colored, are also varied. Some are specialized to consume fruit (parrots, for example) and others to consume nectar (hummingbirds and sunbirds, for example). Most rainforest mammals, such as sloths and monkeys, are adapted to live only in the trees and rarely climb down to the ground, although some large, ground-dwelling mammals, including elephants, are also found in rain forests.

Human population growth and industrial expansion in tropical countries threaten the survival of tropical rain forests. (See Chapter 13 for more discussion of the ecological impacts of rainforest destruction.)

CHAPARRAL

Some hilly temperate environments have mild winters with abundant rainfall combined with very dry summers. Such *Mediterranean climates*, as they are called, occur not only in the area around the Mediterranean Sea but also in the North American Southwest, southwestern and southern Australia, central Chile, and southwestern South Africa. On the mountain slopes of southern California, this Mediterranean-type biome is known as **chaparral** (**FIGURE 6.9**). Chaparral soil is thin and often not very fertile. Wildfires occur naturally in this environment and are particularly frequent in late summer and autumn.

> ■ **chaparral** A biome with mild, moist winters and hot, dry summers; vegetation is typically small-leafed evergreen shrubs and small trees.

Chaparral FIGURE 6.9

Chaparral vegetation consists mainly of drought-resistant evergreen shrubs and small trees. Hot, dry summers and mild, rainy winters characterize the chaparral. Photographed on the Mount Tamalpais in the Marin Hills, California. Climate graph shows monthly temperatures and precipitation for Culver City, California.

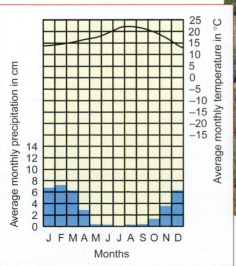

California has about 6,000 wildfires each year, and they are becoming increasingly expensive and dangerous to manage because so many people are building homes and living in the fire-vulnerable chaparral. Yet the topography of chaparral is so steep that firefighters often cannot use mechanized equipment but must transport equipment to fires with helicopters. Afraid that prescribed burns will get out of control, local governments are increasingly trying an effective, low-tech method to reduce the fuel load: During the 6-month fire season, goats are clearing hills around Oakland, Berkeley, Monterey, and Malibu.

A herd of 350 goats can denude an entire acre of heavy brush in about a day, but their use entails a lot of advance organization and support. Before goats can clear hazardous dry fuels from surrounding hillsides, botanists must walk the terrain to put fences around any small trees or other plants that are rare or endangered; the fencing keeps the goats from eating those plants. A portable home site for goatherds is then installed, as are electric fencing and water troughs for the goats. The goatherds typically use dogs to help herd the goats.

Goats are an excellent tool for fire management because they preferentially browse woody shrubs and thick undergrowth—the exact fuel that causes disastrous fires. Fires that have occurred in areas after goats have browsed there are much easier to contain.

Goats prefer woody and weedy species, such as those common to chaparral.

Chaparral vegetation looks strikingly similar in different parts of the world, even though the individual species differ by location. A dense thicket of evergreen shrubs—often short, drought-resistant pine or scrub oak trees that grow 1 to 3 m (3.3 to 9.8 ft) tall—usually dominates chaparral. These plant species have evolved adaptations that equip them to live where precipitation is seasonal. During the rainy winter season, the environment may be lush and green, and during the hot, dry summer, the plants lie dormant. The hard, small, leathery leaves of trees and shrubs resist water loss.

Many plants are also specifically fire adapted and grow best in the months following a fire. Such growth is possible because fire releases into the soil the nutrient minerals present in the aboveground parts of the plants that burned. The seeds and underground parts of plants that survive fire make use of the newly available nutrient minerals and sprout vigorously during winter rains. Mule deer, wood rats, chipmunks, lizards, and many species of birds are common animals of the chaparral. (For more on the role fire plays in nature and on how humans have disrupted this role, see the case study at the end of this chapter.)

Temperate grassland FIGURE 6.10

A black-tailed prairie dog stands next to its grassland burrow. This short-grass prairie contains a mixture of grasses and other herbaceous flowering plants. Photographed in Custer State Park, South Dakota. Climate graph shows monthly temperatures and precipitation for Lawrence, Kansas.

TEMPERATE GRASSLAND

Summers are hot, winters are cold, and rainfall is often uncertain in **temperate grassland**. Average annual precipitation ranges from 25 to 75 cm (10 to 30 in). Grassland soil has considerable organic material because the aboveground portions of many grasses die off each winter and contribute to the organic content of the soil, while the roots and rhizomes (underground stems) survive underground. Many grasses are sod formers—that is, their roots and rhizomes form a thick, continuous underground mat.

Moist temperate grasslands, also known as *tallgrass prairies*, occur in the United States in parts of Illinois, Iowa, Minnesota, Nebraska, Kansas, and other midwestern states. Trees grow sparsely except near rivers and streams, but grasses taller than a person grow in great profusion in the deep, rich soil. Periodic wildfires help to maintain grasses as the dominant vegetation in grasslands.

> **temperate grassland**
> A grassland with hot summers, cold winters, and less rainfall than is found in the temperate deciduous forest biome.

More than 90 percent of the North American grassland encountered by European settlers has been converted to farmland, and the remaining prairie is highly fragmented. Today, the tallgrass prairie is considered North America's rarest biome. Tallgrass prairie formerly supported large herds of grazing animals, such as bison and pronghorn elk. The principal predators were wolves, although in sparser, drier areas coyotes took their place. Smaller animals included prairie dogs and their predators (foxes, black-footed ferrets, and various birds of prey), grouse, reptiles such as snakes and lizards, and great numbers of insects.

Short-grass prairies are temperate grasslands that receive less precipitation than moist temperate grasslands but more precipitation than deserts. In the United States, short-grass prairies occur in parts of Montana, Wyoming, South Dakota, and other midwestern states (**FIGURE 6.10**). Grasses that grow knee high or lower dominate short-grass prairies. Plants grow less abundantly than in the moister grasslands, and bare soil is occasionally exposed.

SAVANNA

Savanna, a tropical grassland, is found in areas of low rainfall or, more commonly, in areas of intense seasonal rainfall with prolonged dry periods. Temperatures in tropical savannas vary little throughout the year. Precipitation is the overriding climate factor: Annual precipitation is 85 to 150 cm (34 to 60 in). Savanna soil is somewhat low in essential nutrient minerals, in part because it is heavily leached during rainy periods—that is, nutrient minerals filter out of the topsoil. Although the African savanna is best known, savanna also occurs in South America, western India, and northern Australia.

Savanna has wide expanses of grasses interrupted by occasional trees like the acacia, which bristles with thorns to provide protection against herbivores. Both trees and grasses have fire-adapted features, such as

savanna A tropical grassland with widely scattered trees or clumps of trees.

extensive underground root systems, that let them survive seasonal droughts as well as periodic fires.

Spectacular herds of herbivores such as wildebeest, antelope, giraffe, zebra, and elephants occur in the African savanna (**FIGURE 6.11**). Large predators, such as lions and hyenas, kill and scavenge the herds. In areas of seasonally varying rainfall, the herds and their predators may migrate annually.

Savanna in many places is being converted into rangeland for cattle and other domesticated animals. The problem is particularly serious in Africa, where human populations are growing rapidly.

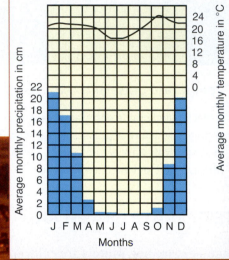

Savanna FIGURE 6.11
Tropical grasslands such as this one, with widely scattered acacia trees, support large herds of grazing animals and their predators. Photographed in Serengeti National Park, Tanzania. Climate graph shows monthly temperatures and precipitation for Lusaka, Zambia.

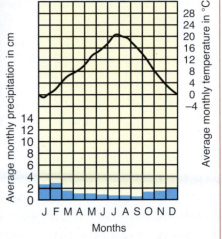

Desert FIGURE 6.12

This desert landscape includes tough-leaved yuccas and spine-covered prickly pear cacti. Desert inhabitants are strikingly adapted to the demands of their environment. Photographed in the Chuahuan Desert, Big Bend National Park, Texas. Climate graph shows monthly temperatures and precipitation for Reno, Nevada.

DESERT

Desert consists of dry areas found in both temperate (cold deserts) and subtropical or tropical regions (warm deserts). The low water vapor content of the desert atmosphere results in daily temperature extremes of heat and cold, so that a major change in temperature occurs in each 24-hour period. Desert environments vary greatly depending on the amount of precipitation they receive, which is generally less than 25 cm (10 in) per year. As a result of sparse vegetation, desert soil is low in organic material but is often high in mineral content, particularly salts.

Plant cover is sparse in deserts, and much of the soil is exposed. Plants in North American deserts include cacti, yuccas, Joshua trees, and sagebrush (**FIGURE 6.12**). Desert plants are adapted to conserve water and as a result tend to have few, small, or no leaves. Cactus leaves are modified into spines, which discourage herbivores. Other desert plants shed their leaves for most of the year, growing only during the brief moist season.

Desert animals are typically small. During the heat of the day, they remain under cover or return to shelter periodically, emerging at night to forage or hunt. In addition to desert-adapted insects and arachnids (such as tarantulas and scorpions), there are a few desert-adapted amphibians (frogs and toads) and many reptiles, such as the desert tortoise, Gila monster, and Mojave rattlesnake. Desert mammals in North America include rodents such as kangaroo rats, as well as mule deer and jackrabbits. Birds of prey, especially owls, live on the rodents and jackrabbits, and even the scorpions. During the driest months of the year, many desert animals tunnel underground, where they remain inactive.

In many areas, human development encroaches on deserts. Off-road vehicle use damages vegetation, and expansion of farms, cities, and residential areas places severe demands on limited groundwater.

desert A biome in which the lack of precipitation limits plant growth; deserts are found in both temperate and tropical regions.

CONCEPT CHECK STOP

What is a biome?

How do you distinguish between temperate rain forest and tropical rain forest?

What are the overriding climate factors for savanna?

Aquatic Ecosystems

LEARNING OBJECTIVES

Summarize the important environmental factors that affect aquatic ecosystems.

Briefly describe the various aquatic ecosystems, giving attention to the environmental characteristics of each.

The most fundamental division in aquatic ecology is probably between freshwater and saltwater environments. Salinity, which is the concentration of dissolved salts (such as sodium chloride) in a body of water, affects the kinds of organisms present in aquatic ecosystems, as does the amount of dissolved oxygen. Water greatly interferes with the penetration of light, so floating aquatic organisms that photosynthesize must remain near the water's surface, and vegetation anchored to lake floors or streambeds will grow only in relatively shallow water. In addition, low levels of essential nutrient minerals limit the number and distribution of organisms in certain aquatic environments. In this section, we discuss freshwater ecosystems only; because the immense marine environment is so critical to the environmental well-being of Earth, we devote an entire chapter to it (see Chapter 11).

Aquatic ecosystems contain three main ecological categories of organisms: free-floating plankton, strongly swimming nekton, and bottom-dwelling benthos. **Plankton** are usually small or microscopic organisms. They tend to drift or swim feebly, so, for the most part, they are carried about at the mercy of currents and waves. **Nekton** are larger, more strongly swimming organisms such as fishes, turtles, and whales. **Benthos** are bottom-dwelling organisms that fix themselves to one spot (sponges and oysters), burrow into the sand (worms and clams), or simply walk about on the bottom (crawfish and aquatic insect larvae).

FRESHWATER ECOSYSTEMS

Freshwater ecosystems include lakes and ponds (standing-water ecosystems), rivers and streams (flowing-water ecosystems), and marshes and swamps (freshwater wetlands). Specific abiotic conditions and characteristic organisms distinguish each freshwater ecosystem. Although freshwater ecosystems occupy only about 2 percent of Earth's surface, they play an important role in the hydrologic cycle: They help recycle precipitation that flows into the ocean as surface runoff. (See Chapter 5 for a detailed explanation of the hydrologic cycle.) Large bodies of fresh water help moderate daily and seasonal temperature fluctuations on nearby land regions, and freshwater habitats provide homes for many species.

Zonation is characteristic of **standing-water ecosystems**. A large lake has three zones: the littoral, limnetic, and profundal zones (see What a Scientist Sees). The *littoral zone* is a shallow-water area along the shore of a lake or pond. Emergent vegetation, such as cattails and bur reeds, as well as several deeper-dwelling aquatic plants and algae, live in the littoral zone.

> **standing-water ecosystem** A body of fresh water surrounded by land and whose water does not flow; a lake or a pond.

The *limnetic zone* is the open water beyond the littoral zone—that is, away from the shore. The limnetic zone extends down as far as sunlight penetrates to permit photosynthesis. The main organisms of the limnetic zone are microscopic plankton. Larger fishes also spend most of their time in the limnetic zone, although they may visit the littoral zone to feed and reproduce. The deepest zone, the *profundal zone*, is beneath the limnetic zone of a large lake; smaller lakes and ponds typically lack a profundal zone. Because light does not penetrate effectively to this depth, plants and algae do not live there. Detritus drifts into the profundal zone from the littoral and limnetic zones; bacteria decompose this detritus. This marked zonation is accentuated by **thermal stratification**, in which the temperature changes sharply with depth.

Zonation in a large lake

The zonation in Bear Lake, in Rocky Mountain National Park, Colorado, is not apparent to a visitor.

(Inset, below) A lake is a standing-water ecosystem surrounded by land. The littoral zone is the shallow-water area around the lake's edge. The limnetic zone is the open, sunlit water away from the shore. The profundal zone, under the limnetic zone, is below where light penetrates.

Littoral zone

Limnetic zone

Profundal zone

Flowing-water ecosystems are highly variable. The surrounding environment changes greatly between a river's source and its mouth (**FIGURE 6.13**). Certain parts of the stream's course are shaded by forest, while other parts are exposed to direct sunlight. Groundwater may well up through sediments on the bottom in one particular area, making the water temperature cooler in summer or warmer in winter than adjacent parts of the stream or river. The kinds of organisms found in flowing water vary greatly from one stream to another, depending primarily on the strength of the current. In streams with fast currents, some inhabitants have adaptations such as suckers, with which they attach themselves to rocks to prevent being swept away. Some stream inhabitants have flattened bodies to slip under or between rocks. Other inhabitants such as fish are streamlined and muscular enough to swim in the current.

> ■ **flowing-water ecosystem**
> A freshwater ecosystem such as a river or stream in which water flows in a current.

Features of a typical river FIGURE 6.13

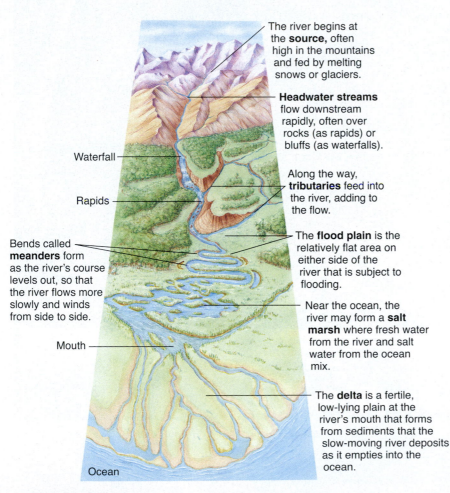

The river begins at the **source,** often high in the mountains and fed by melting snows or glaciers.

Headwater streams flow downstream rapidly, often over rocks (as rapids) or bluffs (as waterfalls).

Along the way, **tributaries** feed into the river, adding to the flow.

The **flood plain** is the relatively flat area on either side of the river that is subject to flooding.

Near the ocean, the river may form a **salt marsh** where fresh water from the river and salt water from the ocean mix.

The **delta** is a fertile, low-lying plain at the river's mouth that forms from sediments that the slow-moving river deposits as it empties into the ocean.

Waterfall

Rapids

Bends called **meanders** form as the river's course levels out, so that the river flows more slowly and winds from side to side.

Mouth

Ocean

A A river flows from its source to the ocean.

Meander

Flood plain

B Aerial view of meanders in the Tambopata River, Peru.

FRESHWATER WETLANDS

Freshwater wetlands include **marshes**, dominated by grasslike plants, and **swamps**, dominated by woody trees or shrubs (**FIGURE 6.14**). Wetland soils are waterlogged for variable periods and are therefore anaerobic (without oxygen). They are rich in accumulated organic materials, partly because anaerobic conditions discourage decomposition.

Wetlands provide excellent wildlife habitat for migratory waterfowl and other bird species, as well as for beaver, otters, muskrats, and game fish. In addition to providing unique wildlife habitat, wetlands serve other important environmental functions, known as **ecosystem services**. When rivers flood their banks, wetlands are capable of holding or even absorbing the excess water, thereby helping to control flooding. The floodwater then drains slowly back into the rivers, providing a steady flow of water throughout the year. Wetlands also serve as groundwater recharging areas. One of their most important roles is to trap and hold pollutants in the flooded soil, thereby cleansing and purifying the water. Although wetlands are afforded some legal protection, they are still threatened by pollution, development, agriculture, and dam construction.

> **freshwater wetlands** Lands that shallow fresh water covers for at least part of the year; wetlands have a characteristic soil and water-tolerant vegetation.

> **ecosystem services** Important environmental benefits, such as clean air to breathe, clean water to drink, and fertile soil in which to grow crops, that the natural environment provides.

Freshwater swamp FIGURE 6.14

Freshwater swamps are inland areas covered by water and dominated by trees, such as baldcypress. Photographed in Lake Verret, Atchafalaya National Wildlife Refuge, Louisiana.

BRACKISH ECOSYSTEMS: ESTUARIES

estuary A coastal body of water, partly surrounded by land, with access to the open ocean and a large supply of fresh water from a river.

Where the ocean meets the land, there may be one of several kinds of ecosystems: a rocky shore, a sandy beach, an intertidal mud flat, or a tidal **estuary**. Water levels in an estuary rise and fall with the tides; salinity fluctuates with tidal cycles, the time of year, and precipitation. Salinity also changes gradually within the estuary, from fresh water at the river entrance, to *brackish* (somewhat salty) water, to salty ocean water at the mouth of the estuary. Because estuaries undergo significant daily, seasonal, and annual variations in physical factors such as temperature, salinity, and depth of light penetration, estuarine organisms must have a high tolerance for changing conditions.

Temperate estuaries usually feature **salt marshes**, shallow wetlands in which salt-tolerant grasses grow (**FIGURE 6.15A**). Salt marshes perform many ecosystem services, including providing biological habi-

Estuaries FIGURE 6.15

A A salt marsh on Assateague Island, Virginia.

B A mangrove forest in the Caroline Islands, Micronesia. Mangrove roots grow into deeper water as well as into mudflats that are exposed at low tide. Many animals live among the complex root systems of mangrove forests.

CAROLINE IS.

Global Locator

tats, trapping sediment and pollution, supplying groundwater, and buffering storms by absorbing their energy, which prevents flood damage elsewhere.

Mangrove forests, the tropical equivalent of salt marshes, cover perhaps 70 percent of tropical coastlines (**Figure 6.15B**). Like salt marshes, mangrove forests provide valuable ecosystem services. Their interlacing roots are breeding grounds and nurseries for several commercially important fishes and shellfish, such as crabs, shrimp, mullet, and spotted sea trout. Mangrove branches are nesting sites for many species of birds, such as pelicans, herons, egrets, and roseate spoonbills. Mangrove roots stabilize the submerged soil, thereby preventing coastal erosion and providing a barrier against the ocean during storms.

Both salt marsh and mangrove forest ecosystems have experienced significant losses due to coastal devel-

opment. Salt marshes have been polluted and turned into dumping grounds; mangrove forests have been logged and used as aquaculture sites.

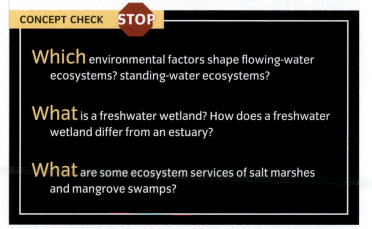

CONCEPT CHECK STOP

Which environmental factors shape flowing-water ecosystems? standing-water ecosystems?

What is a freshwater wetland? How does a freshwater wetland differ from an estuary?

What are some ecosystem services of salt marshes and mangrove swamps?

Population Responses to Changing Conditions over Time: Evolution

LEARNING OBJECTIVES

Define *evolution*.

Explain how evolution by natural selection consists of four observations of the natural world.

Describe various types of evidence that supports evolution.

S cientists think all of Earth's remarkable variety of organisms descended from earlier species by a process known as **evolution**. The concept of evolution dates back to the time of Aristotle, but **Charles Darwin** (1809–1882), a 19th-century naturalist, proposed the mechanism of evolution that today's scientific community still accepts. As you will see, the environment plays a crucial role in Darwin's theory of evolution.

It occurred to Darwin that in a population, inherited traits favorable to survival in a given environ-

ment tended to be preserved over successive generations, whereas unfavorable traits were eliminated. The result is *adaptation*, an evolutionary modification that improves a species' chance of survival and reproductive success in a given environment. Eventually the accumulation of adaptive modifications might result in a new species.

evolution
The cumulative genetic changes in populations that occur during successive generations.

Darwin proposed the theory of evolution by natural selection in his monumental book *The Origin of Species by Means of Natural Selection*, which was published in 1859. Since that time, scientists have accumulated an enormous body of observations and experiments that support Darwin's theory. Although biologists still do not agree completely on some aspects of the evolutionary process, the concept that evolution by natural selection has taken place and is still occurring is now well documented.

NATURAL SELECTION

Evolution occurs through the process of **natural selection**. As favorable traits increase in frequency in successive generations, and as unfavorable traits decrease or disappear, the collection of characteristics of a given population changes. Natural selection is the process by which successful traits are passed on to the next generation and unsuccessful ones are weeded out. It consists of four observations about the natural world:

1. *Overproduction.* Each species produces more offspring than will survive to maturity. Natural populations have the reproductive potential to increase their numbers continuously over time (**Figure 6.16**).

2. *Variation.* The individuals in a population exhibit variation. Each individual has a unique combination of traits, such as size, color, and ability to tolerate harsh environments. Some traits improve the chances of an individual's survival and reproductive success, whereas others do not. It is important to remember that the variation necessary for evolution by natural selection must be inherited so that it can be passed to offspring.

3. *Limits on population growth, or a struggle for existence.* There is only so much food, water, light, growing space, and so on available to a population, and organisms compete with one another for the limited resources available to them. Because there are more individuals than the environment can support, not all of an organism's offspring will survive to reproductive age. Other limits on population growth include predators and diseases.

4. *Differential reproductive success.* Individuals that possess the most favorable combination of characteristics (those that make individuals better adapted to their environment) are more likely than others to survive, reproduce, and pass their traits to the next generation. Sexual reproduction is the key to natural selection: The best-adapted individuals are those that reproduce most successfully, whereas less-fit individuals die

prematurely or produce fewer or inferior offspring. Over time, enough changes may accumulate in geographically separated populations (often with slightly different environments) to produce new species (**Figure 6.17**).

One premise on which Darwin based his theory of evolution by natural selection is that individuals transmit traits to the next generation. However, Darwin could not explain *how* this occurs or *why* individuals within a population vary. Beginning in the 1930s and 1940s, biologists combined the principles of genetics with Darwin's theory of natural selection. The resulting unified explanation of evolution is known as the **modern synthesis** (where *synthesis* refers to combining parts of previous theories). The modern synthesis explains Darwin's observation of variation among offspring in terms of **mutation**, or changes in DNA. Mutations provided the genetic variability on which natural selection acts during evolution.

> **■ natural selection**
> The tendency of better-adapted individuals—those with a combination of genetic traits best suited to environmental conditions—to survive and reproduce, increasing their proportion in the population.

Overproduction FIGURE 6.16

If each breeding pair of elephants were to produce six offspring that lived and reproduced, in 750 years a single pair of elephants would have given rise to more than 15 million elephants! Yet elephants have not overrun the planet. Photographed in Botswana.

Darwin's finches FIGURE 6.17

Charles Darwin was a ship's naturalist on a five-year voyage around the world. During an extended stay in the Galápagos Islands off the coast of Ecuador, he studied the plants and animals of each island, including 14 species of finches.

1 Ancestral species begins in Ecuador.

Grassquit finch (seeds).

There is only one finch species now in Ecuador.

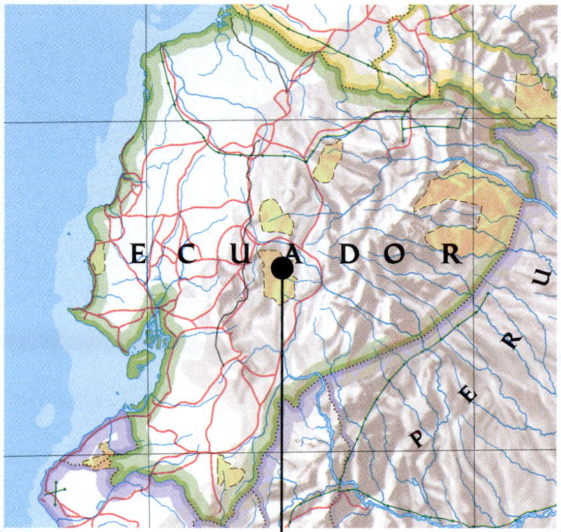

2 Ancestral species reaches the Galápagos Islands.

Small ground finch (soft seeds)

Cactus finch (cactus)

3 Modern species descend from ancestral species.

Large ground finch (hard seeds)

Warbler finch (insects)

I. Darwin
I. Wolf
Isla Pinta
Isla Marchena
Isla Genovesa
I. Isabela
I. San Salvador
I. Fernandina
Isla Santa Cruz
Isla San Cristóbal
I. Santa María
Isla Española

GALÁPAGOS ISLANDS
(Archipiélago de Colón)
Ecuador

4 The apparently related species on the Galápagos Islands have different beak shapes and different diets. Darwin reasoned that finches that colonized from the mainland had changed as the birds adapted to different diets.

Woodpecker finch (insects)

Medium ground finch (moderate seeds)

Atlantic Ocean
ECUADOR
EQUATOR
Galápagos Islands (ECUADOR)
SOUTH AMERICA
Pacific Ocean

ECUADOR
PERU

PACIFIC OCEAN

A vast body of evidence supports evolution, most of which is beyond the scope of this text. This evidence includes observations from the fossil record, comparative anatomy, biogeography (the study of the geographic locations of organisms), and molecular biology (**Figure 6.18**). In addition, evolutionary hypotheses are tested experimentally.

On the basis of these kinds of evidence, virtually all biologists accept the principles of evolution by natural selection, although they don't agree on all the details. They try to better understand certain aspects of evolution, such as the role of chance and how quickly new species evolve. As discussed in Chapter 1, science is an ongoing process, and information obtained in the future may require modifications to certain parts of the theory of evolution by natural selection.

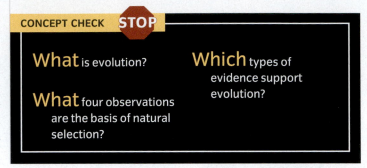

CONCEPT CHECK **STOP**

What is evolution?

What four observations are the basis of natural selection?

Which types of evidence support evolution?

Community Responses to Changing Conditions over Time: Succession

LEARNING OBJECTIVES

Define *ecological succession*.

Distinguish between primary and secondary succession.

A community of organisms does not spring into existence full-blown. By means of **ecological succession**, a given community develops gradually through a sequence of species. Certain organisms colonize an area; over time others replace them, and eventually the replacements are themselves replaced by still other species.

The actual mechanisms that underlie succession are not clear. In some cases, it may be that a resident species modified the environment in some way, thereby making it more suitable for a later species to colonize. It is also possible that prior residents lived there in the first place because there was little competition from other species. Later, as more invasive species arrived, the original species were displaced.

Ecologists initially thought that succession inevitably led to a stable and persistent community, known as a *climax community*, such as a forest. But more recently, this traditional view has fallen out of favor. The apparent stability of a "climax" forest is probably the result of how long trees live relative to the human life span. It is now recognized that mature climax communities are not in a state of stable equilibrium but rather in a state of continual disturbance. Over time, a mature community changes in species composition and in the relative abundance of each species, despite the fact that it retains an overall uniform appearance.

Succession is usually described in terms of the changes in the plant species growing in a given area, although each stage of the succession may also have its own kinds of animals and other organisms. Ecological succession is measured on the scale of tens, hundreds, or thousands of years, not the millions of years involved in the evolutionary time scale.

ecological succession

The process of community development over time, which involves species in one stage being replaced by different species.

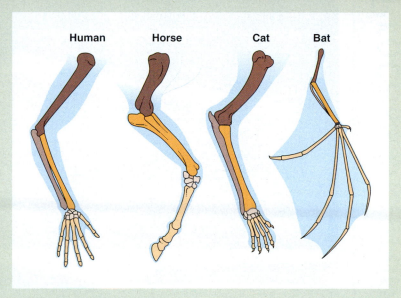

A **The fossil record** Fossils deposited in rock layers, which can be dated, show how organisms evolved over time. These fish fossils from Liaoning Province, China, date from 120 million years ago.

B **Comparative anatomy** Similarities among organisms demonstrate how they are related. These similarities among four vertebrate limbs illustrate that, while proportions of bones have changed in relation to each organism's way of life, the forelimbs have the same basic bone structure.

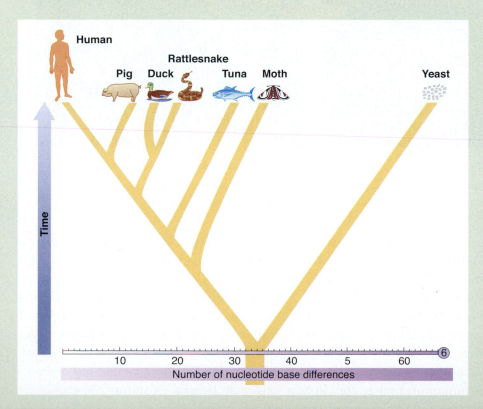

C **Molecular biology** The organisms pictured here all share a particular enzyme, but in the course of evolution, mutations have resulted in changes in the gene that codes for that enzyme. This diagram shows the nucleotide base differences in this gene among humans and other organisms. Note that organisms thought to be more closely related to humans have fewer differences than organisms that are more distantly related to humans.

Primary succession on glacial moraine FIGURE 6.19

A After a glacier's retreat, lichens initially colonize the barren landscape, followed by mosses and small shrubs.

During the past 200 years, glaciers have retreated in Glacier Bay, Alaska. Although these photos were not taken in the same area, they show some of the stages of primary succession on glacial moraine (rocks, gravel, and sand that a glacier deposits).

B At a later time, dwarf trees and shrubs colonize the area.

C Still later, spruces dominate the community.

Primary succession is the change in species composition over time in a previously uninhabited environment (**FIGURE 6.19**). No soil exists when primary succession begins. Bare rock surfaces, such as recently formed volcanic lava and rock scraped clean by glaciers, are examples of sites where primary succession may take place. Details vary from one site to another, but on bare rock, lichens are often the most important element in the *pioneer community*, which is the initial community that develops during primary succession. Lichens secrete acids that help break the rock apart, beginning the process of soil formation. Over time, mosses and drought-resistant ferns may replace the lichen community, followed in turn by tough grasses and herbs. Once soil accumulates, low shrubs may replace the grasses and herbs; over time, forest trees in several distinct stages would replace the shrubs. Primary succession on bare rock from a pioneer community to a forest community often occurs in this sequence: lichens → mosses → grasses → shrubs → trees.

Secondary succession on an abandoned field in North Carolina FIGURE 6.20

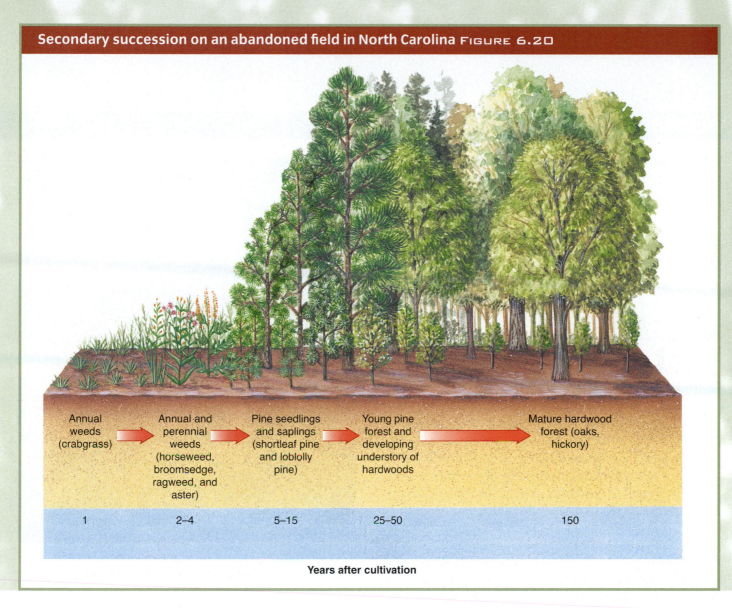

Annual weeds (crabgrass)	Annual and perennial weeds (horseweed, broomsedge, ragweed, and aster)	Pine seedlings and saplings (shortleaf pine and loblolly pine)	Young pine forest and developing understory of hardwoods	Mature hardwood forest (oaks, hickory)
1	2–4	5–15	25–50	150

Years after cultivation

Secondary succession is the change in species composition that takes place after some disturbance destroys the existing vegetation; soil is already present (FIGURE 6.20). Abandoned farmland or an open area caused by a forest fire are common examples of sites where secondary succession occurs. Biologists have studied secondary succession on abandoned farmland extensively. Although it takes more than 100 years for secondary succession to occur at a single site, a single researcher can study old-field succession in its entirety by observing different sites undergoing succession in the same general area. (The biologist may examine county tax records to determine when each field was abandoned.) Secondary succession on abandoned farmland in the southeastern United States proceeds in this sequence: crabgrass → horseweed, broomsedge, and other weeds → pine trees → hardwood trees.

CONCEPT CHECK **STOP**

What is ecological succession?

How does primary succession differ from secondary succession?

WILDFIRES

A wildfire is any unexpected—and unwanted—fire that burns in grass, shrub, and forest areas. Whether started by lightning or by humans, wildfires are an important environmental force in many geographic areas, especially places with wet seasons followed by dry seasons, such as chaparral. Vegetation that grows during the wet season dries to tinder during the dry season. After fire ignites the dry organic material, wind spreads the fire through the area.

At the peak of the wildfire season in the American West, an area prone to wildfires, hundreds of new wildfires can break out each day. In southern California alone, devastating wildfires in 2007 and 2008 consumed well over 260,000 hectares (650,000 acres).

Fires have several effects on the environment. First, combustion frees minerals locked in dry organic matter. The ashes left by fire are rich in potassium, phosphorus, calcium, and other nutrient minerals essential for plant growth. Thus, vegetation flourishes after a fire. Second, fire removes plant cover and exposes the soil, which stimulates the germination of seeds that require bare soil and the growth of shade-intolerant plants. Third, fire increases soil erosion because it removes plant cover, leaving soil more vulnerable to wind and water.

Fires were a part of the natural environment long before humans appeared, and many terrestrial ecosystems have adapted to fire. Grasses adapted to wildfire have underground stems and buds. After fire kills the aerial parts, the untouched underground parts send up new sprouts. Fire-adapted trees such as bur oak and ponderosa pine have thick, fire-resistant bark; others, such as jack pines, depend on fire for successful

A NASA satellite image of the southern California wildfires, October 24, 2007. Red pixels were added to indicate fires.

reproduction because the fire's heat opens the cones and releases the seeds.

Human interference also affects the frequency and intensity of wildfires, even when the actual goal is fire prevention. When fire is excluded from a fire-adapted ecosystem, organic litter accumulates. As a result, when a fire does occur, it burns hotter and is much more destructive than ecologically helpful. Decades of fire suppression in the West are partly responsible for the massively destructive fires that have occurred there in recent years. *Prescribed burning* is an ecological management tool that allows for controlled burning to reduce organic litter and suppress fire-sensitive trees in fire-adapted areas.

A deer escapes from wildfire near Malibu, California, November 2007.

SUMMARY

1 Earth's Major Biomes

1. A **biome** is a large, relatively distinct terrestrial region with characteristic climate, soil, plants, and animals, regardless of where it occurs; a biome encompasses many interacting ecosystems. Near the poles, temperature is generally the overriding climate factor in determining biome distribution, whereas in temperate and tropical regions, precipitation is more significant.

2. **Tundra** is the treeless biome in the far north that consists of boggy plains covered by lichens and small plants such as mosses; it has harsh, very cold winters and extremely short summers. **Boreal forest** is a region of coniferous forest in the Northern Hemisphere, located just south of the tundra. **Temperate rain forest** is a coniferous biome with cool weather, dense fog, and high precipitation. **Temperate deciduous forest** is a forest biome that occurs in temperate areas where annual precipitation ranges from about 75 cm to 126 cm. **Tropical rain forest** is a lush, species-rich forest biome that occurs where the climate is warm and moist throughout the year. **Chaparral** is a biome with mild, moist winters and hot, dry summers; vegetation is typically small-leafed evergreen shrubs and small trees. **Temperate grassland** is grassland with hot summers, cold winters, and less rainfall than is found in the temperate deciduous forest biome. **Savanna** is tropical grassland with widely scattered trees or clumps of trees. **Desert** is a biome in which the lack of precipitation limits plant growth; deserts are found in both temperate and tropical regions.

2 Aquatic Ecosystems

1. In aquatic ecosystems, important environmental factors include salinity, amount of dissolved oxygen, and availability of light for photosynthesis.

2. Freshwater ecosystems include standing-water, flowing-water, and freshwater wetlands. A **standing-water ecosystem** is a body of fresh water surrounded by land and whose water does not flow, such as a lake or pond. A **flowing-water ecosystem** is a freshwater ecosystem such as a river or stream in which the water flows in a current. **Freshwater wetlands** are marshes and swamps—lands that shallow fresh water covers for at least part of the year; wetlands have a characteristic soil and water-tolerant vegetation. An **estuary** is a coastal body of water, partly surrounded by land, with access to the open ocean and a large supply of fresh water from a river. Water in an estuary is brackish rather than truly fresh. Temperate estuaries usually contain **salt marshes**, whereas tropical estuaries are lined with **mangrove forests**.

3 Population Responses to Changing Conditions over Time: Evolution

1. **Evolution** is the cumulative genetic changes in populations that occur during successive generations.

2. **Natural selection** is the tendency of better-adapted individuals—those with a combination of genetic traits best suited to environmental conditions—to survive and reproduce, increasing their proportion in the population. Natural selection is based on four observations established by Charles Darwin: (1) Each species produces more offspring than will survive to maturity. (2) The individuals in a population exhibit inheritable variation in their traits. (3) Organisms compete with one another for the resources needed to survive. (4) Individuals with the most favorable combination of traits are most likely to survive and reproduce, passing their genetic traits to the next generation.

3. Scientific evidence supporting evolution comes from the fossil record, comparative anatomy, biogeography, and molecular biology.

4 Community Responses to Changing Conditions over Time: Succession

1. **Ecological succession** is the process of community development over time, which involves species in one stage being replaced by different species.

2. **Primary succession** is the change in species composition over time in an environment that was not previously inhabited by organisms; examples include bare rock surfaces, such as recently formed volcanic lava and rock scraped clean by glaciers. **Secondary succession** is the change in species composition that takes place after some disturbance destroys the existing vegetation; soil is already present. Examples include abandoned farmland and open areas caused by forest fires.

KEY TERMS

- biome p. 128
- tundra p. 132
- boreal forest p. 133
- temperate rain forest p. 134
- temperate deciduous forest p. 135
- tropical rain forest p. 136

- chaparral p. 137
- temperate grassland p. 139
- savanna p. 140
- desert p. 141
- standing-water ecosystem p. 142
- flowing-water ecosystem p. 144

- freshwater wetlands p. 145
- ecosystem services p. 145
- estuary p. 146
- evolution p. 147
- natural selection p. 148
- ecological succession p. 150

CRITICAL AND CREATIVE THINKING QUESTIONS

1. What two climate factors are most important in determining an area's characteristic biome?

2. What climate and soil factors produce each of the major terrestrial biomes?

3. In which biome do you live? Where would you place your biome in the figure below? How would that compare with your placement of the biome in northern Siberia or the biome dominating northern Africa and Saudi Arabia?

4. If your biome does not match the description given in this book, how do you explain the discrepancy?

5. Which biomes are best suited for agriculture? Explain why each of the biomes you did not specify is less suitable for agriculture.

6. What environmental factors are most important in determining the kinds of organisms found in aquatic environments?

7. Distinguish between freshwater wetlands and estuaries and between flowing-water and standing-water ecosystems.

8. During the mating season, male giraffes slam their necks together in fighting bouts to determine which male is stronger and can therefore mate with females. Explain how the long necks of giraffes may have evolved, using Darwin's theory of evolution by natural selection.

9. Describe the stages in old-field succession.

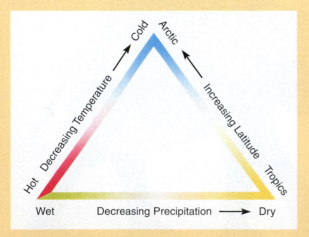

CRITICAL AND CREATIVE THINKING QUESTIONS

10. Although most salamanders have four legs, the aquatic salamander shown below resembles an eel. It lacks hind limbs and has very tiny forelimbs. Propose a hypothesis to explain how limbless salamanders evolved according to Darwin's theory of natural selection.

11. How could you test the hypothesis you proposed in question 10?

12. Which biome discussed in this chapter is depicted by the information in the graph below? Explain your answer.

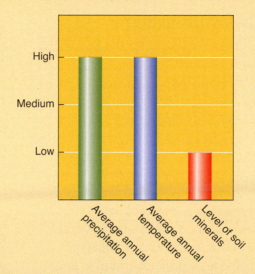

What is happening in this picture ?

■ This picture shows expensive homes built in the chaparral of the Santa Monica Mountains. Based on what you have learned in this chapter, what environmental problem might threaten these homes?

■ Sometimes people have removed the chaparral vegetation to prevent fires from damaging their homes. Where that has occurred, the roots no longer hold the soil in place. What could happen when the winter rains come?

Human Population Change and the Environment

7

SLOWING POPULATION GROWTH IN CHINA

VIEW THIS IN ACTION
in your WileyPLUS course

C hina, with an estimated population of 1.3 billion people, has the largest population in the world. Recognizing that its rate of population growth had to decrease or the quality of life for everyone in China would be compromised, in 1971 the Chinese government began to pursue birth control seriously. It urged couples to marry later, increase spacing between children, and have fewer children.

In 1979 China instigated a more aggressive plan to reduce the birth rate, announcing incentives to promote later marriages and one-child families (see inset). A couple who signed a pledge to limit themselves to a single child might be eligible for incentives like medical care and schooling for the child, cash bonuses, preferential housing, and retirement funds. Penalties were instituted, including fines and the surrender of all of these privileges, if a second child was born.

China's aggressive plan brought about the most rapid and drastic reduction in fertility in the world, from 5.8 births per woman in 1970 to 1.6 births per woman today. However, the plan was controversial and unpopular because it compromises individual freedom of choice. In some instances, social pressures caused women who were pregnant with a second child to get an abortion.

In China, sons are valued more highly than daughters because sons carry on the family name and traditionally provide old-age security for their parents. A disproportionate number of male babies have been born in recent years, suggesting that some expectant parents determine the sex of their fetus and abort it if it is female. In the past, parents required to conform to the one-baby policy abandoned or killed thousands of newborn baby girls because they wanted a boy.

In 1984 the one-child family policy was relaxed in rural China, where 70 percent of all Chinese live. China's recent population control program has relied on education, publicity campaigns (see larger photo), and fewer penalties to achieve its goals.

158

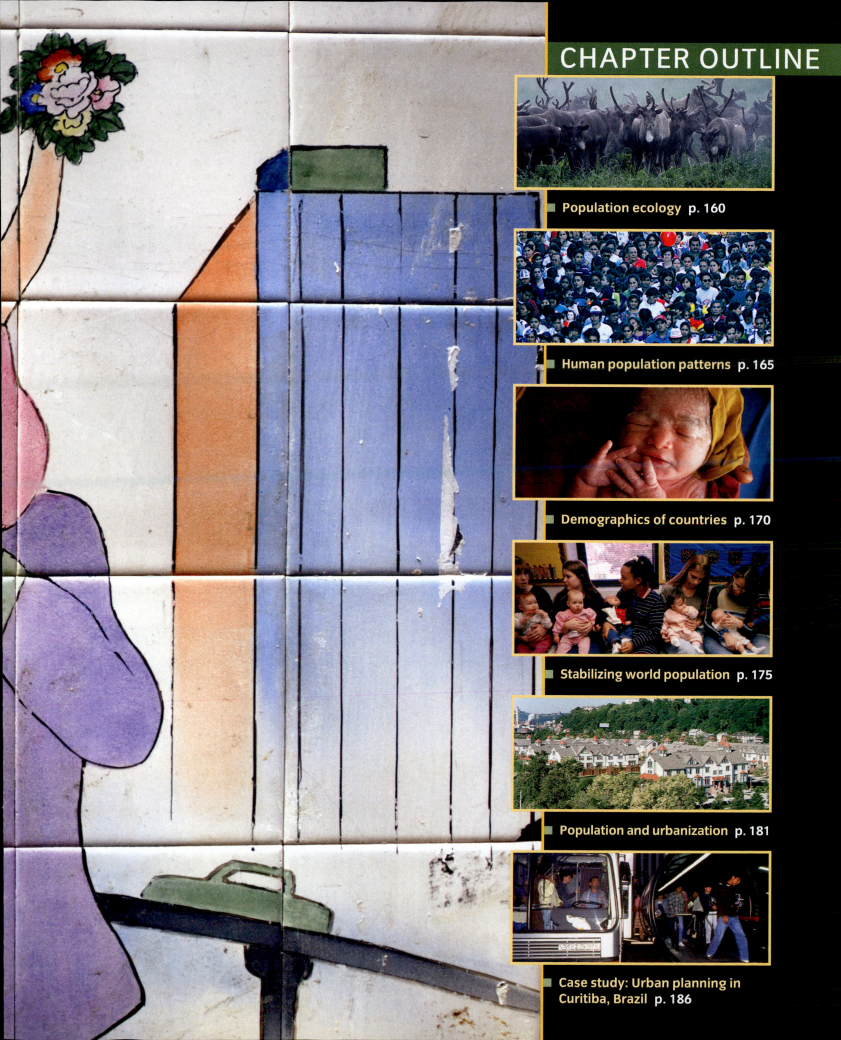

Population Ecology

LEARNING OBJECTIVES

Define *population ecology*.

Explain the four factors that produce changes in population size.

Define *biotic potential* and *carrying capacity*.

Individuals of a given species are part of a larger organization called a *population*. Populations exhibit characteristics that are distinct from those of the individuals in them. Some of the features characteristic of populations but not of individuals are birth and death rates, growth rates, and age structure. Studying populations of other species provides insight into some of the processes that affect the growth of human populations. Understanding human population change is important because the size of the human population is central to most of Earth's environmental problems and their solutions.

Scientists who study **population ecology** try to determine the processes common to all populations (**FIGURE 7.1**). Population ecologists study how a population responds to its environment—such as how individuals in a given population compete for food or other resources, and how predation, disease, and other environmental pressures affect that population. Environmental pressures such as these prevent populations—whether of bacteria or maple trees or giraffes—from increasing indefinitely.

> **population ecology** The branch of biology that deals with the number of individuals of a particular species found in an area and why those numbers increase or decrease over time.

What we learn about one population helps us make predictions about other populations FIGURE 7.1

At first glance, the two populations shown here appear to have little in common, but they share many characteristics.

A A cluster of soft corals. Photographed in the Western Pacific Ocean.

B A herd of impala in a woodland meadow. Photographed in Rwanda.

HOW DO POPULATIONS CHANGE IN SIZE?

Populations of organisms, whether sunflowers, eagles, or humans, change over time. On a global scale, this change is due to two factors: the rate at which individual organisms produce offspring (the birth rate) and the rate at which individual organisms die (the death rate) (FIGURE 7.2A). In humans, the birth rate (b) is usually expressed as the number of births per 1000 people per year, and the death rate (d) as the number of deaths per 1000 people per year. The **growth rate (r)** of a population is the birth rate (b) minus the death rate (d), or $r = b - d$. Growth rate is also referred to as *natural increase* in human populations.

If organisms in the population are born faster than they die, the growth rate is more than zero, and population size increases. If organisms in the population die faster than they are born, the growth rate is less than zero, and population size decreases. If the growth rate is equal to zero, births and deaths match, and population size is stationary, despite continued reproduction and death.

> **growth rate (r)** The rate of change (increase or decrease) of a population's size, expressed in percentage per year.

In addition to birth and death rates, **dispersal**—movement from one region or country to another—affects local populations. There are two types of dispersal: **immigration (i)**, in which individuals enter a population and increase its size, and **emigration (e)**, in which individuals leave a population and decrease its size. The growth rate (r) of a local population must take into account birth rate (b), death rate (d), immigration (i), and emigration (e) (FIGURE 7.2B).

Factors that interact to change population size FIGURE 7.2

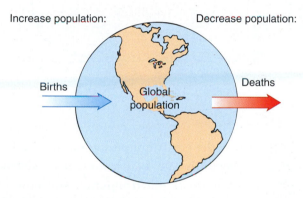

A On a global scale, the change in a population is due to the number of births and deaths.

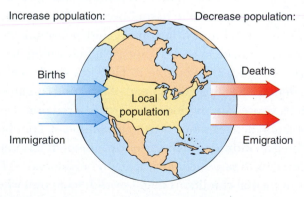

B In local populations, such as the population of the United States, the number of births, deaths, immigrants, and emigrants affect population size.

MAXIMUM POPULATION GROWTH

Different species have different **biotic potentials**. Several factors influence the biotic potential of a species: the age at which reproduction begins, the fraction of the life span during which an individual can reproduce, the number of reproductive periods per lifetime, and the number of offspring produced during each period of reproduction. These factors, called *life history characteristics*, determine whether a particular species has a large or a small biotic potential.

> **biotic potential** The maximum rate at which a population could increase under ideal conditions.

Generally, larger organisms such as blue whales have the smallest biotic potentials, whereas microorganisms have the greatest biotic potentials. Under ideal conditions (that is, in an environment with unlimited resources), certain bacteria reproduce by dividing in half

Exponential population growth FIGURE 7.3

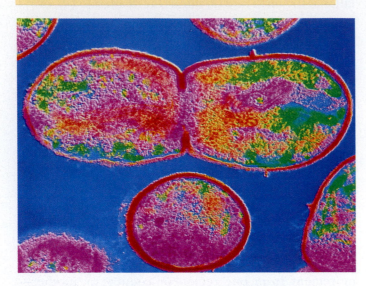

A *Streptococcus* bacterium in the process of dividing.

Time (hours)	Number of bacteria
0	1
1	4
2	16
3	64
4	256
5	1,024
6	4,096
7	16,384
8	65,536
9	262,144
10	1,048,576

B When bacteria divide at a constant rate, their number increases exponentially.

C When their numbers are graphed, the curve of exponential population growth has a characteristic J shape. ▶

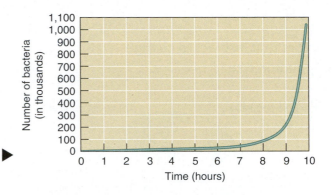

every 30 minutes (**FIGURE 7.3A**). At this rate of growth, a single bacterium increases to a population of more than 1 million in just 10 hours and exceeds 1 billion in 15 hours. If you plot population number versus time, the graph takes on the characteristic J shape of **exponential population growth** (**FIGURE 7.3B** and **C**). When a population grows exponentially, the larger the population gets, the faster it grows. Regardless of species, whenever a population grows at its biotic potential, population size plotted versus time gives the same J-shaped curve. The only variable is time.

ENVIRONMENTAL RESISTANCE AND CARRYING CAPACITY

Certain populations may exhibit exponential population growth for a short period. However, organisms don't reproduce indefinitely at their biotic potentials be-

> **exponential population growth**
>
> The accelerating population growth that occurs when optimal conditions allow a constant reproductive rate.

cause the environment sets limits, which are collectively called **environmental resistance**. Examples of environmental resistance include limited food, water, shelter, and other essential resources, as well as increased disease and predation.

Using the earlier example, we find that bacteria never reproduce unchecked for an indefinite period because they run out of food and living space, and poisonous body wastes accumulate in their vicinity. With crowding, bacteria become more susceptible to parasites (high population densities facilitate the spread of infectious organisms such as viruses among individuals) and predators (high population densities increase the likelihood of a predator catching an individual). As the environment deteriorates, bacterial birth rate declines and death rate increases. The environmental conditions might worsen to a point where the death rate exceeds the birth rate, and as a result, the population decreases. Thus, the environment controls population size: As the population increases, so does environmental resistance, which limits population growth.

Over longer periods, the rate of population growth may decrease to nearly zero. This leveling out occurs at or near the environment's **carrying capacity (K)**. In nature, carrying capacity is dynamic and changes in response to environmental changes. An extended drought, for example, might decrease the amount of vegetation growing in an area, and this change, in turn, would lower the carrying capacity for deer and other herbivores in that environment.

G.F. Gause, a Russian ecologist who conducted experiments in the 1930s, grew a population of *Paramecium* in a test tube (**FIGURE 7.4A**). He supplied a limited amount of food daily and replenished the media to eliminate the buildup of wastes. Under these conditions, the population increased exponentially at first, but then its growth rate declined to zero, and the population size leveled off.

> **■ carrying capacity**
> **(K)** The largest population a particular environment can support sustainably (long term), if there are no changes in that environment.

When a population influenced by environmental resistance is graphed over a long period (**FIGURE 7.4B**), the curve has an **S** shape. The curve shows the population's initial exponential increase (note the curve's **J** shape at the start, when environmental resistance is low). Then the population size levels out as it approaches the carrying capacity of the environment. Although the **S** curve is an oversimplification of how most populations change over time, it fits some populations studied in the laboratory, as well as a few studied in nature.

A population rarely stabilizes at *K* (carrying capacity), as shown in Figure 7.4, but its size may temporarily rise higher than *K*. It will then drop back to, or below, the carrying capacity. Sometimes a population that overshoots *K* will experience a *population crash*, an abrupt decline from high to low population density when resources are exhausted. Such an abrupt change is commonly observed in bacterial cultures, zooplankton, and other populations whose resources are exhausted.

Population growth as carrying capacity is approached FIGURE 7.4

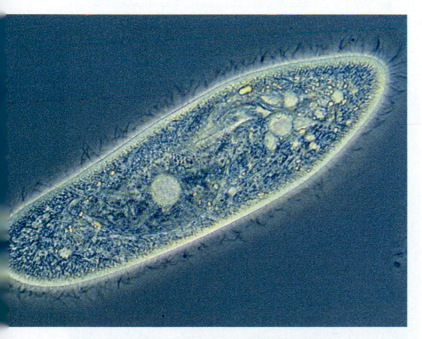

A *Paramecium* is a unicellular microorganism.

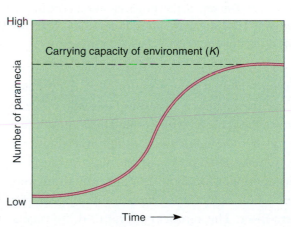

B In many laboratory studies, including Gause's work with *Paramecium*, population growth increases exponentially when the population is low but slows as the carrying capacity of the environment is approached. This produces a curve with a characteristic S shape.

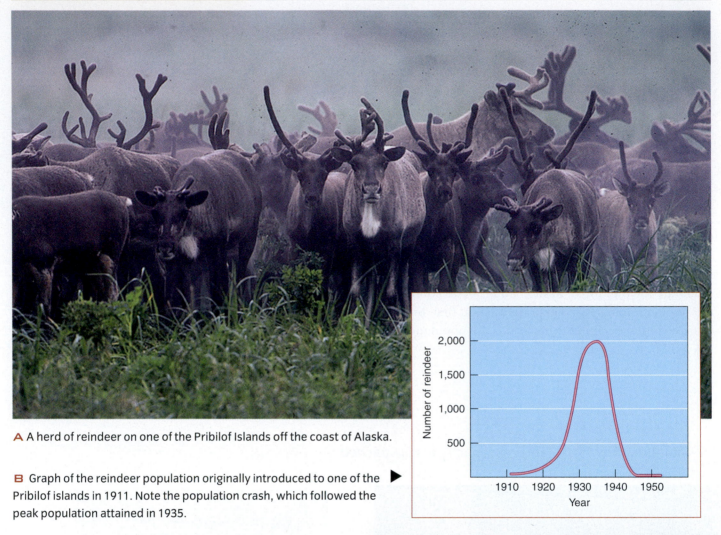

A A herd of reindeer on one of the Pribilof Islands off the coast of Alaska.

B Graph of the reindeer population originally introduced to one of the Pribilof islands in 1911. Note the population crash, which followed the peak population attained in 1935.

The availability of winter forage largely determines the carrying capacity for reindeer, which live in cold northern habitats. In 1911, a small herd of reindeer was introduced on one of the Pribilof Islands in the Bering Sea (FIGURE 7.5A). The herd's population increased exponentially for about 25 years until there were many more reindeer than the island could support, particularly in winter. The reindeer overgrazed the vegetation until the plant life was almost wiped out. Then, in slightly over a decade, as reindeer died from starvation, the number of reindeer plunged to less than 1 percent of the population at its peak (FIGURE 7.5B). If reindeer overgraze the vegetation, it takes 15 to 20 years for it to recover. During that period, the carrying capacity for reindeer is greatly reduced.

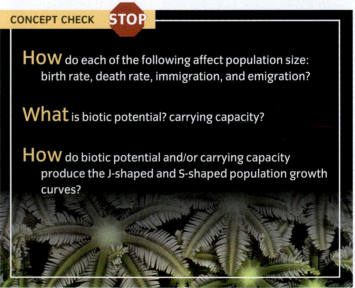

CONCEPT CHECK STOP

How do each of the following affect population size: birth rate, death rate, immigration, and emigration?

What is biotic potential? carrying capacity?

How do biotic potential and/or carrying capacity produce the J-shaped and S-shaped population growth curves?

Human Population Patterns

N ow that you have examined some of the basic concepts of population ecology, let's apply those concepts to the human population. **FIGURE 7.6** shows the increase in human population since 1800. The characteristic **J** curve of exponential population growth reflects the decreasing amount of time it has taken to add each additional billion people to our numbers. It took thousands of years for the human population to reach 1 billion, a milestone that took place around 1800. It took 130 years to reach 2 billion (in 1930), 30 years to reach 3 billion (in 1960), 15 years to reach 4 billion (in 1975), 12 years to reach 5 billion (in 1987), and 12 years to reach 6 billion (in 1999). The United Nations projects that the human population will reach 7 billion by 2012. Population experts predict that the population will level out during the 21st century, possibly forming an **S** curve as observed in other species.

One of the first people to recognize that the human population can't increase indefinitely was **Thomas Malthus** (1766–1834), a British economist. He pointed out that human population growth is not always desirable—a view contrary to the beliefs of his day and to those of many people even today. Noting that human population can increase faster than its food supply, he warned that the inevitable consequences of population growth would be famine, disease, and war. Since Malthus's time, the human population has increased from about 1 billion to more than 6 billion. On the surface, it seems that Malthus was wrong. Our population has grown dramatically because scientific advances have allowed food production to keep pace with population growth. Malthus's ideas may ultimately be proved correct, however, because we don't know whether this increased food production is sustainable. Have we achieved this increase in food production at the environmental cost of reducing the planet's ability to meet the needs of future populations?

Our world population was 6.6 billion in 2007, an increase of about 70 million from 2006. This increase isn't due to a rise in the birth rate (*b*). In fact, the world birth rate has declined slightly during the past 200 years. The population growth is due instead to a dramatic *decrease* in the death rate (*d*) because greater food

Human population numbers, 1800 to present
FIGURE 7.6

Until recently, the human population has been increasing exponentially. There are now indications that the human population is beginning to level out, forming an S curve.

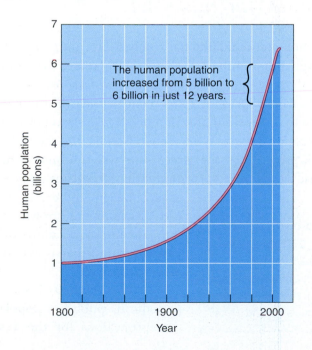

The human population increased from 5 billion to 6 billion in just 12 years.

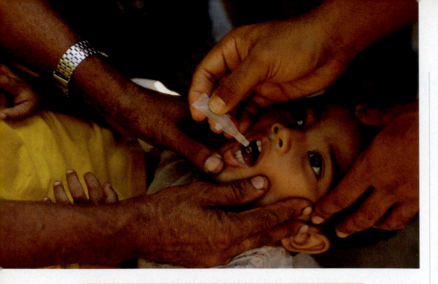

Advances in global health FIGURE 7.7

A child in Bangladesh receives a dose of oral polio vaccine. At one time, polio killed or crippled millions of children each year. Polio is still endemic (constantly present) in Nigeria, India, Afghanistan, and Pakistan, and it sometimes spreads from those countries to other countries.

production, better medical care, and improvements in water quality and sanitation practices have increased life expectancy for a great majority of the global population (**FIGURE 7.7**).

PROJECTING FUTURE POPULATION NUMBERS

The human population has reached a turning point. Although our numbers continue to increase, the world growth rate (*r*) has declined slightly over the past several years, from a peak of 2.2 percent per year in the mid-1960s to the current growth rate of 1.2 percent per year. Population experts at the United Nations and the World Bank project that the growth rate will continue to decrease slowly until **zero population growth** is attained toward the end of the 21st century. Exponential growth of the human population will end, and the **S** curve may replace the **J** curve.

zero population growth The state in which the population remains the same size because the birth rate equals the death rate.

The United Nations periodically publishes population projections for the 21st

century. The latest (2004) U.N. figures forecast that the human population will total between 7.7 billion (their "low" projection) and 10.6 billion (their "high" projection) in the year 2050, with 9.1 billion thought to be most likely (**FIGURE 7.8**). The estimates vary depending on fertility changes, particularly in less developed countries, because that is where almost all of the growth will take place.

Population projections must be interpreted with care because they vary depending on what assumptions are made. In projecting that the world population will be 7.7 billion (their low projection) in the year 2050, U.N. population experts assume that the average number of children born to each woman in all coun-

Population projections to 2050 FIGURE 7.8

In 2004 the United Nations made three projections, each based on different fertility rates.

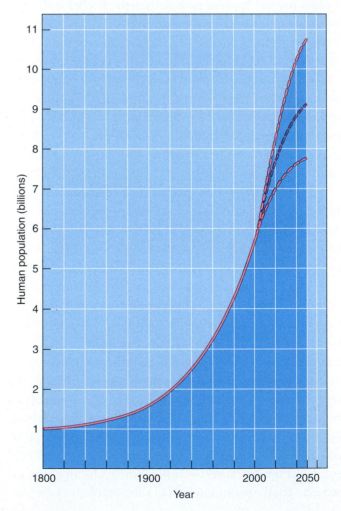

tries will have declined to 1.5 in the 21st century. The average number of children born to each woman on Earth is currently 2.7. If the decline to 1.5 doesn't occur, our population could be significantly higher. If the average number of children born to each woman declines to only 2.5 instead of 1.5, the 2050 population will be 10.6 billion (the U.N. high projection). Small differences in fertility, then, produce large differences in population forecasts.

The main unknown factor in any population growth scenario is Earth's *carrying capacity*. Most published estimates of how many people Earth can support range from 4 billion to 16 billion. For example, in 2004, environmental economists in the Netherlands performed a detailed analysis of 69 recent studies of Earth's carrying capacity for humans. Based on current technology, they estimated that 7.7 billion is the upper limit of human population that the world can support.

These estimates vary widely depending on what assumptions are made about standard of living, resource consumption, technological innovations, and waste generation. If we want all people to have a high level of material well-being equivalent to the lifestyles in highly developed countries, then Earth will support far fewer humans than if everyone lives just above the subsistence level. Unlike with other organisms, environmental constraints aren't the exclusive determinant of Earth's carrying capacity for humans. Human choices and values must be factored into the assessment.

What will happen to the human population when it approaches Earth's carrying capacity? Optimists suggest that a decrease in the birth rate will stabilize the human population. Some experts take a more pessimistic view and predict that our ever-expanding numbers will cause widespread environmental degradation and make Earth uninhabitable for humans as well as other species (**FIGURE 7.9**). These experts contend that a massive wave of human suffering and death will occur. This view doesn't mean we will go extinct as a species, but it projects severe hardship for many people. Some experts think the human population has already exceeded the carrying capacity of the environment, a potentially dangerous situation that threatens our long-term survival as a species.

Human population trends are summarized in **FIGURE 7.10**.

Environmental degradation on a cattle ranch in Brazil FIGURE 7.9

Part of the rain forest in the background was cleared for a cattle pasture. After a few years, the pasture became unproductive, and erosion degraded the land further. Photographed in Amazonas State in the Amazon River Basin.

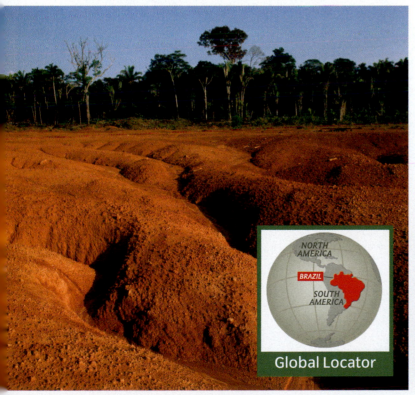

Global Locator

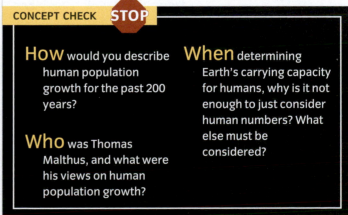

CONCEPT CHECK STOP

How would you describe human population growth for the past 200 years?

Who was Thomas Malthus, and what were his views on human population growth?

When determining Earth's carrying capacity for humans, why is it not enough to just consider human numbers? What else must be considered?

The human population FIGURE 7.10

Human population issues are both local and global in scope.

▶ **POPULATION DENSITY**
Population density can be measured as the average number of people per square unit in a given area (arithmetic density). Populations, however, are not evenly distributed. Often, they're gathered around arable land. By comparing populations to farmland (physiologic density), statistics can be more meaningful.

Egypt, for example, where nearly 90% of the citizens are clustered in the Nile Valley, has a modest overall density of 74 people per sq km (191 people per sq mi), but a physiologic density of 3,089 people per agricultural sq km (8,000 people per agricultural sq mi), among the world's highest.

Population growth rate as a percent of total population, 2005 estimate
(1 block=1 million people)

- More than 3%
- 2.0%–2.9%
- 1.0%–1.9%
- 0%–0.9%
- Negative growth

2005 population (millions) in parentheses

Not all countries or territories shown

▲ **POPULATION CARTOGRAM**
The world appears quite different when countries are sized proportional to their populations and mapped. Underpopulated Canada, the world's second largest country, is reduced to a small strip above the United States, while small, crowded Japan looms large. India and China become the global giants.

▶ **POPULATION PYRAMIDS**
When population is expressed in bars representing age and gender and stacked up (males left, females right), country profiles emerge that have ramifications for the future. Countries with high birthrates and high percentages of young, such as Nigeria, look like pyramids. Countries such as Italy, whose birth rate is below the replacement fertility level of 2.1 children per couple, show bulges in the higher age brackets. The United States clearly shows the "baby boom" of children born in the years after World War II.

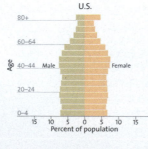

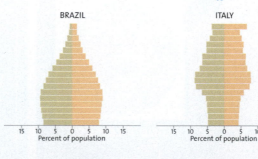

U.S.   BRAZIL   ITALY   NIGERIA

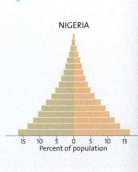

NATIONAL GEOGRAPHIC

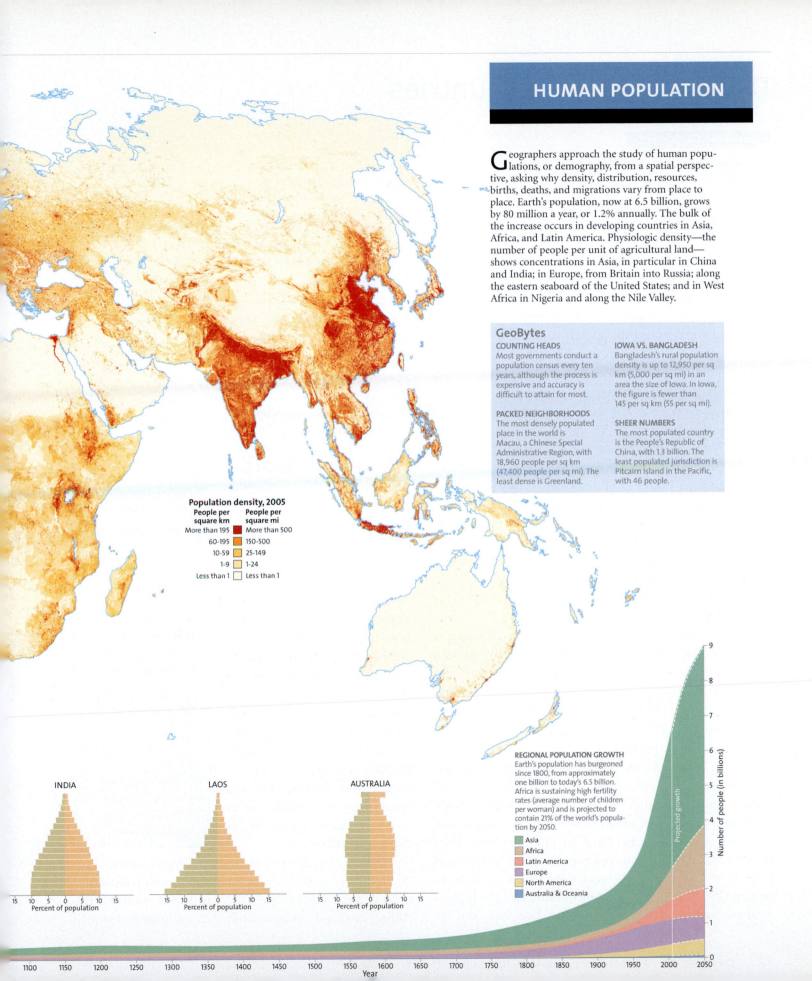

HUMAN POPULATION

Geographers approach the study of human populations, or demography, from a spatial perspective, asking why density, distribution, resources, births, deaths, and migrations vary from place to place. Earth's population, now at 6.5 billion, grows by 80 million a year, or 1.2% annually. The bulk of the increase occurs in developing countries in Asia, Africa, and Latin America. Physiologic density—the number of people per unit of agricultural land—shows concentrations in Asia, in particular in China and India; in Europe, from Britain into Russia; along the eastern seaboard of the United States; and in West Africa in Nigeria and along the Nile Valley.

GeoBytes

COUNTING HEADS
Most governments conduct a population census every ten years, although the process is expensive and accuracy is difficult to attain for most.

PACKED NEIGHBORHOODS
The most densely populated place in the world is Macau, a Chinese Special Administrative Region, with 18,960 people per sq km (47,400 people per sq mi). The least dense is Greenland.

IOWA VS. BANGLADESH
Bangladesh's rural population density is up to 12,950 per sq km (5,000 per sq mi) in an area the size of Iowa. In Iowa, the figure is fewer than 145 per sq km (55 per sq mi).

SHEER NUMBERS
The most populated country is the People's Republic of China, with 1.3 billion. The least populated jurisdiction is Pitcairn Island in the Pacific, with 46 people.

Population density, 2005

People per square km	People per square mi
More than 195	More than 500
60–195	150–500
10–59	25–149
1–9	1–24
Less than 1	Less than 1

INDIA

Percent of population

LAOS

Percent of population

AUSTRALIA

Percent of population

REGIONAL POPULATION GROWTH
Earth's population has burgeoned since 1800, from approximately one billion to today's 6.5 billion. Africa is sustaining high fertility rates (average number of children per woman) and is projected to contain 21% of the world's population by 2050.

- Asia
- Africa
- Latin America
- Europe
- North America
- Australia & Oceania

Projected growth

Number of people (in billions)

Year

Demographics of Countries

LEARNING OBJECTIVES

Define demographics and describe the demographic transition.

Explain how highly developed and developing countries differ in population characteristics such as infant mortality rate, total fertility rate, replacement-level fertility, and age structure.

World population figures illustrate overall trends but don't describe other important aspects of the human population story, such as population differences from country to country. **Demographics** provides interesting information on the populations of various countries. Recall from Chapter 1 that countries are classified into two groups—highly developed and developing—based on population growth rates, degree of industrialization, and relative prosperity.

Highly developed countries have the lowest birth rates in the world. Some countries, such as Germany, have birth rates just below those needed to sustain their populations and are declining slightly in numbers. Highly developed countries also have low **infant mortality rates** (**FIGURE 7.11A**). The infant mortality rate of the United States is 6.5, compared with a world rate of 52. Highly developed countries have longer life expectancies (78 years in the United States versus 68 years worldwide).

Per person GNI PPP is a country's gross national income (GNI) in purchasing power parity (PPP) divided by its population. It indicates the amount of goods and services an average citizen of that particular country could buy in the United States. There is a high average per person GNI PPP in the United States—$44,260—versus the worldwide figure of $9940.

In *moderately developed countries*, birth rates and infant mortality rates are higher than those of highly developed countries, but they are declining. Moderately developed countries have a medium level of industrialization, and their average per person GNI PPPs are lower than those of highly developed countries. *Less developed countries* have the shortest life expectancies, the lowest average per person GNI PPPs, the highest birth rates, and the highest infant mortality rates in the world (**FIGURE 7.11B**).

Replacement-level fertility is usually given as 2.1 children. The number is greater than 2.0 because some infants and children die before they reach reproductive age. Worldwide, the **total fertility rate (TFR)** is currently 2.7, well above the replacement level.

THE DEMOGRAPHIC TRANSITION

Demographers recognize four demographic stages based on their observations of Europe as it became industrialized and urbanized (**FIGURE 7.12**). These stages converted Europe from relatively high birth and death rates to relatively low birth and death rates. All highly developed and moderately developed countries with more advanced economies have gone through this **demographic transition**, and demographers assume that the same progression will occur in less developed countries as they industrialize.

Why has the population stabilized in more than 30 highly developed countries in the fourth (postindustrial) demographic stage? The reasons are complex. Declining birth rate is associated with an improvement in

demographics The applied branch of sociology that deals with population statistics.

infant mortality rate The number of deaths of infants under age 1 per 1000 live births.

replacement-level fertility The number of children a couple must produce to "replace" themselves.

total fertility rate (TFR) The average number of children born to each woman.

demographic transition The process whereby a country moves from relatively high birth and death rates to relatively low birth and death rates.

Infant mortality rates FIGURE 7.11

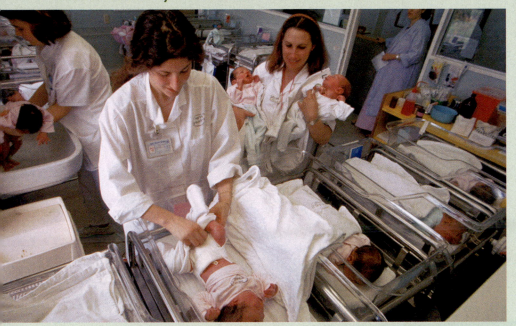

A Nurses care for newborn infants in Israel, a highly developed country with an infant mortality rate of 3.9.

B This infant, just born in the less developed country of Bangladesh, is underweight and therefore at risk. The infant mortality rate in Bangladesh is 65.

The demographic transition FIGURE 7.12

Demographers have identified four stages through which a population progresses as its society becomes industrialized.

VIEW THIS IN ACTION
in your WileyPLUS course

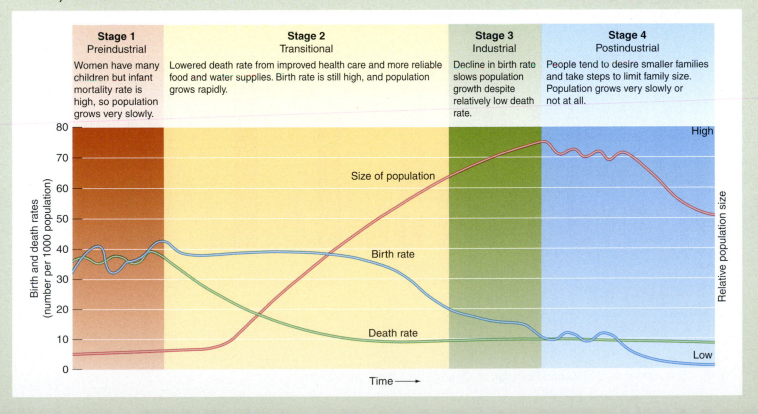

Stage 1	Stage 2	Stage 3	Stage 4
Preindustrial	Transitional	Industrial	Postindustrial
Women have many children but infant mortality rate is high, so population grows very slowly.	Lowered death rate from improved health care and more reliable food and water supplies. Birth rate is still high, and population grows rapidly.	Decline in birth rate slows population growth despite relatively low death rate.	People tend to desire smaller families and take steps to limit family size. Population grows very slowly or not at all.

Birth and death rates (number per 1000 population)

Relative population size

Size of population

Birth rate

Death rate

High

Low

Time ⟶

living standards. It is difficult to say whether improved socioeconomic conditions have resulted in a decrease in birth rate or whether a decrease in birth rate has resulted in improved socioeconomic conditions. Perhaps both are true. Another reason for the decline in birth rate in highly developed countries is the increased availability of family planning services. Other socioeconomic factors that influence birth rate are increased education, particularly of women, and *urbanization* of society (discussed later in this chapter).

Once a country reaches the fourth demographic stage, is it correct to assume that it will continue to have a low birth rate indefinitely? We don't know. Low birth rates may be a permanent response to the socioeconomic factors of an industrialized, urbanized society. On the other hand, low birth rates may be a response to socioeconomic factors, such as the changing roles of women in highly developed countries. Unforeseen changes in the socioeconomic status of women and men in the future may again change birth rates. No one knows for sure.

The population in many developing countries is beginning to approach stabilization (**FIGURE 7.13**).

For example, the TFR in Brazil in 1960 was 6.7 children per woman. Today it is 2.3. Worldwide, the TFR in developing countries has decreased from an average of 6.1 children per woman in 1970 to 2.7 today.

Although fertility rates in these countries have declined, many still exceed replacement-level fertility. Consequently, populations in these countries are still increasing. Even when fertility rates equal replacement-level fertility, population growth will still continue for some time. To understand why this is so, let's examine the age structure of various countries.

AGE STRUCTURE OF COUNTRIES

A population's **age structure** helps predict future population growth. The number of males and the number of females at each age, from birth to death, are represented in an *age structure diagram*. Each diagram is divided vertically in half, the left side represent-

> **age structure**
> The number and proportion of people at each age in a population.

Fertility changes in selected developing countries   FIGURE 7.13

Since the 1960s, fertility levels have dropped dramatically in many developing countries.

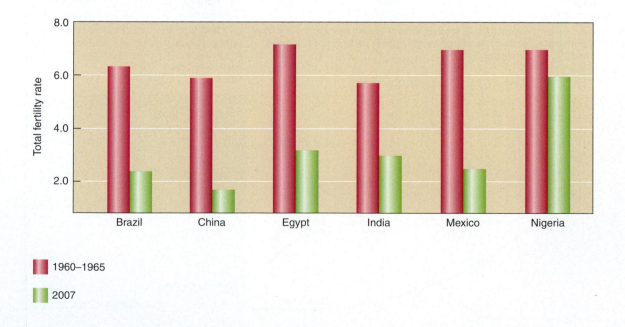

■ 1960–1965

■ 2007

ing the males in a population and the right side the females. The bottom third of each diagram represents prereproductive humans (between 0 and 14 years of age); the middle third, reproductive humans (15 to 44 years); and the top third, postreproductive humans (45 years and older). The widths of these segments are proportional to the population sizes: A broader width implies a larger population. The overall shape of an age structure diagram indicates whether the population is increasing, stable, or shrinking.

The age structure diagram of a country with a high growth rate, based on a high fertility rate—for example, Nigeria or Bolivia—is shaped like a pyramid (**FIGURE 7.14A**). The largest percentage of the population is in the prereproductive age group (0 to 14 years of age), so the probability of future population growth is great. A positive **population growth momentum** exists because when all these children mature, they will become the parents of the next generation, and this group of parents will be larger than the previous group. Even if the fertility rate of such a country has declined to replacement level (couples are having smaller families than their parents did), the population will continue to grow for some time. Population growth momentum explains how a population's present age distribution affects its future growth.

The more tapered bases of the age structure diagrams of countries with slowly growing, stable, or declining populations indicate that a smaller proportion of the population will become the parents of the next generation (**FIGURE 7.14B** and **C**). The age structure diagram of a stable population (neither growing nor shrinking) demonstrates that the numbers of people at prereproductive and reproductive ages are approximately the same. A larger percentage of the population is older—that is, postreproductive—than in a rapidly increasing population. Many countries in Europe have stable populations.

In a shrinking population, the prereproductive age group is smaller than either the reproductive or

Age structure diagrams FIGURE 7.14

Shown are countries with **A** rapid (Nigeria), **B** slow (United States), and **C** declining (Germany) population growth.

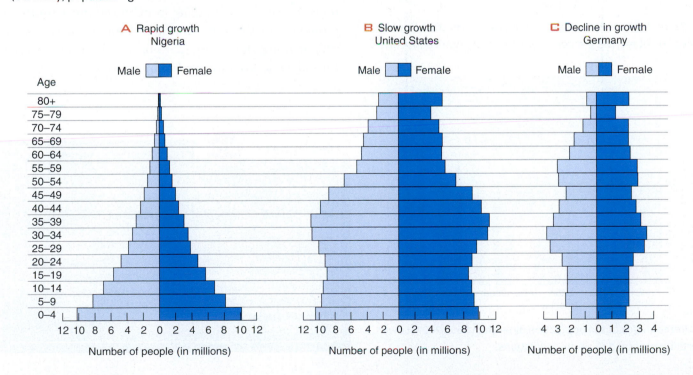

postreproductive age group. Russia, Bulgaria, and Germany are examples of countries with slowly shrinking populations.

Worldwide, 28 percent of the human population is under age 15 (**FIGURE 7.15A**). When these people enter their reproductive years, they have the potential to cause a large increase in the growth rate. Even if the birth rate doesn't increase, the growth rate will increase simply because there are more people reproducing.

Most of the world population increase since 1950 has taken place in developing countries as a result of the younger age structure and the higher-than-replacement-level fertility rates of their populations. In 1950, 67 percent of the world's population was in developing countries in Africa, Asia (minus Japan), and Latin America. Between 1950 and 2000, the world's population more than doubled in size, but most growth occurred in developing countries. As a reflection of this occurrence, the current number of people in developing countries has increased to 82 percent of the world population. Most of the population increase during the 21st century will take place in developing countries, largely as a result of their younger age structures. These countries will have economic difficulty supporting such growth.

Declining fertility rates have profound social and economic implications because as fertility rates drop, the percentage of the population that is elderly increases (**FIGURE 7.15B**). An aging population has a higher percentage of people who are chronically ill or disabled, and these people require more health care and other social services. An aging population reduces a country's productive workforce, increases its tax burden, and strains its social security, health, and pension systems. However, in an aging population, there is often a reduction in the rate of violent crimes. Governments with growing elderly populations may offer incentives to the elderly to work longer before retiring.

Percentages of prereproductive and elderly populations for various regions of the world
FIGURE 7.15

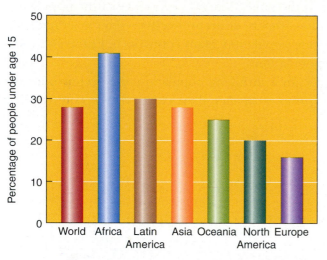

A Percentages of the population under age 15 in 2007. The higher this percentage, the greater the potential for population growth.

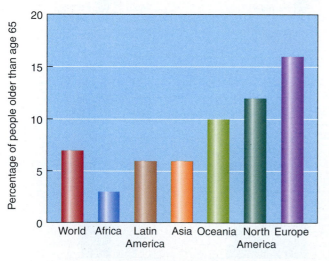

B Percentages of the population older than 65 in 2007. Lower fertility rates lead to aging populations.

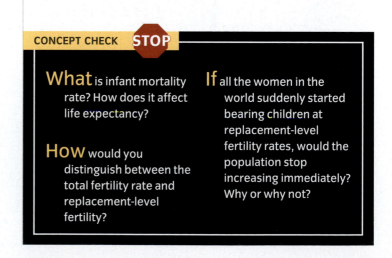

CONCEPT CHECK **STOP**

What is infant mortality rate? How does it affect life expectancy?

How would you distinguish between the total fertility rate and replacement-level fertility?

If all the women in the world suddenly started bearing children at replacement-level fertility rates, would the population stop increasing immediately? Why or why not?

Stabilizing World Population

LEARNING OBJECTIVES

Relate total fertility rates to each of the following: cultural values; social and economic status of women; the availability of family planning services; and government policies.

Explain the link between education and total fertility rates.

Dispersal used to be a solution for overpopulation, but not today. As a species, we humans have expanded our range throughout Earth, and few habitable areas remain with the resources to adequately support a major increase in human population. Nor is increasing the death rate an acceptable means of regulating population size. Clearly, reducing the birth rate is the way to control our expanding population. Cultural traditions, women's social and economic status, family planning, and government policies all influence TFR.

CULTURE AND FERTILITY

The values and norms of a society—what is considered right and important and what is expected of a person—constitute a part of that society's **culture**. Gender—that is, varying roles men and women are expected to fill—is an important part of **culture**. Different societies have different gender expectations (**FIGURE 7.16**). With respect to fertility and culture, a couple is expected to have the number of children traditional in their society.

Varying roles of men and women FIGURE 7.16

A In parts of Latin America, men do the agricultural work. This Argentinian man is harvesting grapes.

B In sub-Saharan Africa, women do most of the agricultural work in addition to caring for their children. Photographed in Mali.

High TFRs are traditional in many cultures. The motivations for having many babies vary from culture to culture, but a major reason for high TFRs is that infant and child mortality rates are high. For a society to endure, it must produce enough children who survive to reproductive age. If infant and child mortality rates are high, TFRs must be high to compensate. Although infant and child mortality rates are decreasing, it will take longer for culturally embedded fertility levels to decline. Parents must have confidence that the children they already have will survive before they stop having additional babies. Another reason for the lag in fertility decline is cultural: Changing anything traditional, including large family size, usually takes a long time.

Higher TFRs in some developing countries are also due to the important economic and societal roles of children. In some societies, children usually work in family enterprises such as farming or commerce, contributing to the family's livelihood. The International Labour Organization estimates that, worldwide, about 218 million children under the age of 15 work full time; more than 95 percent of these children live in developing countries (FIGURE 7.17). When these children be-come adults, they provide support for their aging parents. In contrast, children in highly developed countries have less value as a source of labor because they attend school and because less human labor is required in an industrialized society. Furthermore, highly developed countries provide many social services for the elderly, so the burden of their care doesn't fall entirely on offspring.

Many cultures place a higher value on male children than on female children. In these societies, a woman who bears many sons achieves a high status; thus, the social pressure to have male children keeps the TFR high.

Religious values are another aspect of culture that affects TFRs. Several studies done in the United States point to differences in TFRs among Catholics, Protestants, and Jews. In general, Catholic women have a higher TFR than either Protestant or Jewish women, and women who don't follow any religion have the lowest TFRs of all. The observed differences in TFRs may not be the result of religious differences alone. Other variables, such as ethnicity (certain religions are associated with particular ethnic groups) and residence (certain religions are associated with urban or with rural living), complicate any generalizations that might be made.

Working children FIGURE 7.17

These Indonesian children are making yarn for the textile industry.

ASIA

INDONESIA

Global Locator

THE SOCIAL AND ECONOMIC STATUS OF WOMEN

Gender inequality exists in most societies; in these societies, women don't have the same rights, opportunities, or privileges as men. Because sons are more highly valued than daughters, girls are often kept at home to work rather than being sent to school. In most developing countries, a higher percentage of women are illiterate than men. Fewer women than men attend secondary school (high school). In some African countries only 2 to 5 percent of girls are enrolled in secondary school. Worldwide, some 90 million girls aren't given the opportunity to receive a primary (elementary school) education. Laws, customs, and lack of education often limit women to low-skilled, low-paying jobs. In such societies, marriage is usually the only way for a woman to achieve social influence and economic security.

Evidence suggests that the single most important factor affecting high TFRs may be the low status of women in many societies (**FIGURE 7.18**). A significant way to tackle population growth, then, is to improve the social and economic status of women.

Let's examine how marriage age and educational opportunities, especially for women, affect fertility. The average age at which women marry affects the TFR; in turn, the laws and customs of a given society affect marriage age. Women who marry are more apt to bear children than women who don't marry, and the earlier a woman marries, the more children she is likely to have. Consider Sri Lanka and Bangladesh, two developing countries in South Central Asia. In Sri Lanka the average age at marriage is 25, and the average number of children born per woman is 2.0. In contrast, in Bangladesh the average age at marriage is 17, and the average number of children born per woman is 3.0.

Gender discrimination FIGURE 7.18

Ethiopia is traditionally a male-dominated society in which women and girls have a lower status than men. These young girls are attending an elementary school. Overall, only 17 percent of Ethiopian girls have access to a secondary education, as compared to 28.3 percent of boys.

ETHIOPIA

Global Locator

In nearly all societies women with more education tend to marry later and have fewer children. Providing women with educational opportunities delays their first childbirth, thereby reducing the number of childbearing years and increasing the amount of time between generations. Education provides greater career opportunities and may change women's lifetime aspirations. The What a Scientist Sees feature on the following page discusses this. In the United States, it isn't uncommon for a woman to give birth to her first child in her thirties or forties, after establishing a career.

Education increases the probability that women will know how to control their fertility. It also provides knowledge to improve the health of the women's families, which results in a decrease in infant and child mortality. A study in Kenya showed that 10.9 percent of children born to women with no education died by age 5, as compared with 7.2 percent of children born to women with a primary education, and 6.4 percent of children born to women with a secondary education. Education also increases women's career options and provides ways of achieving status besides having babies.

Education may also have an indirect effect on TFR. Children who are educated have a greater chance of improving their living standards, partly because they have more employment opportunities. Parents who recognize this may be more willing to invest in the education of a few children than in the birth of many children whom they can't afford to educate. The ability of better-educated people to earn more money may be one reason smaller family size is associated with increased family income.

FAMILY PLANNING SERVICES

Socioeconomic factors may encourage people to want smaller families, but fertility reduction won't become a reality without the availability of health and family planning services. The governments of most countries recognize the importance of educating people about basic maternal and child health care. Developing countries that have significantly lowered their TFRs credit many of these results to effective family planning programs. Prenatal care and proper birth spacing make women healthier. In turn, healthier women give birth to healthier babies, leading to fewer infant deaths.

Family planning services provide information on reproductive physiology and contraceptives, as well as the actual contraceptive devices, to people who wish to control the number of children they have or to space out their children's births. Family planning programs are most effective when they are designed with sensitivity to local social and cultural beliefs. Family planning services don't try to force people to limit their family sizes; rather, they attempt to convince people that small families (and the contraceptives that promote small families) are acceptable and desirable.

Contraceptive use is strongly linked to lower TFRs. Research has shown that 90 percent of the decrease in fertility in 31 developing countries was a direct result of increased knowledge and availability of contraceptives (**FIGURE 7.19**). In highly developed countries, where TFRs are at replacement levels or lower, an average of 68 percent of married women of reproductive age use contraceptives. Fertility declines are occurring in developing countries where contraceptives are readily available. During the 1970s, 1980s, and 1990s, use of

Access to contraceptives FIGURE 7.19

A birth control vendor explains condoms to women at the Adjame market in Côte d'Ivoire, a West African country with a TFR of 5.0. Currently, only 7 percent of Ivory Coast women aged 15 to 49 use modern methods of contraception.

Education and Fertility

A Teen mothers gather during lunch at Lincoln High School in Nebraska. A demographer looking at this scene would see the possibility of lower TFRs for the women in this photograph because they are continuing their education.

B Demographers know that the total number of children a woman has during her life (TFR) is affected by the amount of education she has received. The bar graph shows TFRs for 35- to 44-year-old women in the United States by level of education. A similar trend—in which more education leads to lower TFRs—also occurs among women in developing countries. ▶

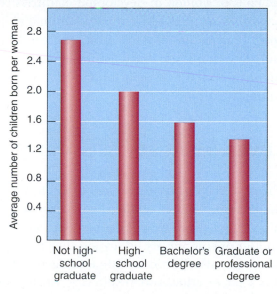

Microcredit programs extend small loans ($50 to $500) to very poor people to help them establish businesses that generate income. The poor use these loans for a variety of projects. Some have purchased used sewing machines to make clothing faster than sewing by hand. Others have opened small grocery stores after purchasing used refrigerators to store food so that it does not spoil.

The **Foundation for International Community Assistance (FINCA)** is a not-for-profit agency that administers a global network of microcredit banks. FINCA uses *village banking*, in which a group of very poor neighbors guarantees one another's loans, administers group lending and saving activities, and provides mutual support. These village banks give autonomy to local people.

FINCA primarily targets women because an estimated 70 percent of the world's poorest people are women. FINCA believes that the best way to alleviate the effects of poverty and hunger on children is to provide their mothers with a means of self-employment. A woman's status in the community is raised as she begins earning income from her business.

contraceptives in East Asia and many areas of Latin America increased significantly, and these regions experienced a corresponding decline in birth rate. In areas where contraceptive use remained low, such as parts of Africa, little or no decline in birth rate took place.

Family planning centers provide information and services primarily to women. As a result, in the male-dominated societies of many developing countries, such services may not be as effective as they could otherwise be. Polls of women in developing countries reveal that many who say they don't want additional children still don't practice any form of birth control. When asked why they don't use birth control, these women frequently respond that their husbands or in-laws want additional children.

GOVERNMENT POLICIES AND FERTILITY

The involvement of governments in childbearing and child rearing is well established. Laws determine the minimum age at which people may marry and the amount of compulsory education they receive. Governments may allot portions of their budgets to family planning services, education, health care, old-age security, or incentives for smaller or larger family size. The tax structure, including additional charges or allowances based on family size, also influences fertility.

In recent years, the governments of at least 78 developing countries—41 in Africa, 19 in Asia, and 18 in Latin America and the Caribbean—have recognized that they must limit population growth. These countries have formulated policies, such as economic rewards and penalties, to achieve this goal. Most countries sponsor family planning projects, which are integrated with health care, education, economic development, and efforts to improve women's status.

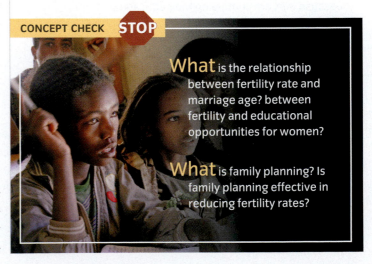

CONCEPT CHECK **STOP**

What is the relationship between fertility rate and marriage age? between fertility and educational opportunities for women?

What is family planning? Is family planning effective in reducing fertility rates?

Population and Urbanization

The geographic distribution of people in rural areas, towns, and cities significantly influences the social, environmental, and economic aspects of population growth. During recent history, the human population has become increasingly urbanized. **Urbanization** involves the movement of people from rural to urban areas as well as the transformation of rural areas into urban areas. When Europeans first settled in North America, the majority of the population consisted of farmers in rural areas. Today, approximately 79 percent of the U.S. population lives in cities.

How many people does it take to make an urban area or city? The answer varies from country to country. According to the U.S. Bureau of the Census, a location with 2500 or more people qualifies as an urban area.

urbanization
A process whereby people move from rural areas to densely populated cities.

One important distinction between rural and urban areas isn't how many people live there but how people make a living. Most people residing in rural areas have occupations that involve harvesting natural resources—such as fishing, logging, and farming. In urban areas, most people have jobs that are not connected directly with natural resources. Cities have traditionally provided more jobs than rural areas because cities are sites of industry, economic development, and educational and cultural opportunities.

Every city is unique in terms of its size, climate, culture, and economic development (**FIGURE 7.20**). Although there is no such thing as a typical city, certain traits are common to city populations in general. One basic characteristic of city populations is their far greater heterogenity with respect to race, ethnicity, religion, and

Vancouver, Canada FIGURE 7.20

Vancouver is Canada's most important port city on the Pacific coast. Noted for its lovely parks and gardens, Vancouver has mild weather and a wet winter season. The greater Vancouver region has a population of more than 2 million, including a large immigrant population.

socioeconomic status than populations in rural areas. People living in urban areas tend to be younger than those living in the surrounding countryside. The young age structure of cities is due to the influx of many young adults from rural areas.

Urban and rural areas often have different proportions of males and females. Cities in developing nations tend to have more males. In cities in Africa, for example, males migrate to the city in search of employment, whereas females tend to remain in the country and tend their farms and children. Cities in highly developed countries often have a higher ratio of females to males. Women in rural areas often have little chance of employment after they graduate from high school, so they move to urban areas.

ENVIRONMENTAL PROBLEMS OF URBAN AREAS

Growing urban areas affect land-use patterns. Suburban sprawl that encroaches into former forest, wetland, desert, or agricultural land destroys or fragments wildlife habitat. Portions of Chicago, Boston, and New Orleans, for example, are former wetlands. Most cities have blocks and blocks of **brownfields**—areas of abandoned, vacant factories, warehouses, and residential sites that may be contaminated from past uses (**FIGURE 7.21**). Meanwhile, the suburbs continue to expand outward, swallowing natural areas and farmland.

Reuse of brownfields is complicated because many have environmental contaminants that must be cleaned up before redevelopment can proceed. Nonetheless, brownfields represent an important potential land resource. Pittsburgh is known for its redevelopment of brownfields that were once steel mills and meatpacking centers. Residential and commercial sites now occupy several of these former brownfields (**FIGURE 7.22**).

Most workers in U.S. cities have to commute dozens of miles through traffic-congested streets from the suburbs where they live to downtown areas where they work. Because development is so spread out in the suburbs, automobiles are a necessity to accomplish everyday chores. This dependence on motor vehicles as our primary means of transportation increases air pollution and causes other environmental problems.

The high density of automobiles, factories, and commercial enterprises in urban areas causes a buildup of airborne emissions, including particulate matter (dust), sulfur oxides, carbon oxides, nitrogen oxides, and volatile organic compounds. Urban areas in developing nations have the worst air pollution in the world. Although we have made progress in reducing air pollution in highly developed nations, the atmosphere in many cities often contains higher levels of pollutants than are acceptable based on health standards.

Cities affect water flow because they cover the rainfall-absorbing soil with buildings and paved roads. Storm sewage systems are built to handle the runoff from rainfall, which is polluted with organic wastes (garbage, animal droppings, and such), motor oil, lawn fertilizers, and heavy metals. In most cities across the United States, urban runoff is cleaned up in sewage treatment plants before being discharged into nearby waterways. In many cities, however, high levels of precipitation can overwhelm the sewage treatment plants, which results in the release of untreated urban runoff. When this occurs, the polluted runoff contaminates water far beyond the boundaries of the city.

Brownfield of vacant warehouses and stores FIGURE 7.21

Pittsburgh's first brownfield redevelopment project, Washington's Landing at Herr's Island, today features upscale housing, recreation, and commerce.

ENVIRONMENTAL BENEFITS OF URBANIZATION

Although our previous discussion may suggest that the concentration of people in cities has a harmful effect overall on the environment, urbanization does have the potential to provide tangible environmental benefits that in many cases outweigh its environmental problems. A well-planned city actually benefits the environment by reducing pollution and preserving rural areas.

One solution to urban growth is **compact development**, which uses land efficiently. Dependence on motor vehicles and their associated pollution are reduced as people walk, cycle, or take public transit such as buses or light rails to work and shopping districts. With compact development, fewer parking lots and highways are needed, so there is more room for parks, open space, housing, and businesses. Compact development makes a city more livable and attractive to people.

Portland, Oregon, provides a good example of compact development. Although Portland is still grap-

> **compact development**
> Design of cities in which tall, multiple-unit residential buildings are close to shopping and jobs, and all are connected by public transportation.

pling with many issues, the city government has developed effective land-use policies that dictate where and how growth will occur. The city looks inward to brownfields rather than outward to the suburbs for new development sites. Since 1975, Portland's population has grown 50 percent, from 0.9 million to 1.8 million (in 2005), yet the urbanized area increased about 2 percent. In contrast, from 1975 to 2005, the population of Chicago grew 22 percent, yet its urbanized area increased more than 50 percent due to sprawl.

Although the automobile is still the primary means of transportation in Portland, the city's public transportation system is an important part of its regional master plan. Public transportation incorporates light-rail lines, bus routes (many of which feature buses arriving every 15 minutes), bicycle lanes, and walkways. Employers are encouraged to provide bus passes to their employees instead of paying for parking. The emphasis on public transportation has encouraged commercial and residential growth along light rail and bus stops instead of in suburbs.

The world's ten largest cities* TABLE 7.1		
1975	**2005**	**2015 (projected)**
Tokyo, Japan, 26.6	Tokyo, Japan, 35.2	Tokyo, Japan, 35.5
New York, USA 15.9	Mexico City, Mexico, 19.4	Mumbai (Bombay), India, 21.9
Mexico City, Mexico, 10.7	New York, USA 18.7	Mexico City, Mexico, 21.6
Osaka-Kobe, Japan, 9.8	São Paulo, Brazil, 18.3	São Paulo, Brazil, 20.5
São Paulo, Brazil, 9.6	Mumbai (Bombay), India, 18.2	New York, USA 19.9
Los Angeles, USA, 8.9	Delhi, India, 15.0	Delhi, India, 18.6
Buenos Aires, Argentina, 8.7	Shangai, China, 14.5	Shangai, China, 17.2
Paris, France, 8.6	Calcutta, India, 14.3	Calcutta, India, 17.0
Calcutta, India, 7.9	Jakarta, Indonesia, 13.2	Dhaka, Bangladesh, 16.8
Moscow, Russian Federation, 7.6	Buenos Aires, Argentina, 12.6	Jakarta, Indonesia, 16.8

*Population in millions

☐ Highly developed countries
☐ Developing countries

URBANIZATION TRENDS

Urbanization is a worldwide phenomenon. Currently, 49 percent of the world population lives in urban areas with populations of 2000 or greater. The percentage of people living in cities compared with rural settings is greater in highly developed countries than in developing countries. In 2007, urban inhabitants composed 75 percent of the total population of highly developed countries but only 43 percent of the total population of developing countries.

Although proportionately more people still live in rural settings in developing countries, urbanization is increasing rapidly. Currently, most urban growth in the world is occurring in developing countries, whereas highly developed countries are experiencing little urban growth. As a result of the greater urban growth of developing nations, most of the world's largest cities are in developing countries. In 1975, four of the world's ten largest cities—Mexico City, São Paulo, Buenos Aires, and Calcutta—were in developing countries. In 2005, eight of the world's ten largest cities—Mexico City, São Paulo, Mumbai (Bombay), Delhi, Shanghai, Calcutta, Jakarta, and Buenos Aires (TABLE 7.1)—were in developing countries. By 2015, eight of the world's ten largest cities will still be from developing countries, although Dhaka, Bangladesh, will have made the list, replacing Buenos Aires (FIGURE 7.23).

According to the United Nations, almost 400 cities worldwide have a population of at least 1 million inhabitants, and 284 of these cities are in developing countries. The number and size of **megacities** (cities with more than 10 million inhabitants) has also increased. In some places, separate urban areas have merged into **urban agglomerations**, urbanized core regions, each of which consists of several adjacent cities or megacities and their surrounding developed suburbs. An example is the Tokyo-Yokohama-Osaka-Kobe agglomeration in Japan, which is home to about 50 million people. However, according to the U.N. Population Division, most of the world's urban population still lives in small or medium-sized cities with populations of less than 1 million.

The fast pace of urban growth in developing nations has outstripped the limited capacity of many cities to provide basic services. It has overwhelmed their economic growth (although cities still offer more job possibilities than rural areas). Consequently, cities in developing nations are generally faced with more serious

The world's ten largest cities in 2015 FIGURE 7.23

In 2015, eight of the ten largest cities will be in developing countries: Mumbai (Bombay), Mexico City, Sao Paulo, Delhi, Shanghai, Calcutta, Dhaka, and Jakarta.

2015 projections

City populations *in millions*
- 4 to 10
- 11 to 16
- 17 to 19
- 20 and over

New York, USA 19.9
Mexico City, Mexico 21.6
São Paulo, Brazil 20.5
Delhi, India 18.6
Dhaka, Bangladesh 16.8
Shanghai, China 17.2
Tokyo, Japan 35.5
Mumbai (Bombay), India 21.9
Calcutta, India 17.0
Jakarta, Indonesia 16.8

The challenge of meeting a fast-growing city's water needs FIGURE 7.24

A woman in Lagos, Nigeria, fills water from a communal tank as another woman walks by with fruit balanced on her head. Lack of access to safe water and basic sanitation services is a problem for many urban residents, particularly the poor, in less developed countries.

challenges than cities in highly developed countries. These challenges include substandard housing (slums and squatter settlements); poverty; exceptionally high unemployment; heavy pollution; and inadequate or nonexistent water, sewage, and waste disposal (FIGURE 7.24). Rapid urban growth also strains school, medical, and transportation systems.

NIGERIA

Global Locator

CONCEPT CHECK STOP

What is urbanization?

Which countries are the most urbanized? the least urbanized? Which countries have the highest rates of urbanization today?

What are some of the problems caused by rapid urban growth in developing countries?

URBAN PLANNING IN CURITIBA, BRAZIL

Livable cities aren't restricted to highly developed countries. Curitiba, a Brazilian city of more than 2.9 million people, provides a good example of *compact development* in a moderately developed country. Curitiba's city officials and planners have had notable successes in public transportation, traffic management, land-use planning, waste reduction and recycling, and community livability.

The city developed an inexpensive, efficient mass transit system that uses clean, modern buses that run in high-speed bus lanes. High-density development was largely restricted to areas along the bus lines, encouraging population growth where public transportation was already available. About 72 percent of commuters use mass transportation (**FIGURE A**). Since the 1970s, Curitiba's population has more than doubled, yet traffic has declined by 30 percent. Curitiba doesn't rely on automobiles as much as comparably sized cities do, so it has less traffic congestion and significantly cleaner air, both of which are major goals of compact development. Instead of streets crowded with vehicular traffic, the center of Curitiba is a *calcadao*, or "big sidewalk," that consists of 49 downtown blocks of pedestrian walkways connected to bus stations, parks, and bicycle paths (**FIGURE B**).

Over several decades, Curitiba purchased and converted flood-prone properties along rivers in the city to a series of interconnected parks crisscrossed with bicycle paths. This move reduced flood damage and increased the per person amount of "green space" from 0.5 m² (5.4 ft²) in 1950 to 50 m² (538 ft²) today, a significant accomplishment considering Curitiba's rapid population growth during the same period.

Another example of Curitiba's creativity is its labor-intensive Garbage Purchase program, in which poor people exchange filled garbage bags for bus tokens, surplus food (eggs, butter, rice, and beans), or school notebooks. This program encourages garbage pickup from the unplanned shantytowns (which garbage trucks can't access) that surround the city. Curitiba supplies more services to these unplanned settlements than most cities do. It tries to provide water, sewer, and bus service for them.

These changes didn't happen overnight. Urban planners can carefully reshape most cities over several decades to make better use of space and to reduce dependence on motor vehicles. City planners and local and regional governments are increasingly adopting measures to provide the benefits of compact development in the future.

Curitiba, Brazil

A Bus passengers pay their fares in advance in the tubular bus stations and then walk directly onto the bus as soon as it arrives.

B The downtown area of Curitiba is filled with open terraces lined with shops and restaurants.

SUMMARY

1 Population Ecology

1. **Population ecology** is the branch of biology that deals with the number of individuals of a particular species found in an area and how and why those numbers change over time.

2. The **growth rate (r)** is the rate of change (increase or decrease) of a population's size, expressed in percentage per year. On a global scale, growth rate is due to the **birth rate (b)** and the **death rate (d)**: $r = b - d$. Emigration (e), the number of individuals leaving an area, and immigration (i), the number of individuals entering an area, also affect a local population's growth rate.

3. **Biotic potential** is the maximum rate a population could increase under ideal conditions. **Exponential population growth** is the accelerating population growth that occurs when optimal conditions allow a constant reproductive rate for limited periods. Eventually, the growth rate decreases to around zero or becomes negative because of **environmental resistance**, unfavorable environmental conditions that prevent organisms from reproducing indefinitely at their biotic potential. The **carrying capacity (K)** is the largest population a particular environment can support sustainably (long term) if there are no changes in that environment.

2 Human Population Patterns

1. It took thousands of years for the human population to reach 1 billion (around 1800). Since then, the population has grown exponentially. The United Nations projects that the population will reach 7 billion by 2013. Although our numbers continue to increase, the growth rate (r) has declined slightly over the past several years. The population should reach **zero population growth**, in which it remains the same size because the birth rate equals the death rate, toward the end of the 21st century.

2. **Thomas Malthus** was a British economist who said that the human population increases faster than its food supply, resulting in famine, disease, and war. Malthus's ideas appear to be erroneous because the human population has grown from about 1 billion in his time to more than 6 billion today, and food production has generally kept pace with population. But Malthus may ultimately be proved correct because we don't know whether our increase in food production is sustainable.

3. Estimates of Earth's carrying capacity for humans vary widely depending on what assumptions are made about standard

of living, resource consumption, technological innovations, and waste generation. In addition to natural environmental constraints, human choices and values determine Earth's carrying capacity for humans.

3 Demographics of Countries

1. **Demographics** is the applied branch of sociology that deals with population statistics. As a country becomes industrialized, it goes through a demographic transition as it moves from relatively high birth and death rates to relatively low birth and death rates.

2. The **infant mortality rate** is the number of deaths of infants under age 1 per 1000 live births. The **total fertility rate (TFR)** is the average number of children born to each woman. **Replacement-level fertility** is the number of children a couple must produce to "replace" themselves. **Age structure** is the number and proportion of people at each age in a population. A country can have replacement-level fertility and still experience population growth if the largest percentage of the population is in the prereproductive years. In contrast to developing countries, highly developed countries have low infant mortality rates, low total fertility rates, and an age structure in which the largest percentage of the population isn't in the prereproductive years.

4 Stabilizing World Population

1. Four factors are most responsible for high total fertility rates: high infant and child mortality rates, the important economic and societal roles of children in some cultures, the low status of women in many societies, and a lack of health and family planning services.

The single most important factor affecting high TFRs is the low status of women. The governments of many developing countries are trying to limit population growth.

2. Education of women decreases the total fertility rate, in part by delaying the first childbirth. Education increases the likelihood that women will know how to control their fertility. Education also increases women's career options, which provide ways of achieving status besides having babies.

5 Population and Urbanization

1. **Urbanization** is the process whereby people move from rural areas to densely populated cities. In developing nations, most people live in rural settings, but their rates of urbanization are rapidly increasing.

2. Rapid urbanization makes it difficult to provide city dwellers with basic services such as housing, water, sewage, and transportation systems.

3. **Compact development** is the design of cities so that tall, multiple-unit residential buildings are close to shopping and jobs, and all are connected by public transportation.

KEY TERMS

- **population ecology** p. 160
- **growth rate (r)** p. 161
- **biotic potential** p. 161
- **exponential population growth** p. 162
- **carrying capacity (K)** p. 163
- **zero population growth** p. 166
- **demographics** p. 170
- **infant mortality rate** p. 170
- **replacement-level fertility** p. 170
- **total fertility rate (TFR)** p. 170
- **demographic transition** p. 170
- **age structure** p. 172
- **urbanization** p. 181
- **compact development** p. 183

CRITICAL AND CREATIVE THINKING QUESTIONS

1. Draw a graph to represent the long-term growth of a population of bacteria cultured in a test tube that contains a nutrient medium that isn't replenished.

2. In Bolivia, 38 percent of the population is younger than 15, and 4 percent is older than 65. In Austria, 16 percent of the population is younger than 15, and 17 percent is older than 65. On the basis of this information, which of these countries will have the higher population growth momentum over the next two decades? Why?

3. Should the rapid increase in world population be of concern to the average citizen in the United States? Why or why not?

4. How is human population growth related to natural resource depletion and environmental degradation?

5. Although the United States ranks third in population number, some experts contend that it is the most overpopulated country in the world. Explain why.

6. Explain the rationale behind this statement: It is better for highly developed countries to spend millions of dollars on family planning in developing countries now than to have to spend billions of dollars on relief efforts later.

7. Discuss this statement: The current human population crisis causes or exacerbates all environmental problems.

8. Discuss some of the ethical issues associated with overpopulation. Is it ethical to have more than two children? Is it ethical to consume so much in the way of material possessions? Is it ethical to try to influence a couple's decision about family size?

9. Tanzania, Argentina, and Poland have about the same population sizes (38 to 39 million each). The current total fertility rates of these countries are Tanzania, 5.4; Argentina, 2.5; and Poland, 1.3. Assuming that fertility rates decline to 2.0 by 2050 in Tanzania and Argentina, will there be a difference in population among the three countries? If so, which country will have the highest population in 2050? Explain your answer.

10. If you were to draw an age structure diagram for Poland, with a total fertility rate of 1.3, which of the following overall shapes would the diagram have? Explain why a country like Poland faces a population decline even if its fertility rate were to start increasing today.

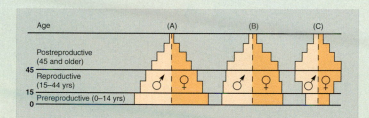

The following graph shows the 10 countries with the largest populations in 2007. Use it to answer questions 11 and 12.

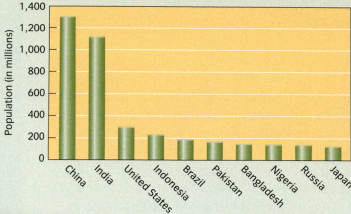

11. Population experts project that in 2050, the Democratic Republic of Congo and Ethiopia will replace Russia and Japan in the top ten. Do Russia and Japan have high, medium, or low fertility rates? Do the Democratic Republic of Congo and Ethiopia have high, medium, or low fertility rates? Explain your answers.

12. Experts project that the United States will have 420 million people in 2050. Given that the U.S. fertility rate is relatively low and stable (2.1), why do you suppose the United States has one of the highest population growth rates in the developed world?

What is happening in this picture ?

The photo shows people—mainly displaced rural workers—picking through trash at the Smoky Mountain Dump in Manila, Philippines. Most of these people moved to Manila looking for work. Why didn't they find better employment?

They are looking mainly for scraps of plastic and metal, which they can sell. What valuable environmental service does such scavenging provide?

Air and Air Pollution

8

VIEW THIS IN ACTION
in your WileyPLUS course

LONG-DISTANCE TRANSPORT OF AIR POLLUTION

Winds distribute certain hazardous air pollutants globally. *Persistent compounds* move through the air from warmer developing countries (where they are still used) to colder, highly developed nations, where they condense and form deposits on land and surface water. This process, in which volatile chemicals enter the atmosphere in warm regions and move to areas at higher, cooler latitudes, is known as the *global distillation effect*.

Dangerous levels of certain persistent toxic compounds are found in the Yukon (in northwestern Canada) and in other pristine arctic regions. These chemicals enter the food chain and become concentrated in the body fat of animals at the top (see *biological magnification* in Chapter 4). Fishes, seals, polar bears, and arctic people such as the Inuit are particularly vulnerable. When an Inuit woman consumes a single bite of raw whale skin, she ingests more PCBs than scientists think should be consumed in a week (*see photograph*). The level of PCBs in the breast milk of that Inuit woman is five times higher than in the milk of women who live in southern Canada.

Examples of air pollution traveling from one continent to another were not well documented until recently. Now we know that atmospheric conditions cause pollutants from Asia to move east across the Pacific Ocean. In 1997 scientists detected pollutants in the air over the western United States that had been produced in Asia six days earlier. In 1998 a major dust storm in China produced a visible cloud of particulate matter that satellites tracked across the Pacific Ocean. When the polluted air reached the United States a few days later, an analysis found it to contain arsenic, copper, lead, and zinc from ore smelters in Manchuria.

No matter where you live, the air you breathe is often dirty and contaminated with pollutants. Air pollution also extends indoors to the air you breathe at home, in school, at work, and in your car. Because air pollution causes many health and environmental problems, most highly developed nations and many developing nations have established air quality standards for numerous air pollutants.

NATIONAL GEOGRAPHIC

The Atmosphere

O
xygen and nitrogen are the predominant gases in the **atmosphere**, accounting for about 99 percent of dry air (**FIGURE 8.1**). Other gases make up the remaining 1 percent. In addition, water vapor and trace amounts of air pollutants are present in the air. The atmosphere becomes less dense as it extends outward into space.

Ulf Merbold, a German space shuttle astronaut, felt differently about the atmosphere after viewing it in space (**FIGURE 8.2**). "For the first time in my life, I saw the horizon as a curved line. It was accen-

> **atmosphere** The gaseous envelope surrounding Earth.

The atmosphere FIGURE 8.2

The "ocean of air" is a thin blue layer that separates the planet from the blackness of space.

tuated by a thin seam of dark blue light—our atmosphere. Obviously, this wasn't the 'ocean' of air I had been told it was so many times in my life. I was terrified by its fragile appearance." The atmosphere is composed of four major concentric layers—the troposphere, stratosphere, mesosphere, and thermosphere (**FIGURE 8.3**). These layers vary in altitude and temperature, depending on the latitude and season.

The atmosphere performs several valuable **ecosystem services**. First, it protects Earth's surface from most of the sun's ultraviolet (UV) radiation and x-rays, and from lethal amounts of cosmic rays from space. Life as we know it would cease to exist without this shielding. Second, the atmosphere allows visible light and some infrared radiation to penetrate, both of which warm Earth's surface and the lower atmosphere. This interaction between solar energy and atmosphere is responsible for our weather and climate.

Organisms depend on the atmosphere for existence, but they also maintain and, in certain instances, modify its composition. Atmospheric oxygen is thought to have increased to its present level as a result of billions of years of photosynthesis. A balance between oxygen-producing photosynthesis and oxygen-using cellular respiration maintains the current level of oxygen.

Composition of the atmosphere FIGURE 8.1

Nitrogen and oxygen form most of the atmosphere. Air also contains water vapor and various pollutants (methane, ozone, dust particles, microorganisms, and chlorofluorocarbons [CFCs]).

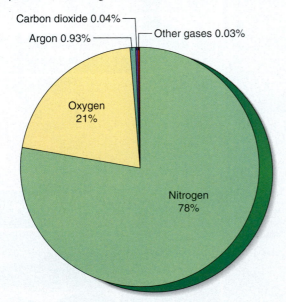

Carbon dioxide 0.04%
Argon 0.93%
Other gases 0.03%

Oxygen 21%

Nitrogen 78%

Layers of Atmosphere

Thermosphere
Extends to 480 km (300 mi)

Mesosphere
Extends to 80 km (50 mi)

Stratosphere
Extends to 50 km (30 mi)

Ozone layer

Troposphere
Average thickness: 12 km (7.5 mi)
16 km (10 mi) thick at equator
8 km (5 mi) thick at poles

Gases in extremely thin air absorb x-rays and short-wave radiation, raising the temperature to 1000°C or more. The thermosphere is important in long-distance communication because it reflects outgoing radio waves back to Earth without the use of satellites. Auroras occur here.

Directly above the stratosphere. Temperatures drop to the lowest in the atmosphere—as low as −138°C. Meteors burn up from friction with air molecules in mesosphere.

Steady wind but no turbulance; commercial jets fly here. Contains a layer of ozone that absorbs much of the Sun's damaging ultraviolet (UV) radiation. Temperature increases with increasing altitude because absorption of UV radiation by ozone layer heats the air.

Layer of atmosphere closest to Earth's surface. Temperature decreases with increasing altitude. Weather, including turbulent wind, storms, and most clouds, occurs here.

A thunderstorm in New Mexico. During a lightning flash, a negative charge moves from the bottom of the cloud to the ground, followed by an upward-moving charge along the same channel. The expansion of air around the lightning stroke produces sound waves, or thunder.

An aurora in the Northern Hemisphere. Electrically charged particles from the sun collide with the gas molecules in the thermosphere, releasing energy visible as light of different colors.

VIEW THIS IN ACTION
in your WileyPLUS course

NATIONAL GEOGRAPHIC

ATMOSPHERIC CIRCULATION

Variations in the amount of solar energy that reaches different areas on Earth cause differences in temperature, which then drive the circulation of the atmosphere. The very warm regions near the equator heat the air, causing it to expand and rise (**Figure 8.4**). As this warm air rises, it cools, and then it sinks again. Much of it recirculates almost immediately to the same areas it has left, but the remainder of the heated air splits and flows in two directions, toward the poles. The air chills enough to sink to the surface at about 30 degrees north and south latitudes. This descending air splits and flows over the surface in two directions.

Similar upward movements of warm air and its subsequent flow toward the poles also occur at higher latitudes farther from the equator. At the poles, the air cools, sinks, and flows back toward the equator, generally be-neath the currents of warm air that simultaneously flow toward the poles. These constantly moving currents transfer heat from the equator toward the poles and cool the land over which they pass on their return. This continuous circulation moderates temperatures over Earth's surface.

In addition to these global circulation patterns, the atmosphere features smaller-scale horizontal movements, or **winds**. The motion of wind, with its eddies, lulls, and turbulent gusts, is difficult to predict. It results partly from fluctuations in atmospheric pressure and partly from the planet's rotation.

The gases that constitute the atmosphere have weight and exert a pressure—about 1013 millibars (14.7 lb per in^2) at sea level. Air pressure is variable, depending on altitude, temperature, and humidity. Winds tend to blow from areas of high atmospheric pressure to areas of low pressure, and the greater the difference between the high- and low-pressure areas, the stronger the wind.

Atmospheric circulation and heat exchange FIGURE 8.4

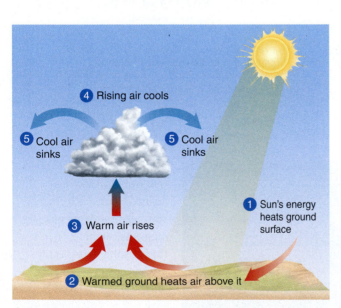

A In atmospheric convection, heating of the ground surface heats the air, producing an updraft of less dense, warm air. The convection process ultimately causes air currents that mix warmer and cooler parts of the atmosphere.

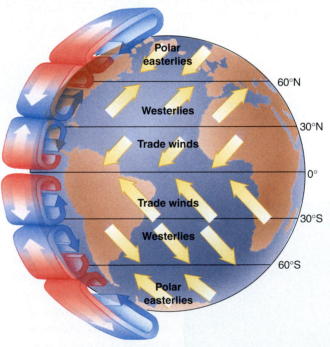

B Atmospheric circulation transports heat from the equator to the poles (left side of figure). The greatest solar energy input occurs at the equator, heating air most strongly in that area. The air rises, travels toward the poles, and cools in the process so that much of it descends again at around 30 degrees latitude in both hemispheres. At higher latitudes, the patterns of air circulation are more complex.

The Coriolis effect FIGURE 8.5

Viewed from the North Pole, the Coriolis effect appears to deflect ocean currents and winds to the right. From the South Pole, the deflection appears to be to the left.

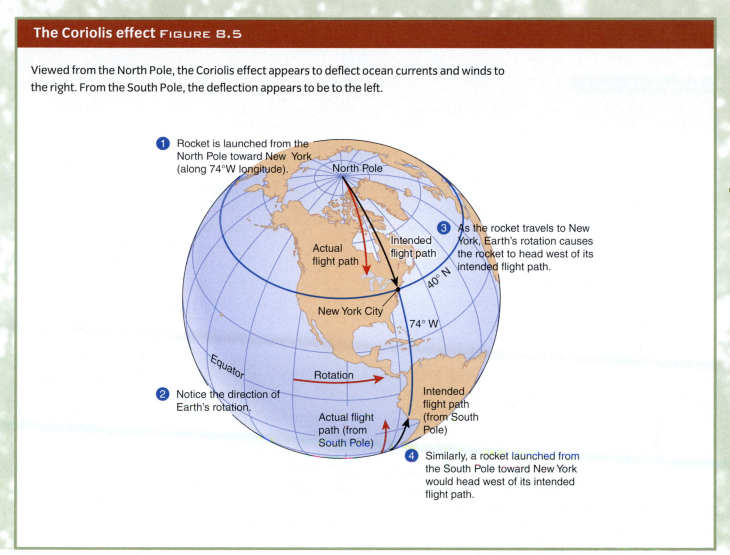

1 Rocket is launched from the North Pole toward New York (along 74°W longitude).

3 As the rocket travels to New York, Earth's rotation causes the rocket to head west of its intended flight path.

North Pole

Actual flight path

Intended flight path

New York City

40° N

74° W

Equator

Rotation

2 Notice the direction of Earth's rotation.

Intended flight path (from South Pole)

Actual flight path (from South Pole)

4 Similarly, a rocket launched from the South Pole toward New York would head west of its intended flight path.

As a result of the **Coriolis effect**, Earth's rotation from west to east also influences the direction of wind. To visualize the Coriolis effect, imagine that a rocket is launched from the North Pole toward New York (**FIGURE 8.5**).

Coriolis effect The tendency of moving air or water to be deflected from its path and swerve to the right in the Northern Hemisphere and to the left in the Southern Hemisphere.

The atmosphere has three **prevailing winds**—major surface winds that blow more or less continually (see Figure 8.4). Prevailing winds from the northeast near the North Pole, or from the southeast near the South Pole, are called *polar easterlies*. Winds that blow in the middle latitudes from the southwest in the Northern Hemisphere or from the northwest in the Southern Hemisphere are called *westerlies*. Tropical winds from the northeast in the Northern Hemisphere or from the southeast in the Southern Hemisphere are called *trade winds*.

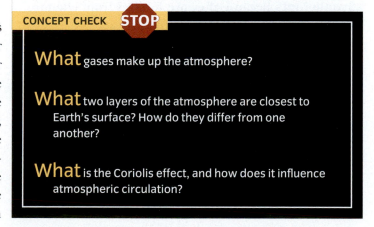

CONCEPT CHECK STOP

What gases make up the atmosphere?

What two layers of the atmosphere are closest to Earth's surface? How do they differ from one another?

What is the Coriolis effect, and how does it influence atmospheric circulation?

Types and Sources of Air Pollution

LEARNING OBJECTIVES

Define *air pollution* and distinguish between primary and secondary air pollutants.

List the seven major classes of air pollutants and describe their characteristics and sources.

ir pollution can come from natural sources, such as smoke from a forest fire ignited by lightning or gases from an erupting volcano. However, human activities release many kinds of substances into the atmosphere and contribute greatly to global air pollution. Some of these substances are harmful when they settle on land and surface waters, and some substances are harmful because they alter the chemistry of the atmosphere.

> **air pollution**
> Various chemicals (gases, liquids, or solids) present in the atmosphere in high enough levels to harm humans, other organisms, or materials.

Although many different air pollutants exist, we focus on the seven most important classes from a regulatory perspective: particulate matter, nitrogen oxides, sulfur oxides, carbon oxides, hydrocarbons, ozone, and air toxics.

Air pollutants are often divided into two categories, primary and secondary (**FIGURE 8.6**). The major **primary air pollutants** are carbon oxides, nitrogen oxides, sulfur dioxide, particulate matter, and hydrocarbons.

Ozone and sulfur trioxide are called **secondary air pollutants** because they are formed from chemical reactions that take place in the atmosphere.

MAJOR CLASSES OF AIR POLLUTANTS

Particulate matter consists of dusts and mists—thousands of different solid and liquid particles suspended in the atmosphere. Particulate matter includes soil particles, soot, lead, asbestos, sea salt, and sulfuric acid droplets. Some particulate matter has toxic or carcinogenic effects.

Particulate matter scatters and absorbs sunlight, reducing visibility. Urban areas receive less sunlight than rural areas, partly as a result of greater quantities of particulate matter in the air. Particulate matter corrodes metals, erodes buildings and sculptures when the air is humid, and soils clothing and draperies.

All particulate matter eventually settles out of the atmosphere, but microscopic particles can remain suspended in the atmosphere for weeks or even years. Trace amounts of hundreds of different chemicals bind to these microscopic particles; inhaling the particles introduces the chemicals, some of which are toxic, into the human body. Microscopic particles are considered more dangerous than larger particles because they are inhaled more deeply into the lungs.

The Environmental Protection Agency (EPA) samples microscopic particulate matter at 1,000 locations around the United States because its composition varies with location and season. To summarize:

> **primary air pollutants** Harmful chemicals that enter directly into the atmosphere due to either human activities or natural processes.

> **secondary air pollutants** Harmful chemicals that form in the atmosphere when primary air pollutants react chemically with one another or with natural components of the atmosphere.

Pollutant	Category	Characteristics
Particulate matter		
Dust particles	Primary	Solid particles
Lead (Pb)	Primary	Solid particles
Sulfuric acid (H_2SO_4)	Secondary	Liquid droplets

Primary and secondary air pollutants FIGURE 8.6

Primary air pollutants are emitted, unchanged, from a source directly into the atmosphere, whereas secondary air pollutants are produced from chemical reactions involving primary air pollutants.

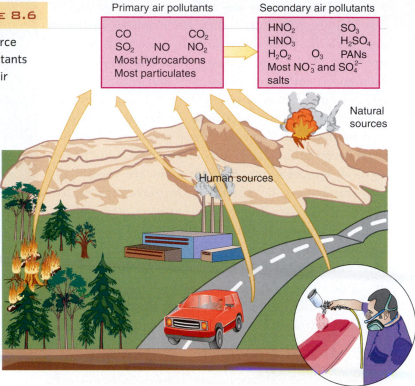

Primary air pollutants
CO $\quad$ CO_2
SO_2 $\quad$ NO $\quad$ NO_2
Most hydrocarbons
Most particulates

Secondary air pollutants
HNO_2 $\qquad$ SO_3
HNO_3 $\qquad$ H_2SO_4
H_2O_2 $\quad$ O_3 $\quad$ PANs
Most NO_3^- and SO_4^{2-} salts

Natural sources

Human sources

Nitrogen oxides are gases produced by chemical interactions between nitrogen and oxygen when a source of energy, such as fuel combustion, produces high temperatures. Collectively known as NO_x, nitrogen oxides consist mainly of nitric oxide (NO), nitrogen dioxide (NO_2), and nitrous oxide (N_2O). Nitrogen oxides inhibit plant growth and, when inhaled, aggravate health problems such as asthma. They are involved in the production of *photochemical smog* (discussed later in the chapter) and acid deposition (see Chapter 9). Nitrous oxide is associated with global warming, and it depletes ozone in the stratosphere (again, see Chapter 9). Nitrogen oxides cause metals to corrode and textiles to fade and deteriorate. As discussed:

Pollutant	Category	Characteristics
Nitrogen oxides		
Nitrogen dioxide (NO_2)	Primary	Reddish-brown gas

Chemical interactions between sulfur and oxygen produce gases called **sulfur oxides**. Sulfur dioxide (SO_2), a colorless, nonflammable gas with a strong, irritating odor, is emitted as a primary air pollutant. Sulfur trioxide (SO_3) is a secondary air pollutant that forms when sulfur dioxide reacts with oxygen in the air. Sulfur trioxide, in turn, reacts with water to form another secondary air pollutant, sulfuric acid. Sulfur oxides play a major role in acid deposition, and they corrode metals and damage stone and other materials. Sulfuric acid and other sulfur oxides damage plants and irritate the respiratory tracts of humans and other animals. As discussed:

Pollutant	Category	Characteristics
Sulfur oxides		
Sulfur dioxide (SO_2)	Primary	Colorless gas with strong odor
Sulfur trioxide (SO_3)	Secondary	Reactive colorless gas

Carbon oxides are the gases carbon monoxide (CO) and carbon dioxide (CO_2). Carbon monoxide is a colorless, odorless, and tasteless gas produced in larger quantities than any other atmospheric pollutant except carbon dioxide. Carbon monoxide is poisonous and reduces the blood's ability to transport oxygen. Carbon dioxide, also colorless, odorless, and tasteless, is associated with global warming. To summarize:

Pollutant	Category	Characteristics
Carbon oxides		
Carbon monoxide (CO)	Primary	Colorless, odorless gas
Carbon dioxide (CO_2)	Primary	Colorless, odorless gas

Hydrocarbons are a diverse group of organic compounds that contain only the elements hydrogen and carbon. Small hydrocarbon molecules, such as methane (CH_4), are gaseous at room temperature. Methane is colorless and odorless and is the principal component of natural gas. (The odor of natural gas comes from sulfur compounds deliberately added so that humans can detect the gas's presence.) Medium-sized hydrocarbons, such as benzene (C_6H_6), are liquids at room temperature, although many are volatile and may evaporate easily. The largest hydrocarbons, such as the waxy fuel paraffin, are solids at room temperature. The many different hydrocarbons have a variety of effects on human and animal health. Some cause no adverse effects, some injure the respiratory tract, and others cause cancer. All except methane contribute to the production of photochemical smog. Methane is linked to global warming. To summarize:

Pollutant	Category	Characteristics
Hydrocarbons		
Methane (CH_4)	Primary	Colorless, odorless gas
Benzene (C_6H_6)	Primary	Liquid with sweet smell

Ozone (O_3) is a form of oxygen considered a pollutant in one part of the atmosphere but an essential component of another. In the stratosphere, oxygen reacts with solar UV radiation to form ozone. Stratospheric ozone protects Earth's surface from receiving harmful levels of solar UV radiation. Unfortunately, certain human-made pollutants, such as chlorofluorocarbons (CFCs), react with stratospheric ozone, breaking it down into molecular oxygen (O_2). As a result, more solar UV reaches Earth's surface.

Unlike stratospheric ozone, ozone in the troposphere—the layer of atmosphere closest to Earth's surface—is a human-made air pollutant. (Ground-level, or tropospheric, ozone does not replenish the ozone depleted from the stratosphere because tropospheric ozone breaks down to form oxygen long before it drifts up to the stratosphere.) Ozone in the troposphere is a secondary air pollutant formed when sunlight triggers

Ozone damage FIGURE 8.7

A scientist measures the effects of ozone on the growth and productivity of plum trees. Plants exposed to ozone generally exhibit damaged leaves, reduced root growth, and reduced productivity. Photographed in the San Joaquin Valley, California.

reactions between nitrogen oxides and volatile hydrocarbons. The most harmful component of photochemical smog, ozone reduces air visibility and causes health problems. Ozone also reduces plant vigor, and chronic ozone exposure (of long duration) lowers crop yields (**FIGURE 8.7**). Chronic exposure to ozone is one possible contributor to forest decline, and ground-level ozone is associated with global warming. As discussed:

Pollutant	Category	Characteristics
Ozone (O_3)	Secondary	Pale blue gas with irritating odor

Most of the hundreds of other air pollutants—such as chlorine (Cl_2), lead, hydrochloric acid, formaldehyde, radioactive substances, and fluorides—are present in very low concentrations, although it is possible to have high local concentrations of specific pollutants. Some of these air pollutants, known as **hazardous air pollutants**, or **air toxics**, are potentially harmful and may pose long-term health risks to people who live and work around chemical factories, incinerators, or other facilities that produce or use them. To limit the release of more than 180 hazardous

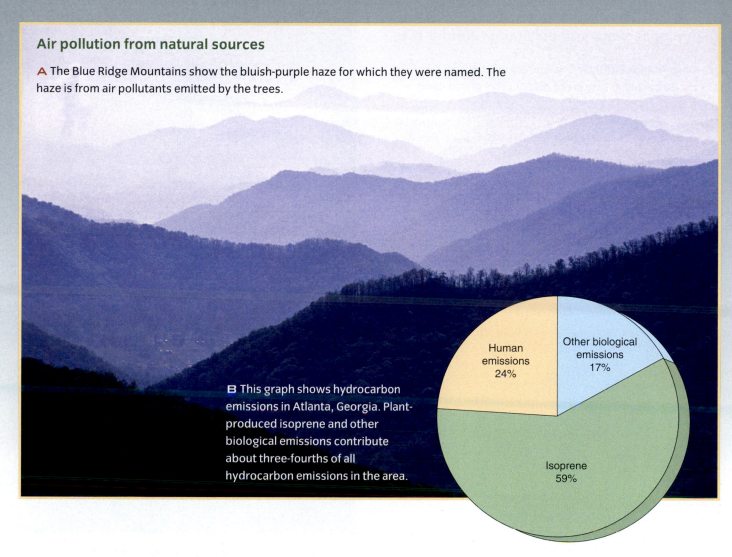

Air pollution from natural sources

A The Blue Ridge Mountains show the bluish-purple haze for which they were named. The haze is from air pollutants emitted by the trees.

B This graph shows hydrocarbon emissions in Atlanta, Georgia. Plant-produced isoprene and other biological emissions contribute about three-fourths of all hydrocarbon emissions in the area.

Human emissions 24%

Other biological emissions 17%

Isoprene 59%

air pollutants, the Clean Air Act Amendments of 1990 (discussed later in this chapter) regulate the pollutant emissions of both large and small businesses. To summarize:

Pollutant	Category	Characteristics
Air toxics		
Chlorine (Cl_2)	Primary	Yellow-green gas
Formaldehyde	Primary	Colorless gas with pungent odor

SOURCES OF OUTDOOR AIR POLLUTION

Not all air pollution is human generated. On a hot summer day in the Blue Ridge Mountains (part of the Appalachians), a bluish-purple haze hangs over the forested hills (see "What a Scientist Sees"). This haze is caused by hydrocarbon emissions from tree leaves. Many plants produce a variety of hydrocarbons in response to heat. The hydrocarbon isoprene, for example, may protect leaves from high temperatures. However, isoprene and other hydrocarbons are volatile and evaporate into the air, where they interact with other substances to affect atmospheric chemistry. These hydrocarbons are highly reactive and contribute to ozone formation, a key ingredient in photochemical smog (discussed shortly). Carbon monoxide, another important air pollutant, is one of the breakdown products of isoprene. Southeastern cities such as Atlanta (which is heavily wooded) contribute substantial biologically generated hydrocarbon emissions.

The two main human sources of primary air pollutants are transportation (mobile sources) and industry

Sources of primary air pollutants FIGURE 8.8

A Transportation and industrial fuel combustion (such as electric power plants) are major contributors of pollutants.

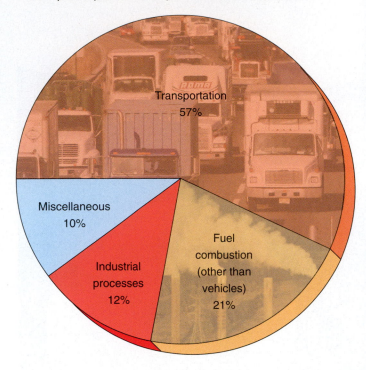

Transportation 57%

Miscellaneous 10%

Industrial processes 12%

Fuel combustion (other than vehicles) 21%

B Mobile source of air pollution. Diesel trucks on the New Jersey Turnpike produce particulate matter and other kinds of air pollution.

C Stationary source of air pollution. Ash is released from smokestacks at this coal-fired electric power plant in Tennessee.

(stationary sources) (FIGURE 8.8A). Cars, trucks, tractors, and heavy construction equipment are known as mobile sources. They release significant quantities of nitrogen oxides, carbon oxides, particulate matter, and hydrocarbons during the combustion of gasoline or diesel fuel. While diesel engines in trucks, buses, trains, and ships consume less fuel than other types of combustion engines, they produce more air pollution (FIGURE 8.8B). One heavy-duty truck emits as much particulate matter as 150 automobiles, whereas one diesel train engine produces, on average, 10 times the particulate matter of a diesel truck.

Electric power plants and other industrial facilities, known as *stationary sources*, emit most of the particulate matter and sulfur oxides released in the United States; they also emit sizable amounts of nitrogen oxides, hydrocarbons, and carbon oxides (FIGURE 8.8C). The combustion of fossil fuels, especially coal, is responsible for most of these emissions. The top three industrial sources of toxic air pollutants are the chemical industry, the metals industry, and the paper industry.

CONCEPT CHECK STOP

What is the difference between primary and secondary air pollutants?

What are the seven main classes of air pollutants, and what are some of their effects?

Effects of Air Pollution

Air pollution injures organisms, reduces visibility, and attacks and corrodes materials such as metals, plastics, rubber, and fabrics. Air pollutants harm the respiratory tracts of animals, including humans, and can worsen existing medical conditions, such as chronic lung disease, pneumonia, and cardiovascular problems. Most forms of air pollution reduce the overall productivity of crop plants. Air pollution is involved in acid deposition, global warming, and stratospheric ozone depletion (all discussed in Chapter 9).

AIR POLLUTION AND HUMAN HEALTH

Generally speaking, exposure to low levels of pollutants irritates the eyes and causes inflammation of the respiratory tract (**TABLE 8.1**). Many air pollutants also suppress the immune system, increasing susceptibility to infection. In addition, exposure to air pollution during respiratory illnesses may result in the development later in life of chronic respiratory diseases, such as emphysema and chronic bronchitis. In *emphysema*, the air sacs (alveoli) in the lungs become irreversibly distended, causing breathlessness and wheezy breathing. *Chronic bronchitis* is a disease in which the air passages (bronchi) of the lungs become permanently inflamed, causing breathlessness and chronic coughing.

URBAN AIR POLLUTION

Air pollution in urban areas is often called **smog**. The term *smog* was coined at the beginning of the 20th century for the smoky fog prevalent in London because of coal combustion. Traditional London-type smog—that is, smoke pollution—is sometimes called **industrial smog**. The principal pollutants in industrial smog are sulfur oxides and particulate matter. The worst episodes of industrial smog typically occur during winter months, when combustion of household fuel such as heating oil or coal is high. Because of air quality laws and pollution-control devices, industrial smog is generally not a significant problem in highly developed countries today, but it is often serious in many developing countries.

Health effects of several major air pollutants TABLE 8.1

Pollutant	Source	Effects
Particulate matter	Industries, motor vehicles	Aggravates respiratory illnesses; long-term exposure may cause chronic conditions such as bronchitis
Sulfur oxides	Electric power plants, industries	Irritate respiratory tract; same effects as particulates
Nitrogen oxides	Motor vehicles, industries, heavily fertilized farmland	Irritate respiratory tract; aggravate respiratory conditions such as asthma and chronic bronchitis
Carbon monoxide	Motor vehicles, industries	Reduces blood's ability to transport oxygen; headache and fatigue at low levels; mental impairment or death at high levels
Ozone	Formed in atmosphere (secondary air pollutant)	Irritates eyes; irritates respiratory tract; produces chest discomfort; aggravates respiratory conditions such as asthma and chronic bronchitis

Another important type of smog is **photochemi-cal smog**. First noted in Los Angeles in the 1940s, photochemical smog is generally worst during the summer months. Both nitrogen oxides and hydrocarbons are involved in its formation. A photochemical reaction occurs among nitrogen oxides, largely from automobile exhaust; volatile hydrocarbons, and oxygen in the atmosphere to produce ozone. This reaction requires solar energy (**FIGURE 8.9**). Ozone formed in this way then reacts with other air pollutants, including hydrocarbons, to form more than 100 different secondary air pollutants (peroxyacyl nitrates [PANs], for example) that injure plant tissues, irritate eyes, and aggravate respiratory illnesses in humans.

The main human source of the ingredients for photochemical smog is automobiles, but bakeries and dry cleaners are also significant contributors. When bread is baked, yeast byproducts that are volatile hydrocarbons are released to the atmosphere, where solar energy powers their interactions with other gases to form ozone. The volatile fumes from dry cleaners also contribute to photochemical smog.

> **photochemical smog** A brownish-orange haze formed by chemical reactions involving sunlight, nitrogen oxides, and hydrocarbons.

Composition of photochemical smog FIGURE 8.9

Photochemical smog is a mixture that includes ozone, peroxyacyl nitrates (PANs), nitric acid, and organic compounds such as formaldehyde.

VIEW THIS IN ACTION in your WileyPLUS course

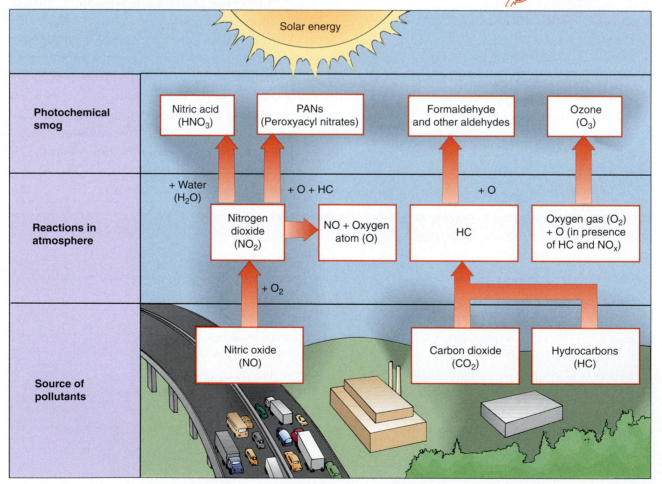

HOW WEATHER AND TOPOGRAPHY AFFECT AIR POLLUTION

Changes in temperature throughout the day produce air circulation patterns that dilute and disperse air pollutants. During a **temperature inversion**, however, polluting gases and particulate matter remain trapped in high concentrations close to the ground, where people live and breathe. Temperature inversions usually persist for only a few hours before solar energy warms the air near the ground. Sometimes a stalled high-pressure air mass allows a temperature inversion to persist for several days, causing atmospheric stagnation.

temperature inversion A layer of cold air temporarily trapped near the ground by a warmer upper layer.

Certain types of topography (surface features) increase the likelihood of temperature inversions. Cities located in valleys, near a coast, or on the leeward side of mountains (the side toward which the wind blows) are prime candidates for temperature inversions. The Los Angeles Basin, for example, lies between the Pacific Ocean on the west and mountains to the north and east. During the summer the sunny climate produces a layer of warm dry air at upper elevations. A region of upwelling occurs just off the Pacific coast, bringing cold ocean water to the surface and cooling the ocean air. As this cool air blows inland over the basin, the mountains block its movement further. Thus, a layer of warm, dry air overlies cool air at the surface, producing a temperature inversion.

EnviroDiscovery   AIR POLLUTION MAY AFFECT PRECIPITATION

For several years, climate scientists have noticed that mountainous areas in the western United States are receiving less precipitation than usual. The effect has been particularly pronounced in mountains located downwind from cities, leading scientists to speculate that air pollution may be altering precipitation patterns. However, long-term data to support this hypothesis were not available until recently.

In 2007 climate scientists evaluated weather data taken atop Mount Hua, a sacred mountain in China that overlooks a plain where several cities (which are a source of air pollution) are located. The data, which include precipitation, visibility, and humidity, have been measured since 1954. By subtracting the effect of humidity on visibility, scientists have been able to estimate the amount of air pollution suspended in the air. The scientists have correlated high visibility—that is, low air pollution—with substantially more precipitation than when air pollution levels were high. They caution, however, that the link between air pollution and precipitation patterns is still tentative and will require more research.

A Taoist Temple on Mount Hua in China. Note the proximity of the mountain to the plain where air pollution is produced.

Photographed in Los Angeles, California, on a day when air pollution exceeded federal air quality standards.

(such as compressed natural gas) for buses to lawn mower emissions to paint vapors. Using the cleanest emission-reduction equipment available significantly reduces emissions from large industrial and manufacturing sources, including oil refineries and power plants. California has no coal-fired power plants, and most of its power plants burn natural gas. To reduce automobile emissions, Los Angeles also has a subway, although it offers limited coverage. Future pollution reductions will come in part from requiring auto manufacturers to sell ultra-low-emission cars.

After several decades devoted to improving its air quality, Los Angeles now has the cleanest skies it has had since the 1950s. Despite the impressive progress, however, Los Angeles still exceeds federal air quality standards on more days than almost any other metropolitan region in the United States. At its current rate of improvement, Los Angeles should attain federal clean-air standards by 2010.

URBAN HEAT ISLANDS AND DUST DOMES

Streets, rooftops, and parking lots in areas of high population density absorb solar radiation during the day and radiate heat into the atmosphere at night. Heat from human activities such as fuel combustion is also highly concentrated in cities. The air in urban areas therefore forms **urban heat islands** in the surrounding suburban and rural areas (FIGURE 8.11).

Urban heat islands also contribute to the buildup of pollutants, especially particulate matter, in the form of **dust domes** over cities (FIGURE 8.12A). Pollutants concentrate in a dust dome because convection (the vertical motion of warmer air) lifts pollutants into the air, where they remain because of the somewhat stable air masses the urban heat island produces. If wind speeds increase, the dust dome moves downwind from the city, and the polluted air spreads over rural areas (FIGURE 8.12B).

urban heat island Local heat buildup in an area of high population.

dust dome A dome of heated air that surrounds an urban area and contains a lot of air pollution.

Air pollution in Los Angeles Los Angeles, California, has some of the worst smog in the world. Its location, in combination with its sunny climate, is conducive to the formation of stable temperature inversions that trap photochemical smog near the ground, sometimes for long periods (FIGURE 8.10). Passenger vehicles, heavy-duty trucks, and buses are the source of more than half of the smog-producing emissions in Los Angeles.

In 1969 California became the first state to enforce emission standards on motor vehicles, largely because of the air pollution problems in Los Angeles. Today Los Angeles has stringent smog controls that regulate everything from low-emission alternative fuels

Urban heat island FIGURE 8.11

This figure shows how temperatures might vary on a summer afternoon. The city stands out as a heat island against the surrounding rural areas.

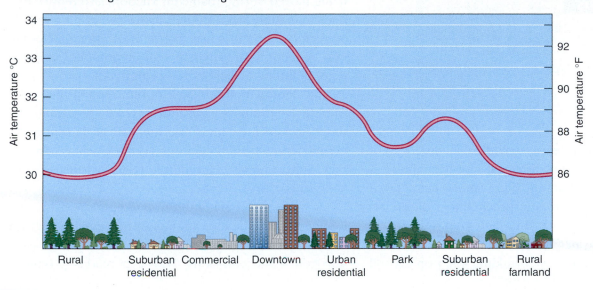

Dust dome FIGURE 8.12

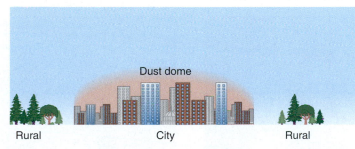

A A dust dome of pollutants forms over a city when the air is somewhat calm and stable.

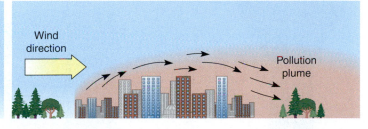

B When wind speeds increase, the pollutants move downwind from the city.

Urban heat islands affect local air currents and weather conditions. For example, urban heat islands may increase the number of thunderstorms over the city during summer months. The uplift of warm air over the city produces a low-pressure cell that draws in cooler air from the surroundings. As the heated air rises, it cools, causing water vapor to condense into clouds and producing thunderstorms.

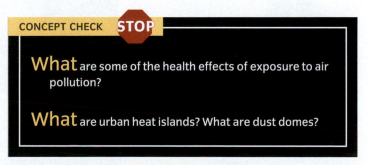

CONCEPT CHECK STOP

What are some of the health effects of exposure to air pollution?

What are urban heat islands? What are dust domes?

Controlling Air Pollutants

Technology exists to control all the forms of air pollution discussed in this chapter except carbon dioxide. Smokestacks fitted with electrostatic precipitators, fabric filters, scrubbers, and other technologies remove particulate matter from the air (**FIGURE 8.13**). Careful land-excavating activities, such as sprinkling water on dry soil being moved during road construction, also reduce particulate matter. Many of the measures that increase energy efficiency and conservation also reduce air pollution. Smaller, more fuel-efficient automobiles produce fewer polluting emissions, for example.

Several methods exist for removing sulfur oxides from flue (chimney) gases, but it is often less expensive simply to switch to a low-sulfur fuel such as natural gas or even to a non–fossil fuel energy source, such as solar energy. Sulfur can also be removed from fuels before they are burned.

Reduction of combustion temperatures in automobiles lessens the formation of nitrogen oxides. Use of mass transit reduces automobile use, thereby decreasing nitrogen oxide emissions. Nitrogen oxides produced

Electrostatic precipitator FIGURE 8.13

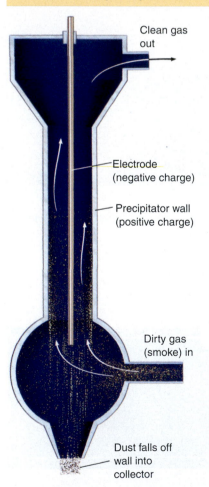

Clean gas out

Electrode (negative charge)

Precipitator wall (positive charge)

Dirty gas (smoke) in

Dust falls off wall into collector

A In an electrostatic precipitator, the electrode imparts a negative charge to particulates in the dirty gas. These particles are attracted to the positively charged precipitator wall, and then fall off into the collector.

B

C

A comparison of emissions from a Delaware Valley steel mill with the electrostatic precipitator turned off (**B**) and on (**C**). The toxic dust must be safely disposed, or it becomes a pollution problem.

during high-temperature combustion processes in industry can be removed from smokestack exhausts.

Modification of furnaces and engines to provide more complete combustion helps control the production of both carbon monoxide and hydrocarbons. Catalytic afterburners, used immediately following combustion, oxidize most unburned gases. The use of catalytic converters to treat auto exhaust reduces carbon monoxide and volatile hydrocarbon emissions about 85 percent over the life of the car. Careful handling of petroleum and hydrocarbons, such as benzene, reduces air pollution from spills and evaporation.

THE CLEAN AIR ACT

There is good news and bad news about air pollution in the United States. The bad news is that many locations throughout the country still have unacceptably high levels of one or more air pollutants. Moreover, most health experts estimate that air pollution causes the premature deaths of thousands of people in the United States each year. The good news is that overall air quality has improved since 1970.

This improvement is largely due to the U.S. **Clean Air Act**, first passed in 1970 and updated and amended in 1977 and 1990. This law authorizes the EPA to set limits on the amount of specific air pollutants permitted everywhere in the United States. Individual states must meet deadlines to reduce air pollution to acceptable levels. States may pass more stringent pollution controls than the EPA authorizes, but they can't mandate weaker limits than those stipulated in the Clean Air Act.

The EPA, which oversees the Clean Air Act, has focused on six air pollutants—lead, particulate matter, sulfur dioxide, carbon monoxide, nitrogen oxides, and ozone—and established maximum acceptable concentrations for each. The most dramatic improvement so far has been in the amount of lead in the atmosphere, which showed a 98 percent decrease between 1970 and 2000, primarily because of the switch from leaded to unleaded gasoline. Atmospheric levels of the other pollutants, with the exception of particulate matter, have also declined (FIGURE 8.14). For example, between 1970 and 2006, sulfur dioxide emissions declined 49 percent. During this same time, the U.S. gross domestic product increased

more than 160 percent, energy consumption increased 48 percent, and vehicle miles increased about 150 percent.

The Clean Air Act of 1970 and its amendments in 1977 and 1990 required progressively stricter controls of motor vehicle emissions. The provisions of the Clean Air Act Amendments of 1990 include the development of "superclean" cars, which emit lower amounts of nitrogen oxides and hydrocarbons, and the use of cleaner-burning gasoline in the most polluted cities in the United States. More recent automobile models do not produce as many pollutants as older models. Yet despite the increasing percentage of newer automobile models on the road, air quality has not improved in some areas of the United States because of the large increase in the number of cars being driven.

The Clean Air Act Amendments of 1990 focus on industrial airborne toxic chemicals in addition to motor vehicle emissions. Between 1970 and 1990, the airborne emissions of only seven toxic chemicals were regulated. In comparison, the Clean Air Act Amendments of 1990 required a 90 percent reduction in the

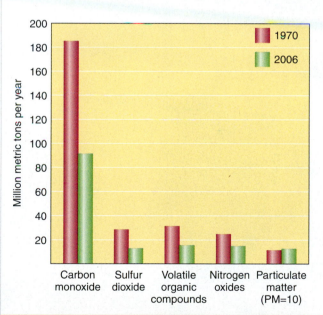

Emissions in the United States, 1970 and 2006
FIGURE 8.14

Carbon monoxide, sulfur dioxide, volatile organic compounds (many of which are hydrocarbons), and nitrogen oxides showed decreases; only particulate matter did not decline. "PM=10" applies to particles less than or equal to 10 μm (10 micrometers). Since 1990 the EPA has also monitored PM=2.5, which are very small particles less than or equal to 2.5 μm.

atmospheric emissions of 189 toxic chemicals. To comply with these requirements, both small businesses (such as dry cleaners) and large manufacturers (such as chemical companies) installed pollution control equipment if they had not already done so.

AIR POLLUTION IN DEVELOPING COUNTRIES

As developing nations become more industrialized, they also produce more air pollution. The leaders of most developing countries believe they must industrialize rapidly to compete economically with more highly developed countries. Environmental quality is usually a low priority in the race for economic development. Outdated technologies are adopted because they are less expensive, and air pollution laws, where they exist, are not enforced. Thus, air quality is deteriorating rapidly in many developing nations.

Many cities and towns in China have so many smokestacks belching coal smoke (coal is burned to heat many homes) that residents see the sun only a few weeks of the year (**FIGURE 8.15**). The rest of the time residents are choked in a haze of orange-colored coal dust. In other developing countries, such as India and Nepal, wood or animal dung is burned indoors, often in poorly designed stoves with little or no outside ventilation, thereby exposing residents to serious indoor air pollution (discussed in the next section).

The growing number of automobiles in developing countries is also contributing to air pollution, particularly in urban areas. Many vehicles in these countries are 10 or more years old and have no pollution control devices. Motor vehicles produce about 60 to 70 percent of the air pollutants in urban areas of Central America,

CHINA

Global Locator

Air pollution in China FIGURE 8.15
Coal smoke pollutes the air in a small town in Shanxi Province, China. All forms of pollution are increasing as China becomes industrialized.

and they produce 50 to 60 percent in urban areas of India. The most rapid proliferation of motor vehicles worldwide is currently occurring in Latin America, Asia, and eastern Europe.

Lead pollution from heavily leaded gasoline is an especially serious problem in developing nations. The gasoline refineries in these countries are generally not equipped to remove lead from gasoline. (The United States was in the same situation until federal law mandated that U.S. refineries upgrade their equipment by 1986.) In Cairo, Egypt, for example, children's blood lead levels are more than two times higher than the level considered at-risk in the United States. Lead can retard children's growth and cause brain damage.

According to the World Health Organization, the five worst cities in the world in terms of exposing chil- dren to air pollution are Mexico City, Mexico; Beijing, China; Shanghai, China; Tehran, Iran; and Calcutta, India. Respiratory disease is now the leading cause of death for children worldwide. More than 80 percent of these deaths occur in children under age 5 who live in cities in developing countries.

CONCEPT CHECK **STOP**

What is the U.S. Clean Air Act and how has it reduced outdoor air pollution?

Where is air pollution worse: in highly developed nations or in developing countries? Why?

Indoor Air Pollution

LEARNING OBJECTIVES

Summarize at least four sources of indoor air pollution.

Describe sick building syndrome.

I f you are reading this chapter indoors, you are probably inhaling air pollution. The air in enclosed places, such as automobiles, homes, schools, and offices, may have significantly higher levels of air pollutants than the air outdoors. Indoor air pollution is of particular concern to urban residents because they may spend as much as 95 percent of their time in enclosed places.

Because illnesses from indoor air pollution usually resemble common ailments such as colds, influenza, or upset stomachs, they are often not recognized. The most common contaminants of indoor air are radon, cigarette smoke, carbon monoxide, nitrogen dioxide (from gas stoves), formaldehyde (from carpet, fabrics, and furniture), household pesticides, cleaning solvents, ozone (from photocopiers), and asbestos. In addition, viruses, bacteria, fungi (yeasts, molds, and mildews), dust mites, pollen, and other organisms are often found in heating, air-conditioning, and ventilation ducts.

Health officials are paying increasing attention to the **sick building syndrome**. The Labor Department estimates that more than 20 million employees are exposed to health risks from indoor air pollution. The EPA estimates that annual medical costs for treating the health effects of indoor air pollution in the United States exceed $1 billion. When lost work time and diminished productivity are added to health care costs, the total annual cost to the economy may be as much as $50 billion. Fortunately, most building problems are relatively inexpensive to alleviate.

sick building syndrome Eye irritations, nausea, headaches, respiratory infections, depression, and fatigue caused by indoor air pollution.

Indoor air pollution FIGURE 8.16

Cooking indoors with open fires or traditional cooking stoves results in dangerous levels of indoor air pollution.

Indoor air pollution is a particularly serious health hazard in developing countries, where many people burn fuels such as firewood or animal dung indoors to cook and heat water (**FIGURE 8.16**). Smoke from indoor cooking contains carbon monoxide, particulates, hydrocarbons, and hazardous air pollutants such as formaldehyde and benzene. Women and children are harmed the most by indoor cooking, which can contribute to acute lower respiratory infections, pneumonia, and lung cancer. The World Health Organization estimates that smoke from indoor cooking kills 1.6 million people each year.

RADON

The most serious indoor air pollutant is probably radon, a colorless, tasteless, odorless radioactive gas produced naturally during the radioactive decay of uranium in Earth's crust. Radon seeps through the ground and enters buildings, where it sometimes accumulates to dangerous levels (**FIGURE 8.17**). Although radon is also emitted into the atmosphere, it gets diluted and dispersed and is of little consequence outdoors.

Only ingested or inhaled radon harms the body. The National Research Council of the National Academy of Sciences estimates that residential exposure to radon causes 12 percent of all lung cancers—between 15,000 and 22,000 lung cancers annually. Cigarette smoking exacerbates the risk from radon exposure; about 90 percent of radon-related cancers occur among current or former smokers.

According to the EPA, about 6 percent of U.S. homes have high enough levels of radon to warrant corrective action—a radon level above 4 picocuries per liter

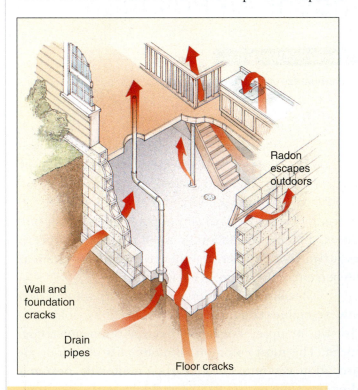

Radon escapes outdoors

Wall and foundation cracks

Drain pipes

Floor cracks

How radon infiltrates a house FIGURE 8.17

Cracks in basement walls or floors, openings around pipes, and pores in concrete blocks provide some of the entries for radon.

Indoor air pollution FIGURE 8.18

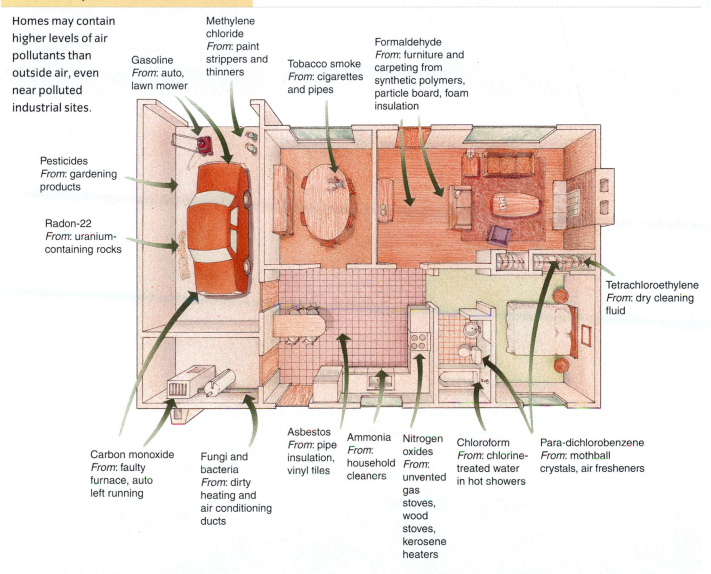

Homes may contain higher levels of air pollutants than outside air, even near polluted industrial sites.

Gasoline
From: auto, lawn mower

Methylene chloride
From: paint strippers and thinners

Tobacco smoke
From: cigarettes and pipes

Formaldehyde
From: furniture and carpeting from synthetic polymers, particle board, foam insulation

Pesticides
From: gardening products

Radon-22
From: uranium-containing rocks

Tetrachloroethylene
From: dry cleaning fluid

Carbon monoxide
From: faulty furnace, auto left running

Fungi and bacteria
From: dirty heating and air conditioning ducts

Asbestos
From: pipe insulation, vinyl tiles

Ammonia
From: household cleaners

Nitrogen oxides
From: unvented gas stoves, wood stoves, kerosene heaters

Chloroform
From: chlorine-treated water in hot showers

Para-dichlorobenzene
From: mothball crystals, air fresheners

---

of air. (As a standard of reference, outdoor radon concentrations range from 0.1 to 0.15 picocuries per liter of air worldwide.) The highest radon levels in the United States are found in homes across southeastern Pennsylvania into northern New Jersey and New York.

Ironically, efforts to make our homes more energy efficient have increased the hazard of indoor air pollutants, including radon. Drafty homes waste energy but allow radon to escape outdoors so it does not build up inside. Every home should be tested for radon because levels vary widely from home to home, even in the same neighborhood. Generally, testing and corrective actions are reasonably priced. However, some corrective actions can cost thousands of dollars, so it's not always inexpensive.

FIGURE 8.18 summarizes many possible sources of air pollution in homes.

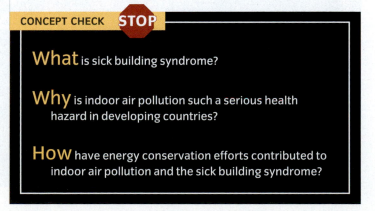

CONCEPT CHECK STOP

What is sick building syndrome?

Why is indoor air pollution such a serious health hazard in developing countries?

How have energy conservation efforts contributed to indoor air pollution and the sick building syndrome?

CURBING AIR POLLUTION IN CHATTANOOGA

During the 1960s, the federal government gave Chattanooga, Tennessee, the dubious distinction of having the worst air pollution in the United States. The air was so dirty in this manufacturing city that sometimes people driving downtown had to turn on their headlights in the middle of the day. The orange air soiled their white shirts so quickly that many businesspeople brought extra ones to work. To compound the problem, the mountains surrounding the city kept the pollutants produced by its inhabitants from dispersing.

Today the air in this scenic midsized city of 200,000 people is clean, and Chattanooga ranks high among U.S. cities in terms of air quality (*see photo*). City and business leaders are credited with transforming Chattanooga's air. Soon after the passage of the federal Clean Air Act of 1970, the city established an air pollution control board to enforce regulations controlling air pollution. New local regulations allowed open burning by permit only, placed limits on industrial odors and particulate matter, outlawed visible automotive emissions, and set a cap on sulfur content in fuel, which controlled the production of sulfur oxides. Businesses installed expensive air pollution control devices. The city started an emissions-free electric bus system. Chattanooga also decided to recycle its solid waste rather than build an emissions-producing incinerator.

In 1984 the EPA declared Chattanooga in attainment for particulate matter; this designation meant particulate levels had been below the federal health limit for one year. The city reached attainment status for ozone in 1989. Since then, the city's levels for all seven EPA-regulated air pollutants have been lower than federal standards require.

In the early 2000s, Chattanoogans continued to move their city toward *environmental sustainability*. The city plans to convert a run-down business district into a community in which people live near their places of work. Businesses located in this district will form an **industrial ecosystem** in which the wastes of one business are raw materials for another business.

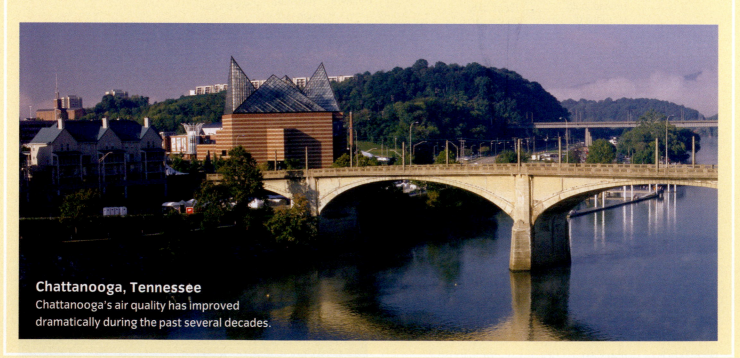

Chattanooga, Tennessee
Chattanooga's air quality has improved dramatically during the past several decades.

1 The Atmosphere

1. Oxygen (21 percent) and nitrogen (78 percent) are the main gases in the **atmosphere**, the gaseous envelope surrounding Earth. Argon, carbon dioxide, other gases, water vapor, and trace amounts of various air pollutants are also present.

2. The **troposphere**, the layer of atmosphere closest to Earth's surface, extends to a height of approximately 12 km. Temperature decreases with increasing altitude, and weather occurs in the troposphere. In the **stratosphere**, there is a steady wind but no turbulence. The stratosphere contains an ozone layer that absorbs much of the Sun's UV radiation. The **mesosphere**, directly above the stratosphere, has the lowest temperatures in the atmosphere. The **thermosphere** has steadily rising temperatures and gases that absorb x-rays and short-wave UV radiation. The thermosphere reflects outgoing radio waves back toward Earth without the aid of satellites.

3. The **Coriolis effect** is the tendency of moving air or water to be deflected from its path and swerve to the right in the Northern Hemisphere and to the left in the Southern Hemisphere.

2 Types and Sources of Air Pollution

1. **Air pollution** consists of various chemicals (gases, liquids, or solids) present in the atmosphere in high enough levels to harm humans, other organisms, or materials. **Primary air pollutants** are harmful chemicals that enter the atmosphere directly due to either human activities or natural processes; examples include carbon oxides, nitrogen oxides, sulfur dioxide, particulate matter, and hydrocarbons. **Secondary air pollutants** are harmful chemicals that form in the atmosphere when primary air pollutants react chemically with each other or with natural components of the atmosphere; ozone and sulfur trioxide are examples.

2. **Particulate matter**—solid particles and liquid droplets suspended in the atmosphere—corrodes metals, erodes buildings, soils fabrics, and can damage the lungs. **Nitrogen oxides** are gases associated with photochemical smog, acid deposition, global warming, and stratospheric ozone depletion; they also corrode metals and fade textiles. **Sulfur oxides** are gases associated with acid deposition; they corrode metals and damage stone and other materials. **Carbon oxides** include the gases carbon monoxide, which is poisonous, and carbon dioxide, which is linked to global warming. **Hydrocarbons** are solids, liquids, or gases associated with photochemical smog and global warming; some are dangerous to human health. **Ozone** is a secondary air pollutant in the lower atmosphere (troposphere) but an essential part of the stratosphere. Tropospheric ozone reduces visibility, causes health problems, stresses plants, and is associated with global warming. Some air pollutants are called **hazardous air pollutants**, or **air toxics**, because they are potentially harmful and may pose long-term health risks to people who are exposed to them; chlorine, lead, hydrochloric acid, formaldehyde, radioactive substances, and fluorides are examples.

3 Effects of Air Pollution

1. Exposure to low levels of air pollutants irritates the eyes and causes inflammation of the respiratory tract. Many air pollutants suppress the immune system, increasing susceptibility to infection. Exposure to air pollution during respiratory illnesses may result in the development of chronic respiratory diseases, such as emphysema and chronic bronchitis.

2. **Industrial smog** refers to smoke pollution. **Photochemical smog** is a brownish-orange haze formed by chemical reactions involving sunlight, nitrogen oxides, and hydrocarbons. A **temperature inversion** is a layer of cold air temporarily trapped near the ground by a warmer upper layer; during a temperature inversion, polluting gases and particulate matter remain trapped in high concentrations close to the ground. An **urban heat island** is local heat buildup in an area of high population. Urban heat islands affect local air currents and weather conditions and contribute to the buildup of pollutants, especially particulate matter, in the form of a **dust dome**, a dome of heated air that surrounds an urban area and contains a lot of air pollution.

4 Controlling Air Pollutants

1. Improvements in U.S. air quality since 1970 are largely due to the **Clean Air Act**, which authorizes the EPA to set limits on specific air pollutants. Individual states must meet deadlines to reduce air pollution to acceptable levels and can't mandate weaker limits than those stipulated in the Clean Air Act.

2. Air quality in the United States has slowly improved since passage of the Clean Air Act. The most dramatic improvement is the decline of lead in the air, although levels of sulfur oxides, ozone, carbon monoxide, volatile compounds, and nitrogen oxides have also declined. Air quality is deteriorating in developing nations as a result of rapid industrialization, growing numbers of automobiles, and a lack of emissions standards.

5 Indoor Air Pollution

1. Indoor air pollution includes radon, cigarette smoke, nitrogen dioxide (from gas stoves), and formaldehyde (from carpet, fabrics, and furniture). Indoor air pollution is a serious health hazard in developing countries.

2. The **sick building syndrome** includes eye irritations, nausea, headaches, respiratory infections, depression, and fatigue caused by indoor air pollution.

KEY TERMS

- **atmosphere** p. 192
- **Coriolis effect** p. 195
- **air pollution** p. 196
- **primary air pollutants** p. 196
- **secondary air pollutants** p. 196
- **photochemical smog** p. 202
- **temperature inversion** p. 203
- **urban heat island** p. 204
- **dust dome** p. 204
- **sick building syndrome** p. 209

CRITICAL AND CREATIVE THINKING QUESTIONS

1. What is the global distillation effect? What kinds of air pollutants are involved in the global distillation effect? Where do these pollutants get permanently deposited? Why?

2. The atmosphere of Earth has been compared to the peel covering an apple. Explain the comparison.

3. What basic forces determine the circulation of the atmosphere? Describe the general directions of atmospheric circulation.

4. Distinguish between primary and secondary air pollutants. Give examples of each that you are likely to encounter.

5. Distinguish between mobile and stationary sources of air pollution.

6. Why might it be more effective to control photochemical smog in heavily wooded areas such as the Atlanta metropolitan area by reducing nitrogen oxides instead of volatile hydrocarbons?

7. These graphs represent air pollutant measurements taken at two different locations. Which location is indoors, and which is outdoors? Explain your answer.

8. One of the most effective ways to reduce the threat of radon-induced lung cancer is to quit smoking. Explain.

9. What air pollutants do the 1990 amendments to the Clean Air Act target?

10. During a formal debate on the hazards of air pollution, one team argues that ozone is helpful to the atmosphere, and the other team argues that it is destructive. Explain why they are both correct.

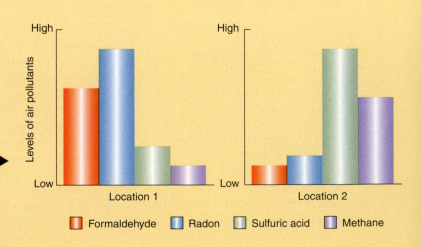

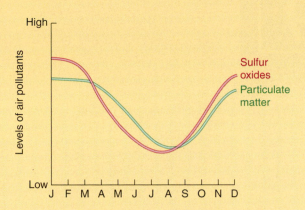

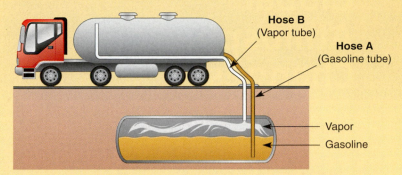

Hose B
(Vapor tube)

Hose A
(Gasoline tube)

Vapor

Gasoline

11. The graph above shows air pollutant levels in a city in the Northern Hemisphere, measured throughout a year. Is this city likely to be found in a developing country or in a developed country? Why?

12. This figure shows phase I vapor recovery from an underground gasoline storage tank. Hose B is used to vent gasoline vapor directly into the air. Now the vapor is vented into the truck and returns to the gasoline depot. What do you think happens to the vapor there? A similar system (phase II vapor recovery) exists to remove gasoline vapor from gas tanks in cars and return it to the underground tank, but this system is not widely used yet. Where does the vapor in a gas tank go when the tank is being filled with gasoline?

What is happening in this picture **?**

■ These scientists are about to launch a balloon attached to meteorological instruments (*in far background*). Judging from the current atmospheric conditions, what processes might the scientists be measuring?

■ During a thunderstorm, the energy produced by a lightning strike chemically combines some nitrogen and oxygen molecules to form the reddish-brown gas nitrogen dioxide (NO_2). Nitrogen dioxide is what the scientists are measuring here. Name at least three environmental effects of NO_2.

■ What are two human sources of NO_2 and other nitrogen oxides?

Global Atmospheric Changes

9

MELTING ICE AND RISING SEA LEVELS

In 2002, an iceberg roughly twice the size of Rhode Island broke off from the Antarctic Peninsula. Studies of the ice-covered ocean in the Arctic from the 1970s to the present show that the ice pack there has retreated, and the remaining pack has thinned rapidly, losing 40 percent of its volume in less than three decades.

Mountain glaciers around the world are melting at accelerating rates. The Muir Glacier in Alaska was formerly enormous, with a huge vertical front from which icebergs calved into Glacier Bay. Today, the Muir Glacier has shrunk to a fraction of its former size (see photograph, taken in August 2004; the inset shows approximately the same location in August 1941).

Sea-level rise is caused not only by the retreat of glaciers and the thawing of polar ice but also by the expansion of ocean water as it warms. During the 20th century, the sea level rose 10 to 20 cm (4 to 8 in). Climate scientists estimate that the sea level will rise an additional 48 cm (19 in) by 2100. Such a rise would flood low-lying coastal areas such as southern Louisiana and South Florida.

Small island nations such as the Maldives, a low-level chain of 1,200 islands in the Indian Ocean, are considered highly vulnerable to a rise in sea level, as storm surges could easily sweep over entire islands. Other vulnerable countries—such as Bangladesh, Egypt, Vietnam, and Mozambique—have dense populations living in low-lying river deltas.

Human-caused climate change is an established phenomenon. Within the scientific community, the question is no longer whether climate change will occur, but at what rate, with what effects, and what can be done about it. The biggest culprit in climate change is an increase in atmospheric carbon dioxide (CO_2), which is generated primarily through burning fossil fuels.

In this chapter we examine the challenges of global atmospheric changes: climate change, ozone depletion, and acid deposition. Changes in economics, politics, energy use, agriculture, and human behavior will be necessary to address these issues.

The Atmosphere and Climate

Weather refers to the conditions in the atmosphere at a given place and time; it includes temperature, atmospheric pressure, precipitation, cloudiness, humidity, and wind. Weather changes from one hour to the next and from one day to the next.

The two most important factors that determine an area's overall **climate** are *temperature*—both average temperature and temperature extremes—and both average and

> **climate** The average weather conditions that occur in a place over a period of years.

Tropical climate FIGURE 9.1

Tropical climates occur in a region that spans the equator, from 15 to 25 degrees latitude north to 15 to 25 degrees latitude south. Photographed at the Na Pali Coast, Kauai, Hawaiian Islands.

seasonal *precipitation*. Latitude, elevation, topography, quantity and quality of vegetation, distance from the ocean, and geographic location all influence temperature, precipitation, and other aspects of climate. Other climate factors include weather conditions such as wind, humidity, fog, cloud cover, and, in some areas, lightning. Unlike weather, which changes rapidly, climate generally changes slowly, over hundreds or thousands of years.

Earth has many different climates, and because each is relatively constant for many years, organisms have adapted to them. The many kinds of organisms on Earth are here in part because of the large number of different climates—from cold, snow-covered polar climates to tropical climates that are hot and have rain almost every day (**FIGURE 9.1**).

SOLAR RADIATION AND CLIMATE

The Sun makes life on Earth possible. It warms the planet, including the atmosphere, to habitable temperatures. Without the Sun's energy, all water on Earth would be frozen, including the ocean. Photosynthetic organisms capture the Sun's energy and use it to make the food molecules almost all forms of life require. Most of our fuels, such as wood and **fossil fuels** (oil, coal, and natural gas), represent solar energy captured by photosynthetic organisms. The amount of solar energy a region receives (in other words, the amount of sunshine a region gets) is the primary determinant of climate.

Clouds, and to a lesser extent snow, ice, and the ocean, reflect away about 31 percent of the solar radiation that falls on Earth (**FIGURE 9.2**). The remaining 69 percent is absorbed and runs the hydrologic cycle, carbon cycle, and other biogeochemical cycles; drives winds and ocean currents; powers photosynthesis; and warms the planet. Ultimately, all this energy returns to space as long-wave **infrared radiation** (heat energy).

> **infrared radiation** Electromagnetic radiation with wavelengths longer than those of visible light but shorter than microwaves; perceived as invisible waves of heat energy.

Fate of solar radiation that reaches Earth FIGURE 9.2

Most of the energy that the Sun produces never reaches Earth. The solar energy that does reach Earth warms the planet's surface, drives the hydrologic cycle and other biogeochemical cycles, produces our climate, and powers almost all life through the process of photosynthesis.

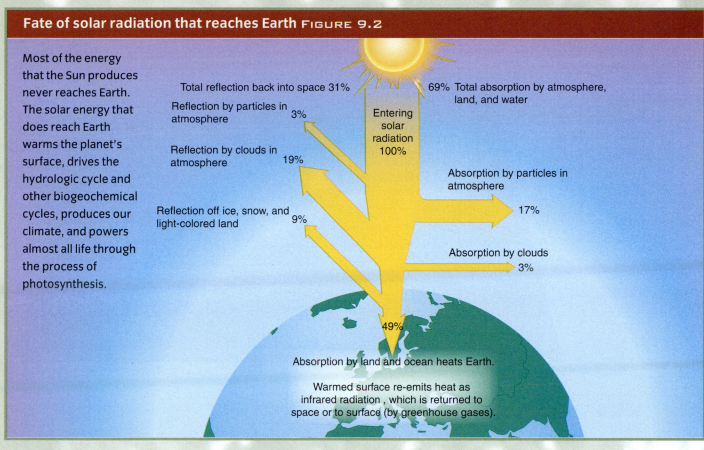

Total reflection back into space 31%

Reflection by particles in atmosphere 3%

Reflection by clouds in atmosphere 19%

Reflection off ice, snow, and light-colored land 9%

Entering solar radiation 100%

69% Total absorption by atmosphere, land, and water

Absorption by particles in atmosphere 17%

Absorption by clouds 3%

49%

Absorption by land and ocean heats Earth.

Warmed surface re-emits heat as infrared radiation , which is returned to space or to surface (by greenhouse gases).

Process Diagram

Temperature changes with latitude and season

Earth's roughly spherical shape and the tilt of its axis produce a great deal of variation in the exposure of the surface to solar energy (**FIGURE 9.3**). Sunlight that shines vertically near the equator (represented by the desk lamp on the left) is concentrated on Earth's surface. As one moves toward the poles, the light hits the surface more and more obliquely (represented by the lamp on the right), spreading the same amount of radiation over larger and larger areas. Because the Sun's energy does not reach all places uniformly, temperature varies locally.

Earth's inclination on its axis (23.5 degrees from a line drawn perpendicular to the orbital plane) determines the seasons. During half the year (March 20 to September 22) the Northern Hemisphere tilts toward the Sun, and during the other half (September 22 to

Solar intensity and latitude FIGURE 9.3

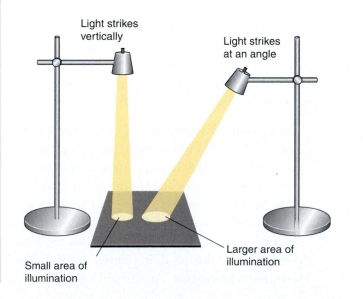

Light strikes vertically

Light strikes at an angle

Larger area of illumination

Small area of illumination

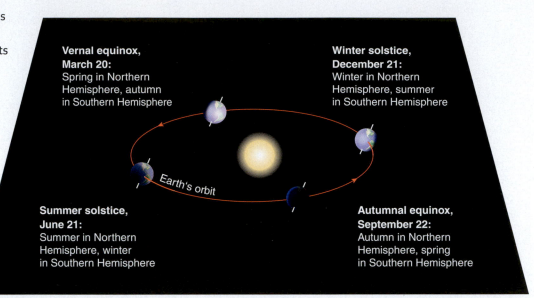

Earth's inclination on its axis remains the same as it travels around the Sun. Thus, the Sun's rays hit the Northern Hemisphere obliquely during its winter months and more directly during its summer. In the Southern Hemisphere, the Sun's rays are oblique during its winter, which corresponds to the Northern Hemisphere's summer. At the equator, the Sun's rays are approximately vertical on March 20 and September 22.

Vernal equinox, March 20:
Spring in Northern Hemisphere, autumn in Southern Hemisphere

Winter solstice, December 21:
Winter in Northern Hemisphere, summer in Southern Hemisphere

Summer solstice, June 21:
Summer in Northern Hemisphere, winter in Southern Hemisphere

Earth's orbit

Autumnal equinox, September 22:
Autumn in Northern Hemisphere, spring in Southern Hemisphere

March 20) it tilts away from the Sun (FIGURE 9.4). The Southern Hemisphere tilts the opposite way, so that summer in the Northern Hemisphere corresponds to winter in the Southern Hemisphere.

PRECIPITATION

Precipitation refers to any form of water, such as rain, snow, sleet, or hail, that falls from the atmosphere. Differences in precipitation depend on three factors:

1. **The amount of water vapor in the atmosphere.** *Equatorial uplift* of warm, moisture-laden air produces heavy rainfall in some areas of the tropics. High surface water temperatures cause vast quantities of water to evaporate from tropical parts of the ocean. Prevailing winds blow the resulting moist air over landmasses. The moist air continues to rise as it is heated by the Sun-warmed land surface. As the air rises, it cools,

which decreases its moisture-holding ability. When the air reaches its saturation point—when it can't hold any additional water vapor—clouds form and water is released as precipitation.

2. **Geographic location.** The rising air from the equator eventually descends to Earth near the Tropic of Cancer and Tropic of Capricorn (latitudes 23.5 degrees north and 23.5 degrees south, respectively). By then most of its moisture has precipitated, and the dry air returns to the equator. Over land, this dry air produces some of the great tropical deserts, such as the Sahara Desert. Air also dries out as it travels long distances over landmasses. Near the windward coasts of continents (the side from which the wind blows), rainfall may be heavy. However, in the temperate zones (the areas between the tropics and the polar zones), continental interiors are usually dry because they are far from the ocean, which replenishes water in the air passing over it.

3. **Topographic features.** When flowing air encounters mountains, it flows up and over them, giving up moisture in the process. The air's temperature cools as it gains altitude; clouds form and precipitation occurs, primarily on the mountains' windward slopes. The air mass is warmed as it moves down on the other side of the mountain, reducing the chance of precipitation of the remaining moisture. This situation exists on the West Coast of North America, where precipitation falls on the western slopes of mountains close to the coast (see What a Scientist Sees). The dry land on the side of the mountains away from the prevailing wind—in this case, east of the mountain range—is called a **rain shadow**. The dry conditions of a rain shadow often occur on a regional scale.

CONCEPT CHECK **STOP**

How do you distinguish between weather and climate?

What are the two most important climate factors?

What are some of the environmental factors that produce areas of precipitation extremes, such as rain forests and deserts?

Rain shadow

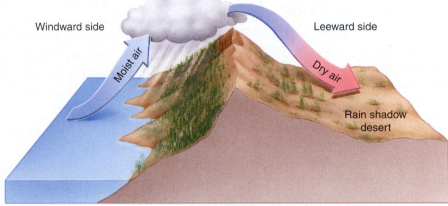

Windward side   Leeward side

Moist air   Dry air

Rain shadow desert

A Proxy Falls is in the Cascade Range, which divides the states of Washington and Oregon into a moist western region and an arid region east of the mountains.

B A rain shadow refers to arid or semiarid land that occurs on the far side (leeward side) of a mountain. Prevailing winds blow warm, moist air from the windward side. Air temperature cools as it rises, releasing precipitation, so dry air descends on the leeward side. Such a rain shadow exists east of the Cascades.

What a Scientist Sees

Global Climate Change

Earth's average temperature is based on daily measurements taken at several thousand land-based meteorological stations around the world, as well as data from weather balloons, orbiting satellites, transoceanic ships, and hundred of sea-surface buoys with temperature sensors. These data show that 11 of the 12 years between 1995 and 2006 were among the 12 warmest years since the mid-1800s. According to the National Oceanic and Atmospheric Administration (NOAA), global temperatures in those years may have been the highest in the last millennium. (Although widespread thermometer records have been assembled only since the mid-19th century, scientists reconstruct earlier temperatures using indirect climate evidence in tree rings, lake and ocean sediments, small air bubbles in ancient ice, and coral reefs.) The last two decades of the 20th century were its warmest (**FIGURE 9.5**).

Other evidence also suggests an increase in global temperature. Several studies indicate that spring in the Northern Hemisphere now comes about six days earlier than it did in 1959, and autumn comes five days later. Since 1949, the United States has experienced an increased frequency of heat waves, resulting in increased heat-related deaths among elderly and other vulnerable people. In the past few decades, the sea level has risen, glaciers worldwide have retreated, and extreme weather events, such as severe storms, have occurred with increasing frequency in certain regions.

Scientists around the world have been researching global climate change for the past 50 years. As the evidence has accumulated, those most qualified to address the issue have reached a strong consensus that the 21st century will experience significant climate change for which human activities are largely responsible.

In response to this growing consensus, governments around the world organized the United Nations Intergovernmental Panel on Climate Change (IPCC). With input from hundreds of climate experts, the IPCC provides the definitive scientific statement about global climate change. In 2007 the IPCC projected a 1.8° to 4.0°C (3.2° to 7.2°F) increase in global temperature by the year 2100. The IPCC predicts that we will observe higher maximum temperatures and more hot days over nearly all land areas, higher minimum temperatures, fewer frost days, fewer cold days, and an increase in the heat index. We may also experience more intense precipitation events over many areas, an increased risk of drought in the continental interiors in the mid-latitudes, and stronger hurricanes in some coastal areas.

Mean annual global temperature, 1960 to present FIGURE 9.5

Data are presented as surface temperatures (°C) for 1960, 1965, and every year thereafter. The measurements, which naturally fluctuate, show the warming trend of the past several decades. (The dip in global temperatures in the early 1990s was caused by the eruption of Mount Pinatubo in 1991.)

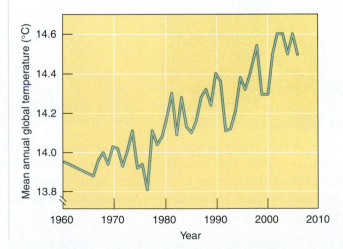

CAUSES OF GLOBAL CLIMATE CHANGE

Carbon dioxide (CO_2) and certain trace gases, including methane (CH_4), nitrous oxide (N_2O), chlorofluorocarbons (CFCs), and tropospheric ozone (O_3), accumulate in the atmosphere as a result of human activities. The concentration of atmospheric carbon dioxide has increased from about 288 parts per million (ppm) approximately 200 years ago (before the Industrial Revolution began) to 382 ppm in 2006 (FIGURE 9.6). Burning carbon-containing fossil fuels accounts for most human-made CO_2. Land conversion, such as when tracts of tropical forests are logged or burned, also releases CO_2. By 2050 the concentration of atmospheric CO_2 may be double what it was in the 1700s.

The combustion of gasoline in your car's engine releases not only CO_2 but also nitrous oxide, which triggers the production of tropospheric ozone (FIGURE 9.7). Various industrial processes, land-use conversion, and the use of fertilizers also produce nitrous oxide. CFCs (discussed later in the chapter as they relate to depletion of the stratospheric ozone layer) are chemicals released into the atmosphere from old, leaking refrigerators and air conditioners. Methane is produced by the decomposition of carbon-containing organic material by anaerobic bacteria in moist places as varied as rice paddies, sanitary landfills, and the intestinal tracts of cattle and other large animals (humans included).

Carbon dioxide (CO_2) in the atmosphere, 1958 to present FIGURE 9.6

Note the steady increase in the concentration of atmospheric CO_2 since 1958, when measurements began at the Mauna Loa Observatory in Hawaii. This location was selected because it is far from urban areas where factories, power plants, and motor vehicles emit CO_2. The seasonal fluctuations correspond to winter (a high level of CO_2), when plants are not actively growing and absorbing CO_2, and summer (a low level of CO_2), when plants are growing and absorbing CO_2.

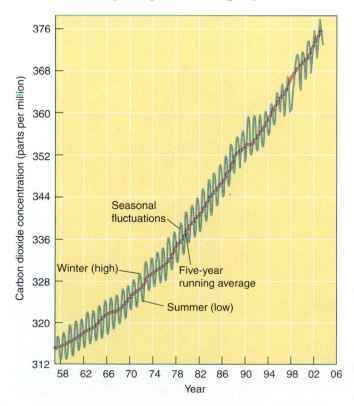

Motor vehicles at a tollbooth FIGURE 9.7

Traffic lines up at the San Francisco Bay Bridge. These motor vehicles produce climate changing gases.

greenhouse gases Gases that absorb infrared radiation, including water vapor, carbon dioxide, methane, and certain other gases.

enhanced greenhouse effect The additional warming that may be produced by human-increased levels of CO_2 and other gases that absorb infrared radiation.

Global climate change occurs because these gases absorb infrared radiation—that is, heat energy—given off by Earth's surface. This absorption slows the natural flow of heat into space, warming the lower atmosphere.

Because CO_2 and other gases trap the Sun's infrared radiation somewhat like glass does in a greenhouse, they are called **greenhouse gases**. Greenhouse gases accumulating in the atmosphere as a result of human activities are causing an **enhanced greenhouse effect** (FIGURE 9.8).

Other pollutants cool the atmosphere

One of the complications that makes the rate and extent of global climate change difficult to predict is that some air pollutants, known as atmospheric aerosols, tend to cool the atmosphere. **Aerosols**, which come from both natural and human sources, are particles so small they remain suspended in the atmosphere for days, weeks, or even months. Sulfur haze is an aerosol that reflects sunlight back into space, reducing the amount of solar energy reaching Earth's surface, and thereby cooling the atmosphere. Sulfur haze significantly moderates warming in industrialized parts of the world. Sulfur emissions come from the same smokestacks that emit CO_2. Volcanic eruptions also eject sulfur particles into the atmosphere (FIGURE 9.9).

Despite their cooling effect, human-produced sulfur emissions should not be viewed as a panacea for the enhanced greenhouse effect. Human-produced sul-

Process Diagram

Enhanced greenhouse effect FIGURE 9.8

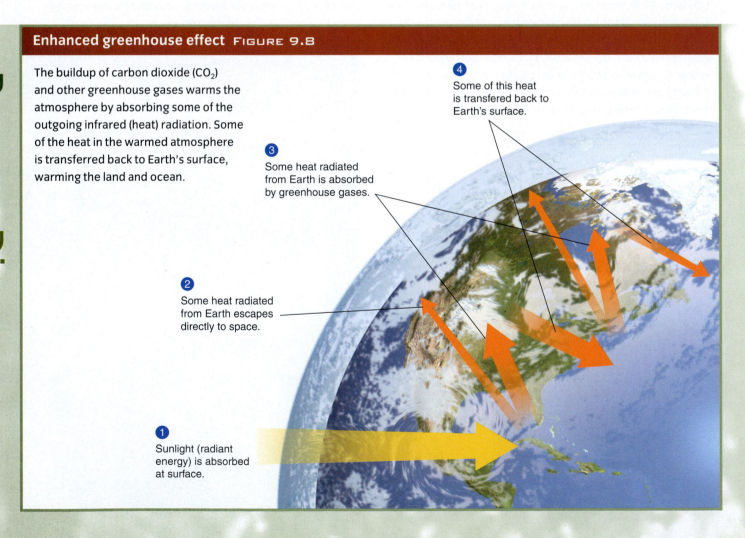

The buildup of carbon dioxide (CO_2) and other greenhouse gases warms the atmosphere by absorbing some of the outgoing infrared (heat) radiation. Some of the heat in the warmed atmosphere is transferred back to Earth's surface, warming the land and ocean.

④ Some of this heat is transfered back to Earth's surface.

③ Some heat radiated from Earth is absorbed by greenhouse gases.

② Some heat radiated from Earth escapes directly to space.

① Sunlight (radiant energy) is absorbed at surface.

Global Locator

PHILIPPINES

Volcanic eruption FIGURE 9.9

The eruption of Mount Pinatubo in the Philippines in 1991 injected massive amounts of sulfur into the atmosphere. Because sulfur haze reduces the amount of sunlight reaching the surface, this eruption caused Earth to cool temporarily. Compared to temperatures during the rest of the 1990s, 1992 and 1993 global temperatures were relatively cool.

fur emissions remain in the atmosphere for only days, weeks, or months. They do not disperse globally and may cause only regional cooling. Greenhouse gases, on the other hand, remain in the atmosphere for hundreds of years. And carbon dioxide and other greenhouse gases help warm the planet 24 hours a day, whereas sulfur haze cools the planet only during the daytime. In addition, sulfur emissions are a respiratory irritant and cause acid deposition (discussed later in this chapter). Most nations are trying to reduce their sulfur emissions, not maintain or increase them.

EFFECTS OF GLOBAL CLIMATE CHANGE

In addition to changes in sea level, already considered in the chapter introduction, some of the observed and potential effects of global climate change include changes in precipitation patterns and impacts on agriculture, human health, and other organisms.

Computer models of weather changes caused by global climate change indicate that precipitation pat-terns will change, causing some areas to have more frequent droughts. At the same time, heavier snow and rainstorms may cause more frequent flooding in other areas. Changes in precipitation patterns could affect the availability and quality of fresh water in many locations, particularly in areas that are currently arid or semiarid, such as the Sahel region just south of the Sahara Desert.

Global climate change will increase problems for agriculture. The rise in sea level will inundate some river deltas, which are fertile agricultural lands. Certain agricultural pests and disease-causing organisms will probably proliferate and reduce crop yields. The likely increase in the frequency and duration of droughts will be a particularly serious problem for countries with limited water resources. Lack of water for drinking and agriculture may force millions of people to relocate.

Currently, most evidence linking climate warming to disease outbreaks is circumstantial, so scientists are reluctant to ascribe cause-and-effect relationships. Nonetheless, data linking climate warming and human health problems are accumulating. More frequent and more severe heat waves during summer may increase the

Sick with malaria FIGURE 9.10

In a climate-warmed world, the mosquito that spreads malaria could expand into temperate areas.
Photographed in Ariquemes, Brazil, where most of the town's inhabitants suffer from malaria.

number of heat-related illnesses and deaths. Climate warming may also affect human health indirectly. Mosquitoes and other disease carriers could expand their range into the newly warm areas and spread malaria, dengue fever, schistosomiasis, and yellow fever (FIGURE 9.10). According to the World Health Organization, during 1998, the second warmest year on record, the incidence of malaria, Rift Valley fever, and cholera surged in developing countries.

An increasing number of studies report measurable changes in the biology of plant and animal species as a result of climate warming. Climate change also affects populations, communities, and ecosystems. In "Visualizing the effects of global climate change" (on facing page), we report on the results of several of the hundreds of studies conducted thus far.

Rising temperatures in the waters around Antarctica have led to a decline in the populations of shrimplike krill and Antarctic silverfish, major food sources for Adélie penguins. This decline has contributed to a reduction in Adélie penguin populations (FIGURE 9.11A). Warmer temperatures also cause higher rates of reproductive failure in these penguins by producing wet, snowy conditions that kill the developing chick embryos at egg-laying sites.

Worldwide, many frog populations have plummeted; these include Puerto Rico's national symbol, the tiny tree frog known as coqui (FIGURE 9.11B). Warmer temperatures and more frequent dry periods have stressed the coqui, making them more vulnerable to infection by a lethal fungus.

Ecosystems considered at greatest risk of climate-change loss are polar seas, coral reefs, mountain ecosystems, coastal wetlands, and tundra. Water temperature increases of 1° to 2°C (1.8° to 3.6°F) cause coral bleaching, which contributes to the destruction of coral reefs (FIGURE 9.11C). In 1998, when tropical waters were some of the warmest ever recorded, about 10 percent of the world's corals died.

As atmospheric CO_2 levels continue to rise, the level of CO_2 that dissolves in the ocean increases. Dissolved CO_2 produces carbonic acid, which acidifies the water (FIGURE 9.12). This acidification could be disastrous for shelled sea animals, particularly zooplankton at the base of the marine food web; the acid would attack and dissolve away their shells.

Global climate change affects organisms
FIGURE 9.11

A Warmer temperatures in Antarctica threaten the Adélie penguin's food supply and reduce its reproductive success.

B A coqui tree frog in Puerto Rico. These once-ubiquitous little frogs have become rarer, an indirect casualty of climate change.

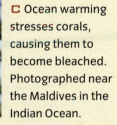

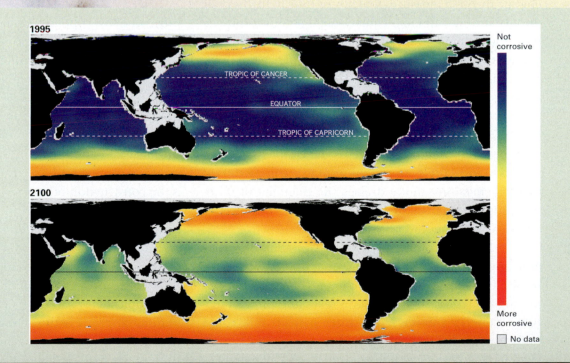

C Ocean warming stresses corals, causing them to become bleached. Photographed near the Maldives in the Indian Ocean.

Acidification of the ocean FIGURE 9.12

Scientific models project that ocean water will become increasingly acidic if human-produced CO_2 levels continue to rise. Shown are computer models for 1995 and 2100 from the NOAA Pacific Marine Environmental Laboratory.

1995

Not corrosive

TROPIC OF CANCER

EQUATOR

TROPIC OF CAPRICORN

2100

More corrosive

No data

NATIONAL GEOGRAPHIC

Species such as weeds, insect pests, and disease-carrying organisms already common in a wide range of environments will come out of global climate change as winners, with greatly expanded numbers and ranges.

SPECIFIC WAYS TO DEAL WITH GLOBAL CLIMATE CHANGE

All greenhouse gases will have to be dealt with as we develop strategies to address global climate change, but we focus on CO_2 because it is produced in the greatest quantity and has the largest total effect. Carbon dioxide has an atmospheric lifetime of more than a century, so emissions produced today will still be around in the 22nd century. The extent and severity of global climate change will depend on the amount of additional greenhouse gas emissions we add to the atmosphere. Many studies make the assumption that atmospheric CO_2 will stabilize at 550 ppm, which is roughly twice the concentration of atmospheric CO_2 in the preindustrial world and almost 50 percent higher than the CO_2 currently in the atmosphere.

There are basically two ways to manage global climate change: mitigation and adaptation. *Mitigation* is the moderation or postponement of global climate change through measures that buy us time to further our understanding of global climate change and to pursue more permanent solutions. *Adaptation* is a response to changes caused by global climate change. Adopting adaptation strategies for climate warming implies that the consequences of climate change are unavoidable.

Mitigation of global climate change
Climate warming is essentially an energy issue. Developing alternatives to fossil fuels offers a solution to warming caused by CO_2 emissions and addresses the dilemma of dwindling supplies of fossil fuels. Alternatives to fossil fuels include solar energy and nuclear energy.

Increasing the energy efficiency of automobiles and appliances—thus reducing the output of CO_2—would help mitigate global climate change. Energy-pricing strategies, such as carbon taxes and the elimination of energy subsidies, are other policies that could mitigate global climate change. Most experts think using current technologies and developing such policies would significantly reduce greenhouse gas emissions with little cost to society.

Planting and maintaining forests also mitigates global climate change. Like other green plants, trees remove carbon dioxide from the air and incorporate the carbon into organic matter through photosynthesis. Reasonable estimates suggest that trees could remove 10 percent to 15 percent of the excess CO_2 in the atmosphere, but only through enormous plantings, so such efforts should not be considered a substitute for cutting emissions of greenhouse gases.

Many countries are investigating **carbon management**. Several power plants currently capture the CO_2 in their flue gases, but the technology is new. Technological innovations that more efficiently trap CO_2 from smokestacks would help mitigate global climate change and yet allow us to continue using fossil fuels (while they last) for energy. The carbon could be sequestered in geologic formations or in depleted oil or natural gas wells on land. **FIGURE 9.13** summarizes several ways to mitigate global climate change.

> ■ **carbon management**
> Ways to separate and capture the CO_2 produced during the combustion of fossil fuels and then sequester (store) it.

Adaptation to global climate change
Because the overwhelming majority of climate experts think human-induced global climate change is inevitable, government planners and social scientists are developing strategies to help various regions and sectors of society adapt to climate warming. One of the most pressing issues is rising sea level. People living in coastal areas could be moved inland, away from the dangers of storm surges, although the societal and economic costs would be great. Another extremely expensive alternative is the construction of massive sea walls to protect coastal land. Rivers and canals that spill into the ocean could be channeled to prevent *saltwater intrusion* into fresh water and agricultural land.

We must also adapt to shifting agricultural zones. Countries with temperate climates are evaluating semitropical crops to determine the best substitutes for traditional crops as the climate warms. Large lumber companies are developing heat- and drought-resistant strains of trees that will be harvested when global climate

Install natural gas–fired plants to replace coal-fired plants.

Develop motor vehicles powered with hydrogen obtained from renewable sources.

Increase use of solar and wind power to replace coal-fired power plants.

Install more nuclear power plants to replace coal-fired power plants.

Increase energy efficiency of coal-fired power plants.

Conserve use of electricity in homes, offices, and stores.

Capture CO_2 at coal–fired power plants and store it.

Increase fuel economy of motor vehicles.

Plant tree on degraded lands.

Stop all deforestation worldwide.

Some of the many strategies suggested for mitigation of global climate change FIGURE 9.13

To be effective these strategies will have to be enacted on a global scale. For example, replacing coal-fired power plants with those powered by natural gas (upper-left corner) would have to involve 1400 or more coal-fired plants around the world.

change may be well advanced. Evaluating such problems and finding and implementing solutions now will ease future stresses of climate warming.

Adaptation to global climate change is under study at 15 locations around the United States. One of the problems identified in the New York City study involves its sewer system. The waterways for storm runoff normally close during high tides. As the sea level rises in response to global climate change, the waterways will have to be shut during many low tides, which will increase the risk of flooding during storms (because excess water will not drain away). City planners will have to rebuild the storm runoff system or find some other way to prevent flooding. Evaluating such problems and implementing solutions now will ease future stresses of climate warming.

CONCEPT CHECK STOP

What is the enhanced greenhouse effect? How does it affect global climate?

What are greenhouse gases?

What are two examples of each of the approaches to global climate change: mitigation and adaptation?

Ozone Depletion in the Stratosphere

LEARNING OBJECTIVES

Describe the importance of the stratospheric ozone layer.

Explain how ozone thinning takes place and relate some of its harmful effects.

Relate how the international community is working to protect the ozone layer.

lthough ozone (O_3) is a human-made pollutant in the troposphere, it is a naturally produced, essential component in the stratosphere, which encircles our planet some 10 to 45 km (6 to 28 mi) above the surface. The ozone layer shields Earth's surface from much of the high-energy **ultraviolet (UV) radiation** coming from the Sun (**FIGURE 9.14** in "Visualizing the ozone layer"). If ozone disappeared from the stratosphere, Earth would become uninhabitable for most forms of life.

A slight **ozone thinning** occurs naturally over Antarctica for a few months each year. In 1985, however, the thinning was first observed to be greater than it should be if natural causes were the only factor inducing it. This increased thinning, which occurs each September, is commonly referred to as the "ozone hole" (**FIGURE 9.15**). There, ozone levels decrease as much as 70 percent each year.

During the 1990s the ozone-thinned area continued to grow, and by 2000 it had reached the record size of 29.2 million km^2 (11.3 million mi^2), which is larger than the North American continent. A smaller thinning was also detected in the stratospheric ozone layer over the Arctic. In addition, world levels of stratospheric ozone have been decreasing for several decades (**FIGURE 9.16**). According to the National Center for Atmospheric Research, ozone levels over Europe and North America have dropped almost 10 percent since the 1970s.

CAUSES OF OZONE DEPLETION

The primary chemicals responsible for ozone loss in the stratosphere are a group of industrial and commercial compounds called **chlorofluorocarbons (CFCs)**. Chlorofluorocarbons such as Freon have been used as propellants for aerosol cans and coolants in air conditioners and refrigerators. Other CFCs have been used as solvents and as foam-blowing agents for insulation and packaging (Styrofoam, for example).

Other compounds that destroy ozone include *halons*, used as fire retardants; *methyl bromide*, a pesticide; *methyl chloroform* and *carbon tetrachloride*, industrial solvents; and *nitrous oxide*, released from the burning of fossil fuels (particularly coal) and from the breakdown of nitrogen fertilizers in the soil.

EFFECTS OF OZONE DEPLETION

With depletion of the ozone layer, higher levels of UV radiation reach the surface of Earth. Increased levels of UV radiation may disrupt ecosystems. For example, the productivity of Antarctic phytoplankton, the microscopic drifting algae that are the base of the Antarctic food web, has declined from increased exposure to UV radiation. (The UV radiation inhibits photosynthesis.) Biologists have documented direct UV damage to natural populations of Antarctic fish. A possible link also exists between the widespread decline of amphibian populations and increased UV radiation levels. Because

ultraviolet (UV) radiation Radiation from the part of the electromagnetic spectrum with wavelengths just shorter than visible light; can be lethal to organisms at high levels of exposure.

ozone thinning The removal of ozone from the stratosphere by human-produced chemicals or natural processes.

chlorofluorocarbons (CFCs) Human-made organic compounds that contain chlorine and fluorine; now banned because they attack the stratospheric ozone layer.

Stratospheric ozone layer FIGURE 9.14

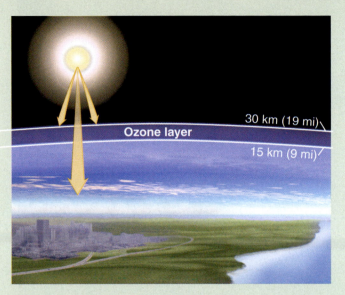

A Stratospheric ozone absorbs about 99 percent of incoming solar ultraviolet (UV) radiation, effectively shielding the surface.

B When stratospheric ozone is present at reduced levels, more high-energy UV radiation penetrates the atmosphere to the surface, where its presence harms organisms.

Ozone depletion FIGURE 9.15

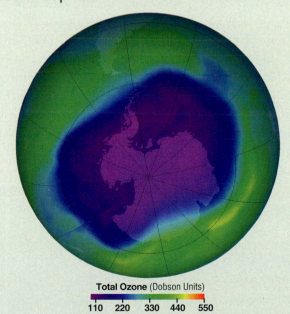

Total Ozone (Dobson Units)

110 220 330 440 550

A computer-generated image of part of the Southern Hemisphere, taken in December 2007 reveals ozone thinning (the purple area over Antarctica). The ozone-thin area is not stationary but moves about as a result of air currents.

World levels of stratospheric ozone, 1980 to 2005
FIGURE 9.16

This graph shows data recorded every five years. Dobson units are used to measure ozone.

VIEW THIS IN ACTION
in your WileyPLUS course

Exposure to UV radiation and skin cancer
FIGURE 9.17

This Australian worked outdoors throughout his career, and it was perhaps inevitable that he contracted skin cancer. Here he attaches an artificial nose where doctors removed a large melanoma.

organisms are interdependent, the negative effect on one species has ramifications throughout the ecosystem.

Excessive exposure to UV radiation is linked to several health problems in humans, including eye cataracts, skin cancer (FIGURE 9.17), and weakened immunity. Malignant melanoma, the most dangerous type of skin cancer, is increasing faster than any other type of cancer.

HELPING THE OZONE LAYER RECOVER

In 1978 the United States, the world's largest user of CFCs, banned the use of CFC propellants in products such as antiperspirants and hair sprays. Although this ban was a step in the right direction, it did not solve the problem. Most nations did not follow suit, and besides, propellants represented only a small portion of all CFC use.

In 1987, representatives from many countries met in Montreal to sign the **Montreal Protocol**, an agreement that originally stipulated a 50 percent reduction of CFC production by 1998. Despite this effort, stratospheric ozone continued to thin over the heavily populated mid-latitudes of the Northern Hemisphere, and the Montreal Protocol was modified to include even stricter limits on CFC production.

Industrial companies that manufacture CFCs quickly developed substitutes, such as hydrofluorocarbons (HFCs) and hydrochlorofluorocarbons (HCFCs). HFCs do not attack ozone, although they are potent greenhouse gases. HCFCs attack ozone but are less destructive than the chemicals they are replacing.

CFC, carbon tetrachloride, and methyl chloroform production was almost completely phased out in the United States and other highly developed countries in 1996, except for a relatively small amount exported to developing countries. Developing countries were on a different timetable and phased out CFC use by 2005. Methyl bromide was phased out in highly developed countries, which were responsible for 80 percent of its global use, by 2005. HCFCs will be phased out in 2030.

Unfortunately, CFCs are extremely stable and will probably continue to deplete stratospheric ozone for several decades. Human-exacerbated ozone thinning will reappear over Antarctica each year, although the area and degree of thinning will gradually decline over time, until full recovery takes place sometime after 2050.

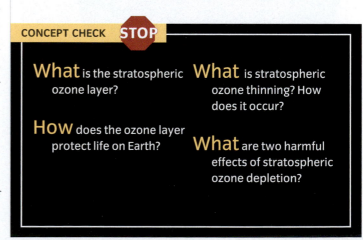

CONCEPT CHECK STOP

What is the stratospheric ozone layer?

How does the ozone layer protect life on Earth?

What is stratospheric ozone thinning? How does it occur?

What are two harmful effects of stratospheric ozone depletion?

Most environmental studies examine a single issue, such as acid deposition, global climate change, or ozone depletion. In the past few years, however, some researchers have been exploring the interactions of all three problems simultaneously. One recent study of such interactions found that North American lakes may be more susceptible to damage from UV radiation than the thinning of the ozone hole would indicate. The reason: The organic matter in the lakes, which absorbs some UV radiation and protects the lakes' plant and fish life, is affected by acid deposition and global climate change. Acid deposition reacts with organic matter in lakes, causing it to settle to the lake floor, where it does not absorb as much of the UV radiation as it once did. And a warmer climate increases evaporation, which reduces the amount of organic matter washed into lakes by streams.

Several studies report a link between human-caused climate warming and polar ozone depletion. Greenhouse gases that warm the troposphere also contribute to stratospheric cooling, presumably because heat trapped in the troposphere is not available to warm the stratosphere. The stratospheric temperature has been dropping for the past several years, and these lower temperatures provide better conditions for ozone-depleting chemicals to attack stratospheric ozone. Record ozone holes over Antarctica are attributed to cooler stratospheric temperatures. Some scientists speculate that if the cooling trend in the stratosphere continues, recovery of the ozone layer may be delayed. This means climate warming could prolong ozone depletion in the stratosphere despite the success of the Montreal Protocol.

Scientists now know environmental problems can't be studied as separate issues because they often interact in surprisingly subtle ways. As global climate change, ozone depletion, and acid deposition are studied further, it is likely that other interactions will be discovered.

Acid Deposition

LEARNING OBJECTIVES

Define *acid deposition* and explain how acid deposition develops.

Relate some of the effects of acid deposition.

What do fishless lakes in the Adirondack Mountains, recently damaged Mayan ruins in southern Mexico, and dead trees in the Czech Republic have in common? All these problems are the result of acid precipitation or, more properly, **acid deposition**. Acid deposition has been around since the Industrial Revolution began. Robert Angus Smith, a British chemist, coined the term *acid rain* in 1872 after he noticed that buildings in areas with heavy industrial activity were being worn away by rain.

Acid precipitation, including acid rain, sleet, snow, and fog, poses a serious threat to the environment. Until recently, industrialized countries in the Northern Hemisphere had been hurt the most, especially the Scandinavian countries, central Europe, Russia, and North America. In the United States alone, the annual damage from acid deposition is estimated at $10 billion. Acid deposition is now recognized as a global problem because it also occurs in developing countries as they become industrialized. For example, Chinese scientists have reported that acid deposition affects 40 percent of their country. In the two decades from 1990 to 2010, the amount of sulfur dioxide released from burning high-sulfur coal will have tripled in China. Sulfur dioxide becomes sulfuric acid in the atmosphere.

> **acid deposition**
> A type of air pollution that includes sulfuric and nitric acids in precipitation, as well as dry acid particles that settle out of the air.

HOW ACID DEPOSITION DEVELOPS

Acid deposition occurs when sulfur dioxide and nitrogen oxides are released into the atmosphere (**FIGURE 9.18**). Motor vehicles are a major source of nitrogen oxides. Coal-burning power plants, large smelters, and industrial boilers are the main sources of sulfur dioxide emissions and produce substantial amounts of nitrogen oxides as well. Wind carries sulfur dioxide and nitrogen oxides, released into the air from tall smokestacks, for long distances. Tall smokestacks allow England to "export" its acid deposition problem to the Scandinavian countries and the midwestern United States to "export" its acid emissions to New England and Canada.

In the atmosphere, sulfur dioxide and nitrogen oxides react with water to produce dilute solutions of sulfuric acid (H_2SO_4), nitric acid (HNO_3), and nitrous acid (HNO_2). Acid deposition returns these acids to Earth's surface in the form of precipitation or particulates.

EFFECTS OF ACID DEPOSITION

Acid deposition corrodes metals, building materials, and statues (**FIGURE 9.19**). It eats away at important monuments, such as the Washington Monument in Washington, DC, and ancient Mayan ruins in southern Mexico.

The link between acid deposition and declining aquatic animal populations, particularly fish, is well established, but other animals are also adversely affected. Birds living in areas with pronounced acid deposition are at increased risk of laying eggs with thin, fragile shells that break or dry out before the chicks hatch. The inability to produce strong eggshells is attributed to reduced calcium in the birds' diets. Calcium is less available to the food chain because in acidic soils calcium becomes soluble and is washed away, with little left for plant roots to absorb.

Acid deposition also has a serious effect on forest ecosystems. In the Black Forest of Germany, for example, up to 50 percent of trees surveyed are dead or severely damaged. This **forest decline** appears to result from a combination of stressors, including tropospheric ozone, UV radiation (which is more intense at higher altitudes), in-

> **forest decline**
> A gradual deterioration and eventual death of many trees in a forest.

sect attack, drought, and acid deposition. When one or more stressors weaken a tree, an additional stressor, such as air pollution, may be decisive in causing the tree's death (**FIGURE 9.20**).

THE POLITICS OF ACID DEPOSITION

Acid deposition is hard to combat because it does not occur only in the locations where acidic gases are emitted. It is entirely possible for sulfur and nitrogen oxides released in one spot to return to Earth's surface hundreds of kilometers from their source.

The United States has wrestled with this issue. Several states in the Midwest and East—Illinois, Indiana, Missouri, Ohio, Pennsylvania, Tennessee, and West Virginia—produce between 50 percent and 75 percent of the acid deposition that contaminates New England and southeastern Canada. Legislation formulated to deal with acid deposition causes arguments about who should pay for the installation of expensive devices to reduce emissions of sulfur and nitrogen oxides. Should the states emitting the gases be required to pay all the expenses to clean up the air, or should the areas that stand to benefit most from the cleaner air absorb some of the cost?

In international disputes, these issues are magnified even more. For example, gases from coal-burning power plants in England move eastward with prevailing winds and return to the surface as acid deposition in Sweden and Norway. Similarly, emissions from mainland China produce acid deposition in Japan, Taiwan, North Korea, and South Korea. Generally, conflicts over international air pollution cannot be resolved by local legislation.

FACILITATING RECOVERY FROM ACID DEPOSITION

Although the science and the politics surrounding acid deposition are complex, the basic concept of control is straightforward: Reducing emissions of sulfur and nitrogen oxides curbs acid deposition. Simply stated, if sulfur and nitrogen oxides are not released into the atmosphere, they cannot come down as acid deposition. Installing scrubbers in the smokestacks of coal-fired power

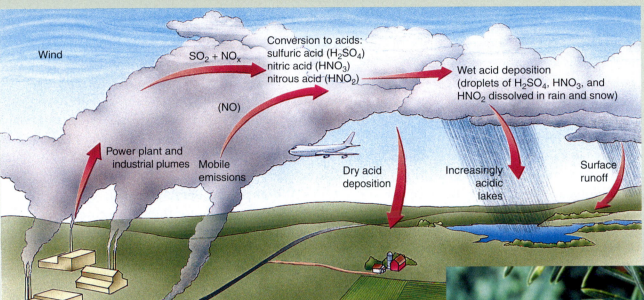

Wind

$SO_2 + NO_x$

Conversion to acids: sulfuric acid (H_2SO_4) nitric acid (HNO_3) nitrous acid (HNO_2)

(NO)

Wet acid deposition (droplets of H_2SO_4, HNO_3, and HNO_2 dissolved in rain and snow)

Power plant and industrial plumes

Mobile emissions

Dry acid deposition

Increasingly acidic lakes

Surface runoff

Acid deposition Figure 9.18

Sulfur dioxide and nitrogen oxide emissions react with water vapor in the atmosphere to form acids that return to the surface as either dry or wet deposition.

A Healthy Sitka spruce branch.

◄ Acid rain damage

Figure 9.19

Acid rain has destroyed much of the detail of this stone angel's face. Photographed in London.

VIEW THIS IN ACTION in your WileyPLUS course

Forest decline Figure 9.20 ▶

Acid deposition is one of several stressors that may interact, contributing to the decline and death of trees.

B Sitka spruce branch exhibiting the effects of forest decline. Photographed in Black Forest, Germany.

plants and using clean-coal technologies to burn coal without excessive emissions effectively diminish acid deposition. In turn, a decrease in acid deposition prevents surface waters and soil from becoming more acidic than they already are.

Rainfall in parts of the Midwest, Northeast, and Mid-Atlantic regions is less acidic today than it was a few years ago as a result of cleaner-burning power plants and the use of reformulated gasoline. Many power plants in the Ohio Valley switched from high-sulfur to low-sulfur coal. However, solving one environmental problem often creates others. While the move to low-sulfur coal reduced sulfur emissions, it contributed to the problem of global climate change. Because low-sulfur coal has a lower heat value than high-sulfur coal, more of it must be burned—and more CO_2 emitted—to generate a given amount of electricity. Low-sulfur coal also contains higher levels of mercury and other trace metals, so burning it adds more of these hazardous pollutants to the air.

Despite the fact that the United States, Canada, and many European countries have reduced sulfur emissions, acid precipitation remains a serious problem. Acidified forests and bodies of water have not recovered as quickly as hoped. Trees in the U.S. Forest Service's Hubbard Brook Experimental Forest in New Hampshire, an area damaged by acid deposition, have grown little since 1987, when emissions started to decline. Many northeastern streams and lakes, such as those in New York's Adirondack Mountains, remain acidic (**FIGURE 9.21**). A likely reason for the slow recovery is that the past 30 or more years of acid rain have profoundly altered soil chemistry in many areas. Essential plant minerals such as calcium and magnesium have washed away from forest and lake soils. Because soils take hundreds or even thousands of years to develop, it may be decades or centuries before they recover from the effects of acid rain.

Many scientists are convinced that ecosystems will not recover from acid rain damage until substantial reductions in nitrogen oxide emissions occur. Nitrogen oxide emissions are harder to control than sulfur dioxide emissions because motor vehicles produce a substantial portion of nitrogen oxides. Engine improvements may help reduce nitrogen oxide emissions, but as the human population continues to grow, the increasing number of motor vehicles will probably offset any engineering gains. Dramatic cuts in nitrogen oxide emissions will require a reduction in high-temperature energy generation, especially in gasoline and diesel engines.

Acidic stream FIGURE 9.21

This mountain stream in the Adirondack Mountains of New York may be acidic because years of acid precipitation have altered soil chemistry.

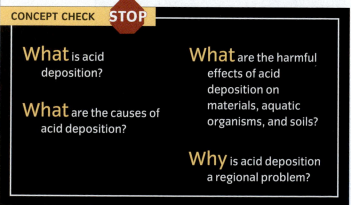

CONCEPT CHECK STOP

What is acid deposition?

What are the causes of acid deposition?

What are the harmful effects of acid deposition on materials, aquatic organisms, and soils?

Why is acid deposition a regional problem?

INTERNATIONAL IMPLICATIONS OF GLOBAL CLIMATE CHANGE

Various social, economic, and political factors complicate international efforts to deal with global climate change. Although highly developed countries are the primary producers of greenhouse gases, many developing countries are rapidly increasing production as they industrialize. But because developing countries have less technical expertise and fewer economic resources, they are less able to respond to the challenges of global climate change.

Tensions exist among nations, especially between the highly developed and developing countries, over their differing self-interests. Most developing countries view fossil fuels as their route to industrial development and resist pressure from highly developed nations to decrease fossil fuel consumption. Developing countries also question the need to curb their CO_2 emissions when rich industrialized nations such as the United States are the main cause of the problem. Currently, highly developed countries produce about five or six times more CO_2 emissions per person than developing countries (see figure). However, according to the U.S. Department of Energy, CO_2 emissions from developing countries will surpass those from highly developed countries by 2020 if current trends in fossil fuel consumption continue.

Despite all the posturing, the international community recognizes that it must stabilize CO_2 emissions. At least 174 nations, including the United States, signed the U.N. Framework Convention on Climate Change developed at the 1992 Earth Summit. At the 1996 U.N. Climate Change Convention held in Geneva, Switzerland, highly developed countries agreed to establish legally binding timetables to cut emissions of greenhouse gases. In 1997 representatives from 160 countries determined these timetables in Kyoto, Japan.

By 2005 enough countries had ratified the Kyoto Protocol for it to come into force. This international treaty provides operational rules on reducing greenhouse gas emissions. It is noteworthy that the United States has not yet ratified the Kyoto Protocol. Most analysts think the Kyoto Protocol will accomplish little without the full participation of the United States. The United States signed the Kyoto Protocol in 1998, but the administration of President George W. Bush withdrew from the commitment in 2001 because of the perceived economic burden that the protocol might inflict. However, many climate experts think that the costs of climate warming—such as damage to agriculture and human health—make enactment of the Kyoto Protocol economically beneficial.

Per-person carbon dioxide (CO_2) emission estimates for selected countries, 1990 and 2003

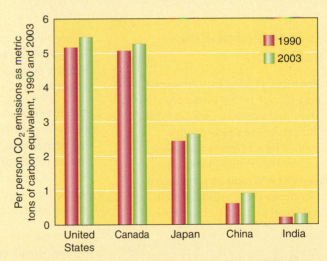

Per person CO_2 emissions have grown from 1990 to 2003. Currently, industrialized nations produce a disproportionate share of CO_2 emissions. As developing nations such as China and India industrialize, however, their per person CO_2 emissions increase.

SUMMARY

1 The Atmosphere and Climate

1. **Weather** is the conditions in the atmosphere at a given place and time; it includes temperature, atmospheric pressure, precipitation, cloudiness, humidity, and wind. **Climate** is the average weather conditions that occur in a place over a period of years. The two most important factors that determine an area's climate are temperature and precipitation.

2. Sunlight is the primary (almost sole) source of energy available in the biosphere. The Sun's energy runs the hydrologic cycle, drives winds and ocean currents, powers photosynthesis, and warms the planet. Of all the solar energy that reaches Earth, 31 percent is immediately reflected away, and the remaining 69 percent is absorbed. Ultimately, all absorbed solar energy is radiated into space as **infrared radiation**, electromagnetic radiation with wavelengths longer than those of visible light but shorter than microwaves; it is perceived as invisible waves of heat energy.

3. **Precipitation** is greatest where warm air passes over the ocean, absorbing moisture, and is then cooled, such as when mountains force humid air upward. Deserts develop in the **rain shadows** of mountain ranges or in continental interiors.

2 Global Climate Change

1. **Greenhouse gases** are gases that absorb infrared radiation; they include carbon dioxide, methane, nitrous oxide, chlorofluorocarbons, and tropospheric ozone. The **enhanced greenhouse effect** is the additional warming that may be produced by human-increased levels of gases that absorb infrared radiation.

2. Global climate change will probably cause a rise in sea level, changes in precipitation patterns, extinction of many species, and problems for agriculture. It could result in the displacement of millions of people, thereby increasing international tensions.

3. Mitigation (slowing down the rate of global climate change) and adaptation (making adjustments to live with climate change) are two ways to address climate change. Mitigation includes developing alternatives to fossil fuels; increasing energy efficiency of automobiles and appliances; planting and maintaining forests; and instigating **carbon management**, by finding ways to separate and capture the CO_2 produced during the combustion of fossil fuels and then sequester it. Adaptation includes developing strategies to help various regions and sectors of society adapt to climate warming.

3 Ozone Depletion in the Stratosphere

1. Ozone (O_3) is a human-made pollutant in the troposphere but a naturally produced, essential component in the stratosphere. The stratosphere contains a layer of ozone that shields the surface from much of the Sun's **ultraviolet (UV) radiation**, that part of the electromagnetic spectrum with wavelengths just shorter than those of visible light; UV radiation is a high-energy form of radiation that can be lethal to organisms at high levels of exposure.

2. **Ozone thinning** is the natural and human-caused removal of ozone from the stratosphere. The primary chemicals responsible for ozone thinning in the stratosphere are **chlorofluorocarbons (CFCs)**, human-made organic compounds that contain chlorine and fluorine. CFCs are now banned because they attack the stratospheric ozone layer. Ozone thinning causes excessive exposure to UV radiation, which may increase cataracts, weaken immunity, and cause skin cancer in humans. Increased levels of UV radiation may also disrupt ecosystems.

3. The Montreal Protocol resulted in an international agreement to phase out CFC production.

4 Acid Deposition

1. **Acid deposition** is a type of air pollution that includes sulfuric and nitric acids in precipitation as well as dry acid particles that settle out of the air. Acid deposition develops when sulfur and nitrogen oxides are released into the air, where they react to form acids and then return to surface waters and soil.

2. Acid deposition kills aquatic organisms, changes soil chemistry, and may contribute to **forest decline**, a gradual deterioration and eventual death of many trees in a forest.

CRITICAL AND CREATIVE THINKING QUESTIONS

1. How does the Sun affect temperature at different latitudes? Why?

2. On the basis of what you know about the nature of science, can we say with absolute certainty that the increased production of greenhouse gases is causing global climate change? Why or why not?

3. Biologists who study plants growing high in the Alps found that plants adapted to cold-mountain conditions migrated up the peaks as fast as 3.7 m (12.1 ft) per decade during the 20th century, apparently in response to climate warming. Assuming that warming continues during the 21st century, what will happen to the plants if they reach the tops of the mountains?

4. Some environmentalists contend that the wisest "use" of fossil fuels is to leave them in the ground. How would this affect air pollution? global climate change? energy supplies?

5. Distinguish between the benefits of the ozone layer in the stratosphere and the harmful effects of ozone at ground level.

6. What is the Montreal Protocol? the Kyoto Protocol?

7. Discuss some of the possible causes of forest decline. How might these factors interact to speed the rate of decline?

8–10. This map shows one model of how warmer global temperatures might alter precipitation in the United States in the next 100 years. Colors indicate the percent change in annual precipitation per century.

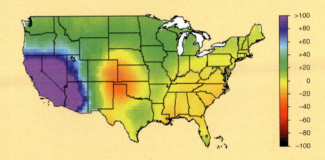

8. Name three states with climates that may become significantly wetter. Name three states with climates that may become significantly drier. Explain your answer.

9. Locate the state where you live. Will it be wetter, drier, or about the same?

10. Why are the projected changes in precipitation in the next 100 years a threat to U.S. agriculture? to the U.S. economy?

What is happening in this picture ?

- These scientists have drilled into a glacier in Greenland to remove an ice core. Do you think the ice deep within the glacier is old or relatively young? Explain your answer.

- Some of the deeper samples were laid down thousands of years ago, when the climate was much cooler. The ice contains bubbles of air. Based on what you have learned in this chapter, do you think the level of carbon dioxide in the air bubbles in the oldest ice is higher or lower than the level in today's atmosphere? Explain your answer.

- If the scientists compared CFCs in the air bubbles in the oldest ice with today's levels of CFCs, what do you think they would find?

Freshwater Resources and Water Pollution

THE MISSOURI RIVER: A BATTLE OVER WATER RIGHTS

VIEW THIS IN ACTION in your WileyPLUS course

The Missouri River flows from Montana to St. Louis, Missouri, where it joins the Mississippi River. The longest river in the United States, the Missouri drains about one-sixth of the country. The six dams on the Missouri offer both benefits and problems for people living along the river.

Since 1987 the Army Corps of Engineers has increased water flow over the northern dams to protect downstream navigation, including 2 million tons of cargo shipped by boat each year. In addition, people who live downstream count on the river water for irrigation, electrical power, and individual water consumption (see inset, of Kansas City, Missouri). On the other hand, the scenic area along the northern Missouri River depends on the river for its multimillion-dollar fishing and tourism industry (see larger photo). The battle over water rights has intensified as both upstream and downstream parties fight to protect their interests.

Complicating the issue is a growing confrontation between environmentalists and farmers. Farmers want to add more dikes and levees to protect their crops from flood damage, such as that experienced in the Upper Midwest (Mississippi River drainage) in 2008, whereas environmentalists want to restore the river to its natural state, in part to protect native fish populations now in decline. Native Americans with claims to water rights want to use the water in a variety of ways, from generating hydroelectric power to irrigating cropland. Such water-use conflicts arise in watersheds around the world.

The Missouri River Basin Association, a coalition of eight river-basin states and two dozen Native American tribes, works with the Corps to meet the demands of the competing interest groups while deciding the river's future.

NATIONAL GEOGRAPHIC

The Importance of Water

LEARNING OBJECTIVES

Describe the structure of a water molecule and explain how hydrogen bonds form between adjacent water molecules.

List the unique properties of water.

Explain how processes of the hydrologic cycle allow water to circulate through the abiotic environment.

L ife on planet Earth would be impossible without water. All life forms, from unicellular bacteria to multicellular plants and animals, contain water. Humans are composed of approximately 60 percent water by body weight. We depend on water for our survival as well as for our convenience: We drink it, cook with it, wash with it (**FIGURE 10.1**), travel on it, and use an enormous amount of it for agriculture, manufacturing, mining, energy production, and waste disposal.

Although Earth has plenty of water, about 97 percent of it is salty and not consumable by most terrestrial organisms. Fresh water is distributed unevenly, resulting in serious regional water supply problems. Water experts predict that by 2025, more than one-third of the human population will live in areas where there isn't enough fresh water for drinking and irrigation.

THE HYDROLOGIC CYCLE AND OUR SUPPLY OF FRESH WATER

In the **hydrologic cycle**, water continuously circulates through the environment, from the ocean to the atmosphere to the land and back to the ocean (see **FIGURE 10.2**; also see Figure 5.9). The result is a balance of the water resources in the ocean, on the land, and in the atmosphere. The hydrologic cycle provides a continual renewal of the supply of fresh water on land, which is essential to terrestrial organisms.

Young brick workers in India bathe with water from an irrigation pipe FIGURE 10.1

Two important components of the hydrologic cycle FIGURE 10.2

A Liquid and solid precipitation continuously falls from the atmosphere to the land and ocean.

B Evaporation continuously moves water vapor from the land and ocean into the atmosphere.

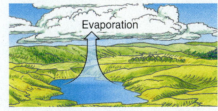

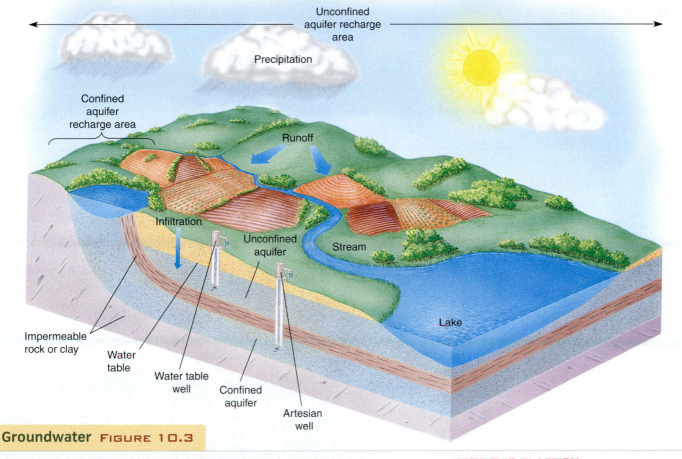

Unconfined aquifer recharge area

Precipitation

Confined aquifer recharge area

Runoff

Infiltration

Unconfined aquifer

Stream

Lake

Impermeable rock or clay

Water table

Water table well

Confined aquifer

Artesian well

Groundwater FIGURE 10.3

Excess surface water seeps downward through soil and porous rock layers until it reaches impermeable rock or clay. An unconfined aquifer has groundwater recharged by surface water directly above it. In a confined aquifer, groundwater is stored between two impermeable layers.

VIEW THIS IN ACTION in your WileyPLUS course

Surface water is water found in streams, rivers, lakes, ponds, reservoirs, and **wetlands** (areas of land covered with water for at least part of the year). The **runoff** of precipitation from the land replenishes surface waters and is considered a renewable, although finite, resource. A **drainage basin**, or **watershed**, is the area of land drained by a single river or stream. Watersheds range in size from less than 1 km² for a small stream to a huge portion of the continent for a major river system such as the Mississippi River.

Earth contains underground formations that collect and store water. This water originates as rain or melting snow that slowly seeps into the soil. It works its way down through cracks and spaces in sand, gravel, or rock until an impenetrable layer stops it; there it accumulates as **groundwater**. Groundwater is eventually discharged into rivers, wetlands, springs, or the ocean. Thus, surface water and groundwater are interrelated parts of the hydrologic cycle. **Aquifers** are underground reservoirs in which groundwater is stored (FIGURE 10.3).

Most groundwater is considered a nonrenewable resource because it has taken hundreds or even thousands of years to accumulate, and usually only a small portion of it is replaced each year by seepage of surface water.

surface water
Precipitation that remains on the surface of the land and does not seep down through the soil.

runoff
The movement of fresh water from precipitation and snowmelt to rivers, lakes, wetlands, and the ocean.

groundwater
The supply of fresh water under Earth's surface that is stored in underground aquifers.

PROPERTIES OF WATER

Water is composed of molecules of H$_2$O, each consisting of two atoms of hydrogen and one atom of oxygen. Water molecules are **polar**—that is, one end of the molecule has a positive electrical charge, and the other end has a negative charge (**FIGURE 10.4**). The negative (oxygen) end of one water molecule is attracted to the positive (hydrogen) end of another water molecule, forming a **hydrogen bond** between the two molecules. Hydrogen bonds are the basis for many of water's physical properties, including its high melting/freezing point (0°C, 32°F) and high boiling point (100°C, 212°F). Because most of Earth has a temperature between 0°C and 100°C, most water exists in the liquid form organisms need.

Water absorbs a great deal of solar heat without substantially increasing in temperature. This high heat capacity allows the ocean to have a moderating influence on climate, particularly along coastal areas.

Water is a *solvent*, meaning that it can dissolve many materials. In nature, water is never completely pure because it contains dissolved gases from the atmosphere and dissolved mineral salts from the land. Water's abilities as a solvent have a major drawback: Many of the substances that dissolve in water cause water pollution.

Chemical properties of water FIGURE 10.4

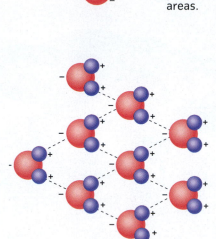

A Water molecules are polar, with positively and negatively charged areas.

B The polarity causes hydrogen bonds (represented by dashed lines) to form between the positive areas of one water molecule and the negative areas of others. Each water molecule forms up to four hydrogen bonds with other water molecules.

CONCEPT CHECK STOP

How do hydrogen bonds form between adjacent water molecules?

What is surface water? groundwater?

What are the unique properties of water?

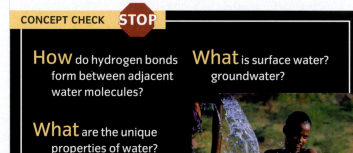

Water Resource Problems

LEARNING OBJECTIVES

Relate some of the problems caused by aquifer depletion, overdrawing surface waters, and salinization of irrigated soil.

Relate the background behind the water problems of the Ogallala Aquifer and the Colorado River Basin.

Briefly describe the role of international cooperation in managing shared water resources.

Water resource problems fall into three categories: too much water, too little water, and poor-quality water. Flooding occurs when a river's discharge cannot be contained within its normal channel. Today's floods are more disastrous in terms of property loss than those of the past because humans often remove water-absorbing plant cover from the soil and construct buildings on flood plains. (A **flood plain** is the area bordering a river channel that has the potential to flood.) These activities increase the likelihood of both floods and flood damage.

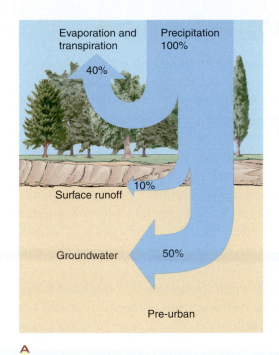

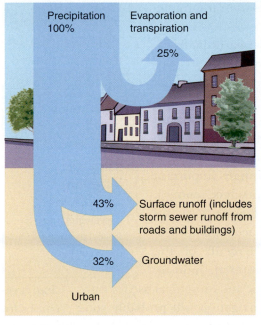

Shown is the fate of precipitation in Ontario, Canada, before (**A**) and after (**B**) urbanization. After Ontario was developed, surface runoff increased substantially, from 10 percent to 43 percent.

How development changes the natural flow of water FIGURE 10.5

When a natural area—that is, an area undisturbed by humans—is inundated with heavy precipitation, the plant-protected soil absorbs much of the excess water. What the soil cannot absorb runs off into the river, which may then spill over its banks onto the flood plain. Because rivers meander, the flow is slowed, and the swollen waters rarely cause significant damage to the surrounding area. (See Figure 6.13 for a diagram of a typical river, including its flood plain.)

When an area is developed for human use, construction projects replace much of this protective plant cover. Buildings and paved roads don't absorb water, so runoff, usually in the form of storm sewer runoff, is significantly greater in developed areas (**FIGURE 10.5**). People who build homes or businesses on the flood plain of a river will most likely experience flooding at some point (**FIGURE 10.6**).

Flooding in California FIGURE 10.6

The swollen Napa Creek roars past a theater in Napa, California, following a storm that caused mudslides and widespread flooding, on December 31, 2005.

Water Resource Problems 245

Arid lands, or deserts, are fragile ecosystems in which plant growth is limited by lack of precipitation. **Semiarid lands** receive more precipitation than deserts but are subject to frequent and prolonged droughts.

Farmers increase the agricultural productivity of arid and semiarid lands with irrigation. Irrigation of arid and semiarid lands has become increasingly important worldwide in efforts to produce enough food for burgeoning populations (**FIGURE 10.7**). In fact, irrigation accounts for the highest percentage, 71 percent, of the world's total water consumption, industry for 20 percent, and domestic and municipal use for only 9 percent. Water use for irrigation will probably continue to increase in the 21st century.

aquifer depletion
The removal of groundwater faster than it can be recharged by precipitation or melting snow.

saltwater intrusion
The movement of seawater into a freshwater aquifer near the coast.

AQUIFER DEPLETION

Aquifer depletion from excessive removal of groundwater lowers the **water table**, the upper surface of the saturated zone of groundwater. Prolonged aquifer depletion drains an aquifer dry, effectively eliminating it as a water resource. Even areas with high rainfall can experience aquifer depletion if humans remove more groundwater than can be recharged. In addition, aquifer depletion from porous sediments causes **subsidence**, or sinking, of the land above it. **Saltwater intrusion** occurs along coastal areas when groundwater is depleted faster than it recharges. Well water in such areas eventually becomes too salty for human consumption or other freshwater uses.

Agricultural use of water FIGURE 10.7

These fields in Kansas use central-pivot irrigation, a method that minimizes evaporative water loss and gives the fields a distinctive circular shape. Each circle is the result of a long irrigation pipe that extends along the radius from the circle's center to its edge and slowly rotates, spraying the crops. This satellite photo was taken in June; wheat fields are bright yellow, corn fields (dark green) are growing vigorously, and the sorghum crop (light green) is just starting to appear.

The Ogallala Aquifer The High Plains cover 6 percent of U.S. land but produce more than 15 percent of the nation's wheat, corn, sorghum, and cotton and almost 40 percent of its livestock. This productivity requires approximately 30 percent of the irrigation water used in the United States. Farmers on the High Plains rely on water from the **Ogallala Aquifer**, the largest groundwater deposit in the world (**Figure 10.8**).

In some areas farmers are drawing water from the Ogallala Aquifer as much as 40 times faster than nature replaces it. This rapid depletion has lowered the water table more than 30 m (100 ft) in some places. Most hydrologists (scientists who study water supplies) predict that groundwater will eventually drop in all areas of the Ogallala to a level uneconomical to pump. Their goal is to postpone that day through water conservation, including the use of water-saving irrigation systems.

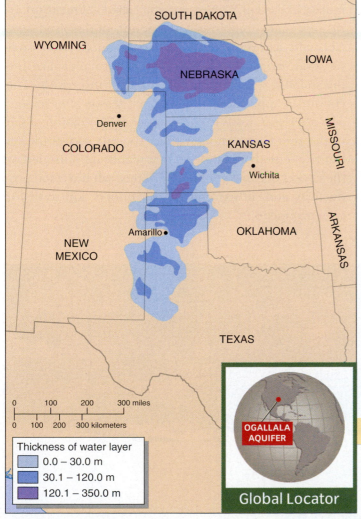

Thickness of water layer
- 0.0 – 30.0 m
- 30.1 – 120.0 m
- 120.1 – 350.0 m

OGALLALA AQUIFER

Global Locator

OVERDRAWING SURFACE WATERS

Removing too much fresh water from a river or lake can have disastrous consequences in local ecosystems. Humans can remove perhaps 30 percent of a river's flow without greatly affecting the natural environment. In some places considerably more is withdrawn for human use. In the arid American Southwest, it is not unusual for 70 percent or more of surface water to be removed.

When surface waters are overdrawn, wetlands dry up. **Estuaries**, where rivers empty into seawater, become saltier when surface waters are overdrawn, which reduces their productivity. Wetlands and estuaries, which serve as breeding grounds for many species of birds and other animals, also play a vital role in the hydrologic cycle. When these resources are depleted, the ensuing water shortages and reduced productivity have economic as well as ecological ramifications.

The increased use of U.S. surface water for agriculture, industry, and personal consumption has caused many water supply and quality problems. Some regions that have grown in population during this period—for example, California, Nevada, Arizona, Georgia (metropolitan Atlanta), and Florida—have had correspondingly greater burdens on their water supplies. If water consumption in these and other areas continues to increase, the availability of surface waters could become a serious regional problem, even in places that have never experienced water shortages.

Nowhere in the country are water problems as severe as they are in the West and Southwest. Much of this large region is arid or semiarid. With the rapid expansion of the population there during the past 25 years, municipal, commercial, and industrial uses now compete heavily with irrigation for available water.

Ogallala Aquifer Figure 10.8

This massive deposit of groundwater lies under eight states, with extensive portions in Texas, Kansas, and Nebraska. Water in the Ogallala Aquifer takes hundreds or even thousands of years to renew after it is withdrawn to grow crops and raise cattle.

The Colorado River Basin One of the most serious water-supply problems in the United States is in the Colorado River Basin. The Colorado River provides water for 25 million people, including those in the cities of Denver, Las Vegas, Salt Lake City, Albuquerque, Phoenix, Los Angeles, and San Diego. It supplies irrigation water for 1.4 million hectares (3.5 million acres) of fruit, vegetable, and field crops worth $1.5 billion per year. The Colorado River has 49 dams, 11 of which produce electricity by hydropower. The river provides $1.25 billion per year in revenues from the recreation industry.

An international agreement with Mexico, along with federal and state laws, severely restricts the use of the Colorado's waters. An important state treaty is the 1922 **Colorado River Compact**. It stipulates an annual allotment of 7.5 million acre-feet of water each to the upper Colorado (Colorado, Utah, and Wyoming) and the lower Colorado (California, Nevada, Arizona, and New Mexico). Each acre-foot equals 326,000 gal, enough for about eight people for 1 year. However, the Colorado River Compact overestimated the average annual flow of the Colorado River, and it locked that estimate into the multistate agreement.

Population growth in the upper Colorado region threatens the lower Colorado region's water supply. Further, people in the states through which the lower Colorado flows take so much water that the remainder is insufficient to meet Mexico's needs, as stipulated by international treaty (**FIGURE 10.9**). To compound the

> **salinization**
> The gradual accumulation of salt in soil, often as a result of improper irrigation methods.

problem, as more and more water is used, the lower Colorado becomes increasingly salty—in some places saltier than the ocean—as it flows toward Mexico.

Some positive steps have been taken. In 2003, California agreed to limit its water withdrawals from the Colorado River to quantities specified in the Colorado River Compact. Also, some California farmers agreed to sell some water they would normally use for irrigation and use the money earned to update their irrigation systems so they make more efficient use of their water.

SALINIZATION OF IRRIGATED SOIL

Although irrigation improves the agricultural productivity of arid and semiarid lands, it often causes salt to accumulate in the soil, a phenomenon called **salinization**. Irrigation water contains small amounts of dissolved salts. Normally, as a result of precipitation runoff, rivers carry salt away. Irrigation water, however, normally soaks into the soil and does not run off the land into rivers. The continued application of such water, season after season, year after year, leads to the gradual accumulation of salt in the soil. Given enough time, the salt concentration can rise to such a high level that plants are poisoned or their roots become dehydrated. Thus, salt hurts soil productivity and, in extreme cases, renders soil unfit for crop production.

Colorado River bed in San Luis Rio Colorado FIGURE 10.9
As a result of diversion for irrigation and other uses in the United States, the Colorado River often dries up before reaching the Gulf of California in Mexico.

GLOBAL WATER ISSUES

As the world's population continues to increase, global water problems will become more serious. In India, where 17 percent of the world's population has access to 4 percent of the world's fresh water, approximately 8,000 villages have no local water supply. Water supplies are precarious in much of China, owing to population pressures: Water table levels are dropping, wells have gone dry, and much of the water in the Yellow River is diverted for irrigation, depriving downstream areas of water. Mexico is facing the most serious water shortages of any country in the Western Hemisphere. The main aquifer supplying Mexico City is dropping rapidly, and the water table is falling fast in Guanajuato, an agricultural state.

As the needs of the growing human population deplete freshwater supplies, less water will be available for crops. Local famines arising from water shortages are a possibility.

Sharing water resources among countries

In the 1950s, the then Soviet Union began diverting water that feeds into the Aral Sea to irrigate nearby desert areas. Since 1960 the Aral Sea has declined more than 50 percent in area (**FIGURE 10.10**). Its total volume is down 80 percent, and much of its biological diversity has disappeared. Millions of people living in the Aral Sea's watershed have developed serious health problems, probably due to storms lifting into the air toxic salts from the receding shoreline.

Following the breakup of the Soviet Union in 1991, responsibility for saving the Aral Sea shifted to the five Asian countries that share the Aral basin—Uzbekistan, Kazakstan, Kyrgyzstan, Turkmenistan, and Tajikistan. Despite recent cooperative restoration efforts made by these nations and backed by the World Bank and the U.N. Environment Program, the Aral Sea will probably never return to its former size and economic importance. Like the Aral Sea, many of Earth's other watersheds cross political boundaries and face management issues associated with their shared use (**FIGURE 10.11**).

Aral Sea FIGURE 10.10

The satellite images show the Aral Sea in 1977, 1989, and 2006. As water was diverted for irrigation, the sea level subsided.

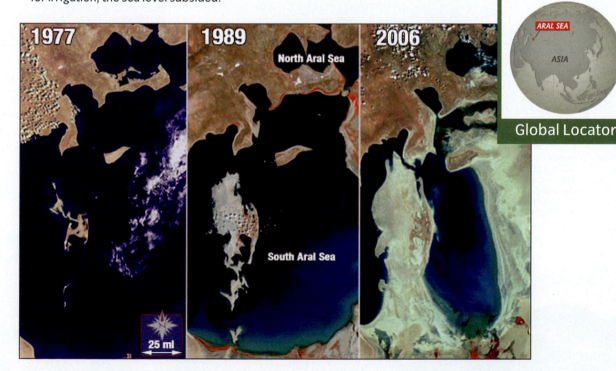

Earth's primary watersheds   FIGURE 10.11

Earth's bodies of fresh water can cover enormous areas and cross many political boundaries.

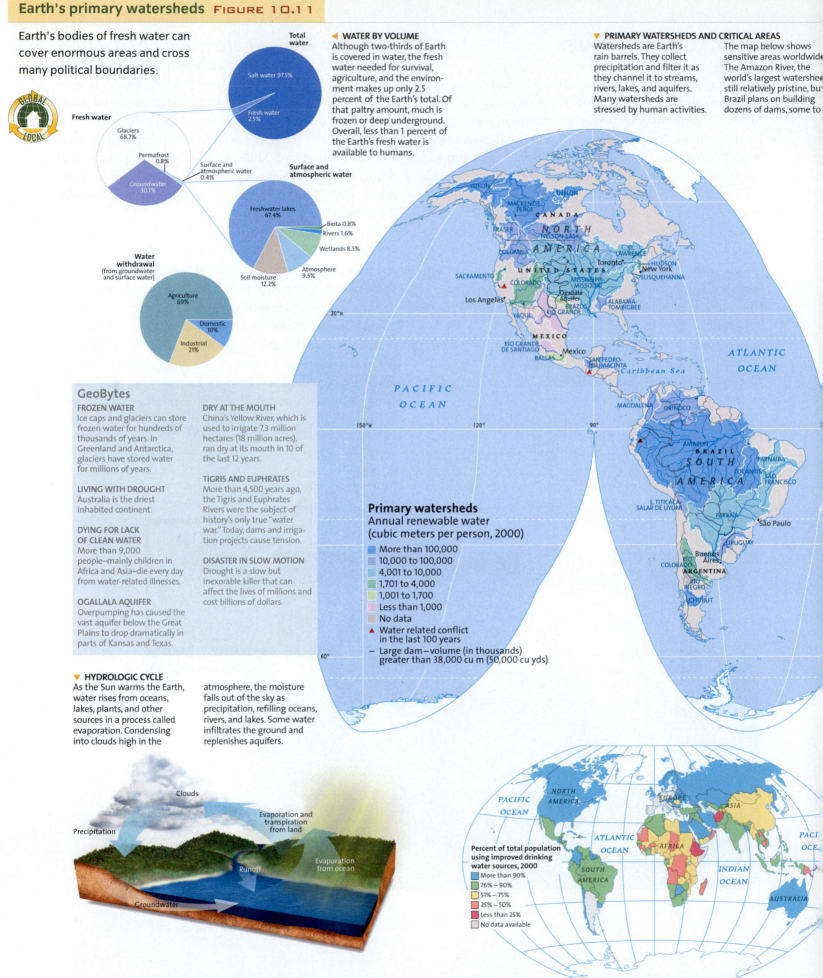

WATER BY VOLUME
Although two-thirds of Earth is covered in water, the fresh water needed for survival, agriculture, and the environment makes up only 2.5 percent of the Earth's total. Of that paltry amount, much is frozen or deep underground. Overall, less than 1 percent of the Earth's fresh water is available to humans.

Total water
Salt water 97.5%
Fresh water 2.5%

Fresh water
Glaciers 68.7%
Permafrost 0.8%
Groundwater 30.1%
Surface and atmospheric water 0.4%

Surface and atmospheric water
Freshwater lakes 67.4%
Biota 0.8%
Rivers 1.6%
Wetlands 8.5%
Atmosphere 9.5%
Soil moisture 12.2%

Water withdrawal (from groundwater and surface water)
Agriculture 69%
Domestic 10%
Industrial 21%

PRIMARY WATERSHEDS AND CRITICAL AREAS
Watersheds are Earth's rain barrels. They collect precipitation and filter it as they channel it to streams, rivers, lakes, and aquifers. Many watersheds are stressed by human activities.

The map below shows sensitive areas worldwide. The Amazon River, the world's largest watershed, still relatively pristine, but Brazil plans on building dozens of dams, some to

GeoBytes

FROZEN WATER
Ice caps and glaciers can store frozen water for hundreds of thousands of years. In Greenland and Antarctica, glaciers have stored water for millions of years.

LIVING WITH DROUGHT
Australia is the driest inhabited continent.

DYING FOR LACK OF CLEAN WATER
More than 9,000 people—mainly children in Africa and Asia—die every day from water-related illnesses.

OGALLALA AQUIFER
Overpumping has caused the vast aquifer below the Great Plains to drop dramatically in parts of Kansas and Texas.

DRY AT THE MOUTH
China's Yellow River, which is used to irrigate 7.3 million hectares (18 million acres), ran dry at its mouth in 10 of the last 12 years.

TIGRIS AND EUPHRATES
More than 4,500 years ago, the Tigris and Euphrates Rivers were the subject of history's only true "water war." Today, dams and irrigation projects cause tension.

DISASTER IN SLOW MOTION
Drought is a slow but inexorable killer that can affect the lives of millions and cost billions of dollars.

Primary watersheds
Annual renewable water
(cubic meters per person, 2000)

- More than 100,000
- 10,000 to 100,000
- 4,001 to 10,000
- 1,701 to 4,000
- 1,001 to 1,700
- Less than 1,000
- No data
- ▲ Water related conflict in the last 100 years
- – Large dam—volume (in thousands) greater than 38,000 cu m (50,000 cu yds)

HYDROLOGIC CYCLE
As the Sun warms the Earth, water rises from oceans, lakes, plants, and other sources in a process called evaporation. Condensing into clouds high in the atmosphere, the moisture falls out of the sky as precipitation, refilling oceans, rivers, and lakes. Some water infiltrates the ground and replenishes aquifers.

Percent of total population using improved drinking water sources, 2000
- More than 90%
- 76% – 90%
- 51% – 75%
- 25% – 50%
- Less than 25%
- No data available

power aluminum smelters. In Africa and Asia, lack of access to water and water-related diseases are the main problems. In Europe and the Middle East, overuse, pollution, and disagreement over diverting water are the major challenges. Hope rests in better planning and community-scale projects.

It's as vital to life as air. Yet fresh water is one of the rarest resources on Earth. Only 2.5 percent of Earth's water is fresh, and of that the usable portion for humans is less than 1 percent of all fresh water, or 0.01 percent of all water on Earth. Water is constantly recycling through Earth's hydrologic cycle. But population growth and pollution are combining to make less and less available per person per year, while global climate change adds new uncertainty.

Efficiency, conservation, and technology can help ensure that the water you absorb today will still be usable and clean hundreds of years from now.

ACCESS TO FRESH WATER

Access to clean fresh water is critical for human health. But, in many regions, potable water is becoming scarce because of heavy demands and pollution. Especially worrisome is the poisoning of aquifers—a primary source of water for nearly one third of the world—by sewage, pesticides, and heavy metals.

▼ GLOBAL IRRIGATED AREAS AND WATER WITHDRAWALS

Since 1970, global water withdrawals have correlated with the rise in irrigated area. Some 70% of withdrawals are for agriculture, mostly for irrigation that helps produce 40 percent of the world's food.

Freshwater withdrawal as a percentage of total water utilization, 2000

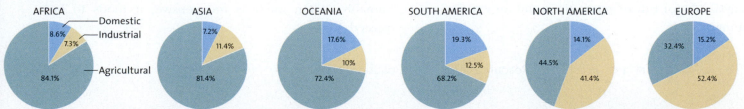

AFRICA — Domestic 8.6% / Industrial 7.3% / Agricultural 84.1%

ASIA — 7.2% / 11.4% / 81.4%

OCEANIA — 17.6% / 10% / 72.4%

SOUTH AMERICA — 19.3% / 12.5% / 68.2%

NORTH AMERICA — 14.1% / 41.4% / 44.5%

EUROPE — 15.2% / 52.4% / 32.4%

Rhine River Basin FIGURE 10.12

The Rhine River drains five European countries. (The green area represents the drainage basin.) Water management of such a river requires international cooperation.

RHINE RIVER BASIN

Global Locator

Three-fourths of the world's 200 or so major watersheds are shared between at least two nations. International cooperation is required to manage rivers that cross international borders. The heavily populated drainage basin for the Rhine River in Europe is in five countries—Switzerland, Germany, France, Luxembourg, and the Netherlands (FIGURE 10.12). All five nations recognize that international cooperation is essential to conserve and protect the supply and quality of the Rhine River. Their efforts have paid off: The main sources of pollution have been eliminated, and water in the Rhine River today is almost as pure as drinking water; long-absent fishes have returned; and projects are under way to restore riverbanks, control flooding, and clean up remaining pollutants.

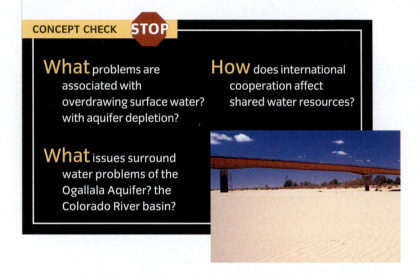

CONCEPT CHECK STOP

What problems are associated with overdrawing surface water? with aquifer depletion?

How does international cooperation affect shared water resources?

What issues surround water problems of the Ogallala Aquifer? the Colorado River basin?

Water Management

LEARNING OBJECTIVES

Define *sustainable water use.*

Contrast the benefits and drawbacks of dams and reservoirs.

Give examples of water conservation in agriculture, industry, and individual homes and buildings.

The main goal of water management is to provide a sustainable supply of high-quality water. **Sustainable water use** means careful human use of water resources so water is available for future generations and for existing non-human needs.

Water supplies are obtained by building dams, diverting water, or removing salt from seawater or salty groundwater. Conservation, which includes reusing water, recycling water, and improving water-use efficiency, augments water supplies and is an important aspect of sustainable water use. Economic policies are also important in managing water sustainably: When water is inexpensive, it tends to be wasted.

sustainable water use
The wise use of water resources, without harming the essential functioning of the hydrologic cycle or the ecosystems on which present and future humans depend.

DAMS AND RESERVOIRS: MANAGING THE COLUMBIA RIVER

Dams generate electricity and ensure a year-round supply of water in areas with seasonal precipitation or snowmelt, but many people think their costs outweigh their benefits. The Columbia River, the fourth-largest river in North America, illustrates the impact of dams on natural fish communities. There are more than 100 dams in the Columbia River system, 19 of which are major generators of hydroelectric power (FIGURE 10.13). The Columbia River system supplies municipal and industrial water to several major urban areas in the northwestern United States and irrigation water for more than 1.2 million hectares (3 million acres) of agricultural land.

As is often the case in natural resource management, one particular use of the Columbia River system may have a negative impact on other uses. The dam impoundments along the Columbia River generate electricity and control floods but have adversely affected fish populations. The salmon population in the Columbia River system is only a fraction of what it was before the watershed was developed. The many dams that impede salmon migrations are widely considered the most significant factor in salmon decline.

Various efforts to assist migrating salmon have not proved particularly successful (FIGURE 10.14). Conservationists and biologists support using a natural approach of releasing water to flush young salmon downstream; they also support adopting a controversial proposal to tear down several dams on the lower Snake River, a tributary of the Columbia River. Farmers and the hydroelectric companies strongly oppose these plans, which they fear would threaten their water supplies.

Grand Coulee Dam on the Columbia River
FIGURE 10.13

Shown are the dam and part of its reservoir, Lake Roosevelt. Dams provide electricity generation, flood control, and water recreation opportunities, but they disrupt or destroy natural river habitats and are expensive to build.

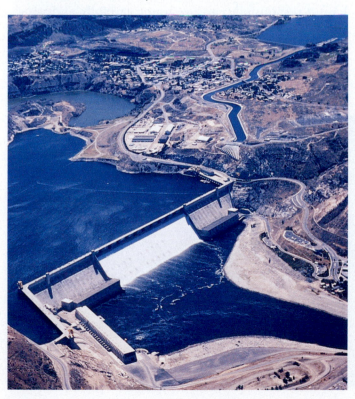

Fish ladder FIGURE 10.14

This ladder is located at the Bonneville Dam along the Oregon side of the Columbia River. Fish ladders help migratory fishes to bypass dams in their migration upstream. Despite the installation of fish ladders, the salmon population remains low.

WATER CONSERVATION

Today there is more competition than ever among water users with different priorities, and water conservation measures are necessary to guarantee sufficient water supplies.

Reducing agricultural water waste

Irrigation generally makes inefficient use of water. Traditional irrigation methods involve flooding the land or diverting water to fields through open channels. Plants absorb about 40 percent of the water that flood irrigation applies to the soil; the rest of the water usually evaporates into the atmosphere or seeps into the ground.

One of the most important innovations in agricultural water conservation is **microirrigation**, also called **drip** or **trickle irrigation**, in which pipes with tiny holes bored in them convey water directly to individual plants (FIGURE 10.15A). Microirrigation substantially reduces the water needed to irrigate crops—usually by 40 percent to 60 percent compared to traditional irrigation—and also reduces the amount of salt that irrigation water leaves in the soil.

> **microirrigation**
> A type of irrigation that conserves water by piping it to crops through sealed systems.

Other measures that could save irrigation water include using lasers to level fields, which allows a more even water distribution, and making greater use of recycled wastewater. A drawback of such techniques is their cost, which makes them unaffordable for most farmers in highly developed countries, let alone subsistence farmers in developing nations.

Reducing water waste in industry

Electric power generators and many industrial processes require water. In the United States, five major industries—chemical products, paper and pulp, petroleum and coal, primary metals, and food processing—consume almost 90 percent of industrial water.

Stricter pollution-control laws provide some incentive for industries to conserve water. Industries usually recapture, purify, and reuse water to reduce their water use and their water treatment costs (FIGURE 10.15B). The U.S. Steel Corporation plant in Granite City, Illinois, for example, recycles approximately two-thirds of the water it uses daily. The Ghirardelli Chocolate Company in San Leandro, California, installed a recycling system to cool large tanks of its chocolate. The potential for industries to conserve water by recycling is enormous.

Reducing municipal water waste

Like industries, regions and cities—and the households within them—recycle or reuse water to reduce consumption (FIGURE 10.15C). For example, homes and other buildings can be modified to collect and store gray water. *Gray water* is water that was already used in sinks, showers, washing machines, and dishwashers. Gray water is recycled to flush toilets, wash cars, or sprinkle lawns. In contrast to water *recycling*, wastewater *reuse* occurs when water is collected and treated before being redistributed. The reclaimed water is generally used for irrigation.

Cities also decrease water consumption by providing consumer education, requiring water-saving household fixtures, developing economic incentives to save water, and repairing leaky water supply systems. Also, increasing the price of water to reflect its true cost promotes water conservation.

The average person in the United States uses 295 L (78 gal) of water per day at home. As a water user, you have a responsibility to use water carefully and wisely. The cumulative effect of many people practicing personal water conservation measures has a significant impact on overall water consumption.

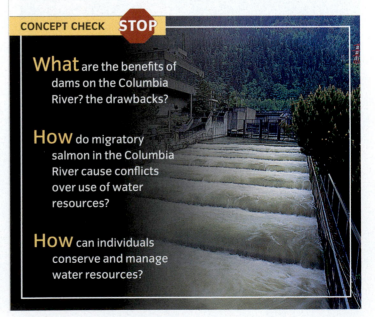

CONCEPT CHECK STOP

What are the benefits of dams on the Columbia River? the drawbacks?

How do migratory salmon in the Columbia River cause conflicts over use of water resources?

How can individuals conserve and manage water resources?

A Microirrigation

Close-up of a drip irrigation on pipe system about to release a drop of water directly over a seedling, eliminating much of the waste associated with traditional methods of irrigation. Photographed in the Negev Desert, Israel.

B Industrial water conservation

The wastewater treatment facilities at General Motors' Orion Assembly Plant in Michigan play a role in the company's efforts to conserve water. The plant has reduced water use by 14.7 percent between 2002 and 2004, a savings of about 1 billion gallons of water annually.

C Conserving water at home

In your bathroom, kitchen, and laundry room, you can take many steps to limit water use. Also, to reuse water, individual homes and buildings can be modified to collect and store "gray water," water already used in sinks, showers, washing machines, and dishwashers. This gray water is used when clean water is not required—for example, in flushing toilets, washing the car, and sprinkling the lawn, or, in the Southwest, irrigating golf courses.*

*Permits to install gray water systems vary from state to state. Arizona and other states with severe water shortages are more flexible than other states about allowing gray water systems.

Install water-saving shower and faucets and low-flush toilets. Or use a water displacement device in the tank of a conventional toilet. Fix leaky fixtures.

Modify personal habits: Avoid leaving faucet running while shaving or brushing teeth. Take shorter showers.

Use a dishwasher. It requires less water than washing dishes by hand with tap running—but only if you run the dishwasher with a full load.

Choose a high-efficiency washing machine. These models use less water or spin more water out of the clothes, which saves on energy costs when using a clothes dryer.

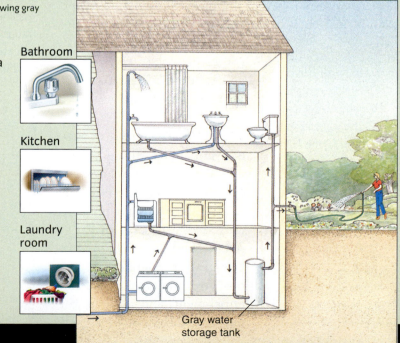

Bathroom

Kitchen

Laundry room

Gray water storage tank

NATIONAL GEOGRAPHIC

Water Pollution

LEARNING OBJECTIVES

Define *water pollution*.

Discuss how sewage is related to eutrophication, biochemical oxygen demand (BOD), and dissolved oxygen.

Distinguish between the two types of pollution sources and give examples of each.

Water pollution is a global problem that varies in magnitude and type of pollutant from one region to another. In many locations, particularly in developing countries, the main water pollution issue is providing individuals with disease-free drinking water.

TYPES OF WATER POLLUTION

As discussed earlier in the chapter, water's chemical properties make it capable of dissolving many substances, including pollutants. Water pollutants are divided into eight categories: sewage, disease-causing agents, sediment pollution, inorganic plant and algal nutrients, organic compounds, inorganic chemicals, radioactive substances, and thermal pollution. Causes and examples of each of these types of water pollution are summarized in TABLE 10.1 (on facing page). Here we explore the pollution threats associated with sewage.

> **water pollution**
> A physical or chemical change in water that adversely affects the health of humans and other organisms.

> **sewage** Wastewater from drains or sewers (from toilets, washing machines, and showers); includes human wastes, soaps, and detergents.

Sewage The release of **sewage** into water causes several pollution problems. First, because sewage may carry disease-causing agents, water polluted with sewage poses a threat to public health (see Chapter 4). Sewage also generates two serious environmental problems: enrichment and oxygen demand. **Enrichment** of a body of water is due to the presence of high levels of plant and algal nutrients such as nitrogen and phosphorus, both of which are sewage products. When an aquatic ecosystem contains high levels of sewage or other organic material, decomposing microorganisms use up most of the dissolved oxygen, leaving little available for fishes or other aquatic animals.

Sewage and other organic wastes are measured in terms of their **biochemical oxygen demand (BOD)**. A large amount of sewage in water generates a high BOD, which robs the water of dissolved oxygen (FIGURE 10.16). BOD measures how fast microorganisms remove oxygen from a body of water. When dissolved oxygen levels are low, anaerobic (without oxygen) microorganisms produce compounds with unpleasant odors, further deteriorating water quality.

> **biochemical oxygen demand (BOD)** The amount of oxygen that microorganisms need to decompose biological wastes into carbon dioxide, water, and minerals.

Effect of sewage on dissolved oxygen and biochemical oxygen demand (BOD) FIGURE 10.16

Note the initial oxygen depletion (blue line) and increasing BOD (red line) close to the sewage spill (at distance 0). The stream gradually recovers as the sewage is diluted and degraded. As indicated by the dashed line, fishes can't live in water that contains less than 4 mg of dissolved oxygen per liter of water.

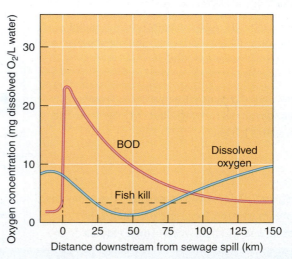

Types of water pollution TABLE 10.1

Type of pollution	Source	Examples	Effects
Sewage	Wastewater from drains or sewers	Human wastes, soaps, detergents	Threatens public health; causes enrichment and high biochemical oxygen demand (BOD)
Disease-causing agents	Wastes of infected individuals	Bacteria, viruses, protozoa, parasitic worms	Spread infectious diseases (cholera, dysentery, typhoid, infectious hepatitis, poliomyelitis, etc.)
Sediment pollution	Erosion of agricultural lands, forest soils exposed by logging, degraded stream banks, overgrazed rangelands, strip mines, construction	Clay, silt, sand, and gravel, suspended in water and eventually settling out	Reduces light penetration, limiting photosynthesis and disrupting food chain; clogs gills and feeding structures of aquatic animals; carries and deposits disease-causing agents and toxic chemicals
Inorganic plant and algal nutrients	Human and animal wastes, plant residues, atmospheric deposition, fertilizer runoff from agricultural and residential land	Nitrogen and phosphorus	Stimulate growth of excess plants and algae, which disrupts natural balance between producers and consumers and causes enrichment, bad odors, and a high BOD; suspected of causing red tides, explosive blooms of toxic pigmented algae that threaten the health of humans and aquatic animals in coastal areas
Organic compounds	Landfills, agricultural runoff, industrial wastes	Synthetic chemicals: pesticides, cleaning solvents, industrial chemicals, plastics	Contaminate groundwater and surface water; threaten drinking water supply; found in some bottled water; some are suspected endocrine disrupters
Inorganic chemicals	Industries, mines, irrigation runoff, oil drilling, urban runoff from storm sewers	Acids, salts, heavy metals such as lead, mercury, and arsenic	Contaminate groundwater and surface water; threaten drinking water supply; found in some bottled water; don't easily degrade or break down
Radioactive substances	Nuclear power plants, nuclear weapons industry, medical and scientific research facilities	Unstable isotopes of radioactive minerals such as uranium and thorium	Contaminate groundwater and surface water; threaten drinking water supply
Thermal pollution	Industrial runoff	Heated water produced during industrial processes, then released into waterways	Depletes water of oxygen and reduces amount of oxygen that water can hold; reduced oxygen threatens fishes

Oligotrophic and Eutrophic Lakes

A

B

A and **B** The average person looking at these two photographs would notice the dramatic differences between them but wouldn't understand the environmental conditions responsible for the differences. **A** shows Crater Lake, an oligotrophic lake in Oregon; **B** shows a small eutrophic lake in western New York.

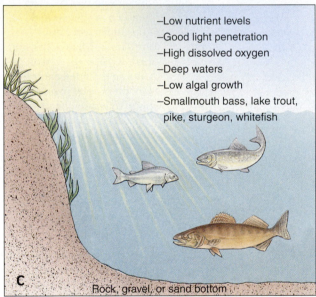

C

–Low nutrient levels
–Good light penetration
–High dissolved oxygen
–Deep waters
–Low algal growth
–Smallmouth bass, lake trout, pike, sturgeon, whitefish

Rock, gravel, or sand bottom

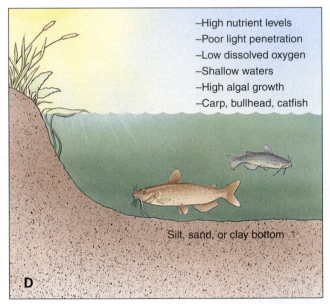

D

–High nutrient levels
–Poor light penetration
–Low dissolved oxygen
–Shallow waters
–High algal growth
–Carp, bullhead, catfish

Silt, sand, or clay bottom

C and **D** Aquatic ecologists understand the characteristics of oligotrophic and eutrophic lakes. **C** An oligotrophic lake has a low level of inorganic plant and algal nutrients. **D** A eutrophic lake has a high level of these nutrients.

Eutrophication: An enrichment problem

Lakes, estuaries, and slow-flowing streams that have minimal levels of nutrients are considered unenriched, or oligotrophic. An **oligotrophic** lake has clear water and supports small populations of aquatic organisms (see What a Scientist Sees, parts **A** and **C**). **Eutrophication** is the enrichment of a lake, estuary, or slow-flowing stream by inorganic plant and algal nutrients such as phosphorus; an enriched body of water is said to be **eutrophic**. The enrichment of water results in an increased photosynthetic productivity. The water in a eutrophic lake is cloudy and usually resembles pea soup because of the presence of vast numbers of algae and cyanobacteria (see parts **B** and **D**).

Over vast periods, oligotrophic lakes, estuaries, and slow-moving streams become eutrophic naturally. However, some human activities greatly accelerate eutrophication. This fast, human-induced process is usually called **artificial eutrophication** to distinguish it from natural eutrophication. Artificial eutrophication results from enrichment of aquatic ecosystems by nutrients found predominantly in fertilizer runoff and sewage.

SOURCES OF WATER POLLUTION

Water pollutants come from both natural sources and human activities. Natural sources of pollution such as mercury and arsenic tend to be local concerns, but human-generated pollution is generally more widespread.

The sources of water pollution are classified into two types: point source pollution and nonpoint source pollution. **Point source pollution** is discharged into the environment through pipes, sewers, or ditches from specific sites such as factories or sewage treatment plants (**FIGURE 10.17**).

Pollutants that enter bodies of water over large areas rather than at a single point cause **nonpoint source pollution**, also called *polluted runoff*. Nonpoint source pollution occurs when precipitation moves over and through the soil, picking up and carrying away pollutants that are eventually deposited in lakes, rivers, wetlands, groundwater, estuaries, and the ocean. Nonpoint source pollution includes agricultural runoff (such as fertilizers, pesticides, livestock wastes, and salt from irrigation), mining wastes (such as acid mine drainage), municipal wastes (such as inorganic plant and algal nutrients), construction sediments, and soil erosion (from fields, logging operations, and eroding stream banks). Although nonpoint sources cover more than one site and can be hard to identify, their combined effect can be huge.

Point source pollution FIGURE 10.17
Industrial runoff pours into a waste pit in the Amazon River basin. Natural gas burns from an adjacent pipe.

According to the EPA, agriculture is the leading source of water quality impairment of surface waters nationwide and is responsible for 72 percent of the water pollution in U.S. rivers. Agricultural practices produce several types of pollutants that contribute to nonpoint source pollution. Fertilizer runoff causes water enrichment. Animal wastes and plant residues in waterways produce high BODs and high levels of suspended solids as well as water enrichment. Highly toxic chemical pesticides may leach into the soil and from there into water or may find their way into waterways by adhering to sediment particles. Soil erosion from fields and rangelands causes sediment pollution in waterways. To address runoff from animal wastes, the EPA unveiled a largely voluntary program in 1998, asking U.S. livestock operations to develop by 2008 plans for managing manure to prevent it from becoming polluted runoff.

Although sewage is the main pollutant produced by cities and towns, municipal water pollution also has a nonpoint source: urban runoff that carries a variety of contaminants (**FIGURE 10.18**). The water quality of urban runoff from the storm sewers of city streets is often worse than that of sewage. Urban runoff carries salt from roadways, untreated garbage, construction sediments, and traffic emissions (via rain carrying air pollutants). When Hurricane Katrina breeched New Orleans's levees in 2005, floodwaters picked up contaminants from houses—such as pesticides, cleaners, paints, and food—and from businesses such as dry-cleaning solvents, lubricants, and medical wastes.

Different industries generate different types of pollutants. Food processing industries produce organic wastes that decompose quickly but have a high BOD. Pulp and paper mills also release wastes with a high BOD and produce toxic compounds and sludge. The paper industry, however, has begun to adopt new manufacturing methods, such as eliminating chlorine as a bleaching agent, that produce significantly less toxic effluents.

Urban runoff FIGURE 10.18

These pollutants may be carried from storm drains on streets to streams and rivers.

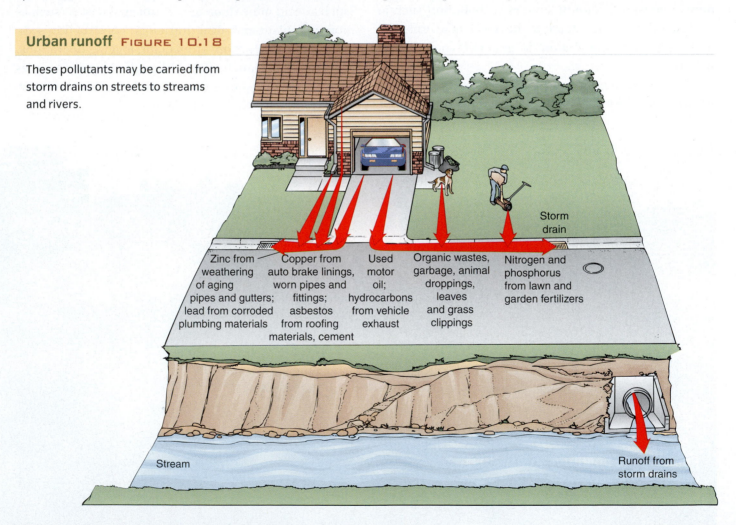

GROUNDWATER POLLUTION

Roughly half the people in the United States obtain their drinking water from groundwater, which is also withdrawn for irrigation and industry. In recent years, the quality of the nation's groundwater has become a concern. The most common pollutants, such as pesticides, fertilizers, and organic compounds, seep into groundwater from municipal sanitary landfills, underground storage tanks, backyards, golf courses, and intensively cultivated agricultural lands (**FIGURE 10.19**).

Currently, most of the groundwater supplies in the United States are of good quality and don't violate standards established to protect human health. However, areas that do experience local groundwater con-

tamination face quite a challenge. Cleanup of polluted groundwater is costly, takes years, and in some cases is not technically feasible.

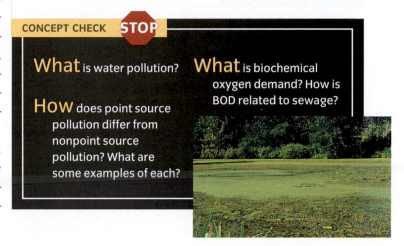

CONCEPT CHECK STOP

What is water pollution?

How does point source pollution differ from nonpoint source pollution? What are some examples of each?

What is biochemical oxygen demand? How is BOD related to sewage?

Sources of groundwater contamination FIGURE 10.19

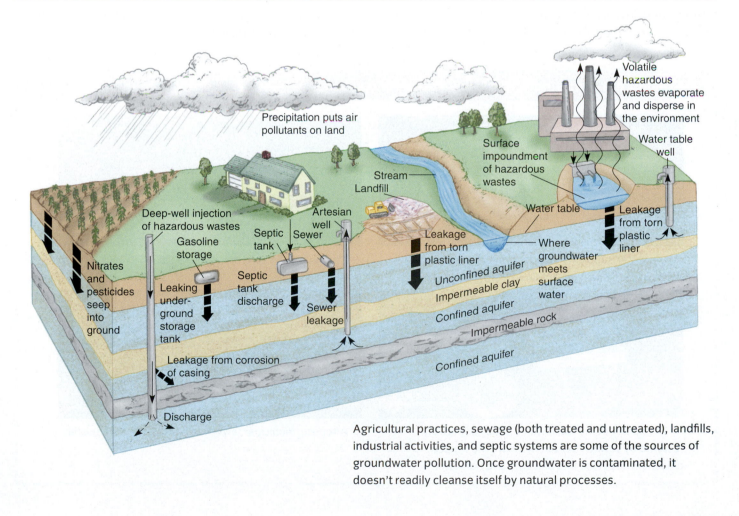

Agricultural practices, sewage (both treated and untreated), landfills, industrial activities, and septic systems are some of the sources of groundwater pollution. Once groundwater is contaminated, it doesn't readily cleanse itself by natural processes.

Improving Water Quality

LEARNING OBJECTIVES

Describe how most drinking water is purified in the United States.

Distinguish among primary, secondary, and tertiary treatments for wastewater.

Compare the goals of the Safe Drinking Water Act and the Clean Water Act.

Water quality is improved by removing contaminants from the water supply before and after it is used. Technology assists in both processes.

PURIFICATION OF DRINKING WATER

Most U.S. municipal water supplies are treated before the water is used so it is safe to drink (**FIGURE 10.20**). Turbid water is treated with a chemical coagulant that causes the suspended particles to clump together and settle out. The water is then filtered through sand to remove remaining suspended materials as well as many microorganisms.

In the final purification step before distribution in the water system, the water is disinfected to kill any remaining disease-causing agents. The most common way to disinfect water is to add chlorine. A small amount of chlorine is left in the water to provide protection during its distribution through many kilometers of pipes. Other disinfection systems use ozone or ultraviolet (UV) radiation in place of chlorine.

Process Diagram

Treatment of water for municipal use FIGURE 10.20

A The water supply for a town may be stored in a reservoir, as shown, or obtained from groundwater.

B The water is treated before use so it is safe to drink.

D The quality of the wastewater is fully or partially restored by sewage treatment before the treated effluent is dispersed into a nearby body of water.

C After use, municipal sewer lines collect the wastewater.

MUNICIPAL SEWAGE TREATMENT

Wastewater, including sewage, usually undergoes several treatments at a sewage treatment plant to prevent environmental and public health problems. The treated wastewater is then discharged into rivers, lakes, or the ocean.

Primary treatment removes suspended and floating particles, such as sand and silt, by mechanical processes such as screening and gravitational settling (FIGURE 10.21, left side). The solid material that settles out at this stage is called **primary sludge**. **Secondary treatment** uses microorganisms (aerobic bacteria) to decompose the suspended organic material in wastewater (Figure 10.21, right side). After several hours of processing, the particles and microorganisms are allowed to settle out, forming **secondary sludge**, a slimy mixture of bacteria-laden solids. Water that has undergone primary and secondary treatment is clear and free of organic wastes such as sewage. About 11 percent of U.S. wastewater treatment facilities have primary treatment only; about 62 percent have both primary and secondary treatments.

> **primary treatment**
> Treatment of wastewater that involves removing suspended and floating particles by mechanical processes.

> **secondary treatment**
> Biological treatment of wastewater to decompose suspended organic material; secondary treatment reduces the water's biochemical oxygen demand.

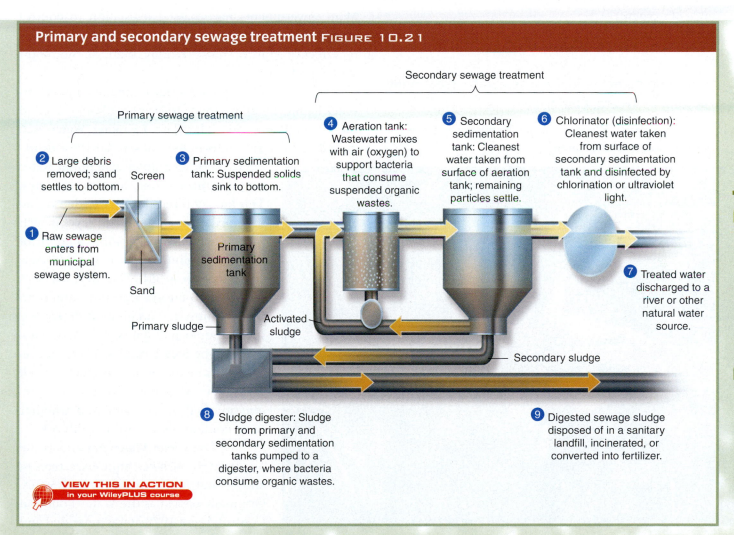

Primary and secondary sewage treatment FIGURE 10.21

Primary sewage treatment

Secondary sewage treatment

1 Raw sewage enters from municipal sewage system.

2 Large debris removed; sand settles to bottom.

Screen

Sand

3 Primary sedimentation tank: Suspended solids sink to bottom.

Primary sedimentation tank

Primary sludge

Activated sludge

4 Aeration tank: Wastewater mixes with air (oxygen) to support bacteria that consume suspended organic wastes.

5 Secondary sedimentation tank: Cleanest water taken from surface of aeration tank; remaining particles settle.

6 Chlorinator (disinfection): Cleanest water taken from surface of secondary sedimentation tank and disinfected by chlorination or ultraviolet light.

7 Treated water discharged to a river or other natural water source.

Secondary sludge

8 Sludge digester: Sludge from primary and secondary sedimentation tanks pumped to a digester, where bacteria consume organic wastes.

9 Digested sewage sludge disposed of in a sanitary landfill, incinerated, or converted into fertilizer.

VIEW THIS IN ACTION in your WileyPLUS course

Process Diagram

Even after primary and secondary treatments, wastewater still contains pollutants, such as dissolved minerals, heavy metals, viruses, and organic compounds. Advanced wastewater treatment methods, or **tertiary treatment**, include a variety of biological, chemical, and physical processes. Tertiary treatment reduces phosphorus and nitrogen, the nutrients most commonly associated with enrichment, and purifies wastewater for reuse in communities where water is scarce.

tertiary treatment Advanced wastewater treatment methods that are sometimes employed after primary and secondary treatments.

Disposal of primary and secondary sludge is a major problem associated with wastewater treatment. Sludge is generally handled by application to soil as fertilizer, incineration, disposal in a sanitary landfill, or anaerobic digestion. (In anaerobic digestion, bacteria break down the organic material in sludge in the absence of oxygen.)

Checking plant life in the wastewater treatment system in Arcata, California FIGURE 10.22

This constructed wetland is a successful solution to treating sewage in a small community.

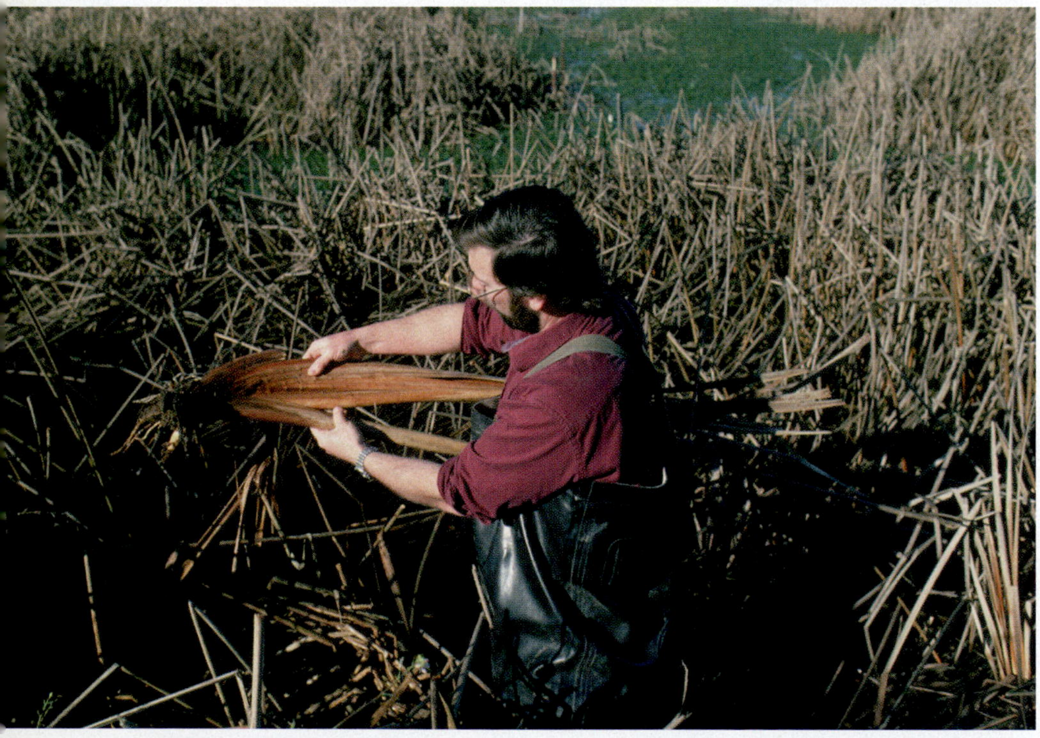

Some communities have adopted an environmentally innovative and economical approach to wastewater treatment (FIGURE 10.22). Beginning in 1978, the small town of Arcata, California, restored and constructed a series of freshwater wetlands in a former industrial area and then routed the town's wastewater through these wetlands. The marshes absorb and assimilate contaminants normally removed through more expensive treatment methods. The highly productive marsh ecosystem—the Arcata Marsh and Wildlife Sanctuary—also provides wildlife habitat for many organisms and opportunities for human recreation.

CONTROLLING WATER POLLUTION

Many governments have passed legislation to control water pollution. Point source pollutants lend themselves to effective control more readily than do nonpoint source pollutants.

The two U.S. laws that have the most impact on water quality today are the Safe Drinking Water Act and the Clean Water Act. The **Safe Drinking Water Act**, passed in 1974, set uniform federal standards for drinking water, to guarantee safe public water supplies throughout the United States. This law required the EPA to determine the **maximum contaminant level**, which is the maximum permissible amount of any water pollutant that might adversely affect human health. The EPA oversees the states to ensure that they adhere to the maximum contaminant levels for specific water pollutants. A 1996 amendment to the Safe Drinking Water Act requires municipal water suppliers to tell consumers what contaminants are present in their city's water and whether these contaminants pose a health risk.

The **Clean Water Act** affects the quality of rivers, lakes, aquifers, estuaries, and coastal waters in the United States. Originally passed as the Water Pollution Control Act of 1972, it was amended and renamed the Clean Water Act of 1977; additional amendments were made in 1981 and 1987. The Clean Water Act has two basic goals: to eliminate the discharge of pollutants in U.S. waterways and to attain water quality levels that make these waterways safe for fishing and swimming. Under the provisions of this act, the EPA is required to set up and monitor **national emission limitations**, the maximum permissible amounts of water pollutants that can be discharged from a sewage treatment plant, factory, or other point source.

Overall, the Clean Water Act has effectively improved the quality of water from point sources. According to the EPA, nonpoint source pollution is a major cause of water pollution yet is much more difficult and expensive to control than point source pollution. The 1987 amendments to the Clean Water Act expanded regulations on nonpoint sources.

The United States has improved its water quality in the past several decades, and demonstrated that the environment recovers once pollutants are eliminated. Much remains to be done, however. The EPA's 2002 *National Water Quality Inventory* indicated that water pollution has increased in U.S. rivers, lakes, estuaries, and coastal areas in recent years. According to the report, 39 percent of the nation's rivers, 45 percent of its lakes, and 51 percent of its estuaries are too polluted for swimming, fishing, or drinking.

Preventing water pollution at home

Although individuals produce little water pollution, the collective effect of municipal water pollution, even in a small neighborhood, can be quite large. There are many things you can do to protect surface waters and groundwater from water pollution (see **TABLE 10.2**).

Preventing water pollution at home TABLE 10.2

Location	What you can do
Bathroom	Never throw unwanted medicines down the toilet.
Kitchen	Use smallest effective amount of toxic household chemicals such as oven cleaners, mothballs, drain cleaners, and paint thinners. Substitute less hazardous chemicals wherever possible. Dispose of unwanted hazardous household chemicals at hazardous waste collection centers.
Driveway/car	Never pour used motor oil or antifreeze down storm drains or on the ground. Recycle these chemicals at service stations or local hazardous waste collection centers. Clean up spilled oil, brake fluid, and antifreeze, and sweep sidewalks and driveways instead of hosing them off. Dispose of dirt properly; don't sweep it into gutters or storm drains. Drive less: Air pollution emissions from automobiles eventually get into surface water and groundwater. Toxic metals and oil byproducts deposited on roads by vehicles are washed into surface waters by precipitation.
Lawn and garden	Pick up pet waste and dispose of it in garbage or toilet. If left on ground, it eventually washes into waterways, where it can contaminate shellfish and enrich water. Replace some grass lawn areas with trees, shrubs, and ground covers, which absorb up to 14 times more precipitation and require little or no fertilizer. To reduce erosion, use mulch to cover bare ground. Use fertilizer sparingly; excess fertilizer leaches into groundwater or waterways. Never apply fertilizer near surface water. Make sure that gutters and downspouts drain onto water-absorbing grass or graveled areas instead of onto paved surfaces.

Water pollution in developing countries

According to the World Health Organization, an estimated 1.4 billion people don't have access to safe drinking water, and about 2.9 billion people don't have access to adequate sanitation systems; most of these people live in rural areas of developing countries. Worldwide, at least 250 million cases of water-related illnesses occur each year, with 5 million or more of them resulting in death.

Municipal water pollution from sewage is a greater problem in developing countries, many of which lack water treatment facilities, than in highly developed nations. Sewage from many densely populated cities in Asia, Latin America, and Africa is dumped directly into rivers or coastal harbors (**FIGURE 10.23**). Other serious sources of water pollution in developing countries include industrial wastes, agricultural chemicals, and even human remains.

China's rapidly developing economy has resulted in the production of such severe water pollution that many of its densely crowded cities face water safety issues (**FIGURE 10.24**). Providing safe drinking water for China's 1.3 billion people is a formidable task, even without taking into account the current increase in water pollution.

Ganges River FIGURE 10.23

Bathing and washing clothes in the Ganges River are common practices in India. The river is contaminated by raw sewage discharged directly into the river at many different locations.

Contaminated water in China FIGURE 10.24

A woman in east China's Jiangxi Province collects bottles from a river polluted with red dye and wastewater released from a nearby paper factory. Widespread water pollution limits supplies of safe drinking water in China's cities.

CONCEPT CHECK STOP

How is most drinking water purified in the United States?

What are the stages in municipal sewage treatment? What happens in each stage?

How has the Safe Drinking Water Act affected U.S. water supplies? How has the Clean Water Act affected them?

WATER POLLUTION IN THE GREAT LAKES

North America's Great Lakes—Superior, Michigan, Huron, Erie, and Ontario—collectively hold about one-fifth of the world's fresh surface water (**FIGURE A**). More than 37 million people live in the Great Lakes watershed, an area that is home to agriculture, trade, industry, and tourism, and at least 26 million people obtain their drinking water from the Great Lakes.

Industrial wastes, sewage, fertilizers, and other pollutants have contaminated the Great Lakes since the mid-1800s. By the 1960s, thousands of toxic chemicals polluted the lakes, eutrophication was pronounced, high bacterial counts caused health hazards, and fish kills were common. Birth defects were observed in almost half of the animal species studied.

Canada and the United States have cooperated to improve the condition of the Great Lakes. Since a joint pollution control program was enacted in 1972, more than $20 billion has been spent on cleanup. As of the late 2000s, the Great Lakes are in better shape than they have been at any other time in recent memory. Levels of dichlorodiphenyl-trichloroethane (DDT) in women's breast milk have declined, as have levels of polychlorinated biphenyls (PCBs) in trout, coho salmon, and herring gulls. Some affected animal populations, such as double-crested cormorants and bald eagles, have rebounded.

However, many problems remain. Zebra mussels, sea lampreys, and at least 160 other invasive aquatic species have proliferated and threaten native species. Shoreline development continues to encroach on natural habitats and contributes to flooding and shoreline erosion during storms. Pesticides and fertilizers from suburban lawns pollute the water.

Persistent toxic compounds, such as mercury and PCBs, remain in the lakes, many of them coming from air pollution or from contaminated lake sediments. The presence of certain toxic chemicals is causing hormonal changes linked to reproductive failures, abnormal development, and abnormal behaviors in some fishes, birds, and mammals (recall the discussion of endocrine disrupters in Chapter 4). Persistent toxic chemicals that have accumulated in Great Lakes food webs threaten the food value of fishes there (**FIGURE B**).

A The eight Great Lakes states and two Canadian provinces signed the Great Lakes Toxic Substance Control Agreement to reduce pollution in the lakes by developing coordinated programs.

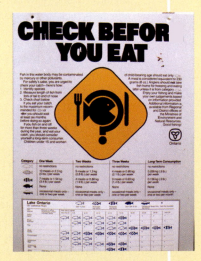

B Health advisories warn people of the potential risks associated with eating Great Lakes fishes.

SUMMARY

1 The Importance of Water

1. Water molecules are **polar**: The negatively charged (oxygen) end of one molecule is attracted to the positively charged (hydrogen) end of another molecule, forming a **hydrogen bond**.

2. Hydrogen bonds are the basis for many of water's properties, including its high melting point, high boiling point, high heat capacity, and dissolving ability.

3. In the **hydrologic cycle**, water continuously circulates through the abiotic environment. **Surface water** is precipitation that remains on the surface. **Runoff** is the movement of fresh water from precipitation and snowmelt to rivers, lakes, wetlands, and the ocean. **Groundwater** is the supply of fresh water that is stored in **aquifers**, underground reservoirs.

2 Water Resource Problems

1. Overdrawing surface water causes **wetlands** to dry up and **estuaries** to become saltier. **Aquifer depletion** is the removal of groundwater faster than it can be recharged. **Saltwater intrusion** is the movement of seawater into a freshwater aquifer near the coast. **Salinization** is the gradual accumulation of salt in soil, often due to improper irrigation.

2. Farmers on the U.S. High Plains are depleting water from the **Ogallala Aquifer** much faster than nature replaces it. In the Colorado River Basin, rapid population growth upstream threatens the water supply of users downstream.

3. Most of the world's major watersheds are shared between at least two nations. International cooperation is often required to manage shared water use.

3 Water Management

1. **Sustainable water use** is the wise use of water resources, without harming the hydrologic cycle or the ecosystems on which humans depend.

2. Dams and reservoirs allow rivers to be tapped for hydroelectric power and used to supply municipal and industrial water, but they are expensive to build and alter the natural environment.

3. **Microirrigation** is an innovative type of irrigation that conserves water by piping it to crops through sealed systems. Industries and cities can employ measures to recapture, purify, and reuse water in homes and buildings.

4 Water Pollution

1. **Water pollution** is any physical or chemical change in water that adversely affects the health of humans and other organisms.

2. **Sewage** is the release of wastewater from drains or sewers. It carries disease-causing agents and causes **enrichment**, the fertilization of a body of water due to high levels of nutrients. **Eutrophication** is the enrichment of a lake, estuary, or slow-flowing stream. **Artificial eutrophication** is overnourishment of an aquatic ecosystem due to human activities. Sewage in water also raises the **biochemical oxygen demand (BOD)**, the amount of oxygen that microorganisms need to decompose biological wastes. A high BOD decreases water quality by reducing levels of dissolved oxygen.

3. **Point source pollution** is water pollution that can be traced to a specific spot, such as wastewater released from a factory or sewage treatment plant. **Nonpoint source pollution** includes pollutants that enter bodies of water over large areas, such as agricultural runoff, mining wastes, municipal wastes, construction sediments, and soil erosion.

5 Improving Water Quality

1. Most U.S. municipal water supplies are treated so the water is safe to drink. A chemical coagulant traps suspended particles, filtration removes suspended materials and microorganisms, and disinfection kills disease-causing agents.

2. Wastewater usually undergoes several treatments at a sewage treatment plant. **Primary treatment** removes suspended and floating particles from wastewater by mechanical processes. **Secondary treatment**, which reduces water's biochemical oxygen demand, treats wastewater biologically to decompose suspended organic material. **Tertiary treatment** reduces pollutants such as phosphorus and nitrogen.

3. The **Safe Drinking Water Act** protects the safety of the nation's drinking water. The **Clean Water Act** affects the quality of U.S. rivers, lakes, aquifers, estuaries, and coastal waters.

KEY TERMS

- **surface water** p. 243
- **runoff** p. 243
- **groundwater** p. 243
- **aquifer depletion** p. 246
- **saltwater intrusion** p. 246
- **salinization** p. 248

- **sustainable water use** p. 252
- **microirrigation** p. 254
- **water pollution** p. 256
- **sewage** p. 256
- **biochemical oxygen demand (BOD)** p. 256
- **artificial eutrophication** p. 259

- **point source pollution** p. 259
- **nonpoint source pollution** p. 259
- **primary treatment** p. 263
- **secondary treatment** p. 263
- **tertiary treatment** p. 264

CRITICAL AND CREATIVE THINKING QUESTIONS

1. What issues complicate water rights in the Missouri River?

2. Are our water supply problems largely the result of too many people? Give reasons for your answer.

3. Briefly describe the complexity of international water use, using the Rhine River or the Aral Sea as examples.

4. Outline a brief water conservation plan for your own daily use. How could you use water more sustainably?

5. Explain why untreated sewage may kill fishes when it is added directly to a body of water.

6. Tell whether each of the following represents point or nonpoint source pollution: fertilizer runoff, thermal pollution from a power plant, urban runoff, sewage from a ship, and erosion sediments from deforestation. Why is nonpoint pollution more difficult to control than point source pollution?

7. Is the Clean Water Act related to the quality of U.S. public drinking water? Explain your answer.

8–9. The graph reflects the monitoring of dissolved oxygen concentrations at six stations along a river. The stations are located 20 m apart, with A the farthest upstream and F the farthest downstream.

8. Where along the river did a sewage spill occur?

9. At which station would you most likely discover dead fish?

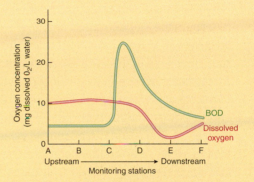

10. Studies have found that some bottled water comes from municipal water supplies and that bottled water frequently contains contaminants. Considering also the waste and energy associated with bottling, under what conditions does it make sense to drink bottled water or to avoid drinking it?

What is happening in this picture ?

- Rain soaks the streets of New York City. Where does the rainwater go?

- Many materials dissolve in water. What types of pollution might be dissolved in or carried by the rainwater?

- Is street runoff an example of point source pollution or nonpoint source pollution?

The Ocean and Fisheries

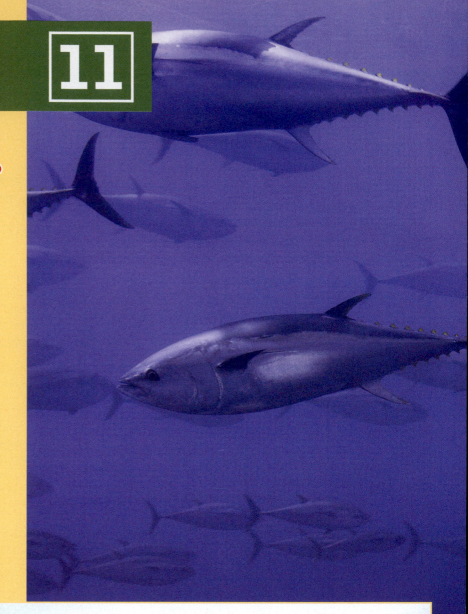

DEPLETING BLUEFIN TUNA STOCKS

VIEW THIS IN ACTION in your WileyPLUS course

Stocks of the giant, or Atlantic, bluefin tuna (*Thunnus thynnus*), a fish highly prized for sushi, are classified as depleted in the Mediterranean Sea by the U.N. Food and Agriculture Organization. Once harvested sustainably through traditional trapping, bluefins in the Mediterranean are now fished—often illegally—at approximately four times the sustainable rate. Spotter aircraft locate fish stocks and alert huge fishing fleets, whose ships (see inset) cast purse seines to net schools. Captured bluefins are fattened in off-shore pens (see large photo) before being butchered for market. The enormous economic value of the bluefin, which as an adult weighs an average of 250 kg (550 lb), places it at great risk. Only recently have Mediterranean nations begun implementing conservation measures to protect the species. In June 2008, the European Union banned several nations from the purse seine harvest of bluefins in Mediterranean and east Atlantic waters.

Overfishing, the harvesting of fishes faster than they can reproduce, is not limited to the Mediterranean. Worldwide, about 30 percent of fish species have been overfished as the world demand for fish has grown and harvesting methods have become more sophisticated. According to the National Research Council, 80 percent of nearly 200 commercially important fish stocks in the United States are fully exploited or overfished. In late 2006, a team of ecologists and economists estimated that, if current trends in overfishing and ocean pollution aren't curbed, populations of virtually all harvested seafood species could collapse by 2048.

The depletion of Mediterranean bluefin tuna stocks and U.S. fish stocks teaches us that as our technologies (such as efficient fishing techniques) advance, so do the impacts that we have on our environment. As a result, we must regulate these impacts (for example, the overharvesting of fishes) more closely than ever.

NATIONAL GEOGRAPHIC

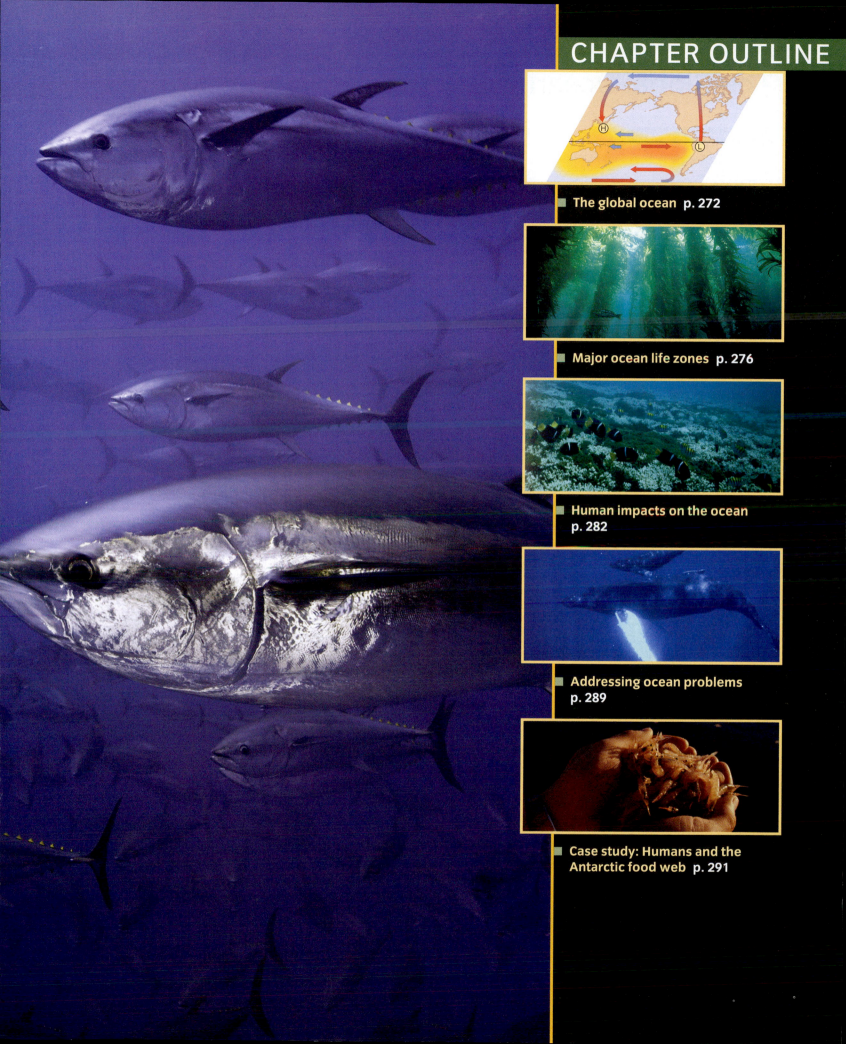

The Global Ocean

LEARNING OBJECTIVES

Describe the global ocean and its significance to life on Earth.

Discuss the roles of winds and the Coriolis effect in producing global water flow patterns, including gyres.

Define El Niño–Southern Oscillation (ENSO) and La Niña and describe some of their effects.

The ocean is a vast wilderness, much of it unknown. It teems with life—from warm-blooded mammals such as whales to soft-bodied invertebrates such as jellyfish. The ocean is essential to the hydrologic cycle that provides us with water. It affects cycles of matter on land, influences our climate and weather, and provides foods that enable millions of people to survive. The ocean dominates Earth, and its condition determines the future of life on our planet. If the ocean dies, then we do as well.

The global ocean is a huge body of salt water that surrounds the continents and covers almost three-fourths of Earth's surface. It is a single, continuous body of water, but geographers divide it into four sections separated by the continents: the Pacific, Atlantic, Indian, and Arctic Oceans. The Pacific is the largest: It covers one-third of Earth's surface and contains more than half of Earth's water.

PATTERNS OF CIRCULATION IN THE OCEAN

The persistent prevailing winds blowing over the ocean produce currents, mass movements of surface–ocean water (**FIGURE 11.1A**). The prevailing winds generate **gyres**, circular ocean currents. In the North Atlantic Ocean, the tropical trade winds tend to blow toward the west, whereas the westerlies in the midlatitudes blow toward the east. This helps establish a clockwise gyre in the North Atlantic.

The **Coriolis effect** (also see Chapter 8) influences the paths of surface, or shallow, ocean currents just as it does the winds. Earth's rotation from west to east causes surface ocean currents to swerve to the right in the Northern Hemisphere, helping establish the circular, clockwise pattern of water currents. In the Southern Hemisphere, ocean currents swerve to the left, thereby moving in a circular, counterclockwise pattern.

> **gyres** Large, circular ocean current systems that often encompass an entire ocean basin.

Vertical mixing of ocean water
Variations in the **density** (mass per unit volume) of seawater affect deep ocean currents. Cold, salty water is denser than warmer, less salty water. Colder, salty ocean water sinks and flows under warmer, less salty water, generating currents far below the surface. Deep ocean currents often travel in different directions and at different speeds than do surface currents, in part because the Coriolis effect is more pronounced at greater depths. **FIGURE 11.1B** shows the present circulation of shallow and deep currents—the **ocean conveyor belt**—that moves cold, salty deep-sea water from higher to lower latitudes, where it warms up. Note that the Atlantic Ocean gets its cold deep water from the Arctic Ocean, whereas the Pacific Ocean and Indian Ocean get theirs from the water surrounding Antarctica.

The ocean conveyer belt affects regional and possibly global climate. As the Gulf Stream and North Atlantic Drift push into the North Atlantic, they deliver an immense amount of heat from the tropics to Europe. (**FIGURE 11.1C**) As this shallow current transfers its heat to the atmosphere, the water becomes denser and sinks. The deep current flowing southward in the North Atlantic is, on average, 8°C cooler than the shallow current flowing northward.

Scientific evidence indicates that the ocean conveyor belt shifts from one equilibrium state to another. Historically, these shifts are linked to major changes in global climate.

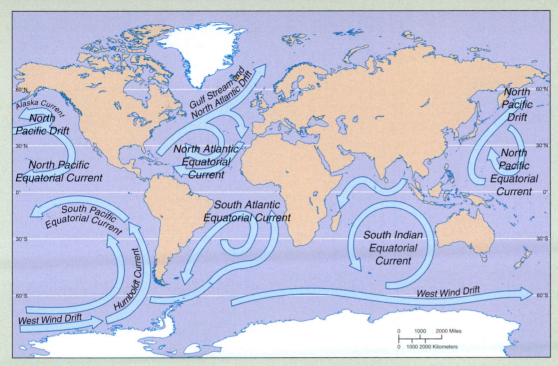

A Surface ocean currents
Winds largely cause the basic pattern of ocean currents. The main ocean current flow—clockwise in the Northern Hemisphere and counterclockwise in the Southern Hemisphere—results partly from the Coriolis effect.

Map labels: North Pacific Drift, Alaska Current, North Pacific Drift, North Pacific Equatorial Current, South Pacific Equatorial Current, Humboldt Current, West Wind Drift, West Wind Drift, Gulf Stream and North Atlantic Drift, North Atlantic Equatorial Current, South Atlantic Equatorial Current, North Pacific Drift, North Pacific Equatorial Current, South Indian Equatorial Current

60°N, 30°N, 0°, 30°S, 60°S

0 1000 2000 Miles
0 1000 2000 Kilometers

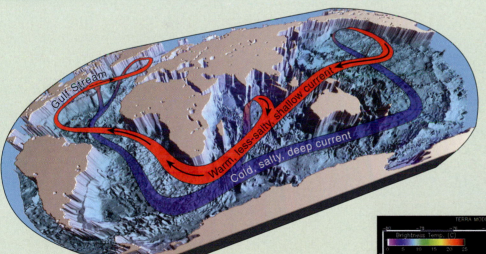

B The ocean conveyer belt This loop consists of deep ocean currents that flow in the opposite direction from surface currents. Vertical motions drive the conveyer: Cold, salty water near Antarctica and the Arctic Ocean sinks and eventually flows northward into the Pacific Ocean, where it wells up, eventually becoming warmer and fresher. The ocean conveyer belt affects regional and possibly global climate.

Diagram labels: Gulf Stream, Warm, less-salty, shallow current, Cold, salty, deep current

C The Gulf Stream The Gulf Stream is a well-known regional link in the ocean conveyor belt. In this satellite image, the colors represent the water's surface temperature: red = warmest, and blue = coolest. The Gulf Stream flows northeast along the North Carolina coast (black, in lower left) and then out to sea, toward Europe. (White areas are clouds.)

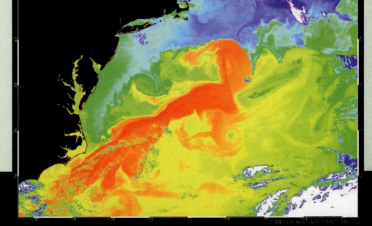

OCEAN–ATMOSPHERE INTERACTION

The ocean and the atmosphere are strongly linked, with wind from the atmosphere affecting the ocean currents and heat from the ocean affecting atmospheric circulation. One of the best examples of the interaction between ocean and atmosphere is the **El Niño–Southern Oscillation (ENSO)** event, which is responsible for much of Earth's interannual (from one year to the next) climate variability. As a result of ENSO, some areas are drier, some wetter, some cooler, and some warmer than usual. Normally, westward-blowing trade winds restrict the warmest waters to the western Pacific near Australia (**FIGURE 11.2A**). Every 3 to 7 years, however, the trade winds weaken, and the warm mass of water expands eastward to South America, increasing surface

El Niño–Southern Oscillation (ENSO) A periodic, large-scale warming of surface waters of the tropical eastern Pacific Ocean that temporarily alters both ocean and atmospheric circulation patterns.

temperatures in the usually cooler east Pacific (**FIGURE 11.2B**). Ocean currents, which normally flow westward in this area, slow down, stop altogether, or even reverse and go eastward. The name for this phenomenon, El Niño (in Spanish, "the boy child"), refers to the Christ child: The warming usually reaches the fishing grounds off Peru just before Christmas. Most ENSOs last between 1 and 2 years.

ENSO devastates the fisheries off South America. Normally, the colder, nutrient-rich deep water is about 40 m (130 ft) below the surface and **upwells** (comes to the surface) along the coast, partly in response to strong trade winds (**FIGURE 11.3A**). During an ENSO event, however, the deep water is about 152 m (500 ft) below the surface, and the warmer surface temperatures and weak trade winds prevent upwelling (**FIGURE 11.3B**). The lack

ENSO FIGURE 11.2

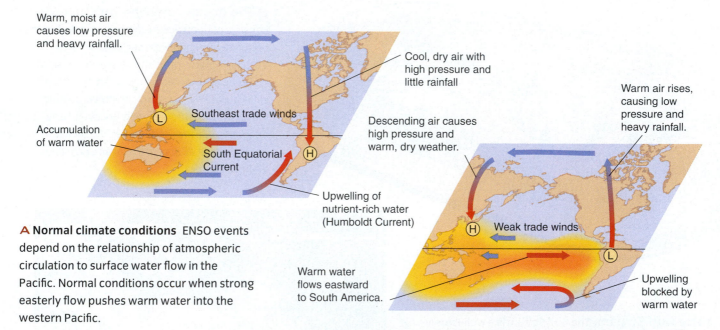

Warm, moist air causes low pressure and heavy rainfall.

Cool, dry air with high pressure and little rainfall

Southeast trade winds

Accumulation of warm water

South Equatorial Current

Descending air causes high pressure and warm, dry weather.

Warm air rises, causing low pressure and heavy rainfall.

Upwelling of nutrient-rich water (Humboldt Current)

Weak trade winds

A Normal climate conditions ENSO events depend on the relationship of atmospheric circulation to surface water flow in the Pacific. Normal conditions occur when strong easterly flow pushes warm water into the western Pacific.

Warm water flows eastward to South America.

Upwelling blocked by warm water

B ENSO conditions An ENSO event occurs when easterly flow weakens, allowing warm water to collect along the South American coast. Note the relationship between precipitation and the location of pressure systems. During an ENSO event, northern areas of the contiguous United States are typically warmer during winter, whereas southern areas are cooler and wetter.

of nutrients in the water results in a severe decrease in the populations of anchovies and many other fishes. During the 1982–1983 El Niño, one of the worst ever recorded, the anchovy population decreased 99%. Other species, such as shrimp and scallops, thrive during an ENSO event.

ENSO alters global air currents, directing unusual, sometimes dangerous, weather to areas far from the tropical Pacific where it originates. By one estimate, the 1997–1998 ENSO, the strongest on record, caused more than 20,000 deaths and $33 billion in property damage worldwide. It resulted in heavy snows in parts of the western United States; ice storms in eastern Canada; torrential rains that flooded Peru, Ecuador, California, Arizona, and western Europe; and droughts in Texas, Australia, and Indonesia. An ENSO-caused drought—the worst in 50 years—particularly hurt Indonesia. Fires that had been deliberately set to clear land for agriculture got out of control and burned an area in Indonesia the size of New Jersey.

Climate scientists observe and monitor sea surface temperatures and winds to better understand and predict the timing and severity of ENSO events. The **TAO/TRITON array** consists of 70 moored buoys in the tropical Pacific Ocean. These instruments collect oceanic and weather data during normal conditions and El Niño events. The data are transmitted to scientists onshore by satellite.

Scientists at the National Oceanic and Atmospheric Administration's Climate Prediction Center forecasted the 1997–1998 ENSO six months in advance, using data from TAO/TRITON. Such forecasts give governments time to prepare for the extreme weather changes associated with ENSO.

Upwelling Figure 11.3

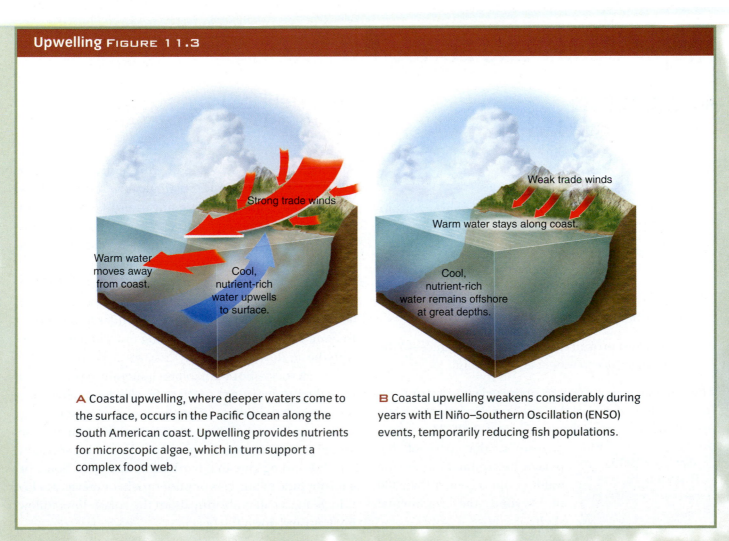

A Coastal upwelling, where deeper waters come to the surface, occurs in the Pacific Ocean along the South American coast. Upwelling provides nutrients for microscopic algae, which in turn support a complex food web.

B Coastal upwelling weakens considerably during years with El Niño–Southern Oscillation (ENSO) events, temporarily reducing fish populations.

Process Diagram

La Niña El Niño isn't the only periodic ocean temperature event to affect the tropical Pacific Ocean. **La Niña** (in Spanish, "the girl child") occurs when the surface water temperature in the eastern Pacific Ocean becomes unusually cool and westbound trade winds become unusually strong. La Niña often occurs after an El Niño event and is considered part of the natural oscillation of ocean temperature.

During the spring of 1998, the surface water of the eastern Pacific cooled 6.7°C (12°F) in just 20 days. Like ENSO, La Niña affects weather patterns around the world, but its effects are more difficult to predict. In the contiguous United States, La Niña typically causes wetter-than-usual winters in the Pacific Northwest, warmer weather in the Southeast, and drought conditions in the Southwest. Atlantic hurricanes are stronger and more numerous during a La Niña event.

CONCEPT CHECK **STOP**

What is the global ocean, and how does it affect Earth's environment?

How are the Coriolis effect, prevailing winds, and surface ocean currents related?

What is the El Niño–Southern Oscillation (ENSO)? What are some of its global effects?

Major Ocean Life Zones

LEARNING OBJECTIVE

Describe and distinguish among the four main ocean life zones.

The immense marine environment is subdivided into several zones (**FIGURE 11.4**):

- The intertidal zone

- The benthic (ocean floor) environment

- The two provinces—neritic and oceanic—of the pelagic (ocean water) environment

The *neritic province* is that part of the pelagic environment from the shore to where the water reaches a depth of 200 m (650 ft). It overlies the continental shelf. The *oceanic province* is that part of the pelagic environment where the water depth is greater than 200 m, beyond the continental shelf.

intertidal zone
The area of shoreline between low and high tides.

THE INTERTIDAL ZONE: TRANSITION BETWEEN LAND AND OCEAN

Although high levels of light, nutrients, and oxygen make the **intertidal zone** a biologically productive habitat, it is a stressful one. On sandy intertidal beaches, inhabitants must contend with a constantly shifting environment that threatens to engulf them and gives them little protection against wave action.

Rocky shores provide fine anchorage for seaweeds and marine animals, but they are exposed to wave action when submerged during high tides and exposed to temperature changes and drying out when in contact with the air during low tides (**FIGURE 11.5**).

A rocky-shore inhabitant generally has some way of sealing in moisture, perhaps by closing its shell (if it has one), and a means of anchoring itself to the rocks. For example, mussels have tough, threadlike anchors secreted by a gland in the foot, and barnacles secrete a tightly bonding glue that hardens under water. Some organisms hide in burrows or under rocks or crevices at low tide. Some small crabs run about the splash line, following it up and down the beach.

Zonation in the ocean FIGURE 11.4

The intertidal zone, the benthic environment, and the pelagic environment make up the ocean. The pelagic environment consists of the neritic and oceanic provinces. (The slopes of the ocean floor aren't as steep as shown; they are exaggerated here to save space.)

Intertidal zone:

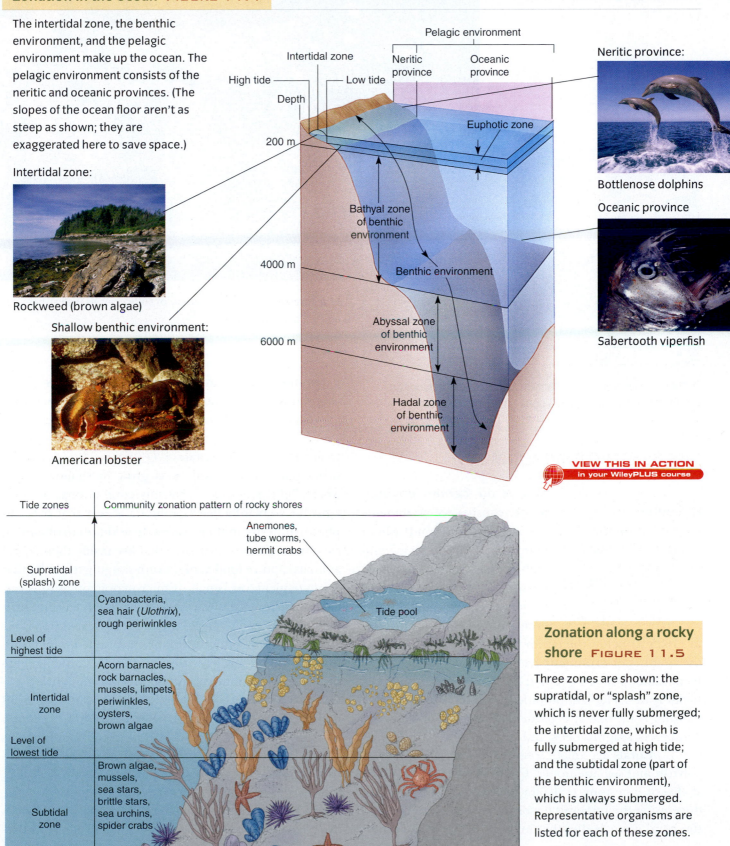

Rockweed (brown algae)

Shallow benthic environment:

American lobster

Pelagic environment

Intertidal zone

Neritic province

Oceanic province

Neritic province:

Bottlenose dolphins

Oceanic province

Sabertooth viperfish

High tide | Low tide

Depth

200 m

Euphotic zone

Bathyal zone of benthic environment

4000 m

Benthic environment

Abyssal zone of benthic environment

6000 m

Hadal zone of benthic environment

VIEW THIS IN ACTION
in your WileyPLUS course

Community zonation pattern of rocky shores

Tide zones

Anemones, tube worms, hermit crabs

Supratidal (splash) zone

Cyanobacteria, sea hair (*Ulothrix*), rough periwinkles

Tide pool

Level of highest tide

Acorn barnacles, rock barnacles, mussels, limpets, periwinkles, oysters, brown algae

Intertidal zone

Level of lowest tide

Brown algae, mussels, sea stars, brittle stars, sea urchins, spider crabs

Subtidal zone

Zonation along a rocky shore FIGURE 11.5

Three zones are shown: the supratidal, or "splash" zone, which is never fully submerged; the intertidal zone, which is fully submerged at high tide; and the subtidal zone (part of the benthic environment), which is always submerged. Representative organisms are listed for each of these zones.

Coral reefs FIGURE 11.6

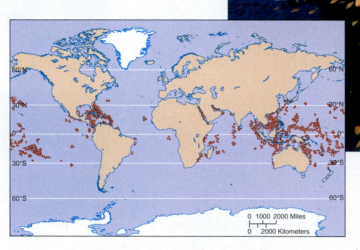

A This map shows the distribution of coral reefs around the world. There are more than 6,000 of them worldwide.

B A coral reef in Fiji has a variety of soft corals as well as basslets and yellow butterfly fish.

THE BENTHIC ENVIRONMENT

benthic environment

The ocean floor, which extends from the intertidal zone to the deep ocean trenches.

Most of the **benthic environment** consists of sediments (mainly sand and mud) where many bottom-dwelling animals, such as worms and clams, burrow. Bacteria are common in marine sediments, found even at depths more than 500 m (1625 ft) below the ocean floor. The deeper parts of the benthic environment are divided into three zones, from shallowest to deepest: the bathyal, abyssal, and hadal zones. The communities in the relatively shallow benthic zone that are particularly productive include coral reefs, seagrass beds, and kelp forests.

Corals are small, soft-bodied animals similar to jellyfish and sea anemones. Corals live in hard cups, or shells, of limestone (calcium carbonate) that they produce using the minerals dissolved in ocean water. When the coral animals die, the tiny cups remain, and a new generation of coral animals grows on top of these. Over thousands of generations, a **coral reef** forms from the accumulated layers of limestone.

Coral reefs are found in warm (usually greater than 21°C), shallow seawater (FIGURE 11.6A). The living portions of coral reefs grow in shallow waters where light penetrates. The tiny coral animals require light for **zooxanthellae** (symbiotic algae) that live and photosynthesize in their tissues. In addition to obtaining food from the zooxanthellae that live inside them, coral animals capture food at night with stinging tentacles that paralyze **plankton** (small or microscopic organisms carried by currents and waves) and small animals that drift nearby. The waters where coral reefs grow are often poor in nutrients, but other factors are favorable for high productivity, including the presence of zooxanthellae, appropriate temperatures, and year-round sunlight.

Coral reef ecosystems are the most diverse of all marine environments (FIGURE 11.6B). They contain hundreds of species of fishes and invertebrates, such as giant clams, snails, sea urchins, sea stars, sponges, flatworms, brittle stars, sea fans, shrimp, and spiny lobsters. The Great Barrier Reef occupies only 0.1 percent of the ocean's surface, but 8 percent of the world's fish species live there. The multitude of relationships and interac-

Sea grass bed FIGURE 11.7

Turtle grasses form underwater meadows that are ecologically important for shelter and food for many organisms. Photographed in the Caribbean Sea, the Cayman Islands.

tions that occur at coral reefs is comparable only to those of the tropical rain forest. As in the rain forest, competition is intense, particularly for light and space to grow.

Coral reefs are ecologically important because they both provide habitat for many kinds of marine organisms and protect coastlines from shoreline erosion. They provide humans with seafood, pharmaceuticals, and recreation and tourism dollars.

Sea grasses are flowering plants adapted to complete submersion in salty ocean water. They occur only in shallow water (to depths of 10 m, or 33 ft) where they receive enough light to photosynthesize efficiently. Extensive beds of sea grasses occur in quiet temperate, subtropical, and tropical waters. Eelgrass is the most widely distributed sea grass along the coast of North America; the world's largest eelgrass bed is in Izembek Lagoon on the Alaska Peninsula. The most common sea grasses in the Caribbean Sea are manatee grass and turtle grass (FIGURE 11.7). Sea grasses have a high primary productivity and are ecologically important: Their roots and rhizomes help stabilize sediments, reducing erosion, and they provide food and habitat for many marine organisms.

In temperate waters, ducks and geese eat sea grasses, and in tropical waters, manatees, green turtles, parrot fish, sturgeon fish, and sea urchins eat them. These herbivores consume only about 5 percent of sea grasses. The remaining 95 percent eventually enters the detritus food web and is decomposed when the sea grasses die. The decomposing bacteria are in turn consumed by animals such as mud shrimp, lugworms, and mullet (a type of fish).

Kelps, known to reach lengths of 60 m (200 ft), are the largest and most complex of all algae commonly called seaweeds (FIGURE 11.8). Kelps, which are brown algae, are common in cooler temperate marine waters of both the Northern and Southern hemispheres. They are especially abundant in relatively shallow waters (depths of about 25 m, or 82 ft) along rocky coastlines. Kelps are photosynthetic and are the primary food producers for the kelp "forest" ecosystem. Kelp forests provide habitats for many marine animals, such as tubeworms, sponges, sea cucumbers, clams, crabs, fishes, and sea otters.

Some animals eat kelp fronds, but kelps are mainly consumed in the detritus food web. Bacteria that decompose kelp provide food for sponges, tunicates, worms, clams, and snails. The diversity of life supported by kelp beds almost rivals that of coral reefs.

Kelp forest FIGURE 11.8

These underwater forests are ecologically important because they support many kinds of aquatic organisms. Photographed off the coast of California.

Sea otters play an important role in their environment. They feed on sea urchins, thereby preventing the urchins from eating kelp, which allows kelp forests to thrive. Now scientists have uncovered an alarming decline in sea otter populations in western Alaska's Aleutian Islands—a stunning 90 percent crash since 1990—that in turn poses wide-ranging threats to the coastal ecosystem there. The population of sea urchins in these areas is exploding, and kelp forests are being devastated. Strong evidence identifies killer whales, or orcas, as the culprits. Recently, orcas were observed for the first time preying on sea otters. Orcas generally feed on sea lions, seals, and fishes of all sizes. Sea otters, the smallest marine mammal species, are more like a snack to the orca than a desirable meal. So why are the orcas now choosing sea otters? Biologists suggest that it is because seal and sea lion populations have collapsed across the north Pacific.

Otters in Alaskan waters

In a scenario that is partly documented and partly speculative, the starting point of this disastrous chain of events is a drop in fish stocks, possibly caused by overfishing or climate change. With their food fish in decline, seal and sea lion populations have suffered, and orcas have looked elsewhere for food. Such a change in the orcas' feeding behavior transforms the food chain of kelp forests, putting orcas rather than otters at the top.

THE NERITIC PROVINCE: FROM THE SHORE TO 200 METERS

neritic province
The part of the pelagic environment that overlies the ocean floor from the shoreline to a depth of 200 m (650 ft).

Organisms that live in the pelagic environment's **neritic province** are all floaters or swimmers. The upper level of the pelagic environment is the **euphotic zone**, which extends from the surface to a maximum depth of 150 m (488 ft) in the clearest open ocean water. Sufficient light penetrates the euphotic zone to support photosynthesis.

Large numbers of phytoplankton (microscopic algae) produce food by photosynthesis and are the base of food webs. Zooplankton, including tiny crustaceans, jellyfish, comb jellies, and the larvae of barnacles, sea urchins, worms, and crabs, feed on phytoplankton. Zooplankton are in turn consumed by plankton-eating nekton (any marine organism that swims freely), such as herring, sardines, squid, baleen whales, and manta rays (**FIGURE 11.9**). These in turn become prey for carnivorous nekton such as sharks, tuna, porpoises, and toothed whales. Nekton are mostly confined to the shallower neritic waters (less than 60 m, or 195 ft, deep), near their food.

Neritic province FIGURE 11.9

A manta ray swims slowly through the water, swallowing vast quantities of microscopic plankton as it swims. The wingspan of a mature manta ray can reach about 6 m (20 ft). Note the remoras that are hitching a ride.

Oceanic province FIGURE 11.10

Found at dark depths of 700 to 3,000 m (2,300 to 3,280 ft), the spiky fanfin anglerfish attracts prey with its glowing lure. Its fin rays allow it to sense movement in the dark water. Photographed in Monterey Bay Canyon, California.

THE OCEANIC PROVINCE: MOST OF THE OCEAN

> **oceanic province**
> The part of the pelagic environment that overlies the ocean floor at depths greater than 200 m (650 ft).

The **oceanic province** is the largest marine environment, representing about 75 percent of the ocean's water. Most of the oceanic province is loosely described as the "deep sea." (The average depth of the ocean is 4000 m, more than 2 mi.) All but the shallowest waters of the oceanic province have cold temperatures, high pressure, and an absence of sunlight. These environmental conditions are uniform throughout the year.

Fishes of the deep waters of the oceanic province are strikingly adapted to darkness and scarcity of food (FIGURE 11.10). Adapted to drifting or slow swimming, animals of the oceanic province often have reduced bone and muscle mass. Many of these animals have light-producing organs to locate one another for mating or food capture.

Most organisms of the deep waters of the oceanic province depend on **marine snow**, organic debris that drifts down into their habitat from the upper, lighted regions of the oceanic province. Organisms of this little-known realm are filter feeders, scavengers, and predators. Many are invertebrates, some of which attain great sizes. The giant squid measures up to 18 m (59 ft) in length, including its tentacles.

CONCEPT CHECK STOP

What are the four main life zones in the ocean?

How do the neritic and oceanic provinces differ?

Which marine environment is transitional between land and ocean?

Human Impacts on the Ocean

LEARNING OBJECTIVES

Contrast fishing and aquaculture and relate the environmental challenges of each activity.

Identify the human activities that contribute to marine pollution and describe their effects.

Explain how global climate change could potentially alter the ocean conveyor belt.

The ocean is so vast, it's hard to imagine that human activities could harm it. Such is the case, however. Fisheries and aquaculture, marine shipping, marine pollution, coastal development, offshore mining, and global climate change all contribute to the degradation of marine environments. Scientists estimate that in 2008, less than 4 percent of the ocean remained unaffected by human activities, and 41 percent had experienced serious harm (**FIGURE 11.11A**)

MARINE POLLUTION AND DETERIORATING HABITAT

One of the great paradoxes of human civilization is that the same ocean that provides food to a hungry world is used as a dumping ground. Coastal and marine ecosystems receive pollution from land, from rivers emptying into the ocean, and from atmospheric contaminants that enter the ocean via precipitation. Offshore mining and oil drilling pollute the neritic province with oil and other contaminants. Pollution increasingly threatens the world's fisheries. Events such as accidental oil spills and the deliberate dumping of litter pollute the water. The World Resources Institute estimates that about 80 percent of global ocean pollution comes from human activities on land. In 2003 the Pew Oceans Commission, composed of scientists, economists, fishermen, and other experts, verified the seriousness of ocean problems in a series of studies. Some of their findings are shown in **FIGURE 11.11B**.

WORLD FISHERIES

The ocean contains a valuable food resource. About 90 percent of the world's total marine catch is fishes, with clams, oysters, squid, octopus, and other molluscs representing an additional 6 percent of the total catch. Crustaceans, including lobsters, shrimp, and crabs, make up about 3 percent, and marine algae constitute the remaining 1 percent.

Fleets of deep-sea fishing vessels obtain most of the world's marine harvest. Numerous fishes are also captured in shallow coastal waters and inland waters. According to the U.N. Food and Agricultural Organization (FAO), the world annual fish harvest increased substantially, from 19 million tons in 1950 to nearly 158 million tons in 2005, the latest year for which data are available.

Problems and challenges for the fishing industry
No nation lays legal claim to the open ocean. Consequently, resources in the ocean are more susceptible to overuse and degradation than land resources, which individual nations own and for which they feel responsible (see the "Tragedy of the Commons" section in Chapter 2).

The most serious problem for marine fisheries is that many species, particularly large predatory fish, have been harvested to the point that their numbers are severely depleted. This generally causes a fishery to become unusable for commercial or sport fishermen, let alone the other marine species that rely on it as part of the food web. Scientists have found that dramatically depleted fish populations recover only slowly. Some show no real increase in population size up to 15 years after the fishery has collapsed.

According to the FAO, at least 76 percent of the world's fish stocks are considered fully exploited, overexploited, or depleted. Fisheries have experienced such pressure for two reasons. First, the growing human population requires protein in its diets, leading to a greater demand for fish. Second, technological advances allow us to fish so efficiently that every single fish is often removed from an area (see What a Scientist Sees on page 284).

A **Mapping human impacts** In 2008, an international team of marine scientists mapped effects of 17 human activities on the ocean. Almost no location remains unaffected, and 41 percent of the ocean has been seriously altered by multiple activities.

VIEW THIS IN ACTION in your WileyPLUS course

- Very low impact
- Low impact
- Medium impact
- Medium high impact
- High impact
- Very high impact

B **Major threats to the ocean**

Climate change
Coral reefs and polar seas are particularly vulnerable to increasing temperatures.

Aquaculture
Fish farms produce wastes that can pollute ocean water and harm marine organisms.

Invasive species
Organisms are transported; released in ballast water from ships contains foreign crabs, mussels, worms, and fishes.

Overfishing
Populations of many commercial fish species are severely depleted.

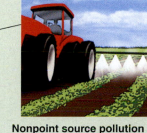

Nonpoint source pollution (runoff from land)
Agricultural runoff (fertilizers, pesticides, and livestock wastes) pollutes water.

Point source pollution
Passenger cruise ships dump sewage, shower and sink water, and oily bilge water.

Bycatch
Fishermen unintentionally kill dolphins, sea turtles, and sea birds.

Habitat destruction
Trawl nets (fishing equipment pulled along the ocean floor) destroy habitat.

Coastal development
Developers destroy important coastal habitat, such as salt marshes and mangrove swamps.

Fishermen tend to concentrate on a few fish species with high commercial value, such as salmon, tuna, and flounder, while other species, collectively called **bycatch**, are unintentionally caught and then discarded. The FAO reports that about 25 percent of all marine organisms caught—some 27 million metric tons (30 million tons)—

bycatch The fishes, marine mammals, sea turtles, seabirds, and other animals caught unintentionally in a commercial fishing catch.

are dumped back into the ocean. Most of these unwanted animals are dead or soon die because they are crushed by the fishing gear or are out of the water too long. The United States and other countries are trying to significantly reduce the amount of bycatch and develop uses for the bycatch that remains.

Modern Commercial Fishing Methods

A A full fishnet is pulled on board a fishing vessel off the coast of Alaska.

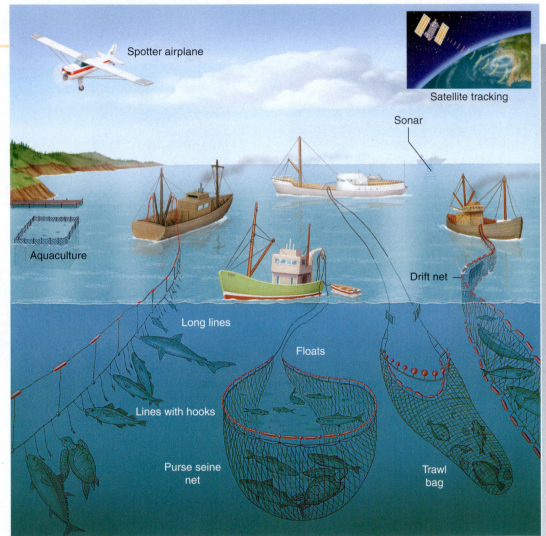

B Scientific evidence indicates that modern methods of harvesting fish are so effective that many fish species have become rare. Sea turtles, dolphins, seals, whales, and other aquatic organisms are accidentally caught and killed in addition to the target fish. The depth of longlines is adjusted to catch open-water fishes such as sharks and tuna or bottom fishes such as cod and halibut. Purse seines catch anchovies, herring, mackerel, tuna, and other fishes that swim near the water's surface. Trawls catch cod, flounder, red snapper, scallops, shrimp, and other fishes and shellfish that live on or near the ocean floor. Drift nets catch salmon, tuna, and other fishes that swim in ocean waters.

In response to harvesting, many nations extended their limits of jurisdiction to 320 km (200 mi) offshore. This action removed most fisheries from international use because more than 90 percent of the world's fisheries are harvested in relatively shallow waters close to land. This policy was supposed to prevent overharvesting by allowing nations to regulate the amounts of fishes and other seafood harvested from their waters. However, many countries also have a policy of **open management**, in which all fishing boats of that country are given unrestricted access to fishes in national waters.

Aquaculture: fish farming

Aquaculture is more closely related to agriculture on land than it is to the fishing industry. Aquaculture is carried out both in fresh and marine water; the cultivation of marine organisms is sometimes called **mariculture**. According to the FAO, world aquaculture production has increased substantially, from 600,000 tons in 1950 to 59.4 million tons in 2004.

> **aquaculture**
>
> The growing of aquatic organisms (fishes, shellfish, and seaweeds) for human consumption.

Aquaculture differs from fishing in several respects. For one thing, although highly developed nations harvest more fishes from the ocean, developing nations produce much more seafood by aquaculture. Developing nations have an abundant supply of cheap labor, which is a requirement of aquaculture because it is labor intensive, like land-based agriculture. Another difference between fishing and aquaculture is that the limit on the size of a catch in fishing is the size of the natural population, whereas the limit on aquacultural production is primarily the size of the area in which organisms can be grown.

In aquacultural "fish farms," fish populations are concentrated in a relatively small area and produce higher-than-normal concentrations of waste that pollute the adjacent water and harm other organisms. Aquaculture also causes a net loss of wild fish because many of the fishes farmed are carnivorous. Sea bass and salmon, for example, eat up to 5 kg (11 lbs) of wild fish to gain 1 kg (2.2 lbs) of weight.

Some aquaculture operations are carried out in deep water (**FIGURE 11.12**). The National Oceanic and Atmospheric Administration (NOAA) is investigating prospects for fish farming in the offshore deep waters of the U.S. Exclusive Economic Zone. Opponents are concerned about the potential for pollution, the spread of disease, and the accidental release of caged species into the deep-water environment. Caged populations are more genetically homogenous than wild ones; if the two groups interbreed, genetic diversity of wild populations could be diminished. The introduced organisms may also outcompete wild species.

An underwater fish farm   FIGURE 11.12

Cobia are raised in deep-water cages off Puerto Rico. The open-ocean circulation reduces the waste problems common in shallow-water aquaculture.

PUERTO RICO

Global Locator

SHIPPING, OCEAN DUMPING, AND PLASTIC DEBRIS

Millions of ships dump oily ballast and other wastes overboard in the neritic and oceanic provinces. The U.N. International Maritime Organization's International Convention for the Prevention of Pollution from Ships (MARPOL) bans marine pollution arising from the shipping industry. MARPOL regulations specifically address six types of marine pollution caused by shipping: oil, noxious liquids, harmful packaged substances, sewage, garbage, and air pollution released by ships. The 2004–2006 revisions to MARPOL regulations included stricter controls on oil tankers and added certain marine sites to the list of special protected areas. Unfortunately, MARPOL is not well enforced in the open ocean.

In the past, U.S. coastal cities such as New York dumped their sewage sludge into the ocean. Disease-causing viruses and bacteria from human sewage contaminated shellfish and other seafood and posed an increasing threat to public health. The **Ocean Dumping Ban Act** barred ocean dumping of sewage and industrial waste, beginning in 1991.

Huge quantities of trash containing plastics are released into the ocean from coastal communities or, sometimes accidentally, from cargo ships. Plastics don't biodegrade; they photodegrade, which means they break down into smaller and smaller pieces yet still exist for an indefinite period. This trash collects in certain areas of the open ocean defined by atmospheric pressure systems. For example, in the north Pacific gyre—halfway between Hawaii and the U.S. mainland—researchers found a continuous array of floating plastics estimated at 1 million pieces per square mile that covered an area the size of Texas.

Not only are marine mammals and birds susceptible to being entangled and/or strangled by larger pieces of plastic (**FIGURE 11.13**), but the many filter-feeding organisms near the bottom of the ocean food chain constantly ingest the smaller degraded pieces. These plastic pieces may absorb and transport hazardous chemicals such as PCBs. Scientists have yet to determine whether these substances are incorporated into marine food webs when organisms ingest the plastic.

Plastic pollution in the ocean FIGURE 11.13

This bottlenose dolphin is unable to fully open its mouth because of the plastic entangled on its snout.

COASTAL DEVELOPMENT

Development of resorts, cities, industries, and agriculture along coasts alters or destroys many coastal ecosystems, including mangrove forests, salt marshes, sea grass beds, and coral reefs. Many coastal areas are overdeveloped, highly polluted, and overfished. Although more than 50 countries have coastal management strategies, their goals are narrow and usually deal only with the economic development of the thin strips of land that directly border the oceans. Coastal management plans generally don't integrate the management of both land and water, nor do they take into account the main cause of coastal degradation—sheer human numbers.

Perhaps as many as 3.8 billion people—about 60 percent of the world's population—live within 150 km (93 mi) of a coastline. Demographers project that three-fourths of all humans—perhaps as many as 6.4 billion—will live in that area by 2025. If the world's natural coastal areas aren't to become urban sprawl or continuous strips of tourist resorts during the 21st century, coastal management strategies must be developed that take into account projections of human population growth and distribution.

HUMAN IMPACTS ON CORAL REEFS

Coral formations, which are important ecosystems, are being degraded and destroyed. Approximately one-fourth of the world's coral reefs are at high risk. In some areas, silt washing downstream from clear-cut inland forests has smothered reefs. High salinity resulting from the diversion of fresh water to supply the growing human population may be killing Florida reefs. Overfishing (particularly the removal of top predators), damage by scuba divers and snorkelers, pollution from ocean dumping and coastal runoff, oil spills, boat groundings, fishing with dynamite or cyanide, hurricane damage, disease, coral bleaching, land reclamation, tourism, and the mining of corals for building material take a heavy toll.

Since the late 1980s, corals in the tropical Atlantic and Pacific have suffered extensive **bleaching** (see What a Scientist Sees), in which stressed corals expel their zooxanthellae. Scientists suspect that several environmental stressors, particularly warmer seawater temperatures, contribute to coral bleaching. Many scientists attribute recent record sea temperatures and large die-offs of corals (more than 70 percent losses in some areas) to El Niño effects, global climate change, or a combination of the two. Impaired coral growth in Australia was recently linked to increased ocean acidification associated with warmer water temperatures. Other potential stressors are pollution and coral diseases.

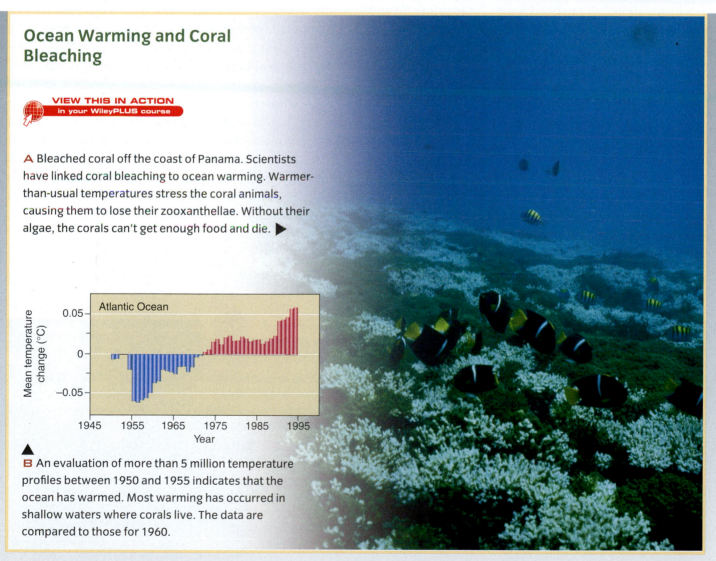

Ocean Warming and Coral Bleaching

VIEW THIS IN ACTION in your WileyPLUS course

A Bleached coral off the coast of Panama. Scientists have linked coral bleaching to ocean warming. Warmer-than-usual temperatures stress the coral animals, causing them to lose their zooxanthellae. Without their algae, the corals can't get enough food and die. ▶

B An evaluation of more than 5 million temperature profiles between 1950 and 1955 indicates that the ocean has warmed. Most warming has occurred in shallow waters where corals live. The data are compared to those for 1960.

What a Scientist Sees

OFFSHORE EXTRACTION OF MINERAL AND ENERGY RESOURCES

Large deposits of minerals, including **manganese nodules**, lie on or below the ocean floor (**FIGURE 11.14**). Dredging manganese nodules from the ocean floor would adversely affect sea life, and the current market value for these minerals wouldn't cover the expense of obtaining them using existing technology. Furthermore, it isn't clear which countries have legal rights to minerals in international waters. Despite these concerns, many experts think that deep-sea mining will be technologically feasible in a few decades, and several industrialized nations have staked claims in a region of the Pacific known for its large number of nodules.

Offshore reserves of oil have long been tapped as a major source of energy. However, obtaining oil and gas resources from the seafloor generally poses a threat to fishing. Fishermen and conservationists worry that Congress may allow oil and gas wells to threaten fisheries such as the Georges Bank fishery, which is already suffering due to decades of overfishing. The environmental concerns associated with extracting offshore energy resources are discussed in Chapter 17.

Manganese nodules on the ocean floor
FIGURE 11.14

These potato-sized nodules have enticed miners, but it isn't yet commercially feasible to obtain them. Photographed in the Pacific Ocean.

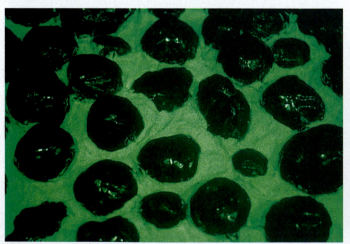

CLIMATE CHANGE, SEA-LEVEL RISE, AND WARMER TEMPERATURES

Our understanding of global climate is so incomplete that unanticipated effects from a globally warmed world will undoubtedly occur. For example, there could be a disruption of the ocean conveyor belt, which transports heat around the globe (see Figure 11.1B). Evidence from seafloor sediments and Greenland ice indicates that the ocean conveyor belt shifts from one equilibrium state to another in a relatively short period (a few years to a few decades). Scientists are concerned that human activities may affect this equilibrium.

Models based on the past behavior of the ocean conveyor belt suggest that climate warming, with its associated freshwater melting off the Greenland ice sheet, could weaken or even shut down the ocean conveyor belt in as short a period as a decade. Such changes in the ocean conveyor belt could cause major cooling in Europe, even while greater climate warming occurs elsewhere.

Until recently, climate scientists couldn't predict whether human-induced global climate change would affect El Niño and La Niña events in the tropical Pacific Ocean. Recent computer models indicate greater extremes of drying and heavy rainfall during El Niño events. Scientists are still uncertain whether El Niño events will occur more frequently with global climate change.

CONCEPT CHECK STOP

What are some of the harmful environmental effects associated with the fishing industry? with aquaculture?

How does the widespread use of plastics contribute to ocean pollution?

How might the effect of global climate change on the ocean alter the global climate?

Addressing Ocean Problems

LEARNING OBJECTIVES

Describe international initiatives that address problems in the global ocean.

Explain goals associated with correcting ocean problems in the future.

The many different threats to the world's ocean are attributed to a range of local, regional, national, and global sources. Problems in the ocean are complex and therefore require complicated solutions.

Industrialized countries' interest in removing manganese nodules from the ocean floor, first expressed in the 1960s, triggered the formation of an international treaty, the **U.N. Convention on the Law of the Sea (UNCLOS)**. UNCLOS, which became effective in 1994, is generally considered a "constitution for the ocean," and its focus is the protection of ocean resources. The provisions of UNCLOS aren't binding for territorial waters, only for international waters.

In 1995 the United Nations approved the **U.N. Fish Stocks Agreement**, the first international treaty to regulate marine fishing. The treaty went into effect in 2001. Because the overfishing problem continues to escalate, the United Nations has sponsored other fishery protection pacts. A plan to reduce fishing capacity by capping the size of fishing fleets and reducing the number of fishing vessels over time is under consideration.

In the United States, the **Magnuson–Stevens Fishery Conservation and Management Act** requires that the National Marine Fisheries Service and eight regional councils regulate fishing under U.S. jurisdiction. These groups are required to protect "essential fish habitat" for more than 600 fish species, reduce overfishing, rebuild the populations of overfished species, and minimize bycatch. Fishing quotas, restrictions of certain types of fishing gear, limits on the number of fishing boats, and closure of fisheries during spawning periods are some of the management tools used to reduce overfishing.

FUTURE ACTIONS

A 2004 report by the U.S. Commission on Ocean Policy, the first comprehensive review of federal ocean policy in 35 years, recommended improving the ocean and coasts in three main ways:

- **Create a new ocean policy to improve decision making**. Currently, an array of agencies and committees manages U.S. waters, and their respective goals often conflict. The commission recommends strengthening and reorganizing the National Oceanic and Atmospheric Administration (NOAA) and consolidating other federal ocean programs under it.

- **Strengthen science and generate information for decision makers**. There is a critical need for high-quality research on how marine ecosystems function and how human activities affect them.

- **Enhance ocean education to instill in citizens a stewardship ethic**. Environmental education should be part of the curriculum at all levels and should include a strong marine component.

Ensuring the recovery of depleted fisheries may require the establishment of networks of "no-take" reserves and a substantial reduction of fishing fleets. Governments will also have to reduce or remove subsidies that help support the fishing industry. (A **subsidy** is a form of government support given to a business or an institution to promote the activity performed by that business or institution.) Government subsidies encourage modernization and expansion of fishing fleets.

Many scientists think the best way to halt and reverse destruction of the ocean is to adopt an *ecosystem-based approach* to manage ocean environments. This means that rather than focus on a single, narrow goal such as reviving a specific fish population, ocean management should focus on preserving the health and function of the entire marine ecosystem.

Marine reserves FIGURE 11.15

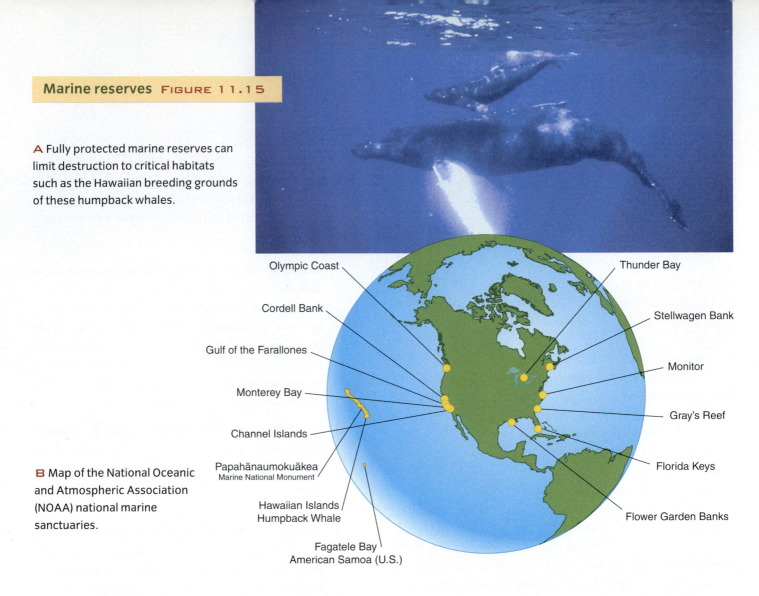

A Fully protected marine reserves can limit destruction to critical habitats such as the Hawaiian breeding grounds of these humpback whales.

B Map of the National Oceanic and Atmospheric Association (NOAA) national marine sanctuaries.

Olympic Coast
Cordell Bank
Gulf of the Farallones
Monterey Bay
Channel Islands
Papahānaumokuākea
Marine National Monument
Hawaiian Islands
Humpback Whale
Fagatele Bay
American Samoa (U.S.)

Thunder Bay
Stellwagen Bank
Monitor
Gray's Reef
Florida Keys
Flower Garden Banks

One proposed approach that would enhance ecosystem-based management would be to establish networks of fully protected marine reserves, within which no habitat destruction or resource extraction would be allowed. Currently less than 5 percent of U.S. marine environments are set aside as fully protected marine reserves, yet these areas have successfully preserved threatened habitats and increased populations of exploited organisms (**FIGURE 11.15**).

In 2006, President George W. Bush established the world's largest protected marine area when he designated the northwestern Hawaiian Islands and surrounding waters—an area almost as large as California—as a national monument. This designation provides permanent funding to manage and preserve the area, now named the Papahānaumokuākea Marine National Monument. This protected area is home to more than 7,000 species, including seabirds, fishes, marine mammals, coral reef colonies, and other organisms, approximately one-quarter of which are found only there.

Initiatives such as the Papahānaumokuākea Marine National Monument will require billions of dollars and years of effort to be fully realized. Like most other countries, the United States recognizes the importance of the ocean to life on this planet. However, it remains to be seen if the United States and other countries will make a strong commitment to protecting and managing the global ocean.

CONCEPT CHECK STOP

Which international treaties aim to protect ocean resources?

What are the main recommendations of the U.S. Commission on Ocean Policy?

HUMANS AND THE ANTARCTIC FOOD WEB

Although the icy waters around Antarctica seem inhospitable, they shelter a complex food web. The base of the web is microscopic algae, which live in vast numbers in the well-lit, nutrient-rich water. A huge population of herbivores—tiny shrimplike **krill**— eat these marine algae (see photo). Krill in turn support a variety of larger animals.

Baleen whales—including blue, humpback, and right whales—are major consumers of krill. Until a 1986 global ban, whaling steadily reduced the numbers of baleen whales in Antarctic waters. As a result, more krill became available for other krill-eating animals, whose populations then increased. Seals, penguins, and smaller baleen whales replaced the large baleen whales as the main eaters of krill. Now that commercial whaling is regulated, it is hoped that the number of large baleen whales will increase, although it is impossible to say whether they will return to their former dominance in the food web in terms of krill consumption.

Scientists are concerned that ozone thinning over Antarctica may damage the algae that form the base of the food web. Increased ultraviolet radiation is penetrating the surface waters around Antarctica, and algal productivity has declined. (The problem of stratospheric ozone depletion is discussed in detail in Chapter 9.)

Another human-induced change that may be responsible for declines in certain Antarctic populations is global climate change. As the water around Antarctica warms, less pack ice forms during winter months. Large numbers of marine algae are found in and around the pack ice, providing a critical supply of food for the krill. Below-average pack ice cover means less algae, which mean less krill. (Global climate change, including its effect on Adélie penguins in Antarctica, is discussed in Chapter 9.)

To complicate matters, commercial fishermen have started harvesting krill to make fishmeal for aquaculture operations. Scientists caution that the human harvest of krill may endanger the marine animals that depend on it for food.

Krill support Antarctic consumers.

SUMMARY

1 The Global Ocean

1. The global ocean is a huge body of salt water that surrounds the continents. It affects the hydrologic and other cycles of matter, influences climate and weather, and provides food to millions.

2. Prevailing winds over the ocean generate **gyres**, large, circular ocean current systems that often encompass an entire ocean basin. The **Coriolis effect** is a force resulting from Earth's rotation that influences the paths of surface ocean currents, which move in a circular pattern, clockwise in the Northern Hemisphere and counterclockwise in the Southern Hemisphere.

3. The ocean and the atmosphere are strongly linked. The **El Niño–Southern Oscillation (ENSO)** event, which is responsible for much of Earth's interannual climate variability, is a periodic, large-scale warming of surface waters of the tropical eastern Pacific Ocean that temporarily alters both ocean and atmospheric circulation patterns. A **La Niña** event occurs when surface water in the eastern Pacific Ocean become unusually cool. Its effects on weather patterns are less predictable than an ENSO event's.

2 Major Ocean Life Zones

1. The vast ocean is subdivided into major life zones. The biologically productive **intertidal zone** is the area of shoreline between low and high tides. The **benthic environment** is the ocean floor, which extends from the intertidal zone to the deep ocean trenches. Most of the benthic environment consists of sediments where many animals burrow. Common benthic habitats include sea grass beds, kelp forests, and coral reefs. The **pelagic environment** is divided into two provinces. The **neritic province** is the part of the pelagic environment from the shore to where the water reaches a depth of 200 m. Organisms that live in the neritic province are all floaters or swimmers. The **oceanic province**, "the deep sea," is the part of the pelagic environment where the water depth is greater than 200 m. The oceanic province is the largest marine environment, comprising about 75 percent of the ocean's water.

3 Human Impacts on the Ocean

1. The most serious problem for marine fisheries is the overharvesting of many species to the point that their numbers are severely depleted. Fishermen usually concentrate on a few fish species with high commercial value. In doing so, they also catch **bycatch**, fishes, marine mammals, sea turtles, seabirds, and other animals caught unintentionally in a commercial fishing catch and then discarded. **Aquaculture** is the growing of aquatic organisms (fishes, shellfish, and seaweeds) for human consumption. Aquaculture is common in developing nations with abundant cheap labor, and it is limited by the size of the space dedicated to cultivation. Aquaculture produces wastes that pollute the adjacent water and also causes a net loss of wild fish because many of the fishes farmed are carnivorous.

2. **Marine pollution** is generated by many human activities, including the release of trash and contaminants through commercial shipping, ocean dumping of sludge and industrial wastes, and discarding of plastics that are potentially harmful to marine organisms. Marine environments are also deteriorated by coastal development and the extraction of offshore minerals.

3. The **ocean conveyor belt** moves cold, salty, deep-sea water from higher to lower latitudes, affecting regional and possibly global climate. Global climate change associated with human activities may alter the link between the ocean conveyor belt and global climate.

4 Addressing Ocean Problems

1. International initiatives aimed at protecting the global ocean include the **U.N. Convention on the Law of the Sea (UNCLOS)**, a "constitution for the ocean" that protects ocean resources, and the **U.N. Fish Stocks Agreement**, the first international treaty to regulate marine fishing.

2. Long-term goals for halting and reversing destruction of the ocean focus on adopting an ecosystem-based approach to management of ocean environments. Consolidating ocean programs, funding research on marine ecosystems, and enhancing ocean education to instill in citizens a stewardship ethic could improve U.S. ocean policy.

KEY TERMS

CRITICAL AND CREATIVE THINKING QUESTION

1. How do ocean currents affect climate on land? In particular, describe the role of the ocean conveyor belt.

2. Compare the different global effects of El Niño with those of La Niña. How are the two events similar? How are they different?

3. Identify which of the ocean life zones would be home to each of the following organisms: giant squid, kelp, tuna, and mussels. Explain your answers.

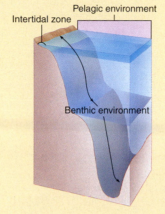

4. Explain how human activities impact coral reefs.

5. How might the production of plastic shopping bags contribute to ocean pollution?

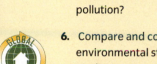

6. Compare and contrast the environmental stresses faced by blue fin tuna populations in the Mediterranean and by krill in Antarctica. Explain the role of human activities in each case.

7. Imagine that you live in a small Atlantic coast community where a company wants to set up an aquaculture facility in a salt marsh. What are its benefits and its environmental drawbacks? Would you support or oppose this proposal? Explain your answer.

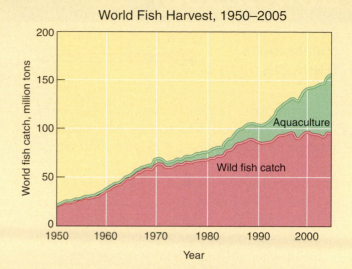

World Fish Harvest, 1950–2005

8. From 1950 to 1980, did most of the growth in the world fish harvest occur in the wild fish catch or in aquaculture?

9. From 1980 to 2005 (most recent data available), did the contribution to the global fish harvest increase or decrease? In 2005, what percentage (approximately) of the global fish catch was from aquaculture?

10. A 2003 study reports that in recent years, tropical ocean waters have become saltier, whereas polar ocean waters have become less salty. What is a possible explanation for these changes?

What is happening in this picture ?

These divers in Florida are scientists conducting underwater experiments to learn more about damage to coral reefs. What climate change are they likely studying?

What controlled differences might there be in the enclosed habitats containing the miniature coral reefs?

Why do you suppose the scientists are conducting their experiments in this habitat rather than in a laboratory?

Mineral and Soil Resources

COPPER BASIN, TENNESSEE

VIEW THIS IN ACTION
in your WileyPLUS course

Copper Basin, Tennessee, provides an example of environmental degradation caused by smelting, a stage of mineral processing. During the 19th century, mining companies in southeastern Tennessee extracted copper ore—rock containing copper—from the ground and dug vast pits to serve as open-air smelters. They cut down the surrounding trees to fire the smelters, producing the high temperatures needed for the separation of copper from other substances in the ore. One of these substances, sulfur, reacted with oxygen in the air to form sulfur dioxide. The sulfur dioxide entered the atmosphere, reacted with water vapor there, and became sulfuric acid that returned to Copper Basin as acid precipitation.

Ecological ruin took only a few short years (see larger photograph). Acid precipitation killed plants. Without plants to hold the soil in place, erosion cut gullies in the rolling hills. The forest animals disappeared along with the plants, their food and shelter destroyed.

State and federal reclamation efforts were only marginally successful until the 1970s, when specialists began using new replanting techniques. The new plants had a greater survival rate, and as they became established, their roots held the soil in place (see inset). Birds and field mice slowly began to return.

Today, reclamation of Copper Basin continues; the goal is to have the entire area under plant cover early in the 21st century. The return of the original forest ecosystem will take at least a century or two. Plant scientists and land reclamation specialists have learned a lot from Copper Basin, and they will put this knowledge to use in future reclamation projects around the world.

NATIONAL GEOGRAPHIC

Pharmaceutical
plant

Greenhouses

Fish farming

Electric
power plant

Plate Tectonics and the Rock Cycle

LEARNING OBJECTIVES

Define *plate tectonics* and explain its relationship to earthquakes and volcanic eruptions.

Diagram a simplified version of the rock cycle.

A The main layers of planet Earth

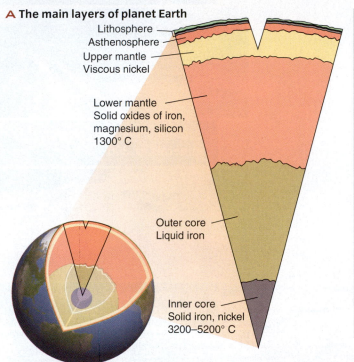

- Lithosphere
- Asthenosphere
- Upper mantle
  Viscous nickel
- Lower mantle
  Solid oxides of iron,
  magnesium, silicon
  1300° C
- Outer core
  Liquid iron
- Inner core
  Solid iron, nickel
  3200–5200° C

G eology is an essential part of environmental science. To better understand the environmental effects of humans on mineral and soil resources, you must first know something about the geologic properties of Earth's crust.

Earth's outermost rigid rock layer (the *lithosphere*) is composed of seven large plates, plus a few smaller ones, that float on the *asthenosphere*, the region of the mantle where rocks become hot and soft (**FIGURE 12.1**). Continents and landmasses are situated on some of these plates. As the plates move across Earth's surface, the continents change their relative positions. **Plate tectonics**, the study of the movement of these plates, explains how most features on Earth's surface originate.

> **plate tectonics**
> The study of the processes by which the lithospheric plates move over the asthenosphere.

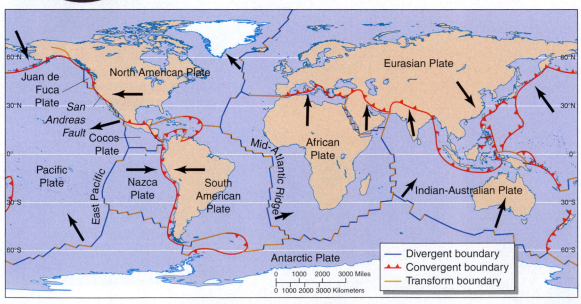

B Plates and plate boundary locations There are seven major independent plates that move horizontally across Earth's surface. Arrows show the directions of plate movements. The three types of plate boundaries are explained in Figure 12.2.

Map labels: North American Plate, Juan de Fuca Plate, San Andreas Fault, Cocos Plate, Pacific Plate, East Pacific, Nazca Plate, South American Plate, Mid-Atlantic Ridge, Eurasian Plate, African Plate, Indian-Australian Plate, Antarctic Plate, 60°N, 30°N, 0°, 30°S, 60°S

Legend:
— Divergent boundary
▲▲ Convergent boundary
— Transform boundary

0 1000 2000 3000 Miles
0 1000 2000 3000 Kilometers

Earth's layers and surface structure FIGURE 12.1

An area where two plates meet—a **plate boundary**—is a site of intense geologic activity (**FIGURE 12.2**). Earthquakes and volcanoes are common in these regions. San Francisco, California (noted for its earthquakes), and the volcano Mount Saint Helens in Washington State are both situated on plate boundaries. Where landmasses meet on the boundary between two plates, mountains may form: When the plate carrying India rammed into the plate carrying Asia, the resulting pressure pushed up the Himalayas. When two plates collide, one of them is sometimes forced under the other, in the process of **subduction**. When two plates move apart, a ridge of molten rock from the mantle wells up between them, continually expanding as the plates move farther apart. The Atlantic Ocean is growing as a result of the buildup of lava along the Mid-Atlantic Rift Zone, where two plates are separating.

VOLCANOES

The movement of tectonic plates on the hot, soft rock of the asthenosphere causes most volcanic activity. In places where the asthenosphere is close to the surface, heat from this part of Earth's mantle melts the surrounding rock, forming pockets of **magma**. When one plate slides under or away from another, this magma may rise to the surface, often forming volcanoes. Magma that reaches the surface is called **lava**.

Volcanoes occur at three kinds of locations: in subduction zones, at spreading centers, and above hot spots. Subduction zones around the Pacific Basin have given rise to hundreds of volcanoes around Asia and the Americas, known as the *Ring of Fire*. Iceland is a volcanic island that formed along the Mid-Atlantic Rift Zone as the adjoining plates there spread apart. The volcanic Hawaiian Islands formed as the Pacific plate moved over a **hot spot**, a rising plume of magma that flowed from an undersea opening in Earth's crust.

The largest volcanic eruption in the 20th century occurred in 1991 when Mount Pinatubo in the Philippines exploded (see Figure 9.9). Despite the evacuation of more than 200,000 people, several hundred deaths occurred, mostly from the collapse of buildings under the thick layer of wet ash that blanketed the area. The lava and ash ejected into the atmosphere by the eruption blocked much of the sun's warmth and caused a slight cooling of global temperatures for a year or so.

Plate boundaries FIGURE 12.2

All three types of plate boundaries occur both in the ocean and on land.

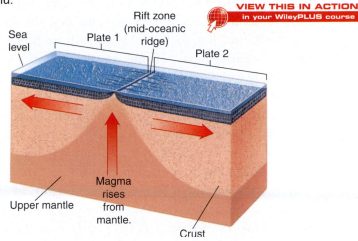

A Two plates move apart at a divergent plate boundary.

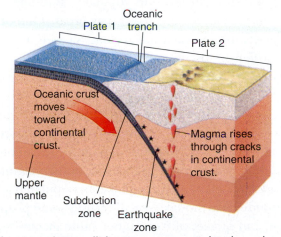

B When two plates collide at a convergent plate boundary in the seafloor, subduction may occur. Convergent collision can also form a mountain range (not shown).

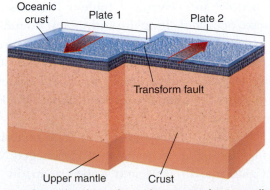

C At a transform plate boundary, plates move horizontally in opposite but parallel directions. On land, such a boundary is often evident as a long, thin valley due to erosion along the fault line.

EARTHQUAKES

Forces inside Earth sometimes push and stretch rocks in the lithosphere. The rocks absorb this energy for a time, but eventually, as the energy accumulates, the stress is too great, and the rocks suddenly shift or break. The energy—released as **seismic waves**, vibrations that rapidly spread through rock in all directions—causes one of the most powerful events in nature, an earthquake. Most earthquakes occur along **faults**, fractures in the crust where rock moves forward and backward, up and down, or from side to side. Fault zones are often found at plate boundaries.

The site where an earthquake begins, often far below the surface, is the **focus** (**FIGURE 12.3**). Directly above the focus, at Earth's surface, is the earthquake's **epicenter**. When seismic waves reach the surface, they cause the ground to shake. Buildings and bridges may collapse, and roads may break. One of the instruments used to measure seismic waves is a seismograph, which helps seismologists (scientists who study earthquakes) determine where an earthquake started, how strong it was, and how long it lasted.

Seismologists record more than 1 million earthquakes each year. Some of these are major, but most are too small for humans to feel, equivalent to readings of about 2 on the *Richter scale*, a measure of the magnitude of energy released by an earthquake. In populated areas, a magnitude 5 earthquake usually causes some property damage, and quakes of 8 or higher cause massive property destruction and kill large numbers of people. In 2005, a 7.6-magnitude earthquake devastated a mountainous region in northern Pakistan, killing as many as 87,000 people and leaving 2 million homeless. The earthquake occurred along the convergent boundary of the Indian and Eurasian plates (Figure 12.1B), a region known for its high tectonic activity. The Indian plate is thrusting northward and being subducted under the Eurasian plate, the same process that formed the Himalayan mountain range.

Not all earthquakes occur at plate boundaries. Some, like the major earthquakes that damaged Northridge, California, in 1994 and Kobe, Japan, in 1995, occur on smaller faults that crisscross the large plates. Such earthquakes pose a seismic hazard because they are difficult to predict. Major quakes may occur on larger

Earthquakes FIGURE 12.3

Earthquakes occur when plates along a fault suddenly move in opposite directions relative to one another. This movement triggers seismic waves that radiate through the crust.

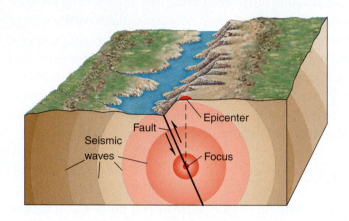

fault lines every century or so, but along any given small fault line, like those in Northridge and Kobe, only every 1,000 to 5,000 years.

Side effects of earthquakes include landslides and tsunamis. A *landslide* is an avalanche of rock, soil, and other debris that slides swiftly down a mountainside. A 1970 earthquake in Peru resulted in a landslide that buried the town of Yungay and killed 17,000 people.

A *tsunami*, a giant sea wave caused by an underwater earthquake or volcanic eruption, may sweep across the ocean at more than 750 km (450 mi) per hour. Although a tsunami may be only about 1 m (3 ft) high in deep ocean water, it can build to a wall of water 30.5 m (100 ft)—as high as a 10-story building—when it comes ashore, often far from where the original earthquake triggered it. Tsunamis have caused thousands of deaths, particularly along the Pacific coast. Although the Pacific Tsunami Warning System monitors submarine earthquakes and warns people of approaching tsunamis, deaths still occur because there is so little time to respond and not all nations are part of the network.

One of the deadliest natural disasters in modern history was the tsunami generated by an Indian Ocean earthquake in 2004 that killed more than 225,000 peo-

The rock cycle FIGURE 12.4

Rocks do not remain in their original form forever. This highly simplified diagram shows how rock cycles from one form to another.

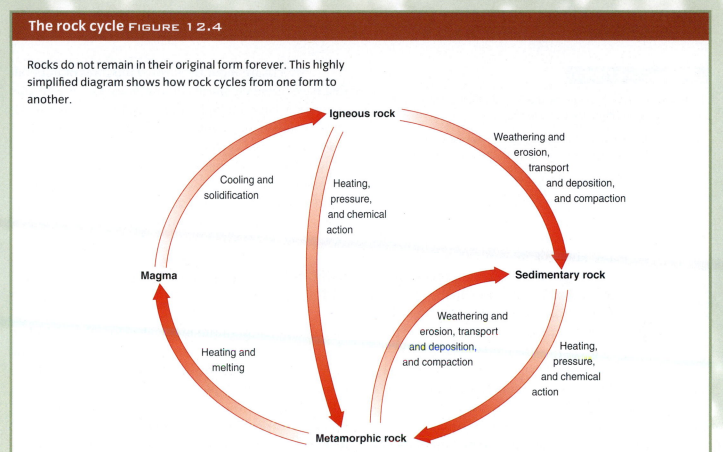

ple along the coasts of South India, Thailand, Sri Lanka, Indonesia, and eastern Africa. The earthquake was recorded as a magnitude 9 or higher, and the waves it generated reached heights of 30.5 m (100 ft). Unfortunately, no gauges or buoys were in place in the Indian Ocean to detect the tsunami as it developed there.

THE ROCK CYCLE

Rocks, which are aggregates of one or more minerals, fall into three categories, based on how they formed: igneous, metamorphic, and sedimentary. **Igneous rocks** form when magma rises from the mantle and cools. **Metamorphic rocks** form when intense heat and pressure alter igneous, sedimentary, or other metamorphic rocks. **Sedimentary rocks** form when small fragments of weathered, eroded rocks (or marine organisms) are deposited, compacted, and cemented together.

Earth's internal structure and the basic geologic processes that we have presented in this chapter result in a **rock cycle**, in which rock moves from one physical state or location to another (**FIGURE 12.4**). The rock cycle is similar to the other cycles of matter, such as the carbon and hydrologic cycles (see Chapter 5). However, rocks are formed and move through the environment much more slowly than the elements of the other cycles.

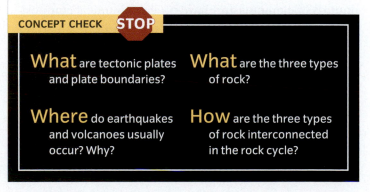

CONCEPT CHECK **STOP**

What are tectonic plates and plate boundaries?

What are the three types of rock?

Where do earthquakes and volcanoes usually occur? Why?

How are the three types of rock interconnected in the rock cycle?

Economic Geology: Useful Minerals

LEARNING OBJECTIVES

Define *minerals* and contrast the consumption of minerals by developing and highly developed countries.

Distinguish between surface mining and subsurface mining, using the terms *overburden* and *spoil bank* in your answer.

Describe briefly the process of smelting.

E arth's outermost layer, the crust, contains many kinds of minerals that are of economic importance. We now focus on the economic and environmental impacts of extracting and using mineral resources. We then consider soil, the part of the crust where biological and physical processes meet.

Minerals are such an integral part of our lives that we often take them for granted. Steel, an essential building material, is a blend of iron and other metals. Beverage cans, aircraft, automobiles, and buildings all contain aluminum. Copper, which readily conducts electricity, is used for electrical and communications wiring. The concrete used in buildings and roads is made from sand and gravel, as well as cement, which contains crushed limestone. Sulfur, a component of sulfuric acid, is an indispensable industrial mineral. It is used to make plastics and fertilizers and to refine oil. Other important minerals include platinum, mercury, manganese, and titanium.

> **minerals** Elements or compounds of elements that occur naturally in Earth's crust.

Earth's minerals are elements or (usually) compounds of elements and have precise chemical compositions. **Sulfides** are mineral compounds in which certain elements are combined chemically with sulfur, and **oxides** are mineral compounds in which elements are combined chemically with oxygen. Minerals are metallic or nonmetallic (**FIGURE 12.5**). **Metals** are minerals such as iron, aluminum, and copper, which are malleable, lustrous, and good conductors of heat and electricity. **Nonmetallic minerals**, such as sand, stone, salt, and phosphates, lack these characteristics.

Rocks are naturally formed mixtures of minerals that have varied chemical compositions. **Ore** is rock that contains a large enough concentration of a particular mineral to be profitably mined and extracted. **High-grade ores** contain relatively large amounts of particular minerals, whereas **low-grade ores** contain lesser amounts. Although some minerals are abundant, all minerals are nonrenewable resources that are not replenished by natural processes on a human timescale.

MINERALS: AN ECONOMIC PERSPECTIVE

At one time, most of the highly developed nations had abundant mineral deposits that enabled them to industrialize. In the process of industrialization, these countries largely deplete their domestic reserves of minerals so that they must increasingly turn to developing countries. This is particularly true for countries in Europe, Japan, and, to a lesser extent, the United States.

As with the consumption of other natural resources, there is a large difference in consumption of minerals between highly developed and developing countries. The United States and Canada, which have about 5.1 percent of the world's population, consume about 25 percent of many of the world's metals. It is too simplistic, however, to divide the world into two groups, the mineral consumers (highly developed countries) and the mineral producers (developing countries). Four of the world's top five mineral producers are highly developed countries: the United States, Canada, Australia, and the Russian Federation. South Africa, a moderately developed, middle-income country, is the other mineral producer in the top five. Many developing countries lack any significant mineral deposits.

Mineral production in China is increasing dramatically, as is China's mineral consumption as the country industrializes. For example, in 2004 China smelted more than 20 percent of the world's primary aluminum (aluminum obtained from ores and not from recycling).

Some important minerals and their uses FIGURE 12.5

Gypsum, silicon, and sulfur are nonmetals. All other minerals shown are metals.

Aluminum	Chromium	Cobalt	Gold
Aircraft, motor vehicles, packaging (cans, foil), water treatment	Chrome plate, dyes and paints, steel alloys (cutlery)	Corrosion and wear-resistant alloys, pigments (cobalt blue)	Jewelry, money, restorative dentistry
Iron	**Magnesium**	**Mercury**	**Molybdenum**
Steel (alloy of iron) buildings and machinery	Beverage cans, electronic devices, firecrackers, flares	Industrial chemicals, electric and electronic applications, batteries	High-temperature alloys for aircraft, industrial motors
Nickel	**Potassium**	**Silver**	**Titanium**
Coins, metal plating, alloys with various uses	Fertilizers, photography	Jewelry, silverware, photography, electronics	Alloy in steel and other industrial alloys, pigment in paints, plastics
Zinc	**Gypsum ($CaSO_4 \cdot 2H_2O$)**	**Silicon**	**Sulfur**
Galvanizing steel, alloys (brass), anode in alkaline batteries	Drywall, plaster of Paris, soil conditioner	Electronic devices, semiconductors, natural stone, glass, concrete	Industrial chemicals, insecticides, gunpowder, vulcanized tires

China also consumed almost all of this aluminum, making it the world's largest producer and largest consumer of primary aluminum.

Because industrialization increases the demand for minerals, developing countries that at one time met their mineral needs with domestic supplies become increasingly reliant on foreign supplies as development occurs.

HOW MINERALS ARE EXTRACTED AND PROCESSED

The process of making mineral deposits available for human consumption occurs in several steps. First, a particular mineral deposit is located. Geologic knowledge of Earth's crust and how minerals are formed is used to estimate locations of possible mineral deposits. Once these sites are identified, geologists drill or tunnel for mineral samples and analyze their composition. Second, mining extracts the mineral from the ground. Third, the mineral is processed, or refined, by concentrating it and removing impurities. Finally, the purified mineral is used to make a product.

Extracting minerals
The depth of a particular mineral deposit determines whether surface or subsurface mining will be used. In **surface mining**, minerals are extracted near the surface. Surface mining is more common because it is less expensive than subsurface mining. Because even surface mineral deposits occur in rock layers beneath Earth's surface, the overlying soil and rock layers, called **overburden**, must first be removed, along with the vegetation growing in the soil. Then giant power shovels scoop out the minerals.

There are two kinds of surface mining, **open-pit surface mining** and strip mining. Iron, copper, stone, and gravel are usually extracted by open-pit surface mining, in which a giant hole, called a quarry, is dug in the ground to extract the minerals (**FIGURE 12.6A**). In **strip**

> **surface mining**
> The extraction of mineral and energy resources near Earth's surface by first removing the soil, subsoil, and overlying rock strata.

> **overburden**
> Soil and rock overlying a useful mineral deposit.

Types of mining operations FIGURE 12.6

A Copper is extracted from this open-pit surface mine near Tucson, Arizona.

B Strip mining removes overburden along narrow strips to reach the ore beneath.

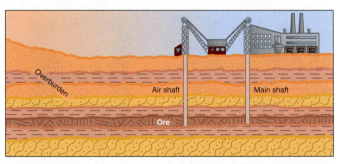

C In a shaft mine, a hole is dug straight through the overburden to the ore, which is removed up through the shaft in buckets.

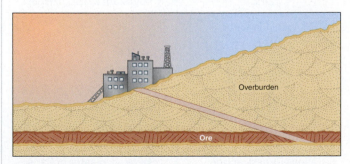

D In a slope mine, an entry to the ore is dug at an angle so that the ore can be hauled out in carts.

spoil bank A hill of loose rock created when the overburden from a new trench is put into the already excavated trench during strip mining.

subsurface mining The extraction of mineral and energy resources from deep underground deposits.

mining, a trench is dug to extract the minerals (**FIGURE 12.6B**). Then a new trench is dug parallel to the old one, and the overburden from the new trench is put into the old one, creating a hill of loose rock called a **spoil bank**.

Subsurface mining extracts minerals too deep in the ground to be removed by surface mining. It disturbs the land less than surface mining, but it is more expensive and more hazardous for miners. There is always a risk of death or injury from explosions or collapsing walls, and prolonged breathing of dust in subsurface mines can result in lung disease.

Subsurface mining may be done with underground shaft mines or slope mines. A **shaft mine**, often used for mining coal, is a direct vertical shaft to the vein of ore (**FIGURE 12.6C**). The ore is broken up underground and then hoisted through the shaft to the surface in buckets. A **slope mine** has a slanting passage that makes it possible to haul the broken ore out of the mine in cars rather than to hoist it up in buckets (**FIGURE 12.6D**). Sump pumps keep a subsurface mine dry, and a second shaft is usually installed for ventilation.

Processing minerals

Processing minerals often involves **smelting**. Purified copper, tin, lead, iron, manganese, cobalt, or nickel smelting is done in a blast furnace. **FIGURE 12.7** shows a blast furnace used to smelt iron. The iron ore reacts with coke (modified coal) to form molten iron and carbon dioxide. The limestone reacts with impurities in the ore to form a molten mixture called **slag**. Note the vent near the top of the iron smelter for exhaust gases. If air pollution control devices are not installed, many dangerous gases are emitted during smelting.

smelting The process in which ore is melted at high temperatures to separate impurities from the molten metal.

Blast furnace FIGURE 12.7

Towerlike furnaces separate metal from impurities in the ore. The energy for smelting comes from a blast of heated air.

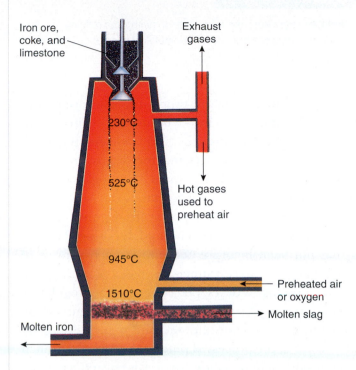

Iron ore, coke, and limestone

Exhaust gases

230°C

525°C

Hot gases used to preheat air

945°C

1510°C

Preheated air or oxygen

Molten slag

Molten iron

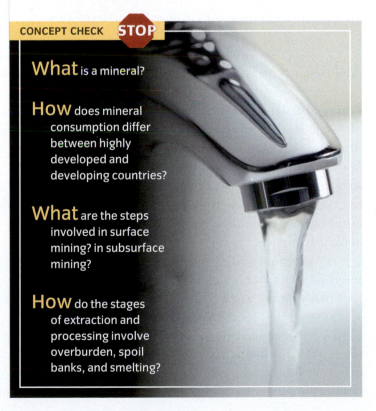

CONCEPT CHECK STOP

What is a mineral?

How does mineral consumption differ between highly developed and developing countries?

What are the steps involved in surface mining? in subsurface mining?

How do the stages of extraction and processing involve overburden, spoil banks, and smelting?

Economic Geology: Useful Minerals **303**

Environmental Implications of Mineral Use

LEARNING OBJECTIVES

Relate the environmental impacts of mining and refining minerals. Include a brief description of acid mine drainage.

Explain how mining lands can be restored.

The extraction, processing, and disposal of minerals clearly harm the environment. Mining disturbs and damages the land, and the processing and disposal of minerals pollute the air, soil, and water. Although pollution can be controlled and damaged lands can be restored, these remedies are costly. Historically, the environmental cost of extracting, processing, and disposing of minerals has not been incorporated into the actual price of mineral products to consumers.

Most highly developed countries have regulatory mechanisms in place to minimize environmental damage from mineral consumption, and many developing nations are in the process of putting them in place. Such mechanisms include policies to prevent or reduce pollution, restore mining sites, and exclude certain recreational and wilderness sites from mineral development.

MINING AND THE ENVIRONMENT

Mining, particularly surface mining, disturbs large areas of land. In the United States, functioning and abandoned metal and coal mines occupy an estimated 9 million hectares (22 million acres). Because mining destroys existing vegetation, this land is particularly prone to erosion, with wind erosion causing air pollution and water erosion polluting nearby waterways and damaging aquatic habitats.

Open-pit mining of gold and other minerals uses huge quantities of water. As miners dig deeper, they eventually hit the water table and pump out the water to keep the pit dry. Farmers and ranchers in open-pit mining areas are concerned about depletion of the groundwater they need for irrigation. Environmentalists and others would like the mining operations to reinject the water into the ground after pumping it out.

Mining has contaminated thousands of kilometers of streams and rivers in the United States. Rocks rich in minerals often contain high concentrations of heavy metals such as arsenic and lead. Rainwater seeping through the sulfide minerals in mine waste produces sulfuric acid, which dissolves the heavy metals and other toxic substances in the spoil banks. These acids, called **acid mine drainage**, are highly toxic and are washed into soil and water, including groundwater, by precipitation runoff (**FIGURE 12.8**). When such acids and toxic compounds make their way into nearby lakes and streams, particularly through "toxic pulses" of thunderstorms or spring snowmelt, they adversely affect aquatic life.

> **acid mine drainage** Pollution caused when sulfuric acid and dangerous dissolved materials such as lead, arsenic, and cadmium wash from mines into nearby lakes and streams.

ENVIRONMENTAL IMPACTS OF REFINING MINERALS

Approximately 80 percent of mined ore consists of impurities that become wastes after processing. These wastes, called **tailings**, are usually left in giant piles on the ground or in ponds near the processing plants (**FIGURE 12.9**). The tailings contain toxic materials such as cyanide, mercury, and sulfuric acid. Left exposed, they contaminate the air, soil, and water.

Smelting plants may emit large quantities of air pollutants during mineral processing, particularly sulfur. Unless expensive pollution control devices are added to smelters, the sulfur escapes into the atmosphere, where it forms sulfuric acid. (The environmental implications of the resulting acid precipitation are discussed in Chapter 9.) Pollution control devices for

Gold is a precious metal used primarily for jewelry and as a medium of exchange in many countries. Annual global gold production in 2006 was about 2,500 tons, up from 1,200 tons in 1980. Worldwide demand for gold is increasing, and the environment is suffering from the increased mining. The waste from mining and processing ore is enormous: Six tons of wastes are produced to yield enough gold to make two wedding rings. One technology, *cyanide heap leaching*, allows profitable mining when minuscule amounts of gold are present, but this process produces up to 3 million pounds of waste for every pound of gold produced. The world's largest gold mine, located in Indonesia but owned by a U.S. company, dumps more than 100,000 tons of cyanide-contaminated waste into the local river each day. The highly toxic cyanide threatens waterfowl and fishes, as well as underground drinking water supplies.

Small-scale miners use other extraction techniques with destructive side effects: soil erosion, production of silt that clogs streams and threatens aquatic organisms, and contamination from mercury used to extract the gold. The environmental hazards of gold mining do not end when the gold is carried away: If not disposed of properly, mining wastes cause long-term problems such as acid mine drainage and heavy-metal contamination.

Acid mine drainage FIGURE 12.8

For more than 100 years, the Leadville mining district in the central Colorado Rockies has been mined for silver, gold, lead, and zinc, which occur as sulfide deposits. Shown is the characteristic orange-red acid runoff that contains sulfuric acid contaminated with lead, arsenic, cadmium, silver, and zinc. The runoff here drains into the Arkansas River from snowmelt and precipitation runoff.

Copper ore tailings FIGURE 12.9

Tailings dumped in mountainous heaps cause air, soil, and water pollution and have serious effects on land use. In highly developed countries, responsible mining companies are developing methods to dispose of tailings in environmentally sustainable ways. However, in developing countries, where there are few environmental regulations, disposal of tailings is a far more serious problem. Photographed in southern Illinois.

smelters are the same as the devices used for the burning of sulfur-containing coal—scrubbers and electrostatic precipitators.

Contaminants in ores include the heavy metals lead, cadmium, arsenic, and zinc. These toxic elements may pollute the atmosphere during the smelting process and cause harm to humans. In addition to airborne pollutants, smelters emit hazardous liquid and solid wastes that can pollute the soil and water.

One of the most significant environmental impacts of mineral production is the large amount of energy required to mine and refine minerals, particularly if they are being refined from low-grade ore. Most of this energy is obtained by burning fossil fuels, which depletes nonrenewable energy reserves and produces carbon dioxide and other air pollutants.

RESTORATION OF MINING LANDS

When a mine is no longer profitable to operate, the land can be reclaimed, or restored to a seminatural condition, as has been done to most of the Copper Basin in Tennessee (see the chapter introduction). Reclamation prevents further degradation and erosion of the land, eliminates or neutralizes local sources of toxic pollutants, and makes the land productive for purposes other than mining (**FIGURE 12.10**). Restoration also makes such areas visually attractive.

Restoring land degraded by mining—called **derelict land**—involves filling in and grading the area to the shape of its natural contours and then planting vegetation to hold the soil in place. Often the topsoil is completely gone or contains toxic levels of metals, so special types of plants that tolerate such a challenging environment must be used.

The **Surface Mining Control and Reclamation Act** of 1977 requires reclamation of areas that were surface mined for coal. However, no federal law is in place to require restoration of derelict lands produced by other kinds of mines. As a result, restoration of mining lands often does not occur.

CONCEPT CHECK **STOP**

What are three harmful environmental effects of mining and processing minerals?

How are mining lands restored?

Restoration of mining lands
FIGURE 12.10

Part of a phosphate mine near Fort Meade, Florida, was reclaimed and is currently used as a pasture (background). The unrestored area that remains is in the foreground. Restoration of mining lands makes them usable once again, or at least stabilizes them so that further degradation does not occur.

Soil Properties and Processes

LEARNING OBJECTIVES

Define *soil* and identify the factors involved in soil formation.

Describe the composition of soil and the organization of soil into horizons.

Relate at least two ecosystem services performed by soil organisms and briefly discuss nutrient cycling.

The relatively thin surface layer of Earth's crust is **soil**, which consists of mineral and organic matter modified by the natural actions of agents such as weather, wind, water, and organisms. It is easy to take soil for granted. We walk on and over it throughout our lives but rarely stop to think about how important it is to our survival. Vast numbers and kinds of organisms, mainly microorganisms, inhabit soil and depend on it for shelter, food, and water. Plants anchor themselves in soil, and from it they receive essential minerals and water. Terrestrial plants could not survive without soil, and because we depend on plants for our food, humans could not exist without soil either (**FIGURE 12.11**).

> **soil** The uppermost layer of Earth's crust, which supports terrestrial plants, animals, and microorganisms.

SOIL FORMATION AND COMPOSITION

Soil is formed from *parent material,* rock that is slowly broken down, or fragmented, into smaller and smaller particles by biological, chemical, and physical **weathering processes**. It takes a long time, sometimes thousands of years, for rock to disintegrate into finer and finer mineral particles. Time is also required for organic material to accumulate in the soil. Soil formation is a continuous process that involves interactions between Earth's solid crust and the biosphere. The weathering of parent material beneath already formed soil continues to add new soil.

Topography, a region's surface features (such as the presence or absence of mountains and valleys), is also involved in soil formation. Steep slopes often have little or no soil on them because soil and rock are contin-ually transported down the slopes by gravity. Runoff from precipitation tends to amplify erosion on steep slopes. Moderate slopes and valleys, on the other hand, may encourage the formation of deep soils.

A farmer plows his soil prior to planting crops
FIGURE 12.11

Soil is an important natural resource that humans and countless soil organisms rely on.

Soil Profile

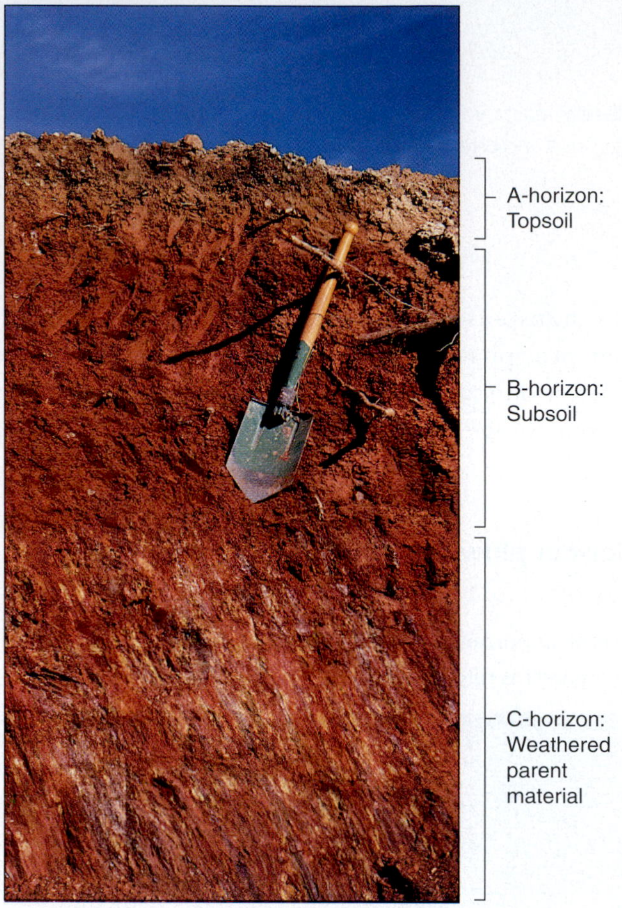

A-horizon: Topsoil

B-horizon: Subsoil

C-horizon: Weathered parent material

A This soil, located on a farm in Virginia, has no O-horizon because it is used for agriculture; the surface litter that would normally compose the O-horizon was plowed into the A-horizon. The shovel gives an idea of the relative depths of each horizon.

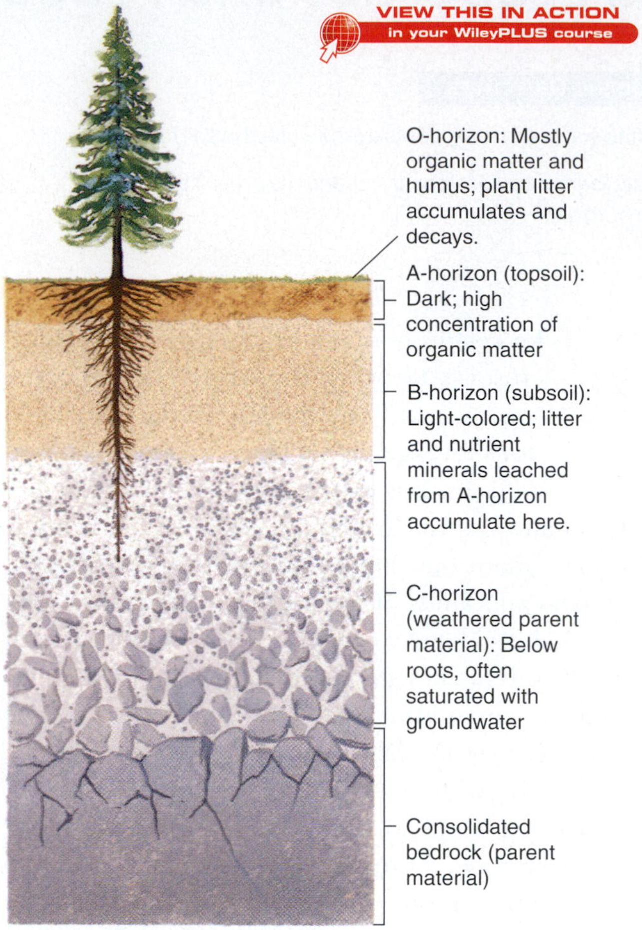

O-horizon: Mostly organic matter and humus; plant litter accumulates and decays.

A-horizon (topsoil): Dark; high concentration of organic matter

B-horizon (subsoil): Light-colored; litter and nutrient minerals leached from A-horizon accumulate here.

C-horizon (weathered parent material): Below roots, often saturated with groundwater

Consolidated bedrock (parent material)

B A "typical" soil profile, as it appears to the trained eye of a soil scientist. Each horizon has its own chemical and physical properties.

Soil is composed of four distinct parts: mineral particles, organic matter, water, and air. The mineral portion, which comes from parent material, is the main component of soil. It provides anchorage and essential nutrient minerals for plants, as well as pore space for water and air. Litter (dead leaves and branches on the soil's surface), animal dung, and the remains of plants, animals, and microorganisms constitute the organic portion of soil. Organisms such as bacteria and fungi gradually decompose this material.

The black or dark brown organic material that remains after extended decomposition is called **humus** (**FIGURE 12.12**). Humus, which is a mix of many organic compounds, binds to nutrient mineral ions and holds water.

Many soils are organized into distinctive horizontal layers called **soil horizons**. A **soil profile** is a vertical section from surface to parent material, showing the soil horizons (see What a Scientist Sees). The topsoil (or A-horizon) is somewhat nutrient poor due to the leaching of many nutrients into deeper soil layers. **Leaching** is the removal of dissolved materials from the soil by water percolating downward.

> **soil horizons**
> Horizontal layers into which many soils are organized, from the surface to the underlying parent material.

Soil rich in humus FIGURE 12.12

Humus is partially decomposed organic material, primarily from plant and animal remains. Soil rich in humus has a loose, somewhat spongy structure with several properties, such as increased water-holding capacity, that are beneficial for plants and other organisms living in it.

SOIL ORGANISMS

Soil organisms, which are usually hidden underground, are remarkably numerous. Organisms that colonize the soil ecosystem include plant roots, insects such as termites and ants, earthworms, moles, snakes, and groundhogs (FIGURE 12.13). Most numerous in soil are bacteria, which number in the hundreds of millions per gram of soil. Other microorganisms that are abundant in soil ecosystems include fungi, algae, microscopic worms such as nematodes, and protozoa.

In a balanced ecosystem, the relationship between soil and the organisms that live in and on it ensure soil fertility. Soil organisms provide several essential **ecosystem services**, such as maintaining soil fertility, preventing soil erosion, breaking down toxic materials, and cleansing water.

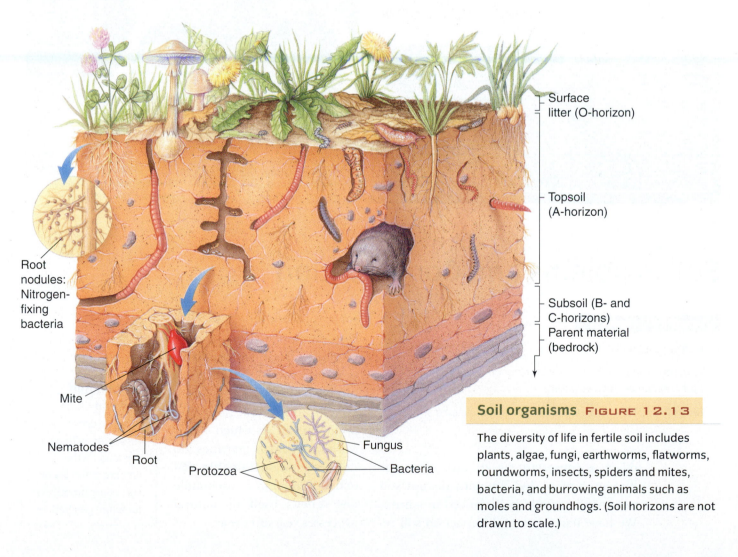

Root nodules: Nitrogen-fixing bacteria

Mite

Nematodes

Root

Protozoa

Fungus

Bacteria

Surface litter (O-horizon)

Topsoil (A-horizon)

Subsoil (B- and C-horizons)

Parent material (bedrock)

Soil organisms FIGURE 12.13

The diversity of life in fertile soil includes plants, algae, fungi, earthworms, flatworms, roundworms, insects, spiders and mites, bacteria, and burrowing animals such as moles and groundhogs. (Soil horizons are not drawn to scale.)

nutrient cycling
The pathway of various nutrient minerals or elements from the environment through organisms and back to the environment.

Essential nutrient minerals such as nitrogen and phosphorus are cycled from the soil to organisms and back to the soil again. Decomposition, another ecosystem service, is part of **nutrient cycling**. Bacteria and fungi decompose plant and animal detritus and wastes, transforming large organic molecules into small inorganic molecules, including carbon dioxide, water, and nutrient minerals; the nutrient minerals are released into the soil to be reused (**FIGURE 12.14**; also see Chapter 5). Nonliving processes are also involved in nutrient cycling: The weathering of the parent material replaces some nutrient minerals lost through erosion or agricultural practices.

VIEW THIS IN ACTION
in your WileyPLUS course

CONCEPT CHECK STOP

How do weathering processes affect soil formation?

What role do soil microorganisms play in nutrient cycling?

What are soil horizons?

Nutrient cycling FIGURE 12.14

In a balanced ecosystem, nutrient minerals cycle from the soil to organisms and then back to the soil.

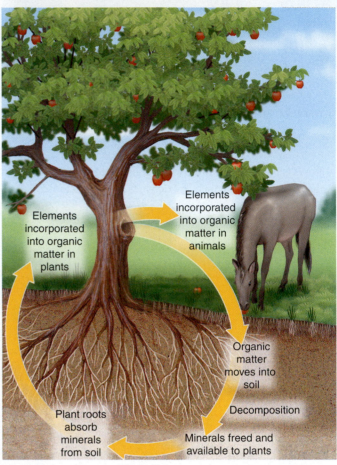

Elements incorporated into organic matter in plants

Elements incorporated into organic matter in animals

Organic matter moves into soil

Decomposition

Plant roots absorb minerals from soil

Minerals freed and available to plants

Soil Problems and Conservation

LEARNING OBJECTIVES

Define *sustainable soil use.*

Explain the impacts of soil erosion on plant growth and on other resources, such as water.

Identify and summarize the major soil conservation methods.

Soil is as important as air and water for human survival. Yet, humans disrupt soil systems that would be balanced in nature. We have had a harmful impact on soil resources worldwide, particularly by intensifying agricultural use. These human activities often cause or exacerbate soil problems such as erosion, mineral depletion, and soil pollution, all of which occur worldwide. Such activities do not promote **sustainable soil use**. Soil used in a sustainable way renews itself by natural processes year after year.

sustainable soil use
The wise use of soil resources, without a reduction in the amount or fertility of soil, so it is productive for future generations.

SOIL EROSION

Water, wind, ice, and other agents promote **soil erosion**, a natural process often accelerated by human activities. Water and wind are particularly effective in moving soil from one place to another. Rainfall loosens soil particles, which are transported by moving water (**FIGURE 12.15**). Wind loosens soil and blows it away, particularly if the soil is barren and dry. Erosion reduces the amount of soil in an area and therefore limits the growth of plants.

> **soil erosion**
> The wearing away or removal of soil from the land.

Humans often accelerate soil erosion with poor soil management. Poor agricultural practices are partly to blame, as are the removal of natural plant communities during road and building construction, and unsound logging practices such as clear-cutting. Soil erosion has an impact on other natural resources as well. Sediment that gets into streams, rivers, and lakes affects water quality and fish habitats (see Chapter 10). If the sediments contain pesticide and fertilizer residues, they further pollute the water.

Sufficient plant cover limits soil erosion. Leaves and stems cushion the impact of rainfall, and roots help to hold soil in place. Although soil erosion is a natural process, the abundant plant cover in many natural ecosystems makes it negligible.

SOIL POLLUTION

Soil pollution is any physical or chemical change in soil that adversely affects the health of plants and other organisms living in or on the soil. Soil pollution is important not only in its own right but because so many soil pollutants tend to also pollute surface water, groundwater, and the atmosphere. For example, selenium, an extremely toxic natural element found in many western soils, leaches off irrigated farmlands and poisons nearby lakes, ponds, and rivers. This has caused death and deformity in thousands of migratory birds and other organisms. Most soil pollutants originate as agricultural chemicals such as fertilizers and pesticides. Other soil pollutants include salts, petroleum products, and heavy metals.

Irrigation of agricultural fields often results in their becoming increasingly saline, an occurrence known as **salinization** (**FIGURE 12.16**). In time, salt concentrations in soil can rise to such a high level that plants are poisoned or their roots become dehydrated.

Soil erosion caused by water FIGURE 12.15

The branching gullies shown here are the most serious form of erosion and will continue to grow unless checked by some type of erosion control. Photographed in Colorado.

Salinized soil FIGURE 12.16

Irrigation water contains small amounts of dissolved salts. Over time, the salt accumulates in the soil. This irrigated soil has become too salty for plants to tolerate.

SOIL CONSERVATION AND REGENERATION

Only 11 percent of the world's soil is suitable for agriculture (**Figure 12.17**). We therefore need to protect the soil we use for agriculture. Although agriculture may cause or accelerate soil degradation, good soil conservation practices promote sustainable soil use. Conservation tillage, crop rotation, contour plowing, strip cropping, terracing, and shelterbelts minimize erosion and mineral depletion of the soil. Badly eroded and depleted land can be restored, but restoration is costly and time-consuming.

Conservation tillage and crop rotation

Conventional methods of tillage, or working the land, include spring plowing, in which the soil is cut and turned in preparation for planting seeds. Although conventional tillage prepares the land for crops, it greatly increases the likelihood of soil erosion. Conventionally tilled fields contain less organic material and generally hold less water than undisturbed soil.

Conservation tillage is one of the fastest-growing trends in U.S. agriculture (**Figure 12.18**). More than one-third of U.S. farmland is currently planted using conservation tillage. In addition to reducing soil erosion, conservation tillage increases the organic material in the soil, which improves the soil's water-holding capacity. Decomposing organic matter releases nutrient minerals more gradually than when conventional tillage methods are employed. However, use of conservation tillage requires new equipment, new techniques, and greater use of herbicides to control weeds. Research in developing alternative methods of weed control for use with conservation tillage is under way. (Chapter 14 discusses *sustainable agriculture*, which includes conservation tillage and the other soil conservation practices presented in this chapter.)

> ■ **conservation tillage** A method of cultivation in which residues from previous crops are left in the soil, partially covering it and helping to hold it in place until the newly planted seeds are established.

Visualizing

Soil conservation

Soil and agriculture Figure 12.17

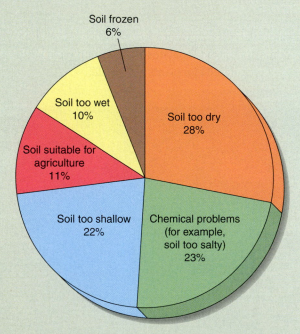

Soil frozen 6%

Soil too wet 10%

Soil suitable for agriculture 11%

Soil too shallow 22%

Soil too dry 28%

Chemical problems (for example, soil too salty) 23%

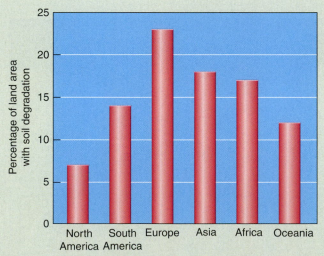

A Only 11 percent of the world's total land area has soil that is naturally suitable for agriculture. Soil that is too dry can be irrigated, whereas soil that is too wet can be drained.

B Degree of soil degradation (eroded, desertified, or salty soil) by continent. North America has the least degraded soils, and Europe, Africa, and Asia have the most degraded soils.

Farmers who practice effective soil conservation measures often use a combination of conservation tillage and **crop rotation**. When the same crop is grown over and over in one place, pests for that crop accumulate to destructive levels, and the essential nutrient minerals for that crop are depleted in greater amounts. This makes the soil more prone to erosion, and it makes the crops less productive as well. Crop rotation is effective in decreasing insect damage and disease, reducing soil erosion, and maintaining soil fertility (Figure 12.18 also shows crop rotation).

A typical crop rotation would be corn → soybeans → oats → alfalfa. Soybeans and alfalfa, both members of the legume family, increase soil fertility through their association with bacteria that fix atmospheric nitrogen into the soil. Thus, planting soybeans and alfalfa as part of crop rotation produces higher yields of the grain crops that are also part of the rotation.

Contour plowing, strip cropping, and terracing

Hilly terrain must be cultivated with care because it is more prone than flatland to soil erosion. Contour plowing, strip cropping, and terracing help control erosion of farmland with variable topography. In **contour plowing**, furrows run around hills rather than in straight rows. Strip cropping, a special type of contour plowing, produces alternating strips of different crops along natural contours (Figure 3.1A on page 50). For example, alternating a row crop such as corn with a closely sown crop such as wheat reduces soil erosion. Even more effective control of soil erosion is achieved when strip cropping is done in conjunction with conservation tillage.

Farming is undesirable on steep slopes, but if it must be done, **terracing** produces level areas and thereby reduces soil erosion from gravity or water runoff (FIGURE 12.19). Nutrient minerals and soil are retained on the horizontal platforms instead of being washed away.

Conservation tillage FIGURE 12.18

Decaying residues from the previous year's crop (rye) surround young soybean plants in a field in Iowa. Conservation tillage reduces soil erosion as much as 70 percent because plant residues from the previous season's crops are left in the soil.

Terracing FIGURE 12.19

Terraces are small earthen embankments placed across a steep hillside or mountain. Terracing hilly or mountainous areas, such as these rice terraces on Bali, Indonesia, curbs water flow and reduces the amount of soil erosion. Farmers must maintain these terraced fields.

Soil reclamation Badly eroded land can be reclaimed by (1) preventing further erosion and (2) restoring soil fertility. To prevent further erosion, the bare ground is seeded with plants; they eventually grow to cover the soil, stabilizing it and holding it in place.

The plants start to improve the quality of the soil almost immediately, as dead material decays into humus. The humus holds nutrient minerals in place, releasing them a little at a time; it also improves the water-holding capacity of the soil. One of the best ways to reduce the effects of wind erosion on soil is to plant **shelterbelts** that reduce the impact of wind (**Figure 12.20**).

> **shelterbelt**
>
> A row of trees planted as a windbreak to reduce soil erosion of agricultural land.

Shelterbelts surrounding kiwi orchards
Figure 12.20

Trees protect the delicate fruits from the wind and reduce wind erosion of farmland soil. Photographed on the North Island, New Zealand.

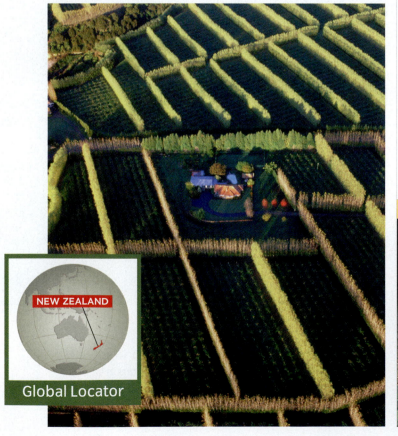

NEW ZEALAND

Global Locator

Restoration of soil fertility is a slow process. The land cannot be farmed or grazed until the soil has completely recovered. But the restriction of land use for an indefinite period may be difficult to accomplish. How can a government tell landowners they may not use their own land? How can land use be restricted when people's livelihoods, and maybe even their lives, depend on it?

Soil conservation policies in the United States

The **Food Security Act (Farm Bill) of 1985** contains provisions for two main soil conservation programs: the Conservation Compliance Program and the Conservation Reserve Program. The Conservation Compliance Program requires farmers with highly erodible land to develop and adopt a 5-year conservation plan for their farms that includes erosion-control measures. If they do not comply, they lose federal agricultural subsidies such as price supports.

The **Conservation Reserve Program (CRP)** is a voluntary subsidy program that pays U.S. farmers to stop producing crops on highly erodible farmland. It requires planting native grasses or trees on such land and then "retiring" it from further use for 10 to 15 years. The CRP has benefited the environment. Annual loss of soil on CRP lands planted with grasses or trees has been reduced more than 90 percent. Because the vegetation is not disturbed once it is established, it provides biological habitat. Small and large mammals, birds of prey, and ground-nesting birds such as ducks have increased in number and kind on CRP lands. The reduction in soil erosion has improved water quality and enhanced fish populations in surrounding rivers and streams.

CONCEPT CHECK STOP

What is sustainable soil use?

How do human activities accelerate soil erosion?

How do conservation tillage, contour plowing, and shelterbelts contribute to soil conservation?

INDUSTRIAL ECOSYSTEMS

Traditional industries operate in a linear fashion: natural resources → products → wastes dumped back into the environment. However, natural resources are finite, and the environment's capacity to absorb waste is limited. The field of **industrial ecology** seeks to address these issues by using resources efficiently, regarding "wastes" as potential products, and creating **industrial ecosystems** that in many ways mimic natural ecosystems.

One pioneering industrial ecosystem in Kalundborg, Denmark (**FIGURE**) consists of an electric power plant, an oil refinery, a pharmaceutical plant, a wallboard factory, a sulfuric acid producer, a cement manufacturer, a fish farm, greenhouses, and area homes and farms. These entities are linked in ways that resemble a food web in a natural ecosystem.

The coal-fired electric power plant originally cooled its waste steam and released it into the local fjord. The steam is now supplied to the oil refinery and pharmaceutical plant, and the surplus heat warms greenhouses, the fish farm, and area homes. Surplus natural gas from the oil refinery is sold to the power plant and the wallboard factory. Before selling the natural gas, the oil refinery removes excess sulfur from it (as required by law) and sells the sulfur to the sulfuric acid producer.

To meet environmental regulations, the power plant installed pollution control equipment to remove sulfur from its coal smoke. This sulfur, in the form of calcium sulfate, is sold to the wallboard plant and used as a gypsum substitute. The fly ash produced by the power plant goes to the cement manufacturer for use in road building.

To fertilize their fields, local farmers use sludge from the fish farm and a high-nutrient sludge generated at the pharmaceutical plant. Most pharmaceutical companies discard this sludge because it contains living microorganisms, but the Kalundborg plant heats the sludge to kill the microorganisms and convert a waste material into a commodity.

It took a decade to develop this entire industrial ecosystem. Although initiated for economic reasons, the industrial ecosystem has distinct environmental benefits, from energy conservation to a reduction of pollution.

The Kalundborg industrial ecosystem

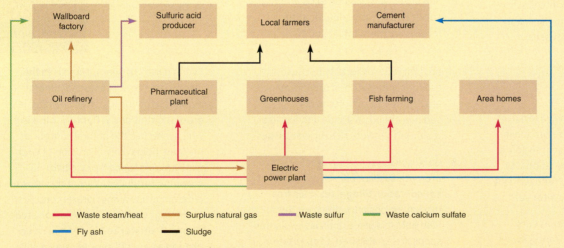

Kalundborg's industrial ecosystem has significantly reduced resource consumption and waste production.

Kalundborg's industrial ecosystem has significantly reduced resource consumption and waste production.

SUMMARY

1 Plate Tectonics and the Rock Cycle

1. The lithosphere, Earth's outermost rigid rock layer, is composed of plates that float on the asthenosphere, the region of the mantle where rocks become hot and soft. **Plate tectonics** is the study of the processes by which the lithospheric plates move over the asthenosphere. **Plate boundaries** are often sites of intense geologic activity: earthquakes, volcanoes, and mountain building.

2. The **rock cycle** shows how rock slowly cycles from one form to another. The three categories of rock are **igneous, metamorphic,** and **sedimentary rock**.

2 Economic Geology: Useful Minerals

1. **Minerals** are metallic or nonmetallic elements or compounds of elements that occur naturally in Earth's crust. Highly developed nations consume a disproportionate share of the world's minerals, but as developing countries become industrialized, their need for minerals increases.

2. Minerals are extracted through surface or subsurface mining. **Surface mining** removes the **overburden**: the overlying soil, subsoil, and rock strata. Strip mining, a type of surface mining, produces a **spoil bank** when the overburden from a new trench is put into an excavated trench. **Subsurface mining** extracts resources from deep underground deposits.

3. Processing minerals often involves **smelting**, melting the ore in a blast furnace to separate impurities from the metal.

3 Environmental Implications of Mineral Use

1. Surface mining destroys vegetation across large areas, increasing erosion. **Open-pit mining** uses huge quantities of water. Mining also affects water quality. **Acid mine drainage** is pollution caused when dissolved toxic materials wash from mines into nearby lakes and streams.

2. **Derelict lands** degraded by mining can be restored by filling in and grading the land to its natural contours and then planting vegetation to hold the soil in place.

4 Soil Properties and Processes

1. **Soil** is the uppermost layer of Earth's crust and supports terrestrial plants, animals, and microorganisms. Soil is formed from parent material—rock that is slowly fragmented into small particles by biological, chemical, and physical **weathering processes**.

2. Soil is composed of mineral particles, organic matter, water, and air. **Soil horizons** are the horizontal layers into which many soils are organized, from the surface to the underlying parent material.

3. Soil organisms provide **ecosystem services** such as maintaining soil fertility and preventing soil erosion.

Soil organisms carry out **nutrient cycling**, the pathway of nutrient minerals or elements from the environment through organisms and back to the environment.

5 Soil Problems and Conservation

1. **Sustainable soil use** is the wise use of soil resources, without a reduction in the amount or fertility of soil, so soil is productive for future generations. Soil used in a sustainable way renews itself by natural processes year after year.

2. Water, wind, ice, and other agents cause **soil erosion**, the wearing away or removal of soil from the land. Soil erosion reduces fertility because essential minerals and organic matter are removed. Erosion causes sediments and pesticide and fertilizer residues to pollute nearby waterways.

3. Good soil conservation practices promote sustainable soil use. In **conservation tillage**, residues from previous crops partially cover the soil to help hold it in place until newly planted seeds are established. **Crop rotation**, the planting of different crops in a field over a period of years, decreases the insect damage, disease, and mineral depletion that occur when one crop is grown continuously. **Contour plowing**, which matches the natural contour of the land, helps control erosion of land with variable topography. **Strip cropping** produces alternating strips of different crops along natural contours. **Terracing** reduces soil erosion on steep slopes. A **shelterbelt** is a row of trees planted as a windbreak to reduce soil erosion.

KEY TERMS

- plate tectonics p. 296
- minerals p. 300
- surface mining p. 302
- overburden p. 302
- spoil bank p. 303
- subsurface mining p. 303
- smelting p. 303
- acid mine drainage p. 304
- soil p. 307
- soil horizons p. 308
- nutrient cycling p. 310
- sustainable soil use p. 310
- soil erosion p. 311
- conservation tillage p. 312
- crop rotation p. 313
- contour plowing p. 313
- shelterbelt p. 314

CRITICAL AND CREATIVE THINKING QUESTIONS

1. How are plate tectonics and tsunamis related?

2. How many minerals have you come in contact with today? Which were metals; which were nonmetals?

3. What is the difference between surface and subsurface mining? open-pit and strip mines? shaft and slope mines? When is each most likely to be used?

4. What are the roles of weathering, organisms, and topography in soil formation?

5. Certain pests that cause plant disease reside in the plant residues left on the ground with conservation tillage. Given that these organisms are often specific for the plants they attack, how can we control them?

6. How did Copper Basin, Tennessee, become an environmental disaster? Are reclamation efforts making a difference there?

7. How does the industrial ecosystem at Kalundborg, Denmark, resemble a natural ecosystem? What are some environmental benefits of this industrial ecosystem?

8-9. The figure below represents the flow of minerals in a low-waste society.

8. Which colors of arrows represent sustainable manufacturing, consumer reuse, and consumer recycling, respectively?

9. Would the flow of minerals be more or less complicated in a high-waste society? How would the wastes generated differ from those of the low-waste society represented here?

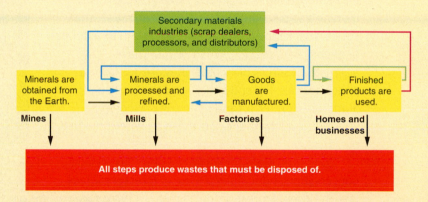

What is happening in this picture ?

- At this former coal mine site in West Virginia, special equipment carries out "hydroseeding," applying a mixture of grass seed, water, and fertilizers. What is the purpose of this effort?

- Why do you think hydroseeding was chosen here rather than applying grass seed using equipment that rolls across the ground?

- Hydroseeding is also used after other activities have disturbed large patches of ground; suggest one or two likely examples.

Land Resources 13

KORUP NATIONAL PARK

Korup National Park in Cameroon has the richest biological diversity in Africa (see larger photograph). It is home to more than 400 tree species, 50 mammal species, and more than 320 bird species.

Korup National Park is a project of the wildlife conservation group World Wildlife Fund (WWF), in cooperation with the government of Cameroon and other organizations. WWF helps protect, maintain, and integrate the park into regional development plans. To alleviate poverty, WWF offers people the opportunity to train in local technical colleges. Individuals receive education about environmental issues and the importance of conservation. Some are employed as park staff members, including game guards. This training also discourages the killing of animals—the park's most serious problem—especially of threatened or endangered species. When villagers have alternative sources of income, they are less likely to poach animals. Recently, the local people and park authorities formed an anti-poaching partnership; villagers now inform the park manager when they find poachers.

Long-term plans for Korup National Park include conservation and sustainable use of forest resources (see the inset of forest products that can be harvested sustainably). The six villages now located inside park boundaries will be voluntarily relocated to fertile land outside them. An established research program develops sustainable management practices for the park. Tourism, which depends on preservation of the area's unique biological diversity, provides income for the local economy. Korup National Park is a model of land conservation that other developing countries, in cooperation with conservation organizations and with foreign aid, might profitably emulate.

Land Use in the United States

Private citizens, corporations, and nonprofit organizations own about 55 percent of the land in the United States, and Native American tribes own about 3 percent. State and local governments own another 7 percent. The federal government owns the rest (about 35 percent).

Government-owned land encompasses all types of ecosystems, from tundra to desert, and includes land that contains important resources such as minerals and fossil fuels, land that possesses historical or cultural significance, and land that provides critical biological habitat. Most federally owned land is in Alaska and 11 western states (**FIGURE 13.1**).

Federal land is managed primarily by four agencies, three in the U.S. Department of the Interior—the Bureau of Land Management (BLM), the Fish and Wildlife Service (FWS), and the National Park Service (NPS)—and one in the Department of Agriculture—the U.S. Forest Service (USFS) (**TABLE 13.1**).

Government-owned lands provide vital **ecosystem services** that benefit humans living far from public

Selected federal lands FIGURE 13.1

Shown are national parks and preserves, national wildlife refuges, national forests and grasslands, and national marine sanctuaries in the United States. Note the preponderance of federal lands in western states and Alaska. Other federal lands, such as military installations and research facilities, aren't shown.

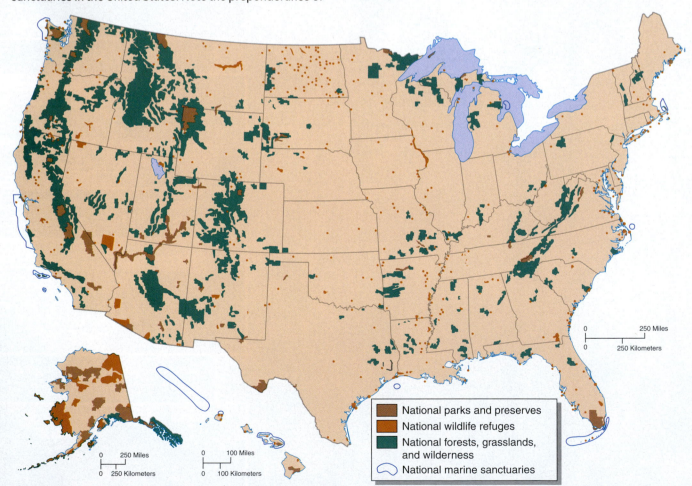

Legend:
- National parks and preserves
- National wildlife refuges
- National forests, grasslands, and wilderness
- National marine sanctuaries

Administration of federal lands TABLE 13.1

Agency	Land held	Primary uses	Area in millions of hectares (acres)
Bureau of Land Management (Dept. of the Interior)	National resource lands	Mining, livestock grazing, oil and natural gas extraction	109 (270)
U.S. Forest Service (Dept. of Agriculture)	National forests	Logging, recreation, conservation of watersheds, wildlife habitat, mining, livestock grazing, oil and natural gas extraction	77 (191)
U.S. Fish and Wildlife Service (Dept. of the Interior)	National wildlife refuges	Wildlife habitat; also logging, hunting, fishing, mining, livestock grazing, oil and natural gas extraction	38 (95)
National Park Service (Dept. of the Interior)	National Park Service	Recreation, wildlife habitat	34 (84)
Other—includes Department of Defense, Corps of Engineers (Dept. of the Army), and Bureau of Reclamation (Dept. of the Interior)	Remaining federal lands	Military uses, wildlife habitat	29 (72)
Total federal lands			**287 (712)**

forests, grasslands, deserts, and wetlands. These services include wildlife habitat, flood and erosion control, groundwater recharge, and the breakdown of pollutants.

Undisturbed public lands are ecosystems that scientists use as a benchmark, or point of reference, to determine the impact of human activity. Geologists, zoologists, botanists, ecologists, and soil scientists are some of the scientists who use government-owned lands for scientific inquiry. These areas provide perfect settings for educational experiences not only in science but also in history, because they can be used to demonstrate the condition of the land when humans originally settled it (**FIGURE 13.2**).

Public lands are important for their recreational value, providing places for hiking, swimming, boating, rafting, sport hunting, and fishing. Wild areas—forest-covered mountains, rolling prairies, barren deserts, and other undeveloped areas—are important to the human spirit. We can escape the tensions of the civilized world by retreating, even temporarily, to the solitude of natural areas.

Not all public lands remain undeveloped. As you will see throughout this chapter, many public lands are developed for uses ranging from logging to cattle grazing to mineral extraction.

Hurricane Ridge in Olympic National Park, Washington FIGURE 13.2

Olympic National Park, one of 58 national parks and nearly 400 total sites in the National Park Service system, includes rugged mountains as well as rain forests and Pacific beaches.

CONCEPT CHECK **STOP**

What percentage of land in the United States is privately owned?

What percentage of land in the United States is owned by the federal government?

Forests

LEARNING OBJECTIVES

Define *sustainable forestry* and explain how monocultures and wildlife corridors are related to it.

Define *deforestation*, including *clear-cutting*, and list the main causes of tropical deforestation.

Describe national forests, stating which government agencies administer them and current issues of concern.

Forests, important ecosystems that provide many goods and services to support human society, occupy less than one-third of Earth's total land area. Timber harvested from forests is used for fuel, construction materials, and paper products. Forests supply nuts, mushrooms, fruits, and medicines. Forests provide employment for millions of people worldwide and offer recreation and spiritual sustenance in an increasingly crowded world.

Forests also provide a variety of beneficial ecosystem services, such as influencing climate conditions. If you walk into a forest on a hot summer day, you will no-

Process Diagram

Role of forests in the hydrologic cycle FIGURE 13.3

Forests return most of the water that falls as precipitation to the atmosphere by transpiration. When an area is deforested, almost all precipitation is lost as runoff.

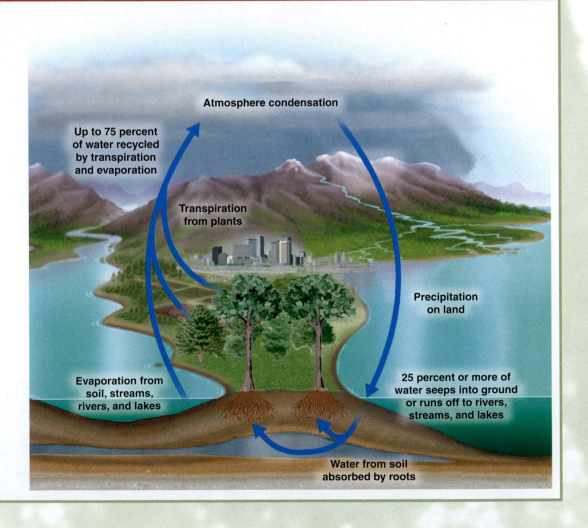

Atmosphere condensation

Up to 75 percent of water recycled by transpiration and evaporation

Transpiration from plants

Precipitation on land

Evaporation from soil, streams, rivers, and lakes

25 percent or more of water seeps into ground or runs off to rivers, streams, and lakes

Water from soil absorbed by roots

tice that the air is cooler and moister than it is outside the forest. This is the result of a biological cooling process called *transpiration*, in which water from the soil is absorbed by roots, transported through plants, and then evaporated from their leaves and stems. Transpiration provides moisture for clouds, eventually resulting in precipitation (see **FIGURE 13.3** on facing page). Thus, forests help maintain local and regional precipitation.

Forests play an essential role in regulating global biogeochemical cycles like those for carbon and nitrogen. Photosynthesis by trees removes large quantities of heat-trapping carbon dioxide from the atmosphere and fixes it into carbon compounds, while releasing oxygen back into the atmosphere. Forests thus act as carbon "sinks," which may help mitigate climate warming, and they produce oxygen, which almost all organisms require for cellular respiration.

Tree roots hold vast tracts of soil in place, reducing erosion and mudslides. Forests protect watersheds because they absorb, hold, and slowly release water; this moderation of water flow provides a more regulated flow of water downstream, even during dry periods, and helps control floods and droughts. Forest soils remove impurities from water, improving its quality. In addition, forests provide a variety of essential habitats for many organisms, such as mammals, reptiles, amphibians, fishes, insects, lichens and fungi, mosses, ferns, conifers, and numerous kinds of flowering plants.

FOREST MANAGEMENT

Management for timber production disrupts a forest's natural condition and alters its species composition and other characteristics. Specific varieties of commercially important trees are planted, and those trees not as commercially desirable are thinned out or removed. Traditional forest management often results in low-diversity forests. In the southeastern United States, many tree plantations of young pine grown for timber and paper production are all the same age and are planted in rows a fixed distance apart (**FIGURE 13.4**). These "forests" are essentially **monocultures**—areas uniformly covered by one crop, like a field of corn. Herbicides are sprayed to kill

Tree plantation FIGURE 13.4

This intensively managed pine plantation in the southern United States is a monoculture, with trees of uniform size and age. Such plantations supplement harvesting of trees in wild forests to provide the United States with the timber it requires.

monoculture

Ecological simplification in which only one type of plant is cultivated over a large area.

shrubs and herbaceous plants between the rows. One of the disadvantages of monocultures is that they are at increased risk of damage from insect pests and disease-causing microorganisms. Consequently, pests and diseases must be controlled in managed forests, usually by applying

Forests **323**

insecticides and fungicides. Also, because managed forests contain few kinds of food, they can't support the variety of organisms typically found in natural forests.

In recognition of the many ecosystem services performed by natural forests, a newer method of forest management, known as **ecologically sustainable forest management**, or simply **sustainable forestry**, is evolving. Sustainable forestry maintains a mix of forest trees, by age and species, rather than imposing a monoculture. This broader approach seeks to conserve forests for the long-term commercial harvest of timber and nontimber forest products. Sustainable forestry also attempts to sustain biological diversity by providing improved habitats for a variety of species, prevent soil erosion and improve soil conditions, and preserve watersheds that produce clean water. Effective sustainable forest management involves cooperation among environmentalists, loggers, farmers, indigenous people, and local, state, and federal governments.

When loggers use sustainable forestry principles, they set aside unlogged areas and **wildlife corridors** as sanctuaries for organisms. The purpose of wildlife corridors is to provide animals with escape routes, should they be needed, and to allow them to migrate so they can interbreed. (Small, isolated, inbred populations may have an increased risk of extinction.) Wildlife corridors may also allow large animals such as the Florida panther to maintain large territories. Some scientists question the effectiveness of wildlife corridors, although recent research on wildlife corridors in fragmented landscapes suggests that wildlife corridors help certain wildlife populations persist. Additional research is needed to resolve the effectiveness of wildlife corridors for all endangered species.

Methods for ecologically sustainable forest management are under development. Such practices vary from one forest ecosystem to another, in response to different environmental, cultural, and economic conditions. In Mexico, many sustainable forestry projects involve communities that are economically dependent on forests. Because trees have such long life spans, scientists and forest managers of the future will judge the results of today's efforts.

sustainable forestry The use and management of forest ecosystems in an environmentally balanced and enduring way.

wildlife corridor A protected zone that connects isolated unlogged or undeveloped areas.

clear-cutting A logging practice in which all the trees in a stand of forest are cut, leaving just the stumps.

Harvesting trees According to the U.N. Food and Agricultural Organization (FAO), 3.5 million cubic meters (123.6 million cubic feet) of wood (for fuelwood, timber, and other products) were harvested in 2005 (the latest available data). The five countries with the greatest tree harvests are the United States, Canada, Russia, Brazil, and China; these countries currently produce more than half the world's timber. About 50 percent of harvested wood is burned directly as fuelwood or used to make charcoal. Most fuelwood and charcoal are used in developing countries (see Chapter 18). Highly developed countries consume more than three-fourths of the remaining 50 percent for paper and wood products.

Loggers harvest trees in several ways—selective cutting, shelterwood cutting, seed tree cutting, and clear-cutting (see What a Scientist Sees on facing page). **Selective cutting**, in which mature trees are cut individually or in small clusters while the rest of the forest remains intact, allows the forest to regenerate naturally.

The removal of all mature trees in an area over an extended period is **shelterwood cutting**. In the first year, undesirable tree species and dead or diseased trees are removed. Subsequent harvests occur at intervals of several years, allowing time for remaining trees to grow.

In **seed tree cutting**, almost all trees are harvested from an area; a scattering of desirable trees is left behind to provide seeds for the regeneration of the forest.

Clear-cutting is harvesting timber by removing all trees from an area and then either allowing the area to reseed and regenerate itself naturally or planting the area with one or more specific varieties of trees. Timber companies prefer clear-cutting because it is the most cost-effective way to harvest trees. However, clear-cutting over wide areas is ecologically unsound. It destroys biological habitats and increases soil erosion, particularly on sloping land, sometimes degrading land so much that reforestation doesn't take place. Clear-cut areas at lower elevations are usually regenerated successfully, whereas those at high elevations are difficult to regenerate. Obviously, most recreational benefits of forests are lost when clear-cutting occurs.

What a Scientist Sees

Harvesting Trees

A Aerial view of a large patch of clear-cut forest in British Columbia, Canada. Clear-cutting is the most common but most controversial type of logging. The obvious line is a road built to haul away the logs.

B As a forest scientist looks at a clear-cut forest, he or she may think about the various kinds of tree harvesting (1 to 3) that are less environmentally destructive than clear-cutting (4).

(1) In **selective cutting**, the older, mature trees are selectively harvested from time to time, and the forest regenerates itself naturally.

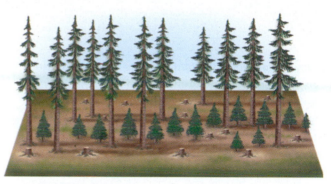

(2) In **shelterwood cutting**, less desirable and dead trees are harvested. As younger trees mature, they produce seedlings, which continue to grow as the now-mature trees are harvested.

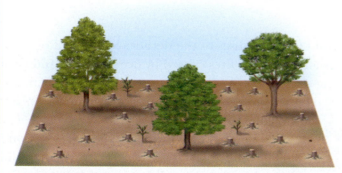

(3) **Seed tree cutting** involves the removal of all but a few trees, which are allowed to remain to provide seeds for natural regeneration.

(4) In **clear-cutting**, all trees are removed from a particular site. Clear-cut areas may be reseeded or allowed to regenerate naturally.

The Forest Stewardship Council ecologically certifies "green" wood.

Often, the consumer pays no additional premium, or only slightly more, for ecologically certified wood, which has become so popular that demand threatens to exceed supply.

Many homebuilders and homeowners are interested in "green" wood for flooring and other building materials (see photograph). Such wood is ecologically certified by a legitimate third party, such as the Germany-based Forest Stewardship Council (FSC), to have come from a forest managed with environmentally sound and socially responsible practices. Although these areas remain a small percentage of world forests, by early 2008 the FSC had certified as well managed more than 98 million hectares (more than 243 million acres) in 79 countries. Certification is based on sustainability of timber resources, socioeconomic benefits provided to local people, and forest ecosystem health, which includes such considerations as preservation of wildlife habitat and watershed stability.

Green forestry has its detractors. Traditional forestry organizations are skeptical about the reliability of FSC investigations and the economic viability of this type of forestry. Trade experts caution that government efforts to specify the purchase of certified timber could violate global free-trade agreements. Still, green timber is gaining market share, pleasing business owners and consumers alike, and offering the promise of better conservation practices in managed forests.

DEFORESTATION

The most serious problem facing the world's forests is **deforestation**. According to latest FAO estimates, world forests shrank by about 36 million hectares (89 million acres) between 2000 and 2005. This estimate represents a total loss of about 1 percent of global forested area and does not include thinned forests or those degraded by overharvesting, declining biological diversity, and reduced

> **deforestation**
> The temporary or permanent clearance of large expanses of forest for agriculture or other uses.

soil fertility. Causes of the decades-long trend of deforestation include fires caused by drought and land clearing practices, expansion of agriculture, construction of roads, tree harvests, insects, disease, and mining.

Most of the world's deforestation is currently taking place in Africa and South America, according to the FAO. Africa lost about 3.2 percent of its forested area from 2000 to 2005, and South America lost about 2.5 percent. Central America and the

Caribbean nations lost about 3.9 percent of their forests during this period. There was a comparatively small loss of forested area in North America, whereas Europe and Asia actually gained forested areas, either by natural regrowth or by increasing forest plantations.

Results of deforestation

Deforestation results in decreased soil fertility, as the essential mineral nutrients found in most forest soils leach away rapidly without trees to absorb them. Uncontrolled soil erosion, particularly on steep deforested slopes, affects the production of hydroelectric power as silt builds up behind dams. Increased sedimentation of waterways caused by soil erosion harms downstream fisheries. In drier areas, deforestation contributes to the formation of deserts. Regulation of water flow is disrupted when a forest is removed, so that the affected region experiences alternating periods of flood and drought.

Deforestation contributes to the extinction of many species. (See Chapter 15 for a discussion of the importance of tropical forests as repositories of biological diversity.) Many tropical species, in particular, have limited ranges within a forest, so they are especially vulnerable to habitat modification and destruction. Migratory species, including birds and butterflies, also suffer because of deforestation.

Deforestation is thought to induce regional and global climate changes. Trees release substantial amounts of moisture into the air; in the hydrologic cycle, about 97 percent of the water that roots absorb from the soil is evaporated directly into the atmosphere and then falls back to Earth. When a large forested area is removed, local rainfall may decline, droughts may become more common in that region, and temperatures may rise slightly. Deforestation may also contribute to an increase in global temperature by releasing carbon originally stored in the trees into the atmosphere as carbon dioxide, which enables the air to retain heat. When an old-growth forest is harvested, researchers estimate that it takes about 200 years for the replacement forest to accumulate the equivalent amount of carbon stored in the original trees.

Boreal forests and deforestation

Extensive deforestation in boreal forests due to logging began in the late 1980s. **Boreal forests** occur in Alaska, Canada, Scandinavia, and northern Russia and are dominated by coniferous evergreen trees such as spruce, fir, cedar, and hemlock. The boreal forest biome is the world's largest, covering about 11 percent of Earth's land. Harvested primarily by clear-cut logging, boreal forests are the primary source of the world's industrial wood and wood fiber. The annual loss of boreal forests is estimated to encompass an area twice as large as the rain forests of Brazil.

About 1 million hectares (2.5 million acres) of forest in Canada—currently the world's biggest timber exporter—are logged annually, and most of Canada's forests are subject to logging contracts (**FIGURE 13.5**). On the basis of current harvest quotas, logging is unsustainable in Canada.

Extensive tracts of Siberian forests in Russia are harvested, although estimates are unavailable. Alaska's boreal forests are at risk because the U.S. government may increase logging on public lands in the future.

Logging in Canada's boreal forest FIGURE 13.5

About 80 percent of Canada's forest products are exported to the United States.

Tropical forests and deforestation There are two types of tropical forests: tropical rain forests and tropical dry forests. **Tropical rain forests** prevail in warm areas that receive 200 cm (at least 79 in) or more of precipitation annually. Tropical rain forests are found in Central and South America, Africa, and Southeast Asia, but almost half of them are in just three countries: Brazil, Democratic Republic of the Congo, and Indonesia (**FIGURE 13.6A**). **Tropical dry forests** occur in other tropical areas where annual precipitation is less but is still enough to support trees. India, Kenya, Zimbabwe, Egypt, and Brazil are a few of the countries that have tropical dry forests.

Most of the remaining undisturbed tropical forests, which lie in the Amazon and Congo river basins of South America and Africa, respectively, are being cleared and burned at a rate unprecedented in human history. Tropical forests are also being destroyed at an extremely rapid rate in southern Asia, Indonesia, Central America, and the Philippines.

Why are tropical rain forests disappearing? Several studies show a strong statistical correlation between population growth and deforestation. More people need more food, and so forests are cleared for agricultural expansion.

However, tropical deforestation can't be attributed simply to population pressures because it is also affected by a variety of interacting economic, social, and government factors that vary from place to place. Government policies sometimes provide incentives that favor the removal of forests. The Brazilian government opened the Amazonian frontier, beginning in the late 1950s, by constructing the Belem–Brasilia Highway, which cut through the Amazon Basin. Such roads open a forest for settlement (**FIGURE 13.6B**). An example of economic conditions encouraging deforestation is a farmer who converts forest to pasture so that he can maintain a larger herd of cattle.

Keeping in mind that the origins of tropical deforestation are complex, three agents—subsistence agriculture, commercial logging, and cattle ranching—are considered its most immediate causes. Subsistence agriculture, in which a family produces just enough food to feed itself, accounts for perhaps 60 percent of tropical deforestation. Subsistence farmers carry out **slash-and-burn agriculture** (discussed further in Chapter 14). Subsistence farmers often follow loggers' access roads until they find a suitable spot. They cut down trees and allow them to dry; then they burn the area and plant crops immediately after burning. The yield from the first crop is often quite high because the nutrients that were in the burned trees are made available in the soil. Soil productivity subsequently declines at a rapid rate, and subsequent crops are poor. In a short time, the people farming the land must move to a new part of the forest and repeat the process. Cattle ranchers often claim the abandoned land for grazing because land not rich enough to support crops can still support livestock.

Slash-and-burn agriculture done on a small scale, with plenty of forest to shift around in so that there are periods of 20 to 100 years between cycles, is sustainable. The forest regrows rapidly after a few years of farming. But when millions of people try to obtain a living in this way, the land is not allowed to lie uncultivated long enough to recover.

Another 20 percent of tropical deforestation is the result of commercial logging. Vast tracts of tropical rain forests are harvested for export abroad. Most tropical countries allow commercial logging to proceed at a much faster rate than is sustainable. Unmanaged logging does not contribute to economic development; rather, it depletes a valuable natural resource faster than it can regenerate for sustainable use.

Approximately 12 percent of tropical deforestation provides open rangeland for cattle. Other causes of tropical rainforest destruction include the development of hydroelectric power, which inundates large areas of forest, and mining, particularly when ore smelters burn charcoal produced from rainforest trees (**FIGURE 13.6C**).

Why are tropical dry forests disappearing? Tropical dry forests are being destroyed at a rapid rate, primarily for fuelwood (**FIGURE 13.6D**). About half of the wood consumed worldwide is used as heating and cooking fuel by much of the developing world. Often the wood cut for fuel is converted to charcoal, which is then used to power steel, brick, and cement factories. Charcoal production is extremely wasteful: 3.6 metric tons (4 tons) of wood produce only enough charcoal to fuel an average-sized iron smelter for 5 minutes.

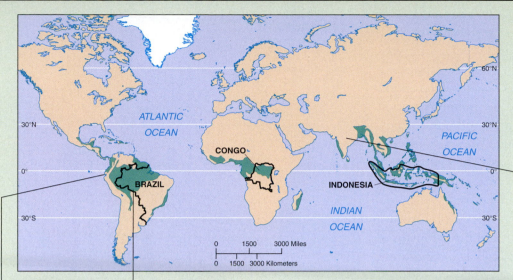

A Distribution of tropical rain forests Rain forests (green areas) occur in Central and South America, Africa, and Southeast Asia. Much of the remaining forested area is highly fragmented. The three countries with the largest area of tropical rain forest are outlined.

B Human settlements along a road in Brazil's tropical rain forest This satellite photograph shows numerous smaller roads extending perpendicularly from the main roads. As farmers settle along the roads, they clear out more and more forest (dark green) for croplands and pastures (tans and pinks).

D Deforestation for fuelwood Women gather firewood in the Ranthambore National Park buffer zone in India. Note the branches trimmed off the trees in the background. About 94 percent of wood removed from Indian forests is burned as fuel.

C Deforestation due to mining Forest was removed for this oil exploration drilling site in Ecuador.

FORESTS IN THE UNITED STATES

In recent years, most temperate forests in the Rocky Mountains, Great Lakes region, and New England and other eastern states have been holding steady or even expanding. In Vermont, the amount of land covered by forests has increased from 35 percent in 1850 to 80 percent today. Expanding forests are the result of *secondary succession* on abandoned farms (see Figure 6.20), the commercial planting of tree plantations on both private and public lands, and government protection. Although these second- and third-growth forests generally don't have the biological diversity of virgin stands, many organisms have successfully become reestablished in the regenerated areas.

Slightly more than one-half of U.S. forests are privately owned (**FIGURE 13.7**), and three-fourths of these private lands are in the Northeast and Midwest. Many private owners are under economic pressure to subdivide the land and develop tracts for housing or shopping malls, as they seek ways to recoup their high property taxes. Projected conversion of forests to agricultural, urban, and suburban lands over the next 40 years will have the greatest potential impact in the South, where more than 85 percent of forest is privately owned and logging is largely unregulated.

U.S. national forests

According to the USFS, the United States has 155 national forests encompassing 77 million hectares (191 million acres) of land, mostly in Alaska and western states. The USFS manages most national forests, and the BLM oversees the remainder. National forests were established for multiple uses, including timber harvesting; mining; livestock foraging; hunting, fishing, and other forms of outdoor recreation; water resources and watershed protection; and habitat for fishes and wildlife. Recreation, which increased dramatically in national forests during the 1990s and early 2000s, ranges from camping at designated campsites to backpacking in the wilderness. Visitors to national forests swim, boat, picnic, and observe nature. With so many possible uses of national forests, conflicts inevitably arise, particularly between timber interests and those who wish to preserve the trees for other purposes.

Road building is a particularly contentious issue, in part because the USFS builds taxpayer-funded roads to allow private logging companies access to forests to

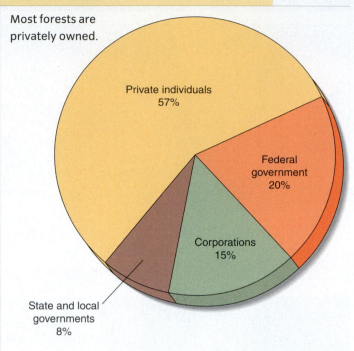

Forest ownership in the United States
FIGURE 13.7

Most forests are privately owned.

- Private individuals 57%
- Federal government 20%
- Corporations 15%
- State and local governments 8%

remove timber. (See the case study at the end of the chapter for an example.) Road building is environmentally destructive when improper construction accelerates soil erosion and mudslides (particularly on steep terrain) and causes water pollution in streams. Biologists are concerned that the many roads that are built fragment wildlife habitat and provide entries for disease organisms and invasive species.

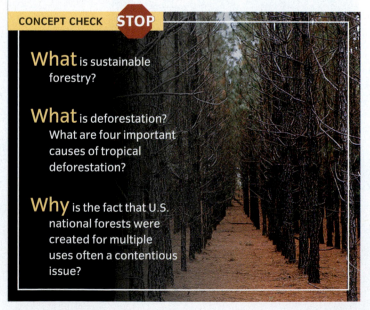

CONCEPT CHECK STOP

What is sustainable forestry?

What is deforestation? What are four important causes of tropical deforestation?

Why is the fact that U.S. national forests were created for multiple uses often a contentious issue?

Rangelands

Rangelands are grasslands, in both temperate and tropical climates, that serve as important areas of food production for humans by providing fodder for livestock such as cattle, sheep, and goats (FIGURE 13.8). Rangelands may be mined for minerals and energy resources, used for recreation, and preserved for biological habitat and for soil and water resources. The predominant vegetation of rangelands includes grasses, *forbs* (small plants other than grasses), and shrubs.

> **rangeland** Land that isn't intensively managed and is used for grazing livestock.

Rangeland FIGURE 13.8

Rangeland is considered a renewable resource when its carrying capacity—the number of animals it can sustain without suffering deterioration—is not exceeded. Photographed along the Salmon River in Idaho.

RANGELAND DEGRADATION AND DESERTIFICATION

Grasses, the predominant vegetation of rangelands, have a *fibrous root system*, in which many roots form a diffuse network in the soil to anchor the plant. Plants with fibrous roots hold the soil in place quite well, thereby reducing soil erosion. Grazing animals eat the leafy shoots of the grass, and the fibrous roots continue to develop, allowing the plants to recover and regrow to their original size.

Carefully managed grazing is beneficial for grasslands. Because rangeland vegetation is naturally adapted to grazing, when grazing animals remove mature vegetation, the activity stimulates rapid regrowth. At the same time, the hooves of grazing animals disturb the soil surface enough to allow rainfall to more effectively reach the root systems of grazing plants. Several studies have reported that moderate levels of grazing encourage greater plant diversity.

The **carrying capacity** of a rangeland is the maximum number of animals the natural vegetation can sustain over an indefinite period without deterioration of the ecosystem. When the carrying capacity of a rangeland is exceeded, **overgrazing** of grasses and other plants occurs. When plants die, the ground is left barren, and the exposed soil is susceptible to erosion. Sometimes plants that do not naturally grow in a rangeland, but which can tolerate the depleted soil, invade an overgrazed area.

Most of the world's rangelands lie in semiarid areas that have natural extended droughts. Under normal conditions, native grasses in these dry lands can survive severe drought. The aboveground portion of the plant dies back, but the extensive root system remains alive and holds the soil in place. But when an extended drought occurs in conjunction with overgrazing, once-fertile rangeland may be converted to desert as reduced grass

overgrazing
A situation that occurs when too many grazing animals consume the plants in a particular area, leaving the vegetation destroyed and unable to recover.

desertification
Degradation of once-fertile rangeland or tropical dry forest into nonproductive desert.

cover allows winds to erode the soil. Even when the rains return, the degradation may be so extensive that the land cannot recover. Water erosion removes the little remaining topsoil, and the sand left behind forms dunes.

Land degradation is a natural or human-induced process that decreases the future ability of the land to support crops or livestock. This progressive degradation, which induces unproductive desert-like conditions on formerly productive rangeland (or tropical dry forest), is **desertification** (**FIGURE 13.9**). It reduces the agricultural productivity of economically valuable land, forces many organisms out, and threatens endangered species. Worldwide, desertification seems to be on the increase. The United Nations estimates that each year since the mid-1990s, 3,560 km² (1,374 mi²)—an area about the size of Rhode Island—has turned into desert.

Desertification in the African Sahel region, Niger FIGURE 13.9

As the goats consume the remaining grass and shrubs, the dunes of the Sahara Desert will inevitably continue to encroach upon the pastureland. There is still much that scientists don't understand about desertification, such as the extent to which desertification results from natural fluctuations in climate versus population pressures and human activities.

Global Locator

RANGELAND TRENDS IN THE UNITED STATES

Rangelands make up approximately 30 percent of the total land area in the United States, mostly in the western states. Of this, approximately one-third is publicly owned and two-thirds are privately owned. Much of the private rangeland is under increasing pressure from developers, who want to subdivide the land into lots for homes and condominiums. To preserve the open land, conservation groups often pay ranchers for **conservation easements** that prevent future owners from developing the land. An estimated 405,000 hectares (1 million acres) of private rangelands are protected by conservation easements.

Excluding Alaska, there are at least 89 million hectares (220 million acres) of public rangelands in the United States. The BLM manages approximately 69 million hectares (170 million acres) of public rangelands, and the USFS manages an additional 20 million hectares (50 million acres).

Overall, the condition of public rangelands in the United States has slowly improved since the low point of the Dust Bowl in the 1930s, when the combined effects of poor agricultural practices, severe winds, and extended drought led to devastating soil erosion and dramatic declines in soil productivity. Much of this improvement is attributed to fewer livestock being permitted to graze the rangelands after the passage of the **Taylor Grazing Act** of 1934, the **Federal Land Policy and Management Act** of 1976, and the **Public Rangelands Improvement Act** of 1978. Better livestock management practices, such as controlling the distribution of animals on a range through fencing or herding, as well as scientific monitoring, have also contributed to rangeland recovery.

But restoration is slow and costly, and more is needed. Rangeland management includes seeding in places where plant cover is sparse or absent, conducting controlled burns to suppress shrubby plants, constructing fences to allow rotational grazing, controlling invasive weeds, and protecting habitats of endangered species. Most livestock operators use public rangelands in a way that results in their overall improvement.

conservation easement A legal agreement that protects privately owned forest, rangeland, or other property from development for a specified number of years.

Issues involving public rangelands The federal government distributes permits that allow private livestock operators to use public rangelands for grazing in exchange for a fee that is much lower than the cost of grazing on private land. The permits are held for many years and are not open to free-market bidding by the general public—that is, only ranchers who live in the local area are allowed to obtain grazing permits. Some environmental groups are concerned about the ecological damage caused by overgrazing of public rangelands and want to reduce the number of livestock animals allowed to graze. They want public rangelands managed for other uses, such as biological habitat, recreation, and scenic value, rather than exclusively for livestock grazing. To accomplish this goal, they would like to purchase grazing permits and set aside the land for nongrazing purposes.

Conservative economists have joined environmentalists in criticizing the management of federal rangelands. According to policy analysts at Taxpayers for Common Sense, in 2003 taxpayers contributed at least $67 million more than the grazing fees collected in order to support grazing on public rangelands. This money is used to manage and maintain the rangelands and to repair damage caused by overgrazing. Taxpayers for Common Sense and other free-market groups want grazing fees increased to cover all costs of maintaining herds on publicly owned rangelands.

CONCEPT CHECK **STOP**

What are rangelands?

How can overgrazing of rangeland lead to desertification?

How do conservation easements help protect privately owned rangelands?

Which agencies manage public rangelands? What management issues do they face?

National Parks and Wilderness Areas

LEARNING OBJECTIVES

State which government agency administers U.S. national parks and describe current issues of concern.

Define *wilderness* and discuss the administration and goals of the National Wilderness Preservation System and the problems faced by wilderness areas.

Explain the differences between the wise-use and environmental movements' views on the use of public lands.

Many acres of federal land are set aside either as national park property or as wilderness areas. Both types of land were established to encourage the protection of the natural environment, yet both experience conflicts associated with how best to use and manage these protected areas.

NATIONAL PARKS

Created in 1916 as a federal bureau in the Department of the Interior, the **National Park Service** (NPS) was originally composed of large, scenic areas in the West such as Yellowstone, the Grand Canyon, and Yosemite Valley (**FIGURE 13.10A**).

Today the NPS has more cultural and historical sites—battlefields and historically important buildings and towns—than places of scenic wilderness. The NPS currently administers 388 sites, 58 of which are national parks. The total acreage that the NPS administers is 34.1 million hectares (84.3 million acres).

Because knowledge and understanding increase enjoyment, one of the primary roles of the NPS is to teach people about the natural environment, management of natural resources, and history of a site by providing nature walks and guided tours of its parks. Exhibits along roads and trails, evening campfire programs, museum displays, and lectures are other common educational tools.

The popularity and success of U.S. national parks (**TABLE 13.2**) have encouraged many other nations to establish national parks. Today about 1,200 national parks exist in more than 100 countries (see the introduction to this chapter). As in the United States, in other countries parks usually have multiple roles, from providing biological habitat to facilitating human recreation.

Threats to U.S. parks Some national parks are overcrowded (**FIGURE 13.10B**). All the problems plaguing urban areas are found in popular national parks during peak seasonal use, including crime, vandalism, litter, traffic jams, and pollution of the soil, water, and air. In addition, thousands of resource violations, from cutting live trees and collecting plants, minerals, and fossils, to defacing historical structures with graffiti, are investigated in national parks each year. Park managers have had to reduce visitor access to park areas that have become degraded from overuse, and in some cases to restrict vehicle traffic (**FIGURE 13.10C**).

Some national parks have imbalances in wildlife populations. Populations of many mammal species are in decline, including bears, white-tailed jackrabbits, and red foxes. For example, grizzly bear populations in national parks of the western United States are threatened. Grizzlies are territorial and require large areas of wilderness as habitat, and the presence of humans in national parks may imperil them. Most importantly, the parks may be too small to support grizzlies. While grizzly bears have survived in sustainable numbers in Alaska and Canada, the populations native to the more restricted spaces of the western continental United States are in considerable danger.

Other mammal populations—notably elk—have proliferated. Elk in Yellowstone National Park's northern range increased from a population of 3,100 in 1968 to a record high of 19,000 in 1994. Ecologists documented that elk reduced the abundance of native vegetation, such as willow and aspen, and seriously eroded

A **Yosemite National Park in California** This winter view shows the Merced River flowing past the rock formation El Capitan.

B **Summer crowd at Grand Canyon National Park in Arizona** The popularity of certain national parks threatens to overwhelm them.

The ten most popular national parks TABLE 13.2

National park	Number of recreational visitors in 2007 (in millions)
Great Smoky Mountains (North Carolina, Tennessee)	9.4
Grand Canyon (Arizona)	4.4
Yosemite (California)	3.5
Yellowstone (Wyoming, Montana, Idaho)	3.1
Olympic (Washington)	3.0
Rocky Mountain (Colorado)	2.9
Zion (Utah)	2.7
Grand Teton (Wyoming)	2.6
Cuyahoga Valley (Ohio)	2.5
Acadia (Maine)	2.2
Total visitors to the national park system	275.6

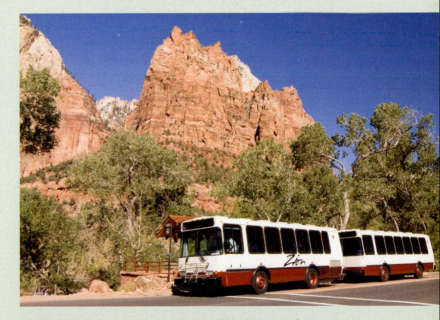

C **Shuttle bus at Zion National Park in Utah** In operation since 2000, the Zion National Park shuttle system eliminates traffic congestion. During peak months, personal vehicles are banned from popular areas; visitors instead ride the propane-powered shuttle.

NATIONAL GEOGRAPHIC

Gray wolves prey on elk in Yellowstone National Park FIGURE 13.11

Since their reintroduction, gray wolf populations have gained a secure foothold in Yellowstone National Park. Early studies of the effects of these predators support scientists' predictions that wolves will help reduce the burgeoning elk population.

stream banks. The reintroduction of gray wolves to Yellowstone, which began in 1995, appears in preliminary studies to be contributing to reductions in the elk population and to improved aspen and willow growth (FIGURE 13.11).

National parks are increasingly becoming islands of natural habitat surrounded by human development. Development on the borders of national parks limits the areas in which wild animals may range, forcing them into isolated populations. Ecologists have found that when environmental stressors occur, several small "island" populations are more likely to become threatened than a single large population occupying a sizable range (see Chapter 15).

WILDERNESS AREAS

Wilderness encompasses regions where the land and its community of organisms are not greatly disturbed by human activities, where humans may visit but don't live permanently. The U.S. Congress recognized that increased human population and expansion into wilderness areas might result in a future where no lands exist in their natural condition. Accordingly, the **Wilderness Act** of 1964 authorized the U.S. government to set aside federally owned land that retains its primeval character and lacks permanent improvements or human habitation, as part of the **National Wilderness Preservation System**. These federal lands range in size from tiny islands of uninhabited land to portions of national parks, national forests, and national wildlife refuges (FIGURE 13.12). Although mountains are the most common wilderness areas, portions of other ecosystems have been set aside, including tundra, desert, and wetlands.

Areas designated as wilderness are given the highest protection of any federal land. These areas are to remain natural and unchanged so they will be unimpaired for future generations to enjoy. The same four government agencies that regulate all publicly owned land—the NPS, USFS, FWS, and BLM—oversee 702 wilderness areas comprising 43.3 million hectares (107 million acres) of land. More than one-half of the lands in the National Wilderness Preservation System lie in Alaska, and western states are home to much of the remainder.

Millions of people visit U.S. wilderness areas each year, and some areas are overwhelmed by this traffic: Eroded trails, soil and water pollution, litter and trash, and human congestion predominate over quiet, unspoiled land. Government agencies now restrict the number of people allowed into each wilderness area at one time so that human use doesn't seriously affect the wilderness.

Some of the most popular wilderness areas may require more intensive future management, such as the development of trails, outhouses, cabins, and campsites. These amenities are not encountered in true wilderness, posing a dilemma between wilderness preservation and human use and enjoyment of wild lands.

Diverse wilderness areas FIGURE 13.12

A The Pelican Island National Wildlife Refuge in Florida is home to the 2.4-hectare (6-acre) Pelican Wilderness, one of the smallest areas protected in the National Wilderness Preservation System.

B At more than 3.7 million hectares (9.1 million acres), the Wrangell–Saint Elias Wilderness in Alaska is the nation's largest designated wilderness area.

Limiting the number of human guests in a wilderness area doesn't control all the factors that threaten wilderness, however. **Invasive species** have the potential to upset the natural balance among native species. For example, white pine blister rust, a foreign (nonnative) fungus that kills white pine trees, has invaded the wilderness in the northern Rocky Mountains. Wilderness managers are concerned that declining white pine populations could affect the population of grizzly bears in the region because pine seeds are a major part of the grizzlies' diet. The Wilderness Act specifies the avoidance of intentional ecological management. In this example, should the white pine populations be scientifically manipulated to help preserve the original wilderness?

Large tracts of wilderness, most of it in Alaska, have been added to the National Wilderness Preservation System since passage of the Wilderness Act in 1964. People who view wilderness as a nonrenewable resource support the designation of additional wilderness areas. They think it is particularly important to preserve additional land in the lower 48 states, where currently less than 2 percent of the total land area is protected as wilderness. Increasing the amount of federal land in the National Wilderness Preservation System is opposed by groups who operate businesses on public lands (such as timber, mining, ranching, and energy companies) and by their political representatives.

Patches of clear-cut forest in Williamette National Forest, Oregon FIGURE 13.13

The wise-use movement favors opening federal lands to logging and other types of economic development.

MANAGEMENT OF FEDERAL LANDS

How do we best manage the legacy of federal lands? Should federal lands be managed under multiple uses, or should they be preserved so that they benefit U.S. citizens for generations to come? These questions have divided many Americans into two groups, each a coalition of several hundred grassroots organizations. Those who wish to exploit resources on federal lands are known collectively as the *wise-use movement*; this group also includes many corporations. Those who wish to preserve the resources on federally owned lands are known collectively as the *environmental movement*. The descriptions of the two movements that follow are mainstream; any given group may not support all of the listed goals of a movement.

In general, people who support the wise-use movement think that the government overregulates environmental protection and that property owners should have more flexibility to use natural resources. They believe that the primary purpose of federal lands is to enhance economic growth (**FIGURE 13.13**).

Some of the goals of the wise-use movement include the following:

1. Put all national forests under timber management, including old-growth forests.
2. Permit mining and commercial development of wilderness areas, wildlife refuges, and national parks, where appropriate.
3. Allow unrestricted development of wetlands.
4. Sell parts of resource-rich federal lands to private interests, such as mining, oil, coal, ranching, and timber groups, for resource extraction.

Many organizations that embrace the wise-use movement have environmentally friendly names. The National Wetlands Coalition, for example, consists primarily of real estate developers and energy companies that want to drain and develop wetlands. Similarly, logging companies support the American Forest Resource Alliance.

In contrast to the wise-use movement, the environmental movement views federal lands as a legacy of U.S. citizens. They think that:

1. The primary purpose of public lands is to protect biological diversity and ecosystem integrity.
2. Those who extract resources from public lands should pay U.S. citizens compensation equal to the fair market value of the resource and not be subsidized by taxpayers.
3. Those who use public lands should be held acountable for any environmental damage they cause.

CONCEPT CHECK STOP

What government agency administers the National Park System? What problems does it face?

What is wilderness, and how does the U.S. National Wilderness Preservation System seek to protect it? What problems do these areas face?

How do the wise-use and environmental movements differ in their views on the use of public lands?

Conservation of Land Resources

Name at least three of the most endangered ecosystems in the United States.

Describe several of the criteria used to evaluate whether an ecosystem is endangered.

Our ancestors considered natural areas an unlimited resource to exploit. They appreciated prairies as valuable agricultural land and forests as immediate sources of lumber and eventual farmland. This outlook was practical as long as there was more land than people needed. But as the population increased and the amount of available land decreased, people began to view land as a limited resource. Thus, exploitation has increasingly shifted to preservation of the remaining natural areas in this country and around the world (**FIGURE 13.14**).

Although all types of ecosystems must be conserved, several are in particular need of protection. The U.S. Geological Survey (USGS) and the Defenders of Wildlife commissioned studies that ranked the most endangered ecosystems in the United States. They used four criteria:

1. The area lost or degraded since Europeans colonized North America

2. The number of present examples of a particular ecosystem, or the total area

3. An estimate of the likelihood that a given ecosystem will lose a significant area or be degraded during the next 10 years

4. The number of threatened and endangered species living in that ecosystem

TABLE 13.3 lists the 15 most endangered U.S. ecosystems based on these criteria. Examples include the South Florida landscape, Southern Appalachian spruce-fir forests, and longleaf pine forests and savannas. As these ecosystems are lost and degraded, the organisms that compose them decline in number and in genetic diversity. Implementing conservation strategies that set aside ecosystems is the best way to preserve an area's biodiversity.

The fifteen most endangered ecosystems in the United States (in order of priority)
TABLE 13.3

South Florida landscape

Southern Appalachian spruce-fir forests

Longleaf pine forests and savannas

Eastern grasslands, savannas, and barrens

Northwestern grasslands and savannas

California native grasslands

Coastal communities in lower 48 states and Hawaii

Southwestern riparian communities

Southern California coastal sage scrub

Hawaiian dry forest

Large streams and rivers in lower 48 states and Hawaii

Cave systems

Tallgrass prairie

California river- and stream-bank communities and wetlands

Florida scrub

As you have seen in this chapter, government agencies, private conservation groups, and private citizens have begun to set aside natural areas for permanent preservation. Such activities ensure that our children and grandchildren will inherit a world with wild places and other natural ecosystems.

CONCEPT CHECK STOP

What are three U.S. ecosystems that need protection?

What are three criteria used to evaluate whether an ecosystem is endangered?

Protected ecosystems around the world FIGURE 13.14

UNPROTECTED AREA
88% of Earth's land surface

PROTECTED AREA
12%

Protected areas worldwide represent 12 percent of the Earth's land surface, according to the U. Environmental Programme World Conservation Monitoring Centre. Only 0.5 percent of the marine environment is within protected areas—an amount considered inadequate by conservationists because of the increasing threats of overfishing and coral reef loss worldwide.

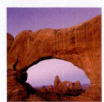

HAWAII VOLCANOES NATIONAL PARK, HAWAII
The park includes two of the world's most active volcanoes, Kilauea and Mauna Loa. The landscape shows the results of 70 million years of volcanism, including calderas, lava flows, and black sand beaches. Lava spreads out to build the island, and seawater vaporizes as lava hits the ocean at 1,149°C (2,100° F). The national park, created in 1916, covers 10% of the island of Hawaii and is a refuge for endangered species, like the hawksbill turtle and Hawaiian goose. It was made a World Heritage site in 1987.

GALÁPAGOS NATIONAL PARK, ECUADOR
Galápago means "tortoise" in Spanish, and at one time 250,000 giant tortoises roamed the islands. Today about 15,000 remain, and three of the original 14 subspecies are extinct—and the Pinta Island tortoise may be extinct soon. In 1959, Ecuador made the volcanic Galápagos Islands a national park, protecting the giant tortoises and other endemic species. The archipelago became a World Heritage site in 1978, and a marine reserve surrounding the islands was added in 2001.

WESTERN UNITED STATES
An intricate public lands pattern—including national forests, wilderness areas, wildlife refuges, and national parks such as Arches (above)—embraces nearly half the surface area of 11 western states. Ten out of 19 World Heritage sites in the United States are found here. It was in the West that the modern national park movement was born in the 19th century, with the establishment of Yellowstone and Yosemite National Parks.

MADIDI NATIONAL PARK, BOLIVIA
Macaws may outnumber humans in Madidi, Bolivia's second largest national park, established in 1995. A complex community of plants, animals, and native Indian groups share this 18,900-sq-km (7,300-sq-mi) reserve, part of the Tropical Andes biodiversity hotspot. Indigenous communities benefit from ecotourism.

AMAZON BASIN, BRAZIL
Indigenous peoples help manage reserves in Brazil that are linked with Jaú National Park. The park and reserves are part of the Central Amazon Conservation Complex, a World Heritage site covering more than 60,000 sq km (23,000 sq mi). It is the largest protected area in the Amazon Basin and one of the most biologically rich regions on the planet.

ARCTIC REGIONS
Polar bears find safe havens in Canadian parks, such as on Ellesmere Island, and in Greenland's huge protected area—Earth's largest—that preserves the island's frigid northeast. In 1996 countries with Arctic lands adopted the Circumpolar Protected Areas Network Strategy and Action Plan to help conserve ecosystems. Today 15% of Arctic land area is protected.

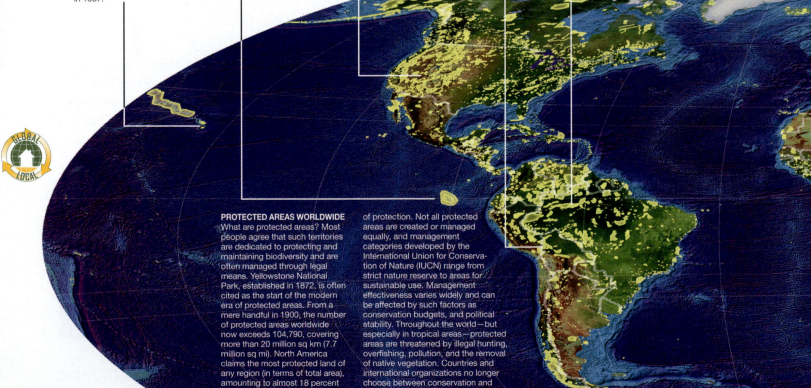

PROTECTED AREAS WORLDWIDE
What are protected areas? Most people agree that such territories are dedicated to protecting and maintaining biodiversity and are often managed through legal means. Yellowstone National Park, established in 1872, is often cited as the start of the modern era of protected areas. From a mere handful in 1900, the number of protected areas worldwide now exceeds 104,790, covering more than 20 million sq km (7.7 million sq mi). North America claims the most protected land of any region (in terms of total area), amounting to almost 18 percent of its territory. South Asia, at about 7 percent, has the least amount of land under some form of protection. Not all protected areas are created or managed equally, and management categories developed by the International Union for Conservation of Nature (IUCN) range from strict nature reserve to areas for sustainable use. Management effectiveness varies widely and can be affected by such factors as conservation budgets, and political stability. Throughout the world—but especially in tropical areas—protected areas are threatened by illegal hunting, overfishing, pollution, and the removal of native vegetation. Countries and international organizations no longer choose between conservation and development; rather the goal for societies is to balance the two for equitable and sustainable resource use.

Wildest biomes

- ■ Wildlands
- □ Ice or snow cover

◄ **WILDEST AREAS**
Although generally far from cities, the world's remaining wild places play a vital role in a healthy global ecosystem. The boreal forests of Canada and Russia, for instance, help cleanse the air we breathe by absorbing carbon dioxide and providing oxygen. With the human population increasing by an estimated 1 billion over the next 15 years, many wild places could fall within reach of the plow or under a cloud of smog.

For millennia, lands have been set aside as sacred ground or as hunting reserves for the powerful. Today, great swaths are protected for recreation, habitat conservation, biodiversity preservation, and resource management. Some groups may oppose protected spaces because they want access to resources now. Yet local inhabitants and governments are beginning to see the benefits of conservation efforts and sustainable use for human health and future generations.

SAREKS NATIONAL PARK, SWEDEN

This remote 1,970-sq-km (760-sq-mi) park, established in 1909 to protect the alpine landscape, is a favorite of backcountry hikers. It boasts some 200 mountains more than 1,800 m (5,900 ft) high, narrow valleys, and about 100 glaciers. Sareks forms part of the Laponian Area World Heritage site and has been a home to the Saami (or Lapp) people since prehistoric times.

AFRICAN RESERVES

Some 120,000 elephants roam Chobe National Park in northern Botswana. Africa has more than 7,500 national parks, wildlife reserves, and other protected areas, covering about 9 percent of the continent. Protected areas are under enormous pressure from expanding populations, civil unrest and war, and environmental disasters.

WOLONG NATURE RESERVE, CHINA

Giant pandas freely chomp bamboo in this 2,000-sq-km (772-sq-mi) reserve in Sichuan Province, near the city of Chengdu. Misty bamboo forests host a number of endangered species, but the critically endangered giant panda—among the rarest mammals in the world—is the most famous resident. Only about 1,600 giant pandas exist in the wild.

KAMCHATKA, RUSSIA

Crater lakes, ash-capped cones, and diverse plant and animal species mark the Kamchatka Peninsula—a World Heritage site—located between the icy Bering Sea and Sea of Okhotsk. The active volcanoes and glaciers form a dynamic landscape of great beauty, known as "The Land of Fire and Ice." Kamchatka's remoteness and rugged landscape help fauna flourish, producing record numbers of salmon species and half of the Steller's sea-eagles on Earth.

GUNUNG PALUNG NATIONAL PARK, INDONESIA

A tree frog's perch could be precarious in this 900-sq-km (347-sq-mi) park on the island of Borneo, in the heart of the Sundaland biodiversity hotspot. The biggest threat to trees and animals in the park and region is illegal logging. Gunung Palung contains a wider range of habitats than any other protected area on Borneo, from mangroves to lowland and cloud forests. A number of endangered species, such as orangutans and sun bears, depend on the dense forests.

AUSTRALIA & NEW ZEALAND

Uluru, a red sandstone monolith (formerly known as Ayers Rock), and the vast Great Barrier Reef, one of the largest marine parks in the world, are outstanding examples of Australia's protected areas—which make up more than 10 percent of the country's area and conserve a diverse range of unique ecosystems. About a third of New Zealand is protected, and it is a biodiversity hotspot because of threats to flightless native birds, such as the kakapo and kiwi. Cats, stoats, and other predators, introduced to New Zealand by settlers, kill thousands of birds each year.

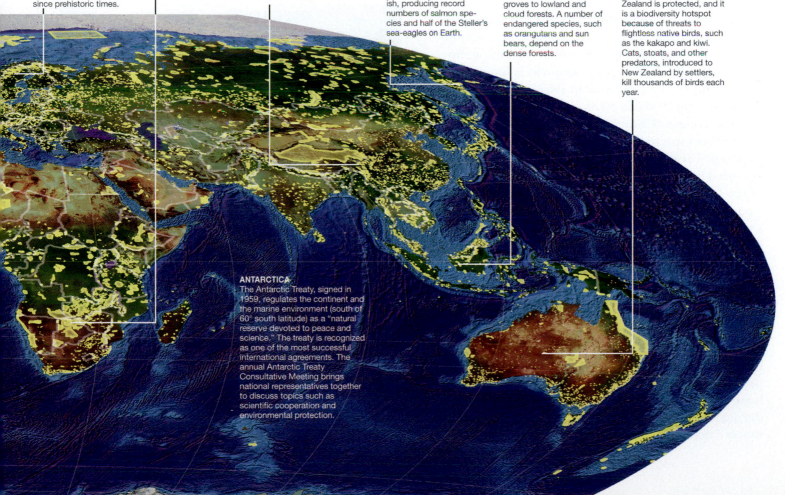

ANTARCTICA

The Antarctic Treaty, signed in 1959, regulates the continent and the marine environment (south of 60° south latitude) as a "natural reserve devoted to peace and science." The treaty is recognized as one of the most successful international agreements. The annual Antarctic Treaty Consultative Meeting brings national representatives together to discuss topics such as scientific cooperation and environmental protection.

THE TONGASS DEBATE OVER CLEAR-CUTTING

Despite its northern location along Alaska's southeastern coast, the Tongass National Forest is one of the world's few temperate rain forests (**FIGURES A** and **B**; also see Chapter 6 for a description of the temperate rainforest biome). It is one of the wettest places in the United States. This moisture supports old-growth forest of giant Sitka spruce, yellow cedar, and western hemlock, some of which are 700 years old. This 6.9-million-hectare (17-million-acre) forest, the largest in the National Forest System, provides habitat for a wealth of wildlife, such as grizzly bears and bald eagles.

The Tongass is a prime logging area because a single large Sitka spruce may yield as much as 10,000 board feet (23.6 cubic meters) of high-quality timber. The logging industry forms the basis of much of the local economy but conflicts with environmental interests seeking to protect the forests from overharvesting. Regeneration of mature forest after it is clear-cut is slow, on the order of several centuries.

As in most other national forests, it is expensive to log in the Tongass. To cover high operating costs, timber interests such as pulp mills rely on obtaining the timber from the federal government at below-market prices. This right was granted in 1954 by a contract that expired in the 1990s. In 1990, congressional efforts to pass the Tongass Timber Reform Act, which would force timber interests to pay market prices, were bitterly opposed. The compromise agreement, reached in 1997, provided timber to the mills at market prices. As a result of this legislation, clear-cut logging continued in the Tongass, but at lower rates than in the past.

In the closing months of the Clinton administration, the USFS officially adopted the *Roadless Area Conservation Rule* to protect roadless national forests from road building and forest harvest. A federal judge blocked the roadless rule in 2001, and shortly thereafter the Bush administration opened to logging and development part of the Alaska forest that had formerly been off-limits (**FIGURE C**).

The take-home message is that the USFS, like other government agencies, takes its lead from current presidential policies. Changes in administrations often leave the USFS and other government agencies floundering as they strive to implement established rulings that are no longer supported.

Alaska's Tongass National Forest

A This temperate rain forest (light green area) is in southeastern Alaska along the Pacific Ocean.

B Aerial view of the Tongass.

C USFS logging road in the Tongass. Roads open the forest to logging and other kinds of development.

SUMMARY

1 Land Use in the United States

1. More than one-half of U.S. land is privately owned. Slightly more than one-third—including many types of ecosystems and land uses—is owned by the federal government. Seven percent belongs to state and local governments.

2 Forests

1. **Sustainable forestry** is the use and management of forest ecosystems in an environmentally balanced and enduring way. Sustainable forestry maintains a mix of forest trees, by age and species, rather than a **monoculture**, in which only one type of plant is cultivated over a large area. Adopting sustainable forestry principles requires setting aside sanctuaries and **wildlife corridors**, protected zones that connect isolated unlogged or undeveloped areas.

2. **Deforestation** is the temporary or permanent clearing of large expanses of forest for agriculture or other uses. **Clear-cutting** is a logging practice in which all the trees in a stand of forest are cut, leaving just the stumps; clear-cutting over a wide area is ecologically unsound. The major causes of tropical deforestation are subsistence farming, commercial logging, and cattle ranching, all accelerated by growing human populations. Increased needs for fuelwood drive deforestation of tropical dry forests.

3. Most U.S. national forests are managed by the U.S. Forest Service (USFS); the rest are overseen by the Bureau of Land Management (BLM). National forests face conflicts associated with supporting multiple uses: timber harvest; livestock forage; water resources and watershed protection; mining; hunting, fishing, and other forms of recreation; and habitat for fishes and wildlife.

3 Rangelands

1. **Rangelands** are grasslands that aren't intensively managed and are used for grazing livestock. Rangelands are also mined for mineral and energy resources, used for recreation, and preserved for biological habitat and for soil and water resources.

2. **Overgrazing** is the destruction of vegetation caused by too many grazing animals consuming the plants in a particular area, leaving them unable to recover. Overgrazing accelerates **land degradation**, which decreases the future ability of the land to support crops or livestock. **Desertification** is the degradation of once-fertile rangeland or tropical dry forest into nonproductive desert.

3. A **conservation easement** is a legal agreement that protects privately owned forest or other property from development for a specified number of years. Conservation groups often pay for conservation easements to preserve open rangeland.

4. The BLM manages more than three-fourths of U.S. public rangelands, excluding Alaska; the USFS manages the remainder. Current issues on public rangelands include conflicts between environmental groups and ranchers over the number of livestock allowed to graze and the potential to manage the areas for uses such as biological habitat, recreation, and scenic value. Conflicts also arise over whether grazing fees paid by livestock operators on public lands should be high enough to cover all costs of maintaining herds, removing taxpayer burden.

4 National Parks and Wilderness Areas

1. The **National Park Service** administers 388 sites in the United States, including 58 national parks. The problems the sites encounter include overcrowding, pollution, crime, resource violations, and imbalanced wildlife populations.

2. **Wilderness** is a protected area of land in which no human development is permitted. The **National Wilderness Preservation System** consists of four U.S. government agencies—the NPS, USFS, FWS, and BLM—that oversee 702 wilderness areas. The problems these areas face include overuse and overcrowding by visitors, pollution, erosion, and the introduction of invasive species.

3. Those who support the **wise-use movement** believe a primary purpose of federal lands is to enhance economic growth. They think that the government overregulates environmental protection and that property owners should have more flexibility to use natural resources. Those who support the **environmental movement** view federal lands as a legacy of U.S. citizens and thus want to preserve resources on federally owned lands.

5 Conservation of Land Resources

1. Endangered U.S. ecosystems include the south Florida landscape, southern Appalachian spruce-fir forests, and longleaf pine forests and savannas.

2. Criteria used to evaluate whether an ecosystem is endangered and to what degree it is threatened include its history of land loss and degradation, its prospects for future loss or degradation, the area the ecosystem occupies, and the number of threatened and endangered species living in that ecosystem.

KEY TERMS

- **monoculture** p. 323
- **sustainable forestry** p. 324
- **wildlife corridor** p. 324
- **clear-cutting** p. 324
- **deforestation** p. 326
- **rangeland** p. 331
- **overgrazing** p. 332
- **desertification** p. 332
- **conservation easement** p. 333
- **wilderness** p. 336

CRITICAL AND CREATIVE THINKING QUESTIONS

1. Why is deforestation a serious global environmental problem?

2. What are the environmental effects of clear-cutting on steep mountain slopes? on tropical rainforest land?

3. Distinguish between rangeland degradation and desertification. Why is moderate grazing beneficial to rangelands, yet overgrazing leads to erosion?

4. Explain the various uses that must be considered in the management of a national park. Use Korup National Park as an example in which the needs of the human population surrounding a park have been addressed successfully.

5. Debate conflicts over logging in Tongass National Forest from two points of view: that of the wise-use movement and that of the environmental movement.

6. Do you think additional federal lands should be added to the wilderness system? Why or why not?

7. Consider the message that the cartoon above sends about the popularity of U.S. national parks. What might park managers do to protect parks while encouraging their use?

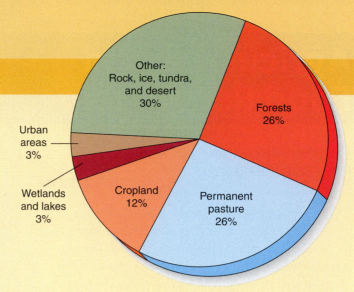

8. Should private landowners have control over what they wish to do to their land? How would you as a landowner handle land-use decisions that may affect the public? Present arguments for both sides of the issue.

9. Explain how economic growth and sustainable use of natural resources can be compatible goals.

10. List at least five ecosystem services provided by forests and other non-urban lands.

11. Suppose that a valley contains a small city surrounded by agricultural land. The land is encircled by a mountain wilderness. Explain why the preservation of the mountain ecosystem would support both urban and agricultural land in the valley.

12. Given the important contributions of forests in providing both timber and ecosystem services, how would you manage U.S. public forests if you were in charge?

Examine the graph of world land use that was assembled by the World Resources Institute and the U.N. Food and Agricultural Organization.

13. How much of the world's total land area is used for agriculture?

14. Approximately 29 percent of the world's land area consists of natural ecosystems that could potentially be developed for human use. What areas were counted in this estimate?

15. Do you think a graph of land uses in your state would look similar to or different from the graph shown? Is the frequency of each land use likely to increase, decrease, or remain the same over the next several decades, both globally and in your state? Explain your answer.

VIEW THIS IN ACTION
in your WileyPLUS course

What is happening in this picture ?

■ Julia "Butterfly" Hill lived in this 600–1,000-year-old, 180-foot-tall California redwood for more than 2 years in the late 1990s, to keep a lumber company from cutting down the tree. Would Hill's perspective on wilderness better fit the wise-use movement or the environmental movement?

■ Explain the likely differences in the perspectives of Hill and the lumber company, especially given that the tree is on the company's land.

■ Why do ecologists and environmentalists believe that the logging of old-growth trees causes such particular damage?

■ See if you can find out what happened to this tree after Julia left it.

Agriculture and Food Resources

14

MAINTAINING GRAIN STOCKPILES FOR FOOD SECURITY

VIEW THIS IN ACTION
in your WileyPLUS course

When people have access at all times to adequate amounts and kinds of food needed for healthy, active lives, they are said to have *food security*. World grain carryover stocks—the amounts of rice, wheat, corn, and other grains remaining from previous harvests—provide a measure of world food security (see larger photo). Stockpiles of grain have decreased since their all-time high in 1987, when carryover stocks were large enough, if evenly distributed, to feed every person in the world for 104 days. In contrast, the amount of grain stockpiled in 2006 would have fed the world's people for only 57 days. According to the United Nations, world grain carryover stocks should not fall below a minimum of 70 days' supply in a given year.

Poor weather is the main reason that the year 2006 had a lower grain harvest than the previous two years. Australia and Europe had major droughts, and North America had an unusually dry spring. Asia experienced droughts, flooding, diseases, and insect infestations. The low 2006 harvest caused grain prices to surge worldwide.

World grain stocks are also under pressure because consumption of beef, pork, poultry, and eggs has increased in developing countries such as China (see inset), where growing affluence has led some people to diversify their diets. This trend represents a global pattern: In highly developed countries, animal products account for nearly half of the calories people consume, compared to only 5 percent of the calories people consume in developing countries. Increased consumption of meat and meat products has prompted a surge in the amount of grain used to feed the world's billions of livestock animals. Thus, the global trend of eating more meat and other animal products is linked to an increased use of grains and other feed crops for livestock.

NATIONAL GEOGRAPHIC

World Food Problems

LEARNING OBJECTIVES

Differentiate between undernutrition and overnutrition.

Define *food insecurity* and relate it to human population, poverty, and world hunger.

The U.N. Food and Agricultural Organization (FAO) estimates that about 852 million people—more than 2.5 times the population of the United States—lack the food needed for healthy, productive lives. Most of these people live in rural areas of the poorest developing countries.

The average adult human must consume enough food to get approximately 2,600 *kilocalories*, or simply *Calories*, per day. People who receive fewer calories than needed are undernourished. Over an extended period of undernourishment, their health and stamina decline, even to the point of death. Worldwide, an estimated 182 million children under age 5 suffer from **undernutrition** (**FIGURE 14.1**) and are seriously underweight, according to the World Health Organization (WHO).

People might receive enough calories in their diets but still be malnourished because they do not receive enough essential nutrients, such as proteins, vitamin A, iodine, or iron. Adults suffering from malnutrition are more susceptible to disease and have less strength to function productively than those who are well fed. In addition to being more susceptible to disease, malnourished children do not grow or develop normally. Because malnutrition affects cognitive development, malnourished children typically do not perform well in school. Currently, WHO estimates that more than 3 billion people worldwide—the greatest number in history—are malnourished. In addition, more than half the deaths in children less than 5 years old in developing countries are associated with malnutrition.

People who eat more food than necessary are overnourished. Generally, a person suffering from **overnutrition** has a diet high in saturated (animal) fats, sugar, and salt. Overnutrition, which is most common among people in highly developed nations such as the United States, results in obesity, high blood pressure, and an increased likelihood of disorders such as diabetes, heart disease, and some cancers. Overnutrition is also emerging in some developing countries, particularly in urban areas, where, as people earn more money, their diets shift from cereal grains to the consumption of more livestock products and processed foods high in fat and sugar.

POPULATION AND WORLD HUNGER

According to the FAO, 86 countries are considered low income and food deficient, which means they cannot produce enough food or afford to import enough food to feed the entire population. South Asia, with an estimated 270 million hungry people, and sub-Saharan Africa, with an estimated 175 million who are hungry, are the regions of the world with the greatest **food insecurity** (**FIGURE 14.2**).

Today, sub-Saharan Africa produces less food per person than it did in 1950. Several factors contribute to the food shortages in Africa; these include civil wars and military actions, HIV/AIDS (which has killed or incapacitated much of the agricultural workforce in some countries), floods, droughts, and soil erosion from hilly, marginal farmlands. People with food insecurity always live under the threat of starvation.

undernutrition
A type of malnutrition in which there is an underconsumption of calories or nutrients that leaves the body weakened and susceptible to disease.

overnutrition
A type of malnutrition in which there is an overconsumption of calories that leaves the body susceptible to disease.

food insecurity
The condition in which people live with chronic hunger and malnutrition.

Effects of hunger FIGURE 14.1

A Millions of children suffer from *kwashiorkor*, caused by severe protein deficiency. Note the characteristic swollen belly, which results from fluid retention. Photographed in Haiti. ▼

B *Marasmus* is progressive emaciation caused by a diet low in both total calories and protein. Symptoms include a pronounced slowing of growth and extreme wasting of muscles. Photographed in Somalia. ▼

C Globally, millions of adult men and women are hungry. This homeless man is suffering from severe malnutrition and starvation. Photographed in New Delhi, India. ▼

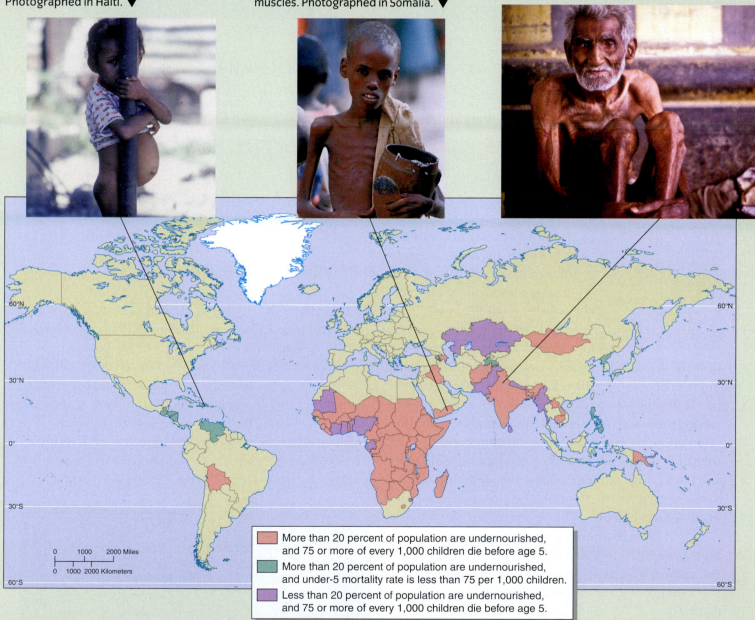

More than 20 percent of population are undernourished, and 75 or more of every 1,000 children die before age 5.

More than 20 percent of population are undernourished, and under-5 mortality rate is less than 75 per 1,000 children.

Less than 20 percent of population are undernourished, and 75 or more of every 1,000 children die before age 5.

Hunger and child mortality FIGURE 14.2

Chronic hunger and malnutrition, which are concentrated in sub-Saharan Africa and South Asia, are linked to an increased mortality for infants and children.

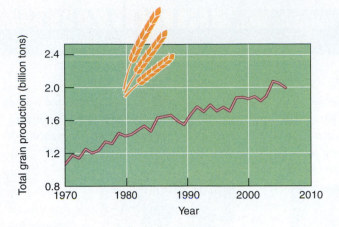

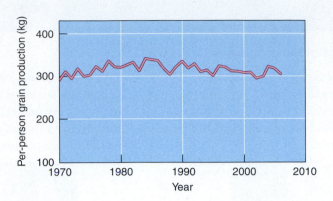

A Total world grain production increased from 1.1 billion tons in 1970 to 2.0 billion tons in 2006.

B The amount of grain produced per person has increased only slightly.

Total world grain production and grain production per person, 1970 to 2006 FIGURE 14.3

---

economic development

An expansion in a government's economy, viewed by many as the best way to raise the standard of living.

The cause of world hunger is not simply the inability of food production to keep pace with population growth. Agriculture currently produces enough food for everyone to have enough if it were distributed evenly. Experts agree that world hunger, population, poverty, and environmental problems are interrelated, but they disagree on the relative importance of each factor. Different groups propose different solutions for resolving the world's food problems, including controlling population growth, promoting the **economic development** of countries that do not produce adequate food, and correcting the inequitable distribution of resources. All experts agree that population pressures exacerbate world food problems.

Although annual grain production almost doubled from 1970 to 2006 (**FIGURE 14.3**), the world population increased so rapidly during that period that the amount of grain *per person* has not changed appreciably.

Producing enough food to feed the world's people is the largest challenge in agriculture today, and it becomes more difficult each year because the human population is continually expanding. Global food production can be increased in the short term, although whether this increase is sustainable is questionable.

POVERTY AND FOOD

The main cause of undernutrition and malnutrition is poverty. Infants, children, and the elderly are most susceptible to poverty and chronic hunger. The world's poorest people—those living in developing countries in Asia, Africa, and Latin America—do not own land on which to grow food and do not have sufficient money to purchase food. Worldwide, hunger is more common in rural areas than in urban areas. Poverty and hunger are not restricted to developing nations, however; poor hungry people are also found in the United States, Europe, and Australia.

CONCEPT CHECK **STOP**

What is the difference between undernutrition and overnutrition?

Where is each type of malnutrition most prevalent in the world?

What is food insecurity and how does it relate to human population, poverty, and world hunger?

The Principal Types of Agriculture

Agriculture can be roughly divided into two types: industrialized agriculture and subsistence agriculture. Most farmers in highly developed countries and some in developing countries practice *high-input agriculture*, or **industrialized agriculture**. It relies on large inputs of capital and energy (in the form of fossil fuels) to make and run machinery, purchase seed, irrigate crops, and produce agrochemicals such as commercial inorganic fertilizers and pesticides (see **FIGURE 14.4**). Industrialized agriculture produces high **yields** (the amount of food produced per unit of land), which allows forests and other natural areas to remain wild instead of being converted to agricultural land. However, the productivity of industrialized agriculture comes with costs, such as

> **industrialized agriculture** Modern agricultural methods, which require a large capital input and less land and labor than traditional methods.

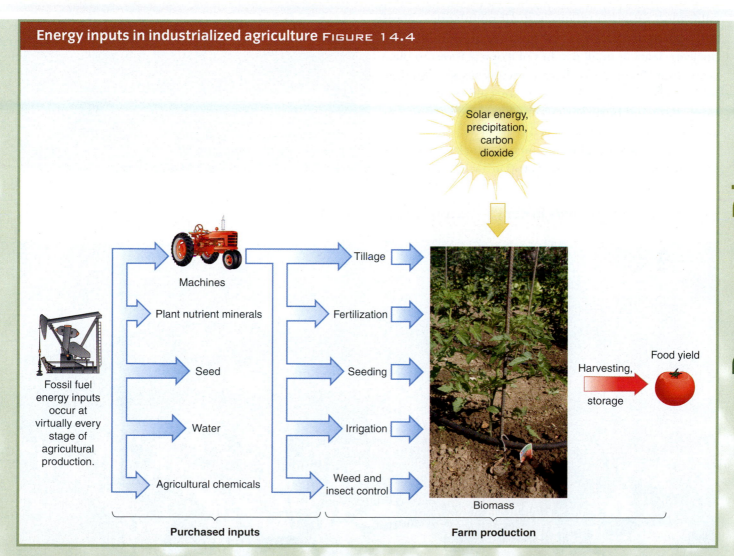

Energy inputs in industrialized agriculture FIGURE 14.4

Solar energy, precipitation, carbon dioxide

Fossil fuel energy inputs occur at virtually every stage of agricultural production.

Machines → Tillage

Plant nutrient minerals → Fertilization

Seed → Seeding

Water → Irrigation

Agricultural chemicals → Weed and insect control

Biomass

Harvesting, storage → Food yield

Purchased inputs   **Farm production**

Process Diagram

soil degradation and increased pesticide resistance in agricultural pests; we discuss these and other problems later in the chapter (and in Chapters 4 and 12).

Most farmers in developing countries practice **subsistence agriculture**, the production of enough food to feed oneself and one's family, with little left over to sell or reserve for hard times. Subsistence agriculture also requires a large input of energy, but from humans and draft animals rather than from fossil fuels.

Some types of subsistence agriculture require large tracts of land. **Shifting cultivation** is a form of subsistence agriculture in which short periods of cultivation are followed by longer periods of fallow (land left uncultivated), during which the land reverts to forest. Shifting cultivation supports relatively small populations. **Slash-and-burn agriculture** is a type of shifting cultivation that involves clearing small patches of tropical forest to plant crops (see Chapter 13). Because tropical soils lose their productivity quickly when they are cultivated, farmers using slash-and-burn agriculture must move from one area of forest to another every 3 years or so.

Nomadic herding, in which livestock is supported by land too arid for successful crop growth, is a similarly

> **subsistence agriculture**
> Traditional agricultural methods, which are dependent on labor and a large amount of land to produce enough food to feed oneself and one's family.

land-intensive form of subsistence agriculture (**FIGURE 14.5**). Nomadic herders must continually move their livestock to find adequate food for the animals.

Intercropping is a form of intensive subsistence agriculture that involves growing a variety of plants on the same field simultaneously. When certain crops are grown together, they produce higher yields than when they are grown as **monocultures**. (A monoculture is the cultivation of only one type of plant over a large area.) **Polyculture** is a type of intercropping in which several kinds of plants that mature at different times are planted together. In polyculture practiced in the tropics, fast- and slow-maturing crops are often planted together so that different crops can be harvested throughout the year.

CONCEPT CHECK STOP

What are some differences between industrialized agriculture and subsistence agriculture?

What are shifting cultivation, nomadic herding, and intercropping?

Nomadic sheep herders in Kenya FIGURE 14.5

Challenges of Agriculture

LEARNING OBJECTIVES

Discuss recent trends in loss of U.S. agricultural land, global declines in domesticated plant and animal varieties, and efforts to increase crop and livestock yields.

Relate the benefits and problems associated with the green revolution.

Describe the environmental impacts of industrialized agriculture, including land degradation and habitat fragmentation.

The United States has more than 121 million hectares (300 million acres) of **prime farmland**, land that has the soil type, growing conditions, and available water to produce food, forage, fiber, and oilseed crops. U.S. agriculture faces a decline in prime farmland. Other challenges include coping with declining numbers of domesticated varieties, improving crop and livestock yields, and addressing environmental impacts.

LOSS OF AGRICULTURAL LAND

There is considerable concern that much of the nation's prime agricultural land is falling victim to urbanization and suburban sprawl by being converted to parking lots, housing developments, and shopping malls (**FIGURE 14.6**). In certain areas of the United States, loss of rural land is a significant problem. According to the American Farmland Trust, more than 162,000 hectares (400,000 acres) of prime U.S. farmland are lost each year.

The 1996 Farm Bill included funding for the establishment of a national *Farmland Protection Program*. This voluntary program lets farmers sell **conservation easements** that prevent their farmland from being converted to nonagricultural uses. The easements are in effect from a minimum of 30 years to forever. As with other conservation easements, the farmers retain full rights to use their property—in this case, for agricultural purposes.

Suburban sprawl onto agricultural land FIGURE 14.6
Virginia housing developments encroach on farmland.

GLOBAL DECLINE IN DOMESTICATED PLANT AND ANIMAL VARIETIES

A global trend is currently under way to replace the many local varieties of a particular crop or domesticated farm animal with just a few kinds. When farmers abandon traditional varieties in favor of more modern ones, which are bred for uniformity and maximum production, the traditional varieties frequently face extinction (**FIGURE 14.7**). This represents a great loss in genetic diversity because each variety's characteristic combination of genes gives it distinctive nutritional value, size, color, flavor, resistance to disease, and adaptability to different climates and soil types.

To preserve older, more diverse varieties of plants and animals, many countries, including the United States, are collecting **germplasm**: seeds, plants, and plant tissues of traditional crop varieties and the sperm and eggs of traditional livestock breeds.

germplasm
Any plant or animal material that may be used in breeding.

Dutch Belted cow FIGURE 14.7

The USDA recognizes Dutch Belts as a viable dairy breed that produces high-quality, flavorful milk. The breed is endangered in the United States and rare in Canada. Photographed in Wisconsin.

INCREASING CROP YIELDS

Until the 1940s, agricultural yields among various countries, whether highly developed or developing, were generally equal. Advances by research scientists since then have dramatically increased food production in highly developed countries (**FIGURE 14.8**). Greater knowledge of plant nutrition has resulted in fertilizers that promote high yields. The use of pesticides to control insects, weeds, and disease-causing organisms has also improved crop yields.

Average U.S. wheat yields, 1950 to 2005
FIGURE 14.8

Each year shown is actually an average of 3 years to minimize the effects that poor weather conditions might have in a given year. Similar increases in yield have occurred in other grain crops.

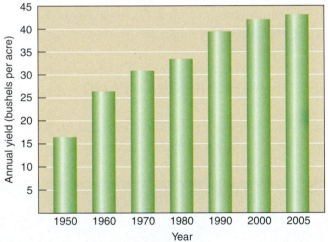

The green revolution By the middle of the 20th century, serious food shortages occurred in many developing countries coping with growing populations. The development and introduction during the 1960s of high-yield varieties of wheat and rice to Asian and Latin American countries gave these nations the chance to provide their people with adequate supplies of food (**FIGURE 14.9**). But the high-yield varieties required intensive industrial cultivation methods, including the use of commercial inorganic fertilizers, pesticides, and mechanized machinery, to realize their potential. These agricultural technologies were passed from highly developed nations to developing nations.

Using modern cultivation methods and the high-yield varieties of certain staple crops to produce more food per acre of cropland is known as the **green revolution**. Some of the success stories of the green revolution are remarkable. During the 1920s, Mexico produced less than 700 kg (0.77 ton) of wheat per hectare annually. During the green revolution years that began in 1965, Mexico's annual wheat production rose to more than 2,400 kg (2.65 tons) per hectare. Indonesia, another green revolution success story, formerly imported more rice than any other country in the world. Today Indonesia produces enough rice to feed its people and export some.

Critics of the green revolution argue that it has made developing countries dependent on imported technologies, such as agrochemicals and tractors, at the expense of traditional agriculture. The two most important problems associated with higher crop production are the high energy costs built into this type of agriculture and the environmental problems caused by the intensive use of fertilizers and pesticides.

Increasing crop yields in the post–green revolution era

In 1999, the International Food Policy Research Institute projected that the world demand for rice, wheat, and corn will increase 40 percent between 2000 and 2020. This rise in demand will require a corresponding rise in grain production to feed the human population, as well as the livestock needed to satisfy the appetites of the increasing numbers of affluent people who can afford to buy meat.

This challenge cannot be met by increasing the amount of land under cultivation, as the best arable lands are already being cultivated. Projected freshwater shortages, rising costs of agricultural chemicals, and deteriorating soil quality caused by intensive agricultural techniques may further constrain productivity. As Figure 14.9 demonstrates, recent progress in coaxing more grain out of crops genetically improved during the green revolution has resulted in diminishing returns. Grain yields have continued to rise since the 1960s, but in recent years the rates of increase have not been as great as they were previously.

Despite these problems, most plant geneticists think we can produce enough food in the 21st century to meet demand if countries spend more money in support of a concerted scientific effort to improve crops. Many scientists think genetic engineering is one of the keys to breeding more productive varieties. In addition, modern agricultural methods, such as water-efficient irrigation, will have to be introduced to developing countries that do not currently have them if we are to continue increasing crop yields.

Development of high-yield rice varieties FIGURE 14.9

Tall conventional plant

Improved high-yielding plant

Low-tillering ideotype (new plant type)

Traditional rice plants on the left are taller and do not yield as much grain (clusters at top of plant) as the more modern varieties shown in the middle and on the right. The rice plant in the middle was developed during the 1960s by crossing a high-yield, disease-resistant variety with a dwarf variety to prevent the grain-heavy plants from falling over. Improvements since the green revolution have been modest, as shown in the rice variety developed during the 1990s (right). Some researchers think rice and certain other genetically improved crops are near their physical limits of productivity.

INCREASING LIVESTOCK YIELDS

The use of hormones and antibiotics, although controversial, increases animal production. **Hormones**, usually administered by ear implants, regulate livestock bodily functions and promote faster growth. Although U.S. and Canadian farmers use hormones, the European Union (EU) currently restricts imports of hormone-treated beef. It cites studies which suggest that these hormones or their breakdown products, both found in trace amounts in meat and meat products, could cause cancer or affect the growth of young children. In 1999 an international scientific committee organized by the FAO and WHO concluded that the trace amounts of hormones found in beef are safe because they are very low compared to the normal hormone concentrations found in the human body.

Modern agriculture has also embraced the addition of low doses of **antibiotics** to the feed for pigs, chickens, and cattle. These animals gain 4 to 5 percent more weight than untreated animals, presumably because they expend less energy fighting infections.

Several studies link the indiscriminate use of antibiotics in humans and livestock to the evolution of bacterial strains that are resistant to antibiotics. Because there is increasing evidence that the use of antibiotics in agriculture reduces their medical effectiveness for humans, WHO recommended in 2003 that routine use of antibiotics in livestock be eliminated. Many European countries have complied, but the United States and many other countries continue the practice.

ENVIRONMENTAL IMPACTS

Industrialized agriculture has many environmental effects (**FIGURE 14.10A**). The agricultural use of fossil fuels and pesticides produces air pollution. Untreated animal wastes and agricultural chemicals such as fertilizers and pesticides cause water pollution, which reduces biological diversity, harms fisheries, and leads to outbreaks of nuisance species. According to the Environmental Protection Agency, agricultural practices are the single largest cause of surface-water pollution in the United States.

Industrialized agriculture has favored the replacement of traditional family farms with large agribusiness conglomerates. In the United States, most cattle, hogs, and poultry are now grown in feedlots and livestock factories (**FIGURE 14.10B**). The large concentrations of animals in livestock factories create many environmental problems, including air and water pollution.

Many insects, weeds, and disease-causing organisms have developed or are developing resistance to **pesticides**. Pesticide resistance forces farmers to apply progressively larger quantities of pesticides (**FIGURE 14.10C**). Pesticide residues contaminate our food supply and reduce the number and diversity of beneficial microorganisms in the soil. Fishes and other aquatic organisms are sometimes killed by pesticide runoff into lakes, rivers, and estuaries.

Land degradation is a reduction in the potential productivity of land. Soil erosion, which is exacerbated by large-scale mechanized operations, causes a decline in soil fertility, and the eroded sediments damage water quality. Other types of degradation are compaction of soil by heavy farm machinery and waterlogging and **salinization** (salting) of soil from improper irrigation methods.

Clearing grasslands and forests and draining wetlands to grow crops result in **habitat fragmentation** that reduces biological diversity. Many species are endangered or threatened as a result of habitat loss to agriculture. The most dramatic example of habitat loss in North America is tallgrass prairie, more than 90 percent of which has been converted to agriculture.

> **pesticide**
> A toxic chemical used to kill pests.

> **degradation (of land)** Natural or human-induced processes that decrease the future ability of the land to support crops or livestock.

> **habitat fragmentation**
> The breakup of large areas of habitat into small, isolated patches.

CONCEPT CHECK STOP

What is the green revolution? What are some of its benefits and problems?

What are the major environmental problems associated with industrialized agriculture?

A Environmental effects of industrialized agriculture

Individual agriculture practices may have multiple environment effects.

Water issues	Air pollution	Land degradation	Loss of biological diversity
• Groundwater depletion from irrigation • Pollution from fertilizer and pesticide runoff • Sediment pollution from eroding soil particles • Pollution from animal wastes (livestock factories) • Enrichment of surface water from fertilizer runoff and livestock wastes	• Pesticide sprays • Soil particles from wind erosion • Odors from livestock factories • Greenhouse gases from combustion of fossil fuels • Other air pollutants from combustion of fossil fuels	• Soil erosion • Loss of soil fertility • Soil salinization • Soil pollution (pesticide residues) • Waterlogged soil from improper irrigation	• Habitat fragmentation (clearing land and draining wetlands) • Monocultures (lack of diversity in croplands) • Stressors from air and water pollution • Stressors from pesticides • Replacement of many traditional crop and livestock varieties with just a few

B Hog factory in Ohio

The hogs remain indoors and are fed and watered by machine at timed intervals throughout the day. Although livestock factories are efficient and produce meat relatively inexpensively, they cause environmental problems such as problems related to sewage disposal. Some livestock factories produce as much sewage as small cities, yet the wastes do not currently fall under water-quality laws and do not go to sewage treatment plants.

C Colorado potato beetle on potato leaf

As a result of being exposed to heavy applications of pesticides over the years, Colorado potato beetles are resistant to most insecticides registered for use on potatoes.

Solutions to Agricultural Problems

LEARNING OBJECTIVES

Define *sustainable agriculture* and contrast it with industrialized agriculture.

Identify the potential benefits and problems of genetic engineering.

The green revolution and industrialized agriculture have unquestionably met the food requirements of most of the human population, even as that population has more than doubled since 1960. But we pay for these food gains with serious environmental problems, and we do not know if industrialized agriculture is sustainable for more than a few decades. To compound the issue, we must continue to increase food production to feed the growing human population—but the resulting damage to the environment may reduce our chances of continuing those food increases into the future.

Fortunately, farming practices and techniques exist that ensure a sustainable output at yields comparable to industrialized agriculture. Farmers who practice industrialized agriculture can adopt these alternative agricultural methods, which cost less and are less damaging to the environment. Advances are also being made in sustainable subsistence agriculture.

MOVING TO SUSTAINABLE AGRICULTURE

In **sustainable agriculture** (also called **alternative** or **low-input agriculture**) certain modern agricultural techniques are carefully combined with traditional farming methods. Sustainable agriculture relies on beneficial biological processes and environmentally friendly chemicals that disintegrate quickly and do not persist as residues in the environment. The sustainable farm consists of field crops, trees that bear fruits and nuts, small herds of livestock, and even tracts of forest (**FIGURE 14.11**).

Such diversification protects the farmer against unexpected changes in the marketplace. The breeding of disease-resistant crop plants and the maintenance of

> **sustainable agriculture**
> Agricultural methods that maintain soil productivity and a healthy ecological balance while having minimal long-term impacts.

Some goals of sustainable agriculture FIGURE 14.11

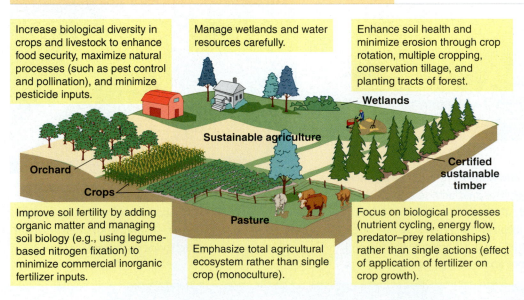

Increase biological diversity in crops and livestock to enhance food security, maximize natural processes (such as pest control and pollination), and minimize pesticide inputs.

Manage wetlands and water resources carefully.

Enhance soil health and minimize erosion through crop rotation, multiple cropping, conservation tillage, and planting tracts of forest.

Wetlands

Sustainable agriculture

Certified sustainable timber

Orchard

Crops

Pasture

Improve soil fertility by adding organic matter and managing soil biology (e.g., using legume-based nitrogen fixation) to minimize commercial inorganic fertilizer inputs.

Emphasize total agricultural ecosystem rather than single crop (monoculture).

Focus on biological processes (nutrient cycling, energy flow, predator–prey relationships) rather than single actions (effect of application of fertilizer on crop growth).

animal health rather than the continual use of antibiotics to prevent disease are important parts of sustainable agriculture. Biological diversity is maintained as a way to minimize pest problems. Water and energy conservation are also practiced in sustainable agriculture.

An important goal of sustainable agriculture is to preserve the quality of agricultural soil. Crop rotation, conservation tillage, and contour plowing help control erosion and maintain soil fertility. Sloping hills converted to mixed-grass pastures erode less than hills planted with field crops, thereby conserving the soil and supporting livestock.

Animal manure added to soil not only cuts costs but decreases the need for high levels of commercial inorganic fertilizers. Legumes such as soybeans, clover, and alfalfa convert atmospheric nitrogen into a form that plants can use in a process called *biological nitrogen fixation,* which reduces the need for nitrogen fertilizers.

Sustainable agriculture is not a single program but a series of programs adapted for specific soils, climates, and farming requirements. Some sustainable farmers—those who practice **organic agriculture**—use no pesticides; others use a system of *integrated pest management (IPM),* which incorporates the limited use of pesticides with pest-controlling biological and cultivation practices.

In growing recognition of the environmental problems associated with industrialized agriculture, more and more mainstream farmers are trying some methods of sustainable agriculture. These methods cause fewer environmental problems to the agricultural ecosystem, or **agroecosystem**, than industrialized agriculture. This trend away from using intensive techniques that produce high yields and toward methods that focus on long-term sustainability of the soil is sometimes referred to as the **second green revolution**.

EnviroDiscovery   A NEW WEAPON FOR LOCUST SWARMS

A non-insect biological control agent targets desert locusts, which periodically increase in number and swarm across the African Sahel, threatening crops across 12 million hectares (30 million acres) located south of the Sahara Desert. The most recent locust outbreak was during the 2004–2005 growing season (see photo). During the last major swarm, in 1988, African farmers sprayed massive quantities of pesticides into the environment to bring the locust population under control. However, many people were concerned about the adverse ecological and health effects of using such large quantities of pesticides. Economists were also concerned about high short-term costs of using pesticides ($300 million for the 1988 swarm) as well as long-term economic costs.

After the 1988 swarm, an international team of scientists worked for 15 years to develop a biological control for locusts that consists of fungal spores. Called Green Muscle, it is environmentally safe and very effective. Unfortunately, Green Muscle wasn't developed in time to prevent the 2004 outbreak, but if used widely in the future, it might prevent the locusts from swarming again.

Global Locator

GENETIC ENGINEERING: A SOLUTION OR A PROBLEM?

Genetic engineering is a controversial technology that has begun to revolutionize medicine and agriculture. The agricultural goals of genetic engineering are not new. Using traditional breeding methods, farmers and scientists have developed desirable characteristics in crop plants and agricultural animals for centuries. It takes time, 15 years or more, to develop such genetically improved organisms using traditional breeding methods. Genetic engineering has the potential to accomplish the same goal in a fraction of that time.

Genetic engineering differs from traditional breeding methods in that desirable genes from any organism can be used, not just those from the species of the

> **genetic engineering**
> The manipulation of genes (for example, taking a specific gene from one species and placing it into an unrelated species) to produce a particular trait.

plant or animal being improved. If a gene for disease resistance found in soybeans would be beneficial to tomatoes, a genetic engineer can splice the soybean gene into the tomato plant's DNA. Traditional breeding methods could not do this because soybeans and tomatoes belong to separate groups of plants and do not interbreed.

Genetic engineering has the potential to produce more nutritious food plants that contain all the essential amino acids (which no single food crop currently does) or that would be rich in necessary vitamins. Crop plants resistant to viral diseases, drought, heat, cold, herbicides, salty or acidic soils, and insect pests are also being developed (**FIGURE 14.12**).

The first **genetically modified (GM)** crops were approved for commercial planting in the United States

Genetically modified cotton FIGURE 14.12

A biologist holds two leaves from different cotton plants grown under identical conditions. The leaf on the left is from a plant genetically modified to resist insect pests, whereas the leaf on the right is from a control plant.

World production of GM crops FIGURE 14.13

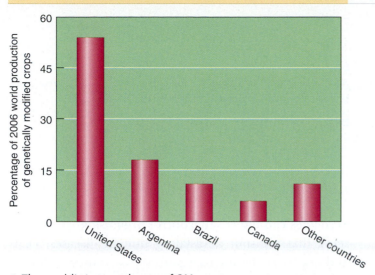

A The world's top producers of GM crops.

B The production of GM crops has increased rapidly.

in the early 1990s. The United States is the world's top producer of GM crops (**FIGURE 14.13A**). Since 2000, however, GM production in developing nations has increased faster than in highly developed countries (**FIGURE 14.13B**).

Genetic engineering has been used to develop more productive farm animals, including rapidly growing hogs and fishes. Perhaps the greatest potential contribution of animal genetic engineering is the production of vaccines against disease organisms that harm agricultural animals. For example, genetically engineered vaccines have been developed to protect cattle against the deadly viral disease rinderpest, which is economically devastating in parts of Asia and Africa.

Concerns about genetically modified foods
During the late 1990s and early 2000s, opposition to genetically engineered crops increased in many countries in Europe and Africa. In 1999 the EU placed a 5-year moratorium on virtually all approvals of GM crops (the moratorium is now lifted). The EU refused to buy U.S. corn because it might be genetically modified. (Currently, about 40 percent of the U.S. corn crop is genetically modified.) One concern is that the inserted genes could spread in an uncontrolled manner from GM crops to weeds or wild relatives of crop plants and possibly harm natural ecosystems in the process. Scientists recognize this concern as legitimate and must take special pre-

cautions to avoid this possibility. Critics also worry that some consumers might develop food allergies to GM foods, although scientists routinely screen new GM crops for allergenicity.

The scientific consensus is that the risks associated with consuming food derived from GM varieties are the same as those associated with consuming food derived from varieties produced by traditional breeding techniques. A growing body of evidence, summarized in the FAO *State of Food and Agriculture 2003–2004*, concludes that current genetically modified (GM) crop plants are as safe for human consumption as crops grown using conventional or organic agriculture. However, more research on the environmental impact of GM crops is required. To that end, strict guidelines exist in areas of genetic engineering research in which there are unanswered questions about possible effects on the environment.

CONCEPT CHECK STOP

What is sustainable agriculture?

What is a GM crop? Give an example.

What are some features of a sustainable farm?

Controlling Agricultural Pests

LEARNING OBJECTIVES

Distinguish between narrow-spectrum and broad-spectrum pesticides.

Relate the benefits of pesticides in disease control and crop protection.

Summarize problems associated with pesticide use, including genetic resistance, ecosystem imbalances, bioaccumulation and biological magnification, and mobility in the environment.

Describe alternative ways to control pests.

A ny organism that interferes in some way with human welfare or activities is a **pest**. Some weeds, insects, rodents, bacteria, fungi, nematodes (microscopic worms), and other pest organisms compete with humans for food; other pests cause or spread disease. People try to control pests, usually by reducing the size of the pest population. Using *pesticides* is the most common way of doing this, particularly in agriculture. Pesticides can be grouped by their target organisms—that is, by the pests they are supposed to eliminate. **Insecticides** kill insects, **herbicides** kill plants, **rodenticides** kill rodents such as rats and mice, and **fungicides** kill fungi (**FIGURE 14.14**).

Oranges being sprayed with fungicide before shipment to grocery stores FIGURE 14.14

BENEFITS OF PESTICIDES

Pesticides can effectively control organisms, such as insects, that transmit devastating human diseases. Fleas and lice carry the microorganism that causes typhus in humans. Malaria, also caused by a microorganism, is transmitted to millions of humans each year by female *Anopheles* mosquitoes (**FIGURE 14.15**). Pesticides help control the population of mosquitoes, thereby reducing the incidence of malaria.

Pesticides also protect crops. It is widely estimated that pests eat or destroy more than one-third of the world's crops. Given our expanding population and world hunger, it is easy to see why control of agricultural pests is desirable. Pesticides reduce the amount of a crop lost through competition with weeds, consumption by insects, and diseases caused by plant **pathogens** (microorganisms, such as fungi and bacteria, that cause disease).

Why are agricultural pests found in such great numbers in our fields? Part of the reason is that agriculture is usually a **monoculture**, in which the field cultivated with a single species represents a simple ecosystem. In contrast, forests, wetlands, and other natural ecosystems are complex and contain many species, including predators and parasites that control pest populations, as well as plant species that pests do not use for food. A monoculture reduces the dangers and accidents that might befall a pest as it searches for food. In the absence of many natural predators and in the presence of plenty of food, a pest population thrives and grows, damaging more of the crop.

Insecticides sprayed to control mosquitoes have saved millions of lives.
(Right) Mosquitoes carry malaria and transfer it to other animals by biting them.

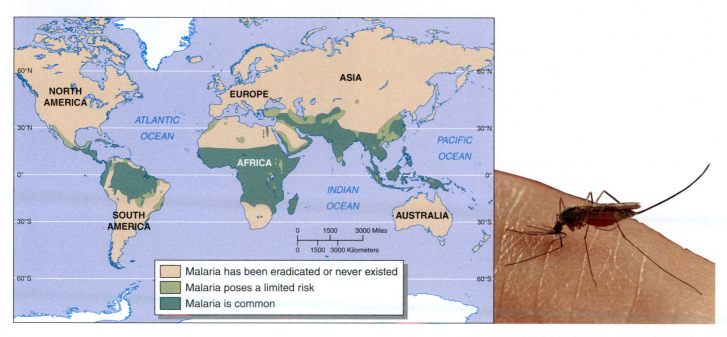

Malaria has been eradicated or never existed
Malaria poses a limited risk
Malaria is common

PROBLEMS WITH PESTICIDES

The ideal pesticide is a **narrow-spectrum pesticide** that kills only the intended organism and does not harm other species. The perfect pesticide would readily break down into safe materials such as water, carbon dioxide, and oxygen. The ideal pesticide would stay exactly where it was put and would not move around in the environment. Unfortunately, no pesticide is perfect. Most pesticides are **broad-spectrum pesticides**. Some pesticides do not degrade readily, or they break down into compounds as dangerous as—if not more dangerous than—the original pesticide. And most pesticides move around the environment quite a bit.

The prolonged use of a particular pesticide can cause a pest population to develop **genetic resistance** to the pesticide. In the 50 years during which pesticides have been widely used, at least 520 species of insects and mites and at least 84 weed species have evolved genetic resistance to certain pesticides.

broad-spectrum pesticide
A pesticide that kills a variety of organisms, including beneficial organisms, in addition to the target pest.

genetic resistance
An inherited characteristic that decreases the effect of a given agent (such as a pesticide) on an organism (such as a pest).

An organism exposed to a chemically stable pesticide that takes years to break down may accumulate high concentrations of the toxin, a phenomenon known as **bioaccumulation**. Organisms at higher levels on food webs tend to have greater concentrations of bioaccumulated pesticide stored in their bodies than those lower on food webs, through a process known as **biological magnification** (see Figure 4.9 on p. 82).

One of the worst problems associated with pesticide use is that pesticides affect more species than the pests for which they are intended. Beneficial insects are killed as effectively as pest insects. Pesticides do not have to kill organisms to harm them. Quite often the stress of carrying pesticides in their tissues makes organisms more vulnerable to predators, diseases, or other stressors in their environments. Because the natural enemies of pests often starve or migrate in search of food after a pesticide is sprayed in an area, pesticides are indirectly responsible for a large reduction in the populations of these natural enemies. Pesticides also kill natural enemies directly because predators may consume lethal amounts of a pesticide while consuming the pests. After a brief period, the pest population rebounds and gets larger than ever, partly because no natural predators are left to keep its numbers in check. In some instances, the use of a pesticide has resulted in a pest problem that did not exist before (see What a Scientist Sees).

Another problem associated with pesticides is that they do not stay where they are applied but tend to move through the soil, water, and air, sometimes traveling long distances (**FIGURE 14.16**). Pesticides applied to agricultural lands wash into rivers and streams, where they can harm fishes. Pesticide mobility is also a problem for humans. About 14 million U.S. residents drink water containing traces of five widely used herbicides, and some people living where the herbicides are commonly used face a slightly elevated cancer risk because of their exposure.

Mobility of pesticides in the environment
FIGURE 14.16

A helicopter sprays pesticides on a crop—and everything else in its pathway—in California.

Pesticide Use and New Pest Species

An infestation of red scale insects on lemons occurred in California after DDT was sprayed to control a different pest. Prior to DDT treatment, red scale did not cause significant economic injury to citrus crops.

A Red scale infestation—shown here on oranges—makes the fruit unappealing and unfit for market.

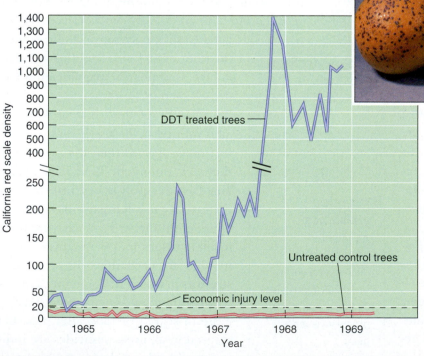

B Scientists understand that the red scale infestation is a direct result of DDT spraying. This graph compares red scale populations on DDT-treated trees (blue line) and untreated trees (red line).

ALTERNATIVES TO PESTICIDES

biological control
A method of pest control that involves the use of naturally occurring disease organisms, parasites, or predators to control pests.

pheromone
A natural substance produced by animals to stimulate a response in other members of the same species.

Given their many problems, pesticides are clearly not the final solution to pest control. Fortunately, pesticides are not the only weapons in our arsenal. Alternative ways to control pests include cultivation methods, **biological controls**, **pheromones** and hormones, reproductive controls, genetic controls, quarantine, and irradiation (**TABLE 14.1** and **FIGURE 14.17**). A combination of these methods in agriculture, often including a limited use of pesticides as a last resort, is the most effective way to control pests.

Pheromone trap FIGURE 14.17

This trap, which was placed in the middle of a pea crop, contains a pheromone to attract pea moth males. Note the dead males in the trap.

Alternative methods of controlling agricultural pests TABLE 14.1

Pest control method	How it works	Disadvantages
Cultivation methods	Interplanting mixes different plants, as by alternating rows; strip cutting alternates crop harvest by portion—remaining portions protect natural predators, parasites of pests	No appreciable disadvantages; more care must be taken in harvest
Biological controls	Naturally occurring predators, parasites, or disease organisms are used to reduce pest populations	Organism introduced for biological control can unexpectedly affect environment or other organisms
Pheromones and hormones	Sexual attractants (pheromones) lure pest species to traps; synthetic regulatory chemicals (hormones) disrupt pests' growth and development	Hormones might affect beneficial species
Reproductive controls	Sterilizing some members of pest population reduces population size	Expensive; must be carried out continually
Genetic controls	Selective breeding develops pest-resistant crops	Plant pathogens evolve rapidly, adapting to disease-resistant host plant; plant breeders forced to constantly develop new strains
Quarantine	Governments restrict importation of foreign pests, diseases	Not foolproof; pests are accidentally introduced
Irradiating foods	Harvested foods are exposed to ionizing radiation that kills potentially harmful microorganisms	Consumers concerned about potential radioactivity (not a true risk); irradiation forms traces of potentially carcinogenic chemicals (free radicals)

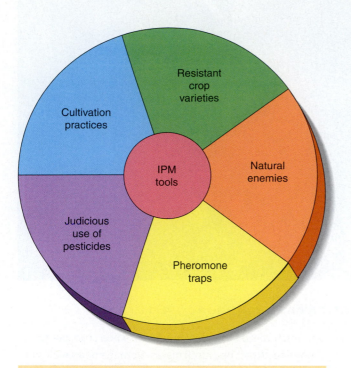

Tools of integrated pest management (IPM)
FIGURE 14.18

INTEGRATED PEST MANAGEMENT

Many pests cannot be controlled effectively with a single technique; a combination of control methods is often more effective. **Integrated pest management (IPM)** combines the use of a variety of biological, cultivation, and pesticide controls tailored to the conditions and crops of an individual farm (**FIGURE 14.18**). Biological and genetic controls, including GM crops designed to resist pests, are used as much as possible, and conven-

> **integrated pest management (IPM)** A combination of pest control methods that, if used in the proper order and at the proper times, keeps the size of a pest population low enough to prevent substantial economic loss.

tional pesticides are used sparingly and only when other methods fail. When pesticides are required, the least toxic pesticides are applied in the lowest possible effective quantities. Thus, IPM allows a farmer to control pests with a minimum of environmental disturbance and often at minimal cost.

To be effective, IPM requires a thorough knowledge

of the life cycles and feeding habits of the pests as well as all their interactions with their hosts and other organisms. The timing of planting, cultivation, and treatments with biological controls is determined by carefully monitoring the concentration of pests. IPM is an important part of **sustainable agriculture**.

IPM is based on two fundamental premises. First, IPM is the *management* rather than the eradication of pests. Farmers who adopt IPM allow a low level of pests in their fields and accept a certain amount of economic damage from the pests. These farmers do not spray pesticides at the first sign of a pest. Instead, they periodically sample the pest population in the field to determine when the benefit of using pesticides exceeds the cost of that action.

Second, IPM requires that farmers be educated to understand what strategies will work best in their particular situations. Managing pests is more complex than trying to eradicate them. The farmer must know what pests to expect on each crop and what to do to minimize their effects.

Adoption of IPM by U.S. farmers has steadily increased since the 1960s, but the overall proportion of farmers using IPM is still small. One reason that IPM is not more widespread is that the knowledge required to use pesticides is relatively simple compared to the sophisticated knowledge needed to implement IPM.

CONCEPT CHECK **STOP**

What is a broad-spectrum pesticide?

What are two important benefits and two potential costs of pesticide use?

How can pests be controlled without pesticides?

DDT AND THE AMERICAN BALD EAGLE

The American bald eagle—the symbol of the United States and an emblem of strength—was a common sight throughout colonial North America (*see photo*). More recently, this species' numbers dropped precipitously to only 417 nesting pairs in the lower 48 states in 1963. In danger of extinction, the bald eagle was listed as an endangered species following enactment of the *Endangered Species Act (ESA)* in 1973.

Several factors contributed to the bald eagle's decline. As European settlers pushed across North America, they cleared many thousands of square kilometers of forest near lakes and rivers, destroying the bald eagle's habitat. Eagles were hunted for sport and because it was thought that they had a significant impact on commercially important fishes. In addition, eagles' numbers dwindled because they could not reproduce at high enough levels to ensure their population growth or their survival. This reproductive failure was the direct result of ingesting food contaminated with the pesticide DDT (dichlorodiphenyltrichloroethane). DDT made the eagles' eggs so thin shelled that they cracked open before the embryos could mature and hatch. Mercury, lead, and selenium were other pollutants that harmed bald eagles.

Banning the use of DDT in the United States in 1972 started the recovery efforts for the bald eagle (see Figure 4.8 on page 82). Conservation efforts involving the U.S. Fish and Wildlife Service, other federal agencies, state and local governments, Native American tribes, conservation organizations, universities, corporations, and individuals have helped the bald eagle make a remarkable comeback.

As a result, the number of nesting pairs in the continental United States increased, and in 1994 the bald eagle was removed from the endangered list and transferred to the less critical threatened list. In 2007, bald eagles had recovered enough—with more than 7,000 pairs in the United States—to be removed from the threatened list.

Nesting pair of bald eagles

SUMMARY

1 World Food Problems

1. **Undernutrition** is a serious underconsumption of calories or nutrients that leaves the body weakened and susceptible to disease. **Overnutrition** is a serious overconsumption of calories.

2. **Food insecurity**, in which people live with chronic hunger, is exacerbated by population growth, environmental problems, and poverty.

2 The Principal Types of Agriculture

1. **Industrialized agriculture** uses modern methods requiring large capital input and less land and labor than traditional methods. **Subsistence agriculture** requires labor and a large amount of land to produce enough food to feed a family. There are three types of subsistence agriculture. In **slash-and-burn agriculture**, small patches of tropical forests are cleared to plant crops. In **nomadic herding**, carried out on arid land, herders move livestock continually to find food for them. **Intercropping** involves growing a variety of plants simultaneously on the same field.

3 Challenges of Agriculture

1. **Prime farmland** in the United States is being lost to urbanization and urban sprawl. Global declines in plant and animal varieties have led many countries to collect **germplasm**, plant and animal material that may be used in breeding. Farmers and ranchers strive to increase yields in many ways, including by administering **hormones** and **antibiotics** to livestock.

2. The **green revolution** introduced modern cultivation methods and high-yield crop varieties to Asia and Latin America. These methods require developing nations to import energy-intensive technologies and to face environmental problems caused by inorganic fertilizers and pesticides.

3. Environmental problems caused by industrialized agriculture include air pollution from the use of fossil fuels and pesticides, water pollution from untreated animal wastes and agricultural chemicals, pesticide-contaminated foods and soils, and increased resistance of pests to pesticides. **Land degradation** decreases the future ability of the land to support crops or livestock. Clearing grasslands and forests and draining wetlands to grow crops have resulted in **habitat fragmentation**, the breakup of large areas of habitat into small, isolated patches.

4 Solutions to Agricultural Problems

1. **Sustainable agriculture** uses methods that maintain soil productivity and a healthy ecological balance while minimizing long-term impacts.

2. **Genetic engineering**, the manipulation of genes to produce a particular trait, can produce more nutritious crops or crop plants that are resistant to pests, diseases, or drought. Concerns about genetic engineering include unknown environmental effects.

5 Controlling Agricultural Pests

1. A **pesticide** is any toxic chemical used to kill pests. Most pesticides are **broad-spectrum pesticides** that kill a variety of organisms in addition to the target pest.

2. Pesticides can effectively control disease-carrying organisms and crop pests. The abundance of pests in agriculture is partly due to the common practice of **monoculture**, the cultivation of only one type of plant over a large area.

3. Pesticide use leads to several problems: pests evolve **genetic resistance**, an inherited characteristic that decreases the effect of a given agent (such as a pesticide) on an organism; ecosystem imbalances occur when pesticides affect species other than the intended pests; and some pesticides exhibit persistence, degrading very slowly. **Bioaccumulation** is the buildup of a persistent pesticide in an organism's body. **Biological magnification** is the increased concentration of certain pesticides in the tissues of organisms at higher levels in food webs. Pesticides also show mobility, moving to places other than where they were applied.

4. Alternatives to pesticides include **biological controls**, which use naturally occurring disease organisms, parasites, or predators to control pests. **Pheromones**, produced by animals to stimulate a response in other members of the same species, are used to attract and trap pest species. **Integrated pest management** is a combination of pest control methods that keep a pest population small enough to prevent substantial economic loss.

CRITICAL AND CREATIVE THINKING QUESTIONS

1. Compare undernutrition and overnutrition. Describe the types of people and the regions of the world most likely to be affected by each.

2. How and why do (a) population growth and (b) the rising consumption of meat affect world grain carryover stocks and world food security?

3. Distinguish between shifting cultivation and slash-and-burn agriculture.

4. What was the green revolution? Describe a few of its successes and shortcomings.

5. Name two environmental problems associated with industrialized agriculture, and give at least three examples of ways that industrialized agriculture could be made more sustainable.

6. Overall, do you think the benefits of pesticide use outweigh its disadvantages? Give at least two reasons for your answer.

7. Is a major goal of integrated pest management (IPM) to eradicate a pest species? Explain your answer.

Use the following graph to answer questions 8 and 9.

School attainment and undernourishment by region, 2000

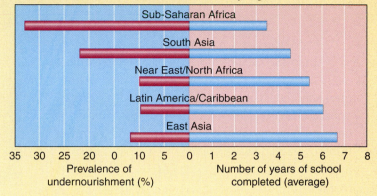

8. Based on the graph, what conclusion can you make about how the level of education people attain in a region is related to that region's prevalence of undernutrition? Give examples of data from specific regions to support your conclusion.

9. How do these results fit in with our understanding of the role that poverty plays in the world's hunger problems?

What is happening in this picture ?

- A pest controller drops insects, young lacewings, over crops. The lacewings attack aphids, which are destructive pests.

- What type of pest control is this?

- Would this practice typically be a part of industrialized agriculture or of sustainable agriculture?

- Can you think of any environmental costs associated with this approach?

Biological Resources

DISAPPEARING FROGS

Amphibians have existed as a group for more than 350 million years. Despite their evolutionary resilience, frogs, toads, and salamanders are remarkably sensitive environmental indicators in both aquatic and terrestrial ecosystems. Most frogs lay gelatinous and unprotected eggs in ponds and other pools of standing water, where the tadpoles metamorphose into adults. As adults, frogs breathe primarily through their moist, permeable skin, which makes them susceptible to environmental contaminants (see larger photo). Amphibians are bellwether species. *Bellwether species*, also called *sentinel species*, provide early warnings of environmental damage that may affect other species.

Since the 1970s, many of the world's frog populations have dwindled or disappeared. More than 150 amphibian species are thought to have gone extinct in the past few decades, and about 1,800 species are threatened, even in remote, pristine locations. Biologists think multiple potential causes may be involved, including pollutants and increased UV radiation, but the two biggest threats are now considered to be global climate change and an emerging fungal disease called chytridiomycosis.

In 1995 Minnesota schoolchildren found that almost half the leopard frogs they caught at a local pond were deformed—with extra legs or toes, eyes located on the shoulder or back, deformed jaws, bent spines, and missing legs, toes, and eyes (see inset). Since then, most states have reported abnormally large numbers of amphibian deformities. Frog deformities have now been found on four continents. Scientists have demonstrated that several factors produce amphibian deformities, including pesticides, which affect development in frog embryos; parasite infestations; and multiple environmental stressors, such as habitat loss, disease, and air and water pollution.

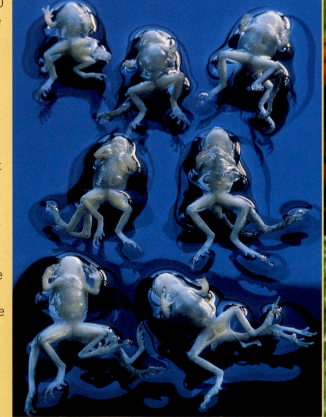

Species Richness and Biological Diversity

LEARNING OBJECTIVES

Describe factors associated with species richness.

Define *biological diversity* and distinguish among species richness, genetic diversity, and ecosystem diversity.

Relate several important ecosystem services provided by biological diversity.

A **species** is a group of distinct organisms that are capable of interbreeding with one another in the wild but that do not interbreed with organisms outside their group. We do not know exactly how many species exist. In fact, biologists now realize how little we know about Earth's diverse organisms.

> **species richness**
> The number of different species in a community.

Earth. To date, about 1.8 million species have been scientifically named and described, including at least 300,000 plant species, 45,000 vertebrate animal species, and some 950,000 insect species. About 10,000 new species are identified each year.

Species richness varies greatly from one community to another. It is related to the abundance of potential ecological niches. A complex community, such as a tropical rain forest or a coral reef, offers a greater variety of potential ecological niches than does a simple community such as mountain chaparral (**FIGURE 15.1**).

HOW MANY SPECIES ARE THERE?

Scientists estimate that there may be as few as 5 million or as many as 100 million different species inhabiting

Species richness is inversely related to the geographic isolation of a community. Isolated island com-

---

Effect of community complexity on species richness FIGURE 15.1

A Chilean dry scrub (chaparral) has low species richness and supports few bird species.

Global Locator

CHILE

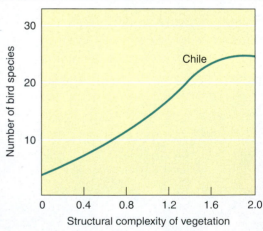

B The structural complexity of vegetation (*x*-axis) is based on vegetation height and density. Note that species richness in birds increases as vegetation becomes more structurally complex.

munities are much less diverse than communities in similar environments found on continents, for two reasons. Many species have difficulty reaching and successfully colonizing an island, and locally extinct species are not readily replaced in isolated environments such as islands or mountaintops.

Species richness is also inversely related to environmental stress. Only species capable of tolerating extreme environmental conditions can live in an environmentally stressed community, such as a polluted stream or a polar region exposed to a harsh climate. Species richness is also reduced when one species enjoys a decided dominance within a community because that species may appropriate a disproportionate share of resources, thus crowding out other species.

Species richness is usually greater at the edges of adjacent communities than in their centers. This is because an **ecotone**—a transitional zone where communities meet—contains all or most of the ecological niches of the adjacent communities as well as some niches unique to the ecotone. The change in species composition produced at ecotones is known as the **edge effect**.

Geologic history greatly affects species richness. Tropical rain forests are probably old, stable communities that have undergone few climate changes over time, which allowed a large number of species to evolve. In contrast, glaciers have repeatedly altered temperate and arctic regions during Earth's history. An area recently vacated by glaciers will have a low species richness because few species will as yet have had a chance to enter it and become established.

> ■ **biological diversity**
>
> The number and variety of Earth's organisms; consists of three components: genetic diversity, species richness, and ecosystem diversity.

WHY WE NEED BIODIVERSITY

The variation among organisms is referred to as **biological diversity**, or **biodiversity**. Biological diversity occurs at all levels of biological organization, from populations to ecosystems (Chapter 5). It takes into account three components: species richness; **genetic diversity**, the genetic variety *within* all populations of that species (**FIGURE 15.2**); and **ecosystem diversity**, the

variety of interactions among organisms in natural communities. For example, a forest community, with its trees, shrubs, vines, herbs, insects, worms, vertebrate animals, fungi, bacteria, and other microorganisms, has greater ecosystem diversity than a cornfield.

Humans depend on the contributions of thousands of species for their survival. For example, insects are instrumental in several ecological and agricultural processes, including pollination of crops, weed control, and insect pest control. Bacteria and fungi provide us with foods, antibiotics and other medicines, and biological processes such as nitrogen fixation (see Chapter 5). However, relatively few species have been evaluated for their potential usefulness to humans. There are at least 300,000 known plant species, but as many as 280,000 of them have yet to be assessed for industrial, medicinal, or agricultural potential. The same is true for most of the millions of microorganisms, fungi, and animals.

Genetic diversity in corn FIGURE 15.2

The variation in corn kernels and ears is evidence of the genetic diversity in the species *Zea mays*.

Ecosystem services and species richness

The living world functions much like a complex machine. Each ecosystem is composed of many parts that are organized and integrated to maintain the ecosystem's overall performance. The activities of all organisms are interrelated; we depend on one another and on the physical environment, often in subtle ways (**FIGURE 15.3**). When one species declines, other species linked to it may either decline or increase in number.

Ecosystems supply human societies with many environmental benefits, or **ecosystem services** (**TABLE 15.1**). Forests are not just a source of lumber; they provide watersheds from which we obtain fresh water, limit the number and severity of local floods, and reduce soil erosion. Many flowering plant species depend on insects to transfer pollen for reproduction. Soil dwellers, from earthworms to bacteria, develop and maintain soil fertility for plants. Bacteria and fungi perform the crucial task of decomposition, which allows nutrients to cycle in the ecosystem. Conservationists maintain that ecosystems with greater species richness supply ecosystem services better than ecosystems with lower species richness.

> **ecosystem services**
> Environmental benefits, such as clean air, clean water, and fertile soil, provided by ecosystems.

You might think that the loss of some species from an ecosystem would not endanger the rest of the organisms, but this is far from true. If enough species are removed, the entire ecosystem will change. Species richness within an ecosystem provides the ecosystem with resilience, the ability to recover from environmental changes or disasters.

Alligators in the environment FIGURE 15.3

The American alligator plays an integral role in its natural ecosystem. Alligators help maintain populations of smaller fishes by eating gar, a fish that preys on them. Alligators dig underwater holes that other aquatic organisms use during droughts when the water level is low. Their nest mounds eventually form small islands colonized by trees and other plants. The trees on these islands support heron and egret populations. The alligator habitat is maintained in part by underwater "gator trails," which help clear out aquatic vegetation that might eventually form a marsh.

Ecosystem services TABLE 15.1	
Ecosystem	**Services provided**
Forests	Purify air and water Produce and maintain soil Absorb carbon dioxide (carbon storage) Provide wildlife habitat Provide humans with wood and recreation
Freshwater systems (rivers, lakes, and groundwater)	Moderate water flow and mitigate floods Dilute and remove pollutants Provide wildlife habitat Provide humans with drinking and irrigation water, food, transportation corridors, electricity, and recreation
Grasslands	Purify air and water Produce and maintain soil Absorb carbon dioxide (carbon storage) Provide wildlife habitat Provide humans with livestock and recreation
Coasts	Provide a buffer against storms Dilute and remove pollutants Provide wildlife habitat Provide humans with food, harbors, transportation routes, and recreation

IMPORTANCE OF GENETIC DIVERSITY

The maintenance of a broad genetic base is critical for the long-term health and survival of each species. Consider economically important crop plants. During the 20th century, plant scientists developed genetically uniform, high-yielding varieties of important food crops such as wheat. However, genetic uniformity resulted in increased susceptibility to pests and disease. By crossing the "super strains" with more genetically diverse relatives, disease and pest resistance can be reintroduced into such plants.

Genetic engineering, the incorporation of genes from one organism into a different species (see Chapter 14), makes it possible to use organisms' genetic resources on a wide scale. Genetic engineering has provided new vaccines, more productive farm animals, and disease-resistant agricultural plants.

Evolution has taken hundreds of millions of years to produce the genetic diversity found on our planet today. This diversity may hold solutions to today's problems and to future problems we have not begun to imagine. It would be unwise to allow such an important part of our heritage to disappear.

Medicinal, agricultural, and industrial importance of organisms
The genetic resources of organisms are vitally important to the pharmaceutical industry, which incorporates hundreds of chemicals derived from plants and other organisms into its medicines (**FIGURE 15.4**). Many of the natural products taken directly from marine organisms are promising anticancer or antiviral drugs. The AIDS (acquired immune deficiency syndrome) drug AZT (azidothymidine), for example, is a synthetic derivative of a compound from a sponge. The 20 best-selling prescription drugs in the United States are either natural products, natural products that are slightly modified chemically, or synthetic drugs whose chemical structures were obtained from organisms.

The agricultural importance of plants and animals is indisputable because we must eat to survive. However, the number of different kinds of foods we eat is limited compared to the total number of edible species available in any given region. Many species are probably nutritionally superior to our common foods.

Medicinal value of the rosy periwinkle
FIGURE 15.4

The rosy periwinkle produces chemicals that are effective against certain cancers. Drugs from the rosy periwinkle have increased the chance of surviving childhood leukemia from about 5 percent to more than 95 percent.

Modern industrial technology depends on a broad range of products from organisms. Plants supply oils and lubricants, perfumes and fragrances, dyes, paper, lumber, waxes, rubber and other elastic latexes, resins, poisons, cork, and fibers. Animals provide wool, silk, fur, leather, lubricants, waxes, and transportation, and they are important in medical research. The armadillo, for example, is used for research in Hansen's disease (leprosy) because it is the only species besides humans known to be susceptible to that disease. Insects secrete a large assortment of chemicals that represent a wealth of potential products. Certain beetles produce steroids with birth-control potential, and fireflies produce a compound that may be useful in treating viral infections. Biologists estimate that perhaps 90 percent of all insects have yet to be identified, and insects represent an important potential biological resource.

Aesthetic, ethical, and spiritual value of organisms Organisms not only contribute to human survival and physical comfort, they provide recreation, inspiration, and spiritual solace. Our natural world is a thing of beauty largely because of the diversity of living forms found in it. Artists have attempted to capture this beauty in drawings, paintings, sculpture, and photography; and poets, writers, architects, and musicians have created works reflecting and celebrating the natural world.

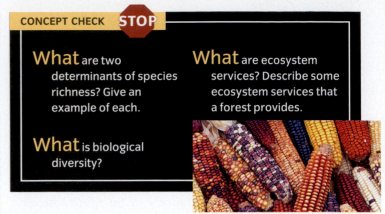

Endangered and Extinct Species

LEARNING OBJECTIVES

Define *extinction* and distinguish between background extinction and mass extinction.

Contrast threatened and endangered species.

Describe four human causes of species endangerment and extinction.

Extinction, the death of a life form, occurs when the last member of a species dies. When a species is extinct, it will never reappear. Biological extinction is the fate of all species, much as death is the fate of all individuals. By one estimate, for every 2,000 species that have ever lived, 1,999 of them are extinct today.

| **extinction** |
| The elimination of a species from Earth. |

During the time in which organisms have occupied Earth, a continuous, low-level extinction of species, or **background extinction**, has occurred. A second kind of extinction, commonly referred to as **mass extinction**, has occurred perhaps five or six times across Earth's history. During a mass extinction event, a large number of species disappear during a relatively short period of geologic time.

The causes of past mass extinctions are not well understood, but biological and environmental factors were probably involved. A major climate change or a catastrophe such as a collision between a large asteroid or comet and Earth could have triggered a mass extinction.

Although extinction is a natural biological process, it is greatly accelerated by human activities. The burgeoning human population has spread into almost all areas of Earth. Whenever humans invade an area, the habitats of many organisms are disrupted or destroyed, which contributes to their extinction.

Earth's biological diversity is disappearing at an unprecedented rate (**FIGURE 15.5**). Conservation biologists estimate that species are now becoming extinct at a rate of 100 to 1,000 times the natural rate of background extinctions. More than 34,000 plant species are currently threatened with extinction.

ENDANGERED AND THREATENED SPECIES

The Endangered Species Act legally defines an **endangered species** as a species in imminent danger of extinction throughout all or a significant portion of

| **endangered species** A species that faces threats that may cause it to become extinct within a short period. |

Representative endangered or extinct species from around the world FIGURE 15.5

Officials at the U.S. Fish and Wildlife Service estimate that more than 500 U.S. species have gone extinct during the past 200 years. Of these, roughly half have become extinct since 1980.

its **range**. (The area in which a particular species is found is its range.) A species is endangered when its numbers are so severely reduced that it is in danger of becoming extinct without human intervention. A species is legally defined as **threatened** when extinction is less imminent but its population is quite low and the species is likely to become endangered in the foreseeable future.

> **threatened species** A species whose population has declined to the point that it may be at risk of extinction.

Endangered and threatened species represent a decline in biological diversity because as their numbers decrease, their genetic variability is severely diminished. Long-term survival and evolution depend on genetic diversity, so a decline in genetic diversity heightens the risk of extinction for endangered and threatened species, as compared to species that have greater genetic variability.

Endangered and Extinct Species **377**

AREAS OF DECLINING BIOLOGICAL DIVERSITY

Declining biological diversity is a concern throughout the United States but is most serious in the states of Hawaii (where 63 percent of species are at risk) and California (where about 29 percent of species are at risk). At least two-thirds of Hawaii's native forests are gone.

As serious as declining biological diversity is in the United States, it is even more serious abroad, particularly in tropical rain forests. Tropical rain forests are being destroyed faster than almost all other ecosystems; approximately 1 percent of these ecosystems are being cleared or severely degraded each year. The forests are making way for human settlements, banana plantations, oil and mineral explorations, and other human activities. (For further discussion of tropical rain forests, see Chapters 6 and 13.)

Tropical rain forests are home to thousands or even millions of species. Many species in tropical rain forests are **endemic** (that is, they are not found anywhere else in the world), and the clearing of tropical rain forests contributes to their extinction.

Perhaps the most unsettling outcome of tropical **deforestation** is its disruptive effect on evolution. In Earth's past, mass extinctions were followed over millions of years by the evolution of new species as replacements for those that died out. In the past, tropical rain forests may have supplied ancestral organisms from which other organisms evolved. Destroying tropical rain forests may reduce nature's ability to replace its species.

> **endemic species**
> Organisms that are native to or confined to a particular region.

EnviroDiscovery IS YOUR COFFEE BIRD FRIENDLY®?

Many species of migratory songbirds, favorites among North American bird lovers, are in decline, and Americans' coffee habits may play a role. In the tropics, high-yield farms cultivating coffee in full sunlight—known as *sun plantations*—are rapidly replacing traditional *shade plantations*. This switch is affecting wintering birds common to southern Mexico, the Caribbean, Costa Rica, and Colombia. Shade plantations grow coffee plants in the shade of tropical rainforest trees. These trees support a vast diversity of songbird species that winter in the tropics (one study counted 150 species in 5 hectares [12.4 acres]), as well as large numbers of other vertebrates and insects. In contrast, sun plantations provide poor bird habitat. Sun-grown varieties of coffee, treated with large inputs of chemical pesticides and fertilizers, outproduce the shade-grown varieties but lack the diverse products that come from the shade trees. About half of the region's shade plantations have been converted to sun plantations since the 1970s.

Songbird populations have declined alarmingly during this period. Researchers counted 94 to 97 fewer bird species on sun plantations in Colombia and Mexico than on shade-grown coffee plantations. Various conservation organizations and development agencies, such as the Smithsonian Migratory Bird Center (SMBC) and the U.S. Agency for International Development (USAID) have initiated programs to certify coffee as "shade grown," which allows consumers the chance to support the preservation of tropical rain forest. Shade-grown coffee typically costs more than sun-grown coffee because it is hand picked, involves more care in selecting only ripe beans, and is often certified organic.

The SMBC allows certified shade-grown and organic coffee farmers to use this label on their product.

The prothonotary warbler spends its summers in eastern North America and its winters in Central and South America.

EARTH'S BIODIVERSITY HOTSPOTS

In the 1980s ecologist Norman Myers of Oxford University coined the term **biodiversity hotspots**. In 2000, using plants as their criteria, Myers and ecologists at Conservation International identified 25 biological hotspots around the world (see What a Scientist Sees). Interestingly, these 25 hotspots for plants contain 29 percent of the world's endemic bird species,

> **biodiversity hotspots** Relatively small areas of land that contain an exceptional number of endemic species and are at high risk from human activities.

27 percent of endemic mammal species, 38 percent of endemic reptile species, and 53 percent of endemic amphibian species. Many humans—nearly 20 percent of the world's population—live in the hotspots. Fifteen of the 25 hotspots are tropical, and 9 are mostly or solely islands. Many biologists recommend that conservation planners focus on preserving land in these hotspots to reduce the mass extinction of species that is currently under way.

Where Is Declining Biological Diversity the Most Serious?

VIEW THIS IN ACTION
in your WileyPLUS course

A If you've visited a zoo and seen a black-and-white ruffed lemur, you may know that lemurs are found in nature only on Madagascar and the small neighboring Comoro Islands. However, most people do not realize that the ruffed lemur comes from a biological hotspot. (Photographed at Folsom Childrens' Zoo in Lincoln, Nebraska.)

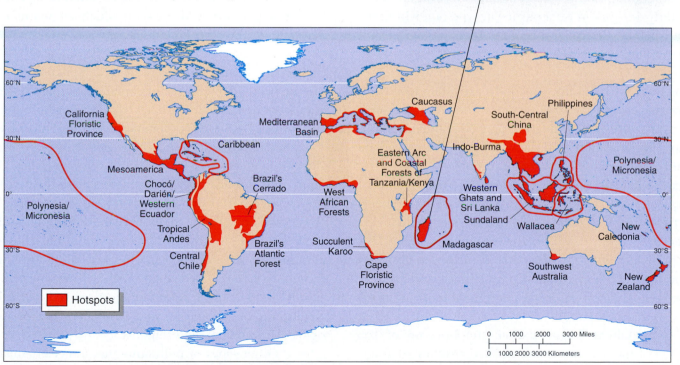

B Ecologists have identified Madagascar and neighboring islands as one of the world's 25 biological hotspots. These hotspots, which are rich in endemic species, are at great risk from human activities. All of Madagascar's 33 lemur species are currently in danger of extinction.

HUMAN CAUSES OF SPECIES ENDANGERMENT

Scientists generally agree that the single greatest threat to biological diversity is loss of habitat. Pollution, the spread of invasive species, and overexploitation are also important.

Habitat destruction, fragmentation, and degradation
Most species facing extinction today are endangered because of the destruction, fragmentation, or degradation of habitats by human activities (**FIGURE 15.6**; also see figures 13.6B and C). We demolish or alter habitats when we build roads, parking lots, bridges, and buildings; clear forests to grow crops or graze domestic animals; and log forests for timber. We drain marshes to build on aquatic habitats, thus converting them to terrestrial ones, and we flood terrestrial habitats when we build dams. Exploration for and mining of minerals, including fossil fuels, disrupt the land and destroy habitats. Habitats are altered by outdoor recreation, including using off-road vehicles, hiking off-trail, golfing, skiing, and camping. Most organisms are utterly dependent on a particular type of environment, and habitat destruction reduces their biological range and ability to survive.

As the human population has grown, the need for increased amounts of food has resulted in a huge conversion of natural lands into croplands and permanent pastures. According to the U.N. Food and Agriculture Organization, total agricultural lands (croplands and pasturelands) currently occupy 38 percent of Earth's land area. Agriculture also has a major impact on aquatic ecosystems because of the diversion of water for irrigation. Little habitat remains for many endangered species. The grizzly bear, for example, occupies about 2 percent of its original habitat in the lower 48 states of the United States. Human population growth and the extraction of resources have destroyed most of the grizzly's wilderness habitat.

Habitat fragmentation, the breakup of large areas of habitat into small, isolated patches (that is, islands), is a major threat to the long-term survival of many species. In ecological terms, *island* refers not only to any land mass surrounded by water (**FIGURE 15.6A**) but also to any isolated habitat surrounded by an expanse of unsuitable territory. Therefore, a small patch of forest surrounded by agricultural and suburban lands is considered an island (**FIGURE 15.6B**). Habitat destruction, fragmentation, and degradation are happening around the world, causing many species to become extinct and reducing the genetic diversity within many surviving species.

Africa provides a vivid example of the conflict over land use between growing human populations and other species. For example, African elephants require a large area of natural landscape in which to forage for the hundreds of kilograms of food that each consumes daily. In Africa, people are increasingly pushing into the elephants' territory to grow crops and graze farm animals. The elephants often trample or devour crops, ruining a year's growth of crops in a single night; they have even killed people. Farmers cannot shoot at or kill elephants because they are a protected species. (Before elephants were listed as a protected species, their numbers had declined precipitously because of overhunting by ivory hunters.) Researchers have found that elephants move out of areas when they become too crowded with people. Unfortunately, the wild areas to which elephants can move are steadily shrinking (**FIGURE 15.6C**).

Pollution
Human-produced acid precipitation, stratospheric ozone depletion, and climate warming degrade even wilderness habitats that are "completely" natural and undisturbed. Acid precipitation is thought to have contributed to the decline of large stands of forest trees and the biological death of many freshwater lakes. Because ozone in the upper atmosphere shields the ground from a large proportion of the sun's harmful ultraviolet (UV) radiation, ozone depletion in the upper atmosphere represents a threat to all terrestrial life. Climate warming, caused in part by an increase in atmospheric carbon dioxide released when fossil fuels are burned, is another threat. Such habitat modifications reduce the biological diversity of species with particularly narrow and rigid environmental requirements.

Other types of pollutants that affect organisms include industrial and agricultural chemicals, organic pollutants from sewage, acid mine drainage, and thermal pollution from heated industrial wastewater.

A Destruction of the world's wildlife habitats This tiny island, located in the Panama Canal, was once a hilltop in a forest that was flooded when the Panama Canal was constructed.

B Isolating wildlife habitats Roads and agricultural lands effectively isolate the scattered remnants, or "islands," of forest. Photographed in Maryland.

AFRICA

TANZANIA

Global Locator

C Crowded refuge Seeking sanctuary from human encroachment, an overpopulation of endangered African elephants in this Tanzanian crater refuge is exploiting its limited resources too quickly for the ecosystem to recover. Elephants crowd into these shrinking habitats.

Invasive species The introduction of a nonnative or foreign species into an ecosystem in which it did not evolve often upsets the balance among the organisms living in that area and interferes with the ecosystem's normal functioning. The foreign species may compete with native species for food or habitat or may prey on them. Generally, an introduced competitor or predator has a greater negative effect on local organisms than do native competitors or predators. Foreign species whose introduction causes economic or environmental harm are called **invasive species** (**FIGURE 15.7**). Although invasive species may be introduced into new areas by natural means, humans are usually responsible for such introductions, either knowingly or unknowingly.

invasive species
Foreign species that spread rapidly in a new area if free of predators, parasites, or resource limitations that may have controlled their population in their native habitat.

Cargo-carrying ocean vessels carry *ballast water* from their ports of origin to increase their stability in the ocean. When they reach their destinations, they discharge this water into local bays, rivers, or lakes. Ballast water may contain clams, mussels, worms, small fishes, and crabs, along with millions of microscopic aquatic organisms. These organisms, if they establish themselves, may threaten the area's aquatic environment and contribute to the extinction of native organisms.

One of North America's greatest biological threats—and its most costly aquatic invader—is the zebra mussel, a native of the Caspian Sea, probably introduced through ballast water flushed into the Great Lakes by a foreign ship in 1985 or 1986. Since then, the tiny freshwater mussel has clustered in extraordinary densi-

Invasive species FIGURE 15.7

Selected examples of the more than 6,500 established foreign species accidentally or deliberately introduced into the United States.

Asian long-horned beetle	Asian tiger mosquito	Brazilian pepper tree	Brown tree snake	Caulerpa
European wild boar	European gypsy moth	Fire ant	Formosan termite	Japanese mud snail
Kudzu	Northern snakehead	Nutria	Puerto Rican frog	Purple loosestrife
Rosy wolf snail	Sea lamprey	Tamarisk	Water hyacinth	Zebra mussel

Zebra mussels FIGURE 15.8

Zebra mussels, here clustered around a pipe opening, have caused billions of dollars in damage, in addition to displacing native clams and mussels.

ties on hulls of boats, piers, buoys, and, most damaging of all, water intake systems (**FIGURE 15.8**).

The zebra mussel's strong appetite for algae and zooplankton reduces the food supply of native fishes, mussels, and clams, threatening their survival. The U.S. Coast Guard estimates that economic losses and control efforts associated with the zebra mussel cost the United States about $5 billion each year.

More than 200 foreign species have been introduced in the United States since 1980, and 25 percent of them have caused significant enough damage to be considered invasive. Worldwide, most regions are estimated to contain 10 percent to 30 percent foreign species.

Overexploitation Sometimes species become endangered or extinct as a result of deliberate efforts to eradicate or control their numbers. Ranchers, hunters, and government agents have reduced populations of large predators such as wolves and grizzly bears. Some animals are killed because they cause problems for humans. The Carolina parakeet, a beautiful green, red, and yellow bird endemic to the southern United States, was exterminated as a pest by farmers because it ate fruit and grain crops. It was extinct by 1920.

Prairie dogs and pocket gophers were poisoned and trapped so extensively by ranchers and farmers that between 1900 and 1960 they disappeared from most of their original geographic range. As a result of sharply decreased numbers of prairie dogs, the black-footed ferret, the natural predator of these animals, became endangered. A successful captive-breeding program has allowed black-footed ferrets to be reintroduced into the wild and reproduce successfully, though some populations have been decimated by disease.

Unregulated hunting, or overhunting, was a factor contributing to the extinction of certain species in the past but is now strictly controlled in most countries. The passenger pigeon was one of the most common birds in North America in the early 1800s, but a century of overhunting resulted in its extinction in the early 1900s. Unregulated hunting was one of several factors that caused the near extinction of the American bison.

Illegal commercial hunting, or **poaching**, endangers many larger animals, such as the tiger, cheetah, and snow leopard, whose beautiful furs are quite valuable. Rhinoceroses are slaughtered primarily for their horns, used for ceremonial dagger handles in the Middle East, and for purported medicinal purposes in Asian medicine. Bears are killed for their gallbladders, used in Asian medicine to treat ailments from indigestion to heart ailments. Caimans (reptiles similar to crocodiles) are killed for their skins and made into shoes and handbags. Although these animals are legally protected, the demand for their products on the black market has led to their being hunted illegally.

In West Africa, poaching has contributed to the decline in lowland gorilla and chimpanzee populations. The meat (called *bushmeat*) of these rare primates and other protected species, such as anteaters, elephants, and mandrill baboons, provides an important source of protein for indigenous people. Bushmeat is also sold to urban restaurants. This demand for a meat source increases the incidence of poaching.

Endangered and Extinct Species   383

ened or endangered, in part because of unregulated commercial trade (**FIGURE 15.9**).

Animals are not the only organisms threatened by excessive commercial harvest. Many unique and rare plants have been collected from nature to the point that they are endangered. These include carnivorous plants, wildflowers, grasses, and ferns, certain cacti, and orchids. On the other hand, carefully monitored and regulated commercial use of animal and plant resources can create an economic incentive to ensure that these resources do not disappear.

Illegal animal trade FIGURE 15.9

These green parrots, captured illegally in the Amazon rain forest, are held for sale at a market in Inquitos, Peru.

Live organisms collected through **commercial harvest** end up in zoos, aquaria, biomedical research laboratories, circuses, and pet stores. Several million birds are commercially harvested each year for the pet trade, but unfortunately many of them die in transit, and many more die from improper treatment after they are in their owners' homes. At least 40 parrot species are now threat-

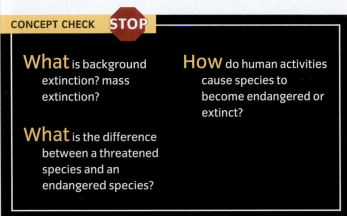

CONCEPT CHECK STOP

What is background extinction? mass extinction?

What is the difference between a threatened species and an endangered species?

How do human activities cause species to become endangered or extinct?

Conservation Biology

LEARNING OBJECTIVES

Define *conservation biology* and compare in situ and ex situ conservation.

Describe restoration ecology.

S tudies in the field of **conservation biology** cover everything from the processes that influence biological diversity to the protection and restoration of endangered species and to the preservation of entire ecosystems and landscapes.

Conservation biology includes two problem-solving techniques to save organisms from extinction: in situ and ex situ conservation. **In situ conservation**, which includes the establishment of parks and reserves, concentrates on preserving biological diversity in nature. With increasing demands on land, in situ conser-

vation cannot guarantee the preservation of all types of biological diversity. Sometimes only ex situ conservation can save a species. **Ex situ conservation** involves conserving biological diversity in human-controlled settings. The breeding of captive species in zoos and the seed storage of genetically diverse plant crops are examples of ex situ conservation.

conservation biology

The scientific study of how humans affect organisms and of the development of ways to protect biological diversity.

PROTECTING HABITATS

Protecting animal and plant habitats—that is, conserving and managing the ecosystem as a whole—is the single best way to preserve biological diversity. Because human activities adversely affect the sustainability of many ecosystems, direct conservation management of protected areas is often required (**FIGURE 15.10**).

Currently, more than 3,000 national parks, sanctuaries, refuges, forests, and other protected areas exist worldwide (see Figure 13.14). Protected areas are not always effective in preserving biological diversity. Many existing protected areas are too small or too isolated from other protected areas to efficiently conserve species. In developing countries where biological diversity is greatest, there is little money or expertise to manage them. Finally, many of the world's protected areas are in lightly populated mountain areas, tundra, and the driest deserts, places that often have spectacular scenery but relatively few species. In contrast, ecosystems in which biological diversity is greatest often receive little protection. Protected areas are urgently needed in tropical rain forests, deserts, the tropical grasslands and savannas of Brazil and Australia, many islands and temperate river basins, and dry forests all over the world.

Wildlife refuges
The **National Wildlife Refuge System**, established in 1903 by President Theodore Roosevelt, is the most extensive network of lands and waters committed to wildlife habitat in the world. The National Wildlife Refuge System contains more than 545 refuges, with at least one in each of the 50 states, and encompasses 38.4 million hectares (95 million acres) of land (see Figure 13.1). The refuges represent all major U.S. ecosystems, from tundra to temperate rain forest to desert, and are home to some of North America's most endangered species, such as the whooping crane. The mission of the National Wildlife Refuge System, which the U.S. Fish and Wildlife Service (FWS) administers, is to preserve lands and waters for the conservation of fishes, wildlife, and plants of the United States.

Some challenges in conservation management FIGURE 15.10

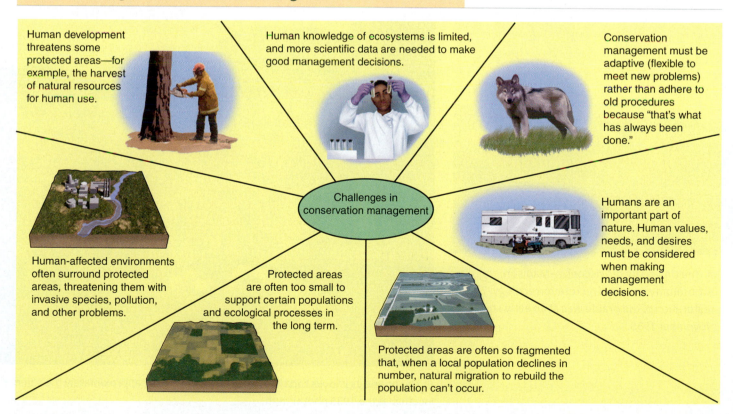

Human development threatens some protected areas—for example, the harvest of natural resources for human use.

Human knowledge of ecosystems is limited, and more scientific data are needed to make good management decisions.

Conservation management must be adaptive (flexible to meet new problems) rather than adhere to old procedures because "that's what has always been done."

Challenges in conservation management

Human-affected environments often surround protected areas, threatening them with invasive species, pollution, and other problems.

Protected areas are often too small to support certain populations and ecological processes in the long term.

Protected areas are often so fragmented that, when a local population declines in number, natural migration to rebuild the population can't occur.

Humans are an important part of nature. Human values, needs, and desires must be considered when making management decisions.

RESTORING DAMAGED OR DESTROYED HABITATS

When preserving habitats is not possible, scientists can reclaim disturbed lands and convert them into areas with high biological diversity. In **restoration ecology**, ecological principles are used to help return a degraded environment to a more functional and sustainable one (**FIGURE 15.11**).

> ■ **restoration ecology**
> The study of the historical condition of a human-damaged ecosystem, with the goal of returning it as close as possible to its former state.

Restoration of disturbed lands creates biological habitats and provides additional benefits, such as the regeneration of soil damaged by agriculture or mining. The disadvantages of restoration include the expense and the time required to restore an area. Even so, restoration is an important aspect of in situ conservation, as restoration may reduce extinction.

CONSERVING SPECIES

Zoos, aquaria, botanical gardens, and other organizations practicing ex situ conservation often play critical roles in saving species on the brink of extinction. Eggs or seeds may be collected from nature, or the few remaining wild animals may be captured and bred in research environments (**FIGURE 15.12A**). But attempting to save a species approaching extinction is expensive, and only a small proportion of endangered species can be saved.

Conservation organizations are an essential part of the effort to maintain biological diversity through species and habitat conservation. These groups help educate policymakers and the public about the importance of biological diversity. In certain instances, they galvanize public support for important biodiversity preservation efforts. They provide financial support for conservation projects, from basic research to the purchase of land that is a critical habitat for a particular species or group of species (**FIGURE 15.12B**).

Prairie restoration FIGURE 15.11

A The University of Wisconsin–Madison Arboretum pioneered restoration ecology. The restoration of the prairie was at an early stage in November 1935.

B The prairie as it looks today. This picture was taken at approximately the same location as the 1935 photograph.

Reintroducing endangered species to nature

The ultimate goal of the captive-breeding programs practiced by zoos, aquaria, and other conservation organizations is to produce offspring in captivity and then release them into nature to restore wild populations. However, only one of every ten reintroductions using animals raised in captivity is successful. Before attempting a reintroduction, conservation biologists make a feasibility study. This includes determining what factors originally caused the species to become extinct in nature, whether these factors still exist, and whether any suitable habitat still remains. Captive-breeding programs are sometimes unsuccessful because it is impossible to teach critical survival skills to animals raised in captivity.

Seed banks

More than 100 seed collections, called **seed banks**, or *gene banks*, exist around the world and collectively hold several million samples at low temperatures (**FIGURE 15.12C**). The Svalbard Global Seed Vault in Norway, opened in February 2008, is designed to store as many as 4.5 million distinct seed samples. Seed banks offer the advantage of storing a large amount of

Visualizing

Efforts to conserve species FIGURE 15.12

◄ **A Captive breeding** Lucky (right) is the first whooping crane born to parents that were raised in captivity and then released, and he is the first in almost 70 years to be born in the wild in the United States. Here Lucky stands with one of his parents at their central Florida home.

▲ **C Seeds from a seed bank** Shown are seeds to be stored in a seed bank in Sussex, England.

B Minimum critical size of ecosystems project When a protected area is set aside, it is important to know what the minimum size of that area must be so it is not affected by encroaching species from surrounding areas. Shown are 1-hectare (2.5-acre) and 10-hectare (25-acre) plots that are part of a long-term study on the effects of habitat fragmentation on Amazonian rain forest by the World Wildlife Fund and Brazil's National Institute for Amazon Research. Preliminary data indicate that the smaller forest fragments do not maintain their ecological integrity.

plant genetic material in a small space. Seeds stored in seed banks are safe from habitat destruction, climate warming, and general neglect. There have even been some instances of using seeds from seed banks to reintroduce to nature a plant species that had become extinct.

Seed banks have some disadvantages. Seeds of many types of plants, such as avocados and coconuts, cannot be stored because they do not tolerate being dried out, a necessary step in freezing the seeds. Seeds do not remain alive indefinitely and must be germinated periodically so new seeds can be collected. Also, growing, harvesting, and returning seeds to storage is expensive. Perhaps the most important disadvantage of seed banks is that plants stored in this manner remain stagnant in an evolutionary sense. Thus, they may be less fit for survival when they are reintroduced into nature. Despite their shortcomings, seed banks are increasingly viewed as important because they safeguard seeds for future generations.

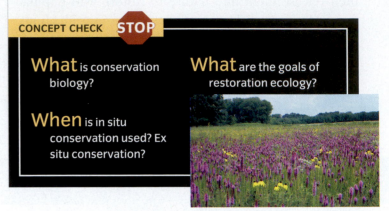

CONCEPT CHECK STOP

What is conservation biology?

When is in situ conservation used? Ex situ conservation?

What are the goals of restoration ecology?

Conservation Policies and Laws

LEARNING OBJECTIVES

Briefly describe the benefits and shortcomings of the U.S. Endangered Species Act.

Relate the purpose of the World Conservation Strategy.

In 1973 the **Endangered Species Act (ESA)** was passed in the United States, authorizing the FWS to protect endangered and threatened species in the United States and abroad. Many other countries now have similar legislation. International laws and policies also seek to conserve Earth's biological resources.

THE ENDANGERED SPECIES ACT

Currently, nearly 1,300 species in the United States are listed as endangered or threatened (TABLE 15.2, FIGURE 15.13). The ESA provides legal protection to listed species to reduce their danger of extinction. The ESA requires the FWS to select critical habitats and design a detailed recovery plan for each species listed. The recovery plan includes an estimate of the current population size, an analysis of the factors contributing to its endangerment, and a list of activities to help the population recover.

The ESA was updated in 1982, 1985, and 1988. It is considered one of the strongest pieces of U.S. environmental legislation, in part because species are designated as endangered or threatened entirely on biological grounds. Currently, economic considerations cannot influence the designation of endangered or threatened species. Biologists generally agree that fewer species have become extinct than would have if the ESA had not been passed.

The ESA is also one of the most controversial pieces of environmental legislation. The ESA does not provide compensation for private property owners who suffer financial losses because they cannot develop their land if a threatened or endangered species lives there. The ESA has also interfered with some federally funded development projects.

The ESA was scheduled for congressional reauthorization in 1992 but has been entangled since then in political wrangling between conservation advocates and supporters of private property rights. Conservation advocates think the ESA does not do enough to save endangered species, whereas those who own the land on which rare species live think the law goes too far and infringes on property rights. Another contentious issue is the financial cost of the law.

Some critics view the ESA as an impediment to economic progress, as when the timber industry was blocked from logging old-growth forests in certain parts of the Pacific Northwest to protect the habitat of the northern spotted owl (see Chapter 3).

Those who defend the ESA point out that of 34,000 past cases of endangered species versus development, only 21 cases were not resolved by some sort of a compromise. When the black-footed ferret was reintroduced on the Wyoming prairie, for example, it was classified as an "experimental, nonessential species" so that its reintroduction would not block ranching and mining in the area. Thus, the ferret release program obtained the

Endangered species FIGURE 15.13

The Florida panther is an endangered subspecies of cougar that exists in small pockets of isolated habitat in southern Florida.

support of local landowners, support that was deemed necessary to the ferrets' survival in nature.

This type of compromise is crucial to the success of saving endangered species because, according to the U.S. General Accounting Office, more than 90 percent of endangered species live on at least some privately owned land. Some critics of the ESA think the law should be changed so that private landowners are given economic incentives to help save endangered species living on their lands. For example, tax cuts for property owners who are good land stewards could make the presence of endangered species on their properties an asset instead of a liability.

Defenders of the ESA agree that it is not perfect. Few endangered species have recovered enough to be delisted—that is, removed from protection of the ESA. However, the FWS says that hundreds of listed species are stable or improving; it expects as many as several dozen species to be delisted in the next decade or so.

Conservationists would like the ESA to be strengthened in such a way as to manage whole ecosystems and maintain complete biological diversity rather than attempt to save endangered species as isolated entities. This approach offers collective protection to many declining species rather than to single species.

U.S. organisms listed as endangered or threatened, 2008 TABLE 15.2		
Type of organism	**Number of endangered species**	**Number of threatened species**
Mammals	69	12
Birds	75	14
Reptiles	13	24
Amphibians	13	10
Fishes	74	65
Snails	64	11
Clams	62	8
Crustaceans	19	3
Corals	0	2
Insects	47	10
Spiders	12	0
Flowering plants	570	143
Conifers	2	1
Ferns and other plants	24	2
Lichens	2	0
TOTAL	1046	305

INTERNATIONAL CONSERVATION POLICIES AND LAWS

The **World Conservation Strategy**, a plan designed to conserve biological diversity worldwide, was formulated in 1980 by the International Union for the Conservation of Nature, the World Wildlife Fund, and the U.N. Environment Program. In addition to conserving biological diversity, the World Conservation Strategy seeks to preserve the vital ecosystem services on which all life depends for survival and to develop sustainable uses of organisms and their ecosystems.

The Convention on Biological Diversity produced by the 1992 Earth Summit requires that each signatory nation must inventory its own biodiversity and develop a **national conservation strategy**, a detailed plan for managing and preserving the biological diversity of that specific country. Currently, 188 nations participate in the Convention on Biological Diversity.

The exploitation of endangered species is somewhat controlled through legislation. At the international level, 160 countries participate in the **Convention on International Trade in Endangered Species of Wild Flora and Fauna (CITES)**, which went into effect in 1975. Originally drawn up to protect endangered animals and plants considered valuable in the highly lucrative international wildlife trade, CITES bans hunting, capturing, and selling of endangered or threatened species and regulates trade of organisms listed as potentially threatened (**FIGURE 15.14**). Unfortunately, enforcement of this treaty varies from country to country. Even where enforcement exists, the penalties are not severe. As a result, illegal trade continues in rare, commercially valuable species.

The goals of CITES often stir up controversy over such issues as who actually owns the world's wildlife and whether global conservation concerns take precedence over competing local interests. These conflicts often highlight socioeconomic differences between wealthy consumers of CITES products and poor people who trade the endangered organisms.

The case of the African elephant, discussed earlier in the chapter, is a good example of these controversies. Listed as an endangered species since 1989 to halt the slaughter of elephants driven by the ivory trade, the species seems to have recovered in southern Africa. Organizations such as the Humane Society in the United

Illegal trade in products made from endangered species FIGURE 15.14

A merchant in Myanmar deals in wildlife products.

States are developing a birth control vaccine to reduce the number of elephant births. However, the African people living near the elephants want to cull the herd periodically and sell elephant meat, hides, and ivory for profit. In the late 1990s and early 2000s, CITES transferred elephant populations in Namibia, Botswana, and South Africa to a less restrictive listing to allow a one-time trade of legally obtained stockpiled ivory (from animals that died of natural causes). The money earned from the sale of ivory is funding elephant conservation programs and community development projects for people living near the elephants.

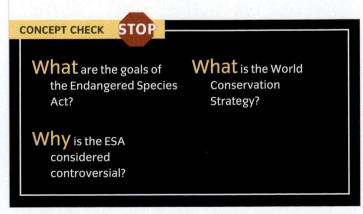

CONCEPT CHECK STOP

What are the goals of the Endangered Species Act?

What is the World Conservation Strategy?

Why is the ESA considered controversial?

REINTRODUCING THE CALIFORNIA CONDOR

A condor perches on the rim of the Grand Canyon in Arizona.

Species that require extremely large territories to survive—often because they are tertiary consumers at the top of the food web—may be threatened with extinction when all or part of their territory is modified by human activity. The California condor, a huge scavenger bird that feeds on carrion and requires a large, undisturbed territory—hundreds of square kilometers—to find adequate food, is slowly recovering from the brink of extinction.

When humans settled North America, the California condor ranged across the continent. Growing human populations and changing climate patterns reduced condor numbers down to approximately 600 by 1890. A combination of causes—the low birth rate and particular mating habits of the species, combined with habitat destruction, poaching, DDT poisoning, contamination of food supplies with lead shot, and power line hazards—rapidly decimated condor populations. By 1982, only 22 condors remained, all in California, and the species faced imminent extinction.

Recovery efforts for the California condor, managed cooperatively by federal, state, and local organizations, took the controversial step of capturing all remaining wild condors to serve as a starting population for captive breeding. The first captively bred condors were released into the wild in 1992. Care and training of young birds has evolved to address reintroduction challenges, such as lead poisoning, collisions with power lines, and contact with humans.

The reintroduction of the California condor cannot yet be called a complete success because no condor population is self-sustaining (able to replace its numbers without captive breeding), and the total numbers of condors fall far short of the recovery program's goal of 450 individuals. But substantial and encouraging gains have been made. As of March 2008, California condors numbered 297, with 146 of those birds living (and breeding) wild at four sites in California, one in Mexico, and one in Arizona.

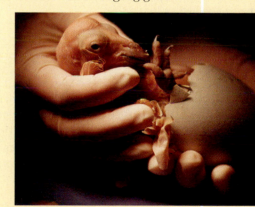

A condor chick hatches at the San Diego Wild Animal Park rearing facility.

SUMMARY

1 Species Richness and Biological Diversity

1. **Species richness** is the number of different species in a community. High species richness is associated with communities that are ecologically complex, not isolated, geologically old and stable, and not subject to environmental stress. Species richness is also higher when no one species dominates the community.

2. **Biological diversity** is the number and variety of Earth's organisms; it consists of three components: genetic diversity, species richness, and ecosystem diversity. **Genetic diversity** is the genetic variety within all populations of a given species. **Ecosystem diversity** is the variety of interactions among organisms in natural communities.

3. Ecosystems with greater species richness are better able to supply **ecosystem services**: environmental benefits, such as clean air to breathe, clean water to drink, and fertile soil in which to grow crops.

2 Endangered and Extinct Species

1. **Extinction** is the elimination of a species from Earth. **Background extinction**, a continuous, low-level extinction of species, has occurred throughout Earth's history. **Mass extinction**, in which many species disappear during a relatively short period of geologic time, has occurred at certain times in Earth's history.

2. An **endangered species** is a species that faces threats that may cause it to become extinct within a short period. A species is defined as **threatened** when extinction is less imminent but its population is quite low.

3. Humans cause species endangerment through habitat destruction, fragmentation, and degradation; pollution; the spread of invasive species; and the overexploitation of biological resources. **Endemic species** are organisms that are native to, or whose range is limited to, a specific place. **Biodiversity hotspots** are areas that contain particularly high numbers of endemic species. **Invasive species** are foreign species, usually introduced by humans, that spread rapidly in a new area where they are free of predators, parasites, or resource limitations that may have controlled their population in their native habitat.

3 Conservation Biology

1. **Conservation biology** is the scientific study of how humans affect organisms and of the development of ways to protect biological diversity. **In situ** conservation includes the establishment of parks and reserves, concentrating on preserving biological diversity in nature; **ex situ conservation** involves conservation of biological diversity in human-controlled settings such as zoos and seed banks.

2. **Restoration ecology** is the study of the historical condition of a human-damaged ecosystem, with the goal of returning it as nearby as possible to its former state.

4 Conservation Policies and Laws

1. The **Endangered Species Act (ESA)** authorizes the U.S. Fish and Wildlife Service (FWS) to protect endangered and threatened species in the United States and abroad. The ESA requires the FWS to select critical habitats and design a detailed recovery plan for each species listed. The act does not compensate private property owners who suffer financial losses when they cannot develop land because it hosts a threatened or endangered species. Species are designated as endangered or threatened entirely on biological grounds, not economic factors.

2. The **World Conservation Strategy**, formulated by the International Union for the Conservation of Nature, the World Wildlife Fund, and the U.N. Environment Program, seeks to conserve biological diversity worldwide, to preserve vital ecosystem services, and to develop sustainable uses of organisms and their ecosystems.

KEY TERMS

CRITICAL AND CREATIVE THINKING QUESTIONS

1. Is biological diversity a renewable or nonrenewable resource? Why could it be seen both ways?

2. List at least five important ecosystem services provided by living organisms.

3. If we preserve species solely on the basis of their potential economic value—such as a source of a novel drug—do they lose their "value" after we have capitalized on a newly discovered chemical? Why or why not?

4. What are the four main causes of species endangerment and extinction? Which cause do biologists consider most important?

5. What are invasive species?

6. Why are frogs and other amphibians considered bellwether species?

7. If you had the assets and authority to take any measure to protect and preserve biological diversity, but could take only one, what would it be?

8. The most recent version of the World Conservation Strategy includes stabilizing the human population. How would stabilizing the human population affect biological diversity?

9. In *A Sand County Almanac* and *Sketches Here and There*, Aldo Leopold wrote, "To keep every cog and wheel is the first precaution of intelligent tinkering." How does his statement relate to this chapter?

10. Why do you suppose the Svalbard Seed Vault in Norway is called the "Doomsday vault"? How is this name reflected in the vault's location, buried in a frozen hillside?

The Nature Conservancy evaluated the extent of human-caused habitat disturbance in the world's various biomes. Use the data in the graph at right to answer the following questions.

11. On which global region have humans had the greatest impact—polar, temperate, or tropical? Suggest why human impact has been greatest in this region.

12. What percentage of tropical rain forests has been disturbed by human activities? what percentage of temperate deciduous forests? Given these two values, why do you think people are more concerned about destruction in tropical rain forests than in temperate deciduous forests?

13. Which biome has had the lowest percentage of habitat disturbance? Suggest two possible reasons why human impacts on this biome may increase greatly in the future.

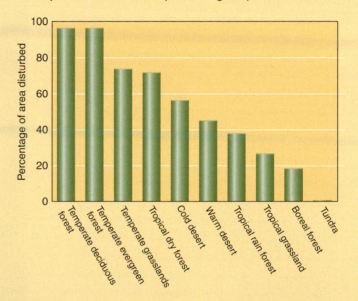

What is happening in this picture ?

- At Zoo Atlanta in the United States, an actor playing an African park guard and a volunteer teach children about ivory poaching.

- What are the goals of this effort?

- Which conservation law bans hunting elephants for their ivory?

- Do you think displaying elephants in zoos helps or hurts these species? Why?

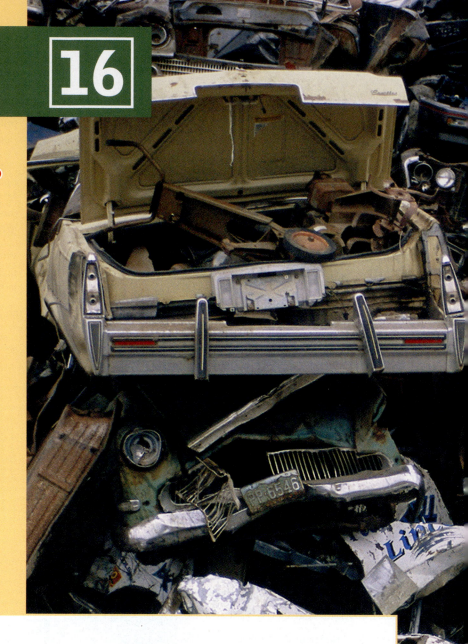

Solid and Hazardous Waste

REUSING AND RECYCLING OLD AUTOMOBILES

VIEW THIS IN ACTION in your WileyPLUS course

In the United States, about 35 million motor vehicles leave service each year. Most are exported to developing countries, but about 11 million cars and trucks are discarded (see photo of a Virginia salvage yard). Although 75 percent of a car can be reused as secondhand parts or recycled, the remaining 25 percent—glass, metals, plastics, fabrics, rubber, foam, and leather—usually ends up in landfills.

About 37 percent of the iron and steel scrap reprocessed in the United States comes from old cars. According to the Environmental Protection Agency, recycling scrap iron and steel produces 86 percent less air pollution and 76 percent less water pollution than mining and refining an equivalent amount of iron ore.

Recycling plastic, which automakers use because it is lightweight and improves fuel efficiency, is one of the biggest challenges in auto recycling. No industry standards currently exist for plastic parts, so the kinds and amounts used in cars vary a great deal.

Auto manufacturers worldwide have begun to address the challenge of recycling old cars. Toyota developed a way to recycle urethane foam and other shredded materials. Chrysler is developing a concept vehicle with completely recyclable body sections. Other manufacturers have started to design cars with completely recyclable or reconditionable parts (see inset of recyclable plastic and composite parts on a Toyota vehicle).

The European Union has mandated that by 2015, 95 percent of each discarded car must be recoverable. U.S. legislators may one day make the same requirement. When they design new models, European auto manufacturers now take into account the car's entire life cycle. This kind of *product stewardship* encourages optimal reuse and recycling when products are returned to manufacturers.

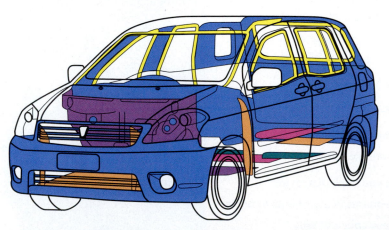

NATIONAL GEOGRAPHIC

Solid Waste

LEARNING OBJECTIVES

Distinguish between municipal and nonmunicipal solid waste.

Describe the features of a modern sanitary landfill and relate some of the problems associated with sanitary landfills.

Describe the features of a mass burn incinerator and relate some of the problems associated with incinerators.

Explain the composting process.

T he United States generates more solid waste per capita than any other country. (Canada is a close second.) Each person in the United States produces an average of 2.1 kg (4.6 lb) of solid waste per day. This amount corresponded to a total of 215 million metric tons (238 million tons) in 2005. The problem worsens each year as the U.S. population increases.

Waste generation is an unavoidable consequence of the prosperous, high-technology, industrial economies of the United States and other highly developed nations. Many products that would be repaired, reused, or recycled in less affluent nations are simply thrown away. Nobody likes to think about solid waste, but it is certainly a concern of modern society—we keep producing it, and places to dispose of it safely are dwindling in number (**FIGURE 16.1**).

TYPES OF SOLID WASTE

Municipal solid waste consists of the combined residential and commercial waste produced in a municipal area. Municipal solid waste is a heterogeneous mixture composed primarily of paper and paperboard; yard waste; plastics; food waste; metals; rubber, leather, and textiles; wood; and glass (**FIGURE 16.2**). The proportions of the major types of solid waste in this mixture change over time. Today's solid waste contains more paper and plastics than in the past, whereas the amounts of glass and steel have declined.

municipal solid waste Solid materials discarded by homes, offices, stores, restaurants, schools, hospitals, prisons, libraries, and other commercial and institutional facilities.

Empty plastic water bottles FIGURE 16.1

Global demand for bottled water is increasing, but most of these empty bottles are not recycled. The average American drinks 104 liters (110 qts) of bottled water annually, which translates into 60 million plastic water bottles per day.

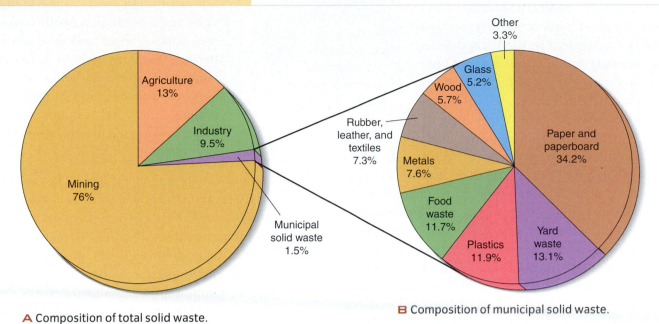

A Composition of total solid waste.

B Composition of municipal solid waste.

Municipal solid waste makes up only a small proportion—less than 2 percent—of the total solid waste produced each day. **Nonmunicipal solid waste**, which includes mining, agricultural, and industrial wastes, is produced in substantially larger amounts. Most solid waste generated in the United States is from nonmunicipal sources.

> **nonmunicipal solid waste** Solid waste generated by industry, agriculture, and mining.

DISPOSAL OF SOLID WASTE

Solid waste has traditionally been regarded as material that is no longer useful and should be disposed of. We can get rid of solid waste in four ways: dump it, bury it, burn it, or recycle or compost it (**FIGURE 16.3**).

Open dumps The old method of solid waste disposal was dumping. Open dumps, which are now illegal, were unsanitary, malodorous places where disease-carrying vermin such as rats and flies proliferated. Methane gas was released into the surrounding air as microorganisms decomposed the solid waste, and fires polluted the air with acrid smoke. Liquid oozed and seeped through the heaps of solid waste, often leaching hazardous materials that then contaminated soil, surface water, and groundwater.

U.S. disposal of municipal solid waste FIGURE 16.3

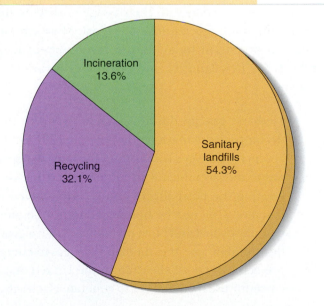

Sanitary landfills Open dumps have been replaced by **sanitary landfills**, which receive about 54 percent of the municipal solid waste generated in the United States today. Sanitary landfills differ from open dumps in that the solid waste is placed in a hole, compacted, and covered with a thin layer of soil every day (see What a Scientist Sees). This process reduces the number of rats and other vermin usually associated with solid waste, lessens the danger of fires, and decreases the amount of odor. If a sanitary landfill is operated in accordance with solid waste management–approved guidelines, it does not pollute local surface water and groundwater. Layers of compacted clay and plastic sheets at the bottom of the landfill prevent liquid waste from seeping into groundwater.

> **sanitary landfill** The most common disposal site for solid waste, where waste is compacted and buried under a shallow layer of soil.

Newer landfills possess a double liner system (plastic, clay, plastic, clay) and use sophisticated systems to collect **leachate** (liquid that seeps through the solid waste) and gases that form during decomposition.

The choice of where to locate a sanitary landfill is based on a variety of factors, including the geology of the area, soil drainage properties, and the proximity of nearby bodies of water and wetlands. A landfill should be far enough away from centers of dense population so it is inoffensive but close enough so as not to require high transportation costs.

Although the operation of sanitary landfills has improved over the years with the passage of stricter and stricter guidelines, few landfills are ideal. Most sanitary landfills in operation today do not meet current legal standards for new landfills and encounter a variety of problems:

- Methane gas, produced by microorganisms that decompose organic material anaerobically (in the absence of oxygen), may seep through the solid waste and accumulate in underground pockets, creating the possibility of an explosion—even in basements of nearby homes. Landfill operators typically collect and burn off methane, but a growing number use the methane for gas-to-energy projects. About 425 landfills in the United States currently use methane gas to generate electricity.

- Leachate that seeps from unlined landfills or through cracks in the lining of lined landfills can potentially contaminate surface water and groundwater. Because even household trash contains toxic chemicals such as heavy metals, pesticides, and organic compounds, the leachate must be collected and treated to neutralize its negative effects.

- Landfills, by their nature, fill up. They are not a long-term remedy for waste disposal. From 1988 to 2006, the number of U.S. landfills in operation decreased from 8000 to 1754; many reached their capacity, and others did not meet state or federal environmental standards. Fewer new sanitary landfills are being opened; many desirable sites are already taken, and people are usually adamantly opposed to the construction of a landfill near their homes.

The special problem of plastic The amount of plastic in our solid waste, more than half of it from packaging, is growing faster than any other component of municipal solid waste. Most plastics are chemically stable and do not readily decompose. This characteristic, although essential in the packaging of products such as food and medicine, causes long-term problems: Most plastic debris disposed of in sanitary landfills will probably last for centuries. In response to concerns about the volume of plastic waste, some areas have banned the use of certain types of plastic, such as the polyvinyl chloride employed in packaging.

Special plastics that have the ability to degrade or disintegrate have been developed. Some of these are **photodegradable**—that is, they break down after being exposed to sunlight—which means they will not break down if buried in a sanitary landfill. Other plastics are **biodegradable**—they are decomposed by microorganisms such as bacteria. Whether biodegradable plastics actually break down under the conditions found in a sanitary landfill is not yet clear, although preliminary studies indicate that they probably do not. (Other waste management options for plastic are discussed later in this chapter.)

Sanitary Landfill

VIEW THIS IN ACTION
in your WileyPLUS course

B Environmental engineers know that sanitary landfills constructed today have protective liners of compacted clay and high-density plastic and sophisticated leachate collection systems that minimize environmental problems such as groundwater contamination. Solid waste is spread in a thin layer, compacted into small sections called "cells," and covered with soil.

A A bulldozer compacts trash at a sanitary landfill in California. To the average person, a sanitary landfill is just a "dump."

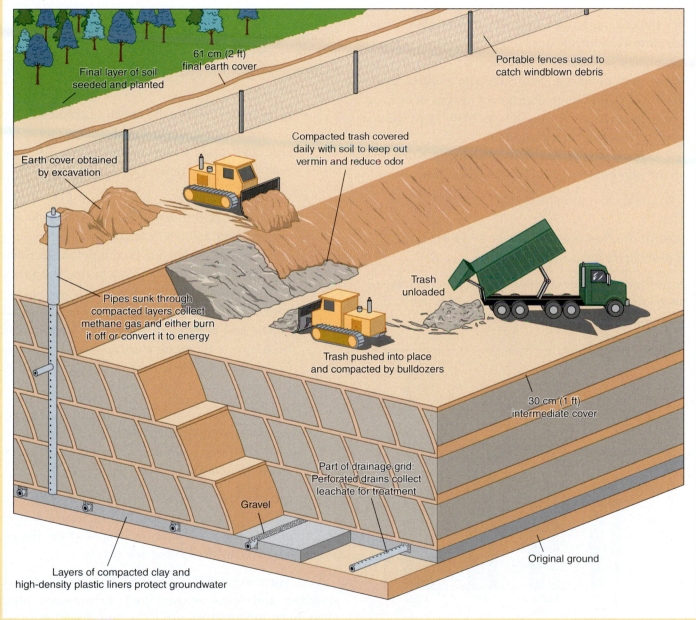

Final layer of soil seeded and planted

61 cm (2 ft) final earth cover

Portable fences used to catch windblown debris

Earth cover obtained by excavation

Compacted trash covered daily with soil to keep out vermin and reduce odor

Pipes sunk through compacted layers collect methane gas and either burn it off or convert it to energy

Trash unloaded

Trash pushed into place and compacted by bulldozers

30 cm (1 ft) intermediate cover

Part of drainage grid: Perforated drains collect leachate for treatment

Gravel

Original ground

Layers of compacted clay and high-density plastic liners protect groundwater

Solid Waste   399

Incineration When solid waste is incinerated, two positive things are accomplished. First, the volume of solid waste is reduced by up to 90 percent: Ash is more compact than unburned solid waste. Second, incineration produces heat that can make steam to warm buildings or generate electricity. In 2005 the United States had 88 waste-to-energy incinerators, which produce substantially less carbon dioxide emissions than power plants that burn fossil fuels (**FIGURE 16.4**). (Recall from Chapter 9 that carbon dioxide is a potent greenhouse gas.)

The best materials for incineration are paper, plastics, and rubber, all of which produce a lot of heat. Paper burns readily, and 1 kg (2.2 lbs) of plastic waste yields almost as much heat as 1 kg of fuel oil.

Tires produce as much heat as coal and often generate less pollution. Some electric utilities in the United States and Canada burn tires instead of or in addition to coal (**FIGURE 16.5**). In 2003, 45 percent of all discarded tires were incinerated.

Most problems associated with incineration arise from the potential for environmental contamination:

Tires that will be burned to generate electricity
FIGURE 16.5

This mountain in Westley, California, contains 4 to 6 million old tires. The power plant that burns them supplies electricity to 3,500 homes. (The person wearing red gives a sense of scale.)

- Incinerators pollute the air with carbon monoxide, particulates, heavy metals such as mercury, and other toxic materials, unless expensive air pollution control devices are used.

- Incinerators produce large quantities of ash, which must be disposed of properly. **Bottom ash**, or slag, is the ash left at the bottom of the incinerator when combustion is completed. **Fly ash** is the ash from the flue (chimney) that is trapped by air pollution control devices. Fly ash usually contains more toxic materials, including heavy metals and possibly dioxins, than bottom ash. Both types of incinerator ash are best disposed of in specially licensed hazardous waste landfills (discussed later in this chapter).

- As with sanitary landfills, site selection for incinerators is controversial. People may recognize the need for an incinerator, but they do not want it near their homes.

- Incinerators are expensive to run. Prices have escalated because costly pollution control devices are now required. Economic factors have also restricted construction of new plants.

The three types of incinerators are mass burn, modular, and refuse-derived fuel incinerators. Most **mass burn incinerators** are large and designed to re-

Carbon dioxide emissions per kilowatt-hour of electricity generated FIGURE 16.4

Waste-to-energy incinerators release less carbon dioxide into the atmosphere than do equivalent power plants that burn fossil fuels.

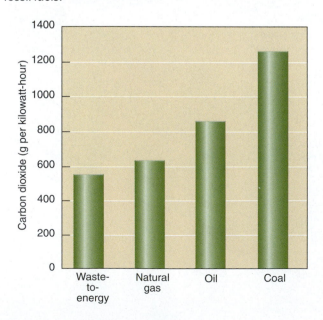

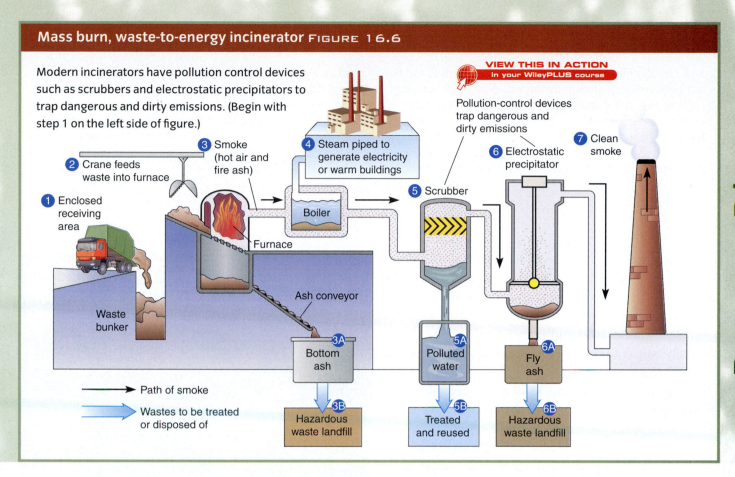

Mass burn, waste-to-energy incinerator FIGURE 16.6

Modern incinerators have pollution control devices such as scrubbers and electrostatic precipitators to trap dangerous and dirty emissions. (Begin with step 1 on the left side of figure.)

VIEW THIS IN ACTION in your WileyPLUS course

Pollution-control devices trap dangerous and dirty emissions

1 Enclosed receiving area
2 Crane feeds waste into furnace
3 Smoke (hot air and fire ash)
4 Steam piped to generate electricity or warm buildings
5 Scrubber
6 Electrostatic precipitator
7 Clean smoke

Boiler
Furnace
Waste bunker
Ash conveyor

3A Bottom ash
5A Polluted water
6A Fly ash

→ Path of smoke
⇒ Wastes to be treated or disposed of

3B Hazardous waste landfill
5B Treated and reused
6B Hazardous waste landfill

Process Diagram

■ **mass burn incinerator** A large furnace that burns all solid waste except for unburnable items such as refrigerators.

cover the energy produced from combustion (**FIGURE 16.6**). **Modular incinerators** are smaller incinerators that burn all solid waste. Assembled at factories, they are less expensive to build. **Refuse-derived fuel incinerators** burn the combustible portion of solid waste. First, noncombustible wastes, such as glass and metals, are removed by machine or by hand. The remaining solid waste, including plastic and paper, is shredded or shaped into pellets and burned.

Composting
Yard waste, such as grass clippings, branches, and leaves, is a substantial component of municipal solid waste (see Figure 16.2). As space in sanitary landfills becomes more limited, other ways to dispose of yard waste are being implemented.

One of the best recovery methods for yard waste is to convert it into soil conditioners such as compost or mulch (**FIGURE 16.7**). Food scraps, sewage sludge,

Home composting of household waste
FIGURE 16.7

The drum of this composting bin can be rotated on its base to mix the decomposing materials, thereby speeding up the microbial decomposition process.

New York City's Rikers Island, site of 10 jails and home to as many as 17,000 inmates, serves as the city's surprising frontrunner in innovative recycling, composting, and gardening. Prisoners compost about 20 tons of food scraps each day, in a facility with a roof made of photovoltaic cells. The food scraps are first mixed with wood chips and placed in a long bin, or bay. Agitators mix the material (to aid decomposition) and move it slowly through

the bay. In about 20 days, the material reaches the end of the bay, and it is placed outside for several months of curing. After curing, the compost is sifted to remove any materials that didn't decompose (such as plastic cutlery). The composting produces high-grade fertilizer used in the prison's garden and farm programs. At the city's largest community garden, inmates raise vegetables for prison meals.

and agricultural manure are other forms of solid waste that can be used to make compost. Compost provides nutrients to the soil and reduces the need for fertilizers and pesticides. Compost and mulch are used for landscaping in public parks and playgrounds and as part of the daily soil cover at sanitary landfills. Compost and mulch are also sold to gardeners.

Composting as a means of managing solid waste first became popular in Europe. Many municipalities in the United States have composting facilities as part of their comprehensive solid waste management plans, and many states have banned yard waste from sanitary landfills. This trend is likely to continue, making composting even more desirable.

CONCEPT CHECK STOP

How do municipal and nonmunicipal solid waste differ?

What are some features of sanitary landfills? What problems are associated with them?

What are the main features of a mass burn

incinerator? What problems are associated with incinerators?

How do composters work?

Reducing Solid Waste

LEARNING OBJECTIVES

Define *source reduction.*

Summarize how source reduction, reuse, and recycling help reduce the volume of solid waste.

Define *integrated waste management.*

G iven the problems associated with sanitary landfills and incinerators, it makes sense to do whatever we can to reduce the wastes we generate. The three goals of waste prevention, in order of priority, are (1) reduce the amount of waste as much as possible; (2) reuse products as much as possible; and (3) recycle materials as much as possible.

Reducing the amount of waste includes purchasing products that have less packaging and that last longer or are repairable (**FIGURE 16.8**). Consumers can also decrease their consumption of products to reduce waste. Before deciding to purchase a product, a consumer should ask, "Do I really *need* this product, or do I merely *want* it?"

SOURCE REDUCTION

source reduction
An aspect of waste management in which products are designed and manufactured in ways that decrease the amount of solid and hazardous waste in the solid waste stream.

The most underutilized aspect of waste management is **source reduction**. Source reduction is accomplished in a variety of ways, including using raw materials that introduce less waste during manufacturing, and reusing and recycling wastes at the plants where they are generated. Innovation and product modifications play a key role. Consider aluminum cans: They are 35 percent lighter now than they were in the 1970s because less material is introduced into their manufacture. Dry-cell batteries are another example: They contain much less mercury today than they did in the early 1980s.

Dematerialization, the progressive decrease in the size and weight of a product as a result of technological improvements, is an example of source reduction only if the new product is at least as durable as the one it replaces. If the smaller, lighter product has a shorter life span and must be replaced more often, source reduction is not achieved.

REUSING PRODUCTS

One example of reuse is refillable glass beverage bottles. Years ago, refillable beverage bottles were used a great deal in the United States. Today they are rarely used. For a glass bottle to be reused, it must be considerably thicker (and heavier) than a single-use bottle. Because of the increased weight, transportation costs are higher. In the past, reuse of glass bottles made sense because there were many small bot-

tlers scattered across the United States, minimizing transportation costs. Today there are approximately one-tenth as many bottlers. The centralization of bottling facilities makes it economically difficult to go back to the days of refillable bottles. Several other countries still reuse glass extensively, including Japan, Ecuador, Denmark, Finland, Germany, the Netherlands, Norway, Sweden, and Switzerland.

RECYCLING MATERIALS

Many materials found in solid waste can be collected and reprocessed into new products. Recycling is preferred over landfill disposal because it conserves natural resources and is more environmentally benign. Every ton of recycled paper saves 17 trees, 26,500 L (7,000 gal) of water, 4,100 kilowatt-hours of energy, and 2.3 cubic meters (3 cubic yards) of landfill space. Recycling also has a positive effect on the economy by generating jobs and revenues from selling the recycled materials. Recycling does have environmental costs: Like all other human activities, it uses energy and generates pollution. For example, the de-inking process in paper recycling requires energy and produces a toxic sludge that contains heavy metals.

The many different materials in municipal solid waste must be separated before recycling. The separation of materials in items with complex compositions is difficult. Some food containers are composed of thin layers of metal foil, plastic, and paper, and trying to separate these layers is a daunting prospect.

The number of communities with recycling programs increased remarkably during the 1990s but leveled off somewhat in the early 2000s. Recycling programs include curbside collection, drop-off centers, buy-back programs, and deposit systems. In 2005 the annual recycling rate of aluminum and steel cans, plastic bottles, glass containers, newspapers, and cardboard was 87 kg (39.4 lb) per person.

Recyclables are usually sent to a materials recovery facility, where they are sorted and prepared for remanufacturing. Currently, the United States recycles about 32 percent of its municipal solid waste, including composting of yard trimmings; this value is higher than in most other highly developed nations (**Figure 16.9A**). (Recall, however, that the United States also generates more municipal solid waste than any other country.)

Most people think recycling involves merely separating certain materials from the solid waste stream, but that is only the first step. For recycling to work, there must be a market for the recycled goods, and the recycled products must be used in preference to virgin products. Prices paid by processors for recyclable materials vary significantly from one year to the next, depending largely on the demand for recycled products. In some places, recycling—particularly curbside collection—is not economically feasible.

Greater recycling generally occurs when the economy is strong, and the amount of recycling nationwide varies from year to year. Thus, although the amount of municipal solid waste generated is fairly constant, the amount of recycling varies.

Recycling paper
The United States currently recycles about 50 percent of its paper and paperboard. Many highly developed countries have higher recycling rates. Denmark, for example, recycles 97 percent of its paper. Part of the reason paper is not recycled more in the United States is that many older paper mills are not equipped to process waste paper. The number of mills that can process waste paper has increased in recent years, in part because of consumer demand. In addition to a slow increase in paper recycling in the United States, there is a growing demand for U.S. waste paper in other countries. China, Mexico, Taiwan, and Korea import large quantities of waste paper and cardboard from the United States.

Recycling glass
The United States currently recycles about 25 percent of its glass containers. Recycled glass costs less than glass made from virgin materials. Glass food and beverage containers are crushed to form **cullet**, which glass manufacturers can use to make new products. Cullet is more valuable when glass containers of different colors are separated before being crushed (**Figure 16.9B**).

Recycling metals
The recycling of aluminum is one of the best success stories in U.S. recycling, largely because of economic factors (**Figure 16.9C**). Making a new aluminum can from a recycled one requires a fraction of the energy it would take to make a new can from raw metal. According to the EPA, in 2005, about 45 percent of discarded aluminum beverage cans were recycled, saving about 15 million barrels of oil.

Other recyclable metals include lead, gold, iron and steel, silver, and zinc. For example, according to the Institute of Scrap Recycling Industries, new steel products contain an average of 56 percent recycled scrap steel. One of the obstacles to recycling metal products discarded in municipal solid waste is that their metallic compositions are often unknown. It is also difficult to extract metal from products such as stoves that contain materials besides metal (plastic, rubber, or glass, for instance).

Recycling plastic
Less than 20 percent of plastic is recycled (see Figure 16.1 on page 396). Depending on the economic situation, it is sometimes less expensive to make plastic from raw materials (petroleum and natural gas) than to recycle it. Some local and state governments support or require the recycling of plastic. According to the EPA, 34 percent of plastic bottles and packaging is recycled to make such diverse products as carpet, automobile parts, tennis ball felt, and polyester cloth. One of the challenges associated with recycling plastic is that there are many different kinds. Forty-six different plastics are common in consumer products, and many products contain multiple kinds of plastic that must be separated or sorted before they can be recycled.

Recycling tires
According to the EPA, 36 percent of tires are currently recycled to make other products. About 290 million tires are discarded in the United States each year, yet recycled tires are used for only a few products: retread tires; playground equipment; trash cans, garden hoses, and other consumer products; and rubberized asphalt for pavement. Most states have established tire-recycling programs. As noted earlier in this chapter, tires are also burned in waste-to-energy incinerators to provide electricity.

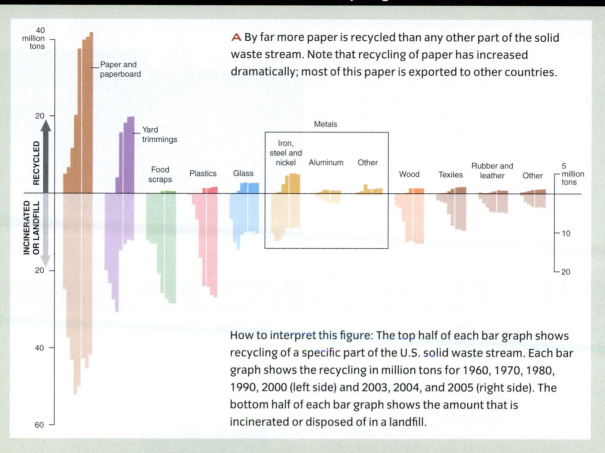

A By far more paper is recycled than any other part of the solid waste stream. Note that recycling of paper has increased dramatically; most of this paper is exported to other countries.

How to interpret this figure: The top half of each bar graph shows recycling of a specific part of the U.S. solid waste stream. Each bar graph shows the recycling in million tons for 1960, 1970, 1980, 1990, 2000 (left side) and 2003, 2004, and 2005 (right side). The bottom half of each bar graph shows the amount that is incinerated or disposed of in a landfill.

B A bin of green glass that has been sorted and crushed at a recycling plant. Separating the different colors of glass in a cost-effective manner is challenging. Note in A that the amount of glass recycling has not increased since 1990.

C Bales of crushed aluminum beverage cans are ready for processing. A typical aluminum can contains about 40 percent recycled aluminum. The amount of aluminum recycling is declining because of the increased use of plastic beverage containers.

INTEGRATED WASTE MANAGEMENT

integrated waste management
A combination of the best waste management techniques into a consolidated program to deal effectively with solid waste.

The most effective way to deal with solid waste is with a combination of techniques. In **integrated waste management**, a variety of waste minimization methods, including the three R's of waste prevention (reduce, reuse, and recycle), are incorporated into an overall waste management plan (**FIGURE 16.10**). Even on a large scale, recycling and source reduction will not entirely eliminate the need for disposal facilities such as incinerators and landfills. However, recycling and source reduction will substantially reduce the amount of solid waste requiring disposal in incinerators and landfills.

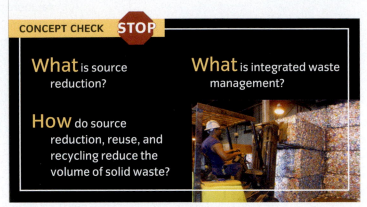

CONCEPT CHECK   STOP

What is source reduction?

How do source reduction, reuse, and recycling reduce the volume of solid waste?

What is integrated waste management?

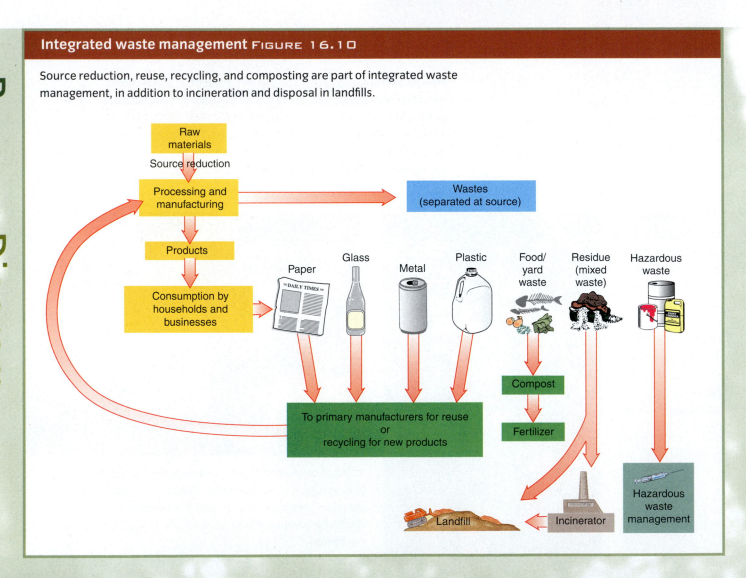

Process Diagram

Integrated waste management FIGURE 16.10

Source reduction, reuse, recycling, and composting are part of integrated waste management, in addition to incineration and disposal in landfills.

Hazardous Waste

Hazardous waste (also called **toxic waste**) accounts for about 1 percent of the solid waste stream in the United States. Hazardous waste includes dangerously reactive, corrosive, ignitable, or toxic chemicals. The chemicals may be solids, liquids, or gases. More than 700,000 different chemicals are known to exist. How many are hazardous is unknown because most have never been tested for toxicity, but without a doubt, there are thousands.

hazardous waste
A discarded chemical that threatens human health or the environment.

Hazardous waste has periodically been in the news since 1977. At that time it was discovered that toxic waste from an abandoned chemical dump had contaminated homes and possibly people in **Love Canal**, a small neighborhood on the edge of Niagara Falls, New York. Love Canal became synonymous with chemical pollution caused by negligent hazardous waste management. In 1978, it became the first location ever declared a national emergency disaster area because of toxic waste; more than 700 families were evacuated (**FIGURE 16.11**).

From 1942 to 1953, a local industry, Hooker Chemical Company, disposed of about 20,000 metric tons (22,000 tons) of toxic chemical waste in the 914-m-long (3,000-ft-long) Love Canal. When the site was filled, Hooker added topsoil and donated the land to the local board of education. A school and houses were built on the site, which began oozing toxic waste several years later. More than 300 chemicals, many of them carcinogenic, have been identified in Love Canal's toxic waste.

In 1990, after almost 10 years of cleanup, the EPA and the New York Department of Health declared the area safe for resettlement. Today, the canal is a 16.2 hectare (40-acre) mound covered by clay and surrounded by a chain-link fence and warning signs. The Love Canal episode resulted in the passage of the federal Superfund Act, which holds polluters accountable for the cost of cleanups (discussed shortly). Its passage generated immediate and ongoing concern about hazardous waste.

Aerial view of Love Canal toxic waste site in the early 1980s FIGURE 16.11

All the homes shown in this photograph were evacuated and demolished.

TYPES OF HAZARDOUS WASTE

Hazardous chemicals include a variety of acids, dioxins, abandoned explosives, heavy metals, infectious waste, nerve gas, organic solvents, polychlorinated biphenyls (PCBs), pesticides, and radioactive substances. Many of these chemicals are discussed in other chapters; see Chapters 4, 8, 10, 14, and 17, which examine endocrine disrupters, air pollution, water pollution, pesticides, and radioactive waste. Here we discuss dioxins and PCBs because they are some of the most persistent hazardous compounds that contaminate our environment.

Dioxins **Dioxins** are a group of 75 similar chemical compounds formed as byproducts during the combustion of chlorine compounds. Incineration of medical and municipal wastes accounts for 70 to 95 percent of known human emissions of dioxins. Some other known sources of dioxins are iron ore mills, copper smelters, cement kilns, metal recycling, coal combustion, pulp and paper plants that use chlorine for bleaching (**FIGURE 16.12**), and chemical accidents. Motor vehicles, out-door grills, and cigarette smoke emit minor amounts of dioxins. Forest fires and volcanic eruptions are natural sources of dioxins. Dioxins also form during the production of some pesticides.

Dioxins are emitted in smoke and then settle on plants, the soil, and bodies of water; from there they are incorporated into the food web. When humans and other animals ingest dioxins—primarily in contaminated meat, dairy products, and fish—they store and accumulate the dioxins in their fatty tissues (see the bioaccumulation and biomagnification discussion in Chapter 4). Because dioxins are so widely distributed in the environment, virtually everyone has dioxins in their body fat.

Dioxins cause several kinds of cancer in laboratory animals, but the data conflict on their cancer-causing ability in humans. A 2001 EPA report suggests that dioxins probably cause several kinds of cancer in humans and likely affect the human reproductive, immune, and nervous systems. Because dioxins are passed through human milk, nursing infants are considered particularly at risk.

EnviroDiscovery   **HANDLING NANOTECHNOLOGY SAFELY**

Nanotechnology is in the news a lot these days. **Nanomaterials**, which are unique materials and devices designed on the ultra-small scale of atoms or molecules, have numerous possible applications. For example, nanoparticles of cadmium selenide might be injected into cancerous tissue, where they would accumulate inside cancer cells; when exposed to ultraviolet radiation, these nanoparticles glow, and surgeons could more easily excise the cancerous tissues and leave the healthy tissues intact. Nanocrystals have the potential to be used in thin-film solar panels to convert solar energy to electricity. Silica nanoparticles embedded in glass make a heat-resistant glass capable of withstanding temperatures of up to 1800° F for several hours (see photograph)

Despite the potential of nanotechnology, particles on the nanometer scale (a nanometer is one-billionth of a meter) might pose health, safety, or environmental risks. No one knows for sure. The EPA has adopted a precautionary approach (see the section on the precautionary principle in Chapter 4) and decided to regulate nanomaterials that might adversely affect the environment. This means that the burden of proof about product safety will fall on companies that sell nanotechnology. Similarly, the Food and Drug Administration will have to oversee regulation of nanotechnology that has potential health and safety risks.

Nanotechnology

This glass has fire-resistant properties because of the addition of silica nanoparticles. Many familiar materials, such as silica, exhibit unusual properties at the nanoparticle level.

Air pollution from a paper mill FIGURE 16.12
These emissions may contain dioxins.

Studying bacteria that break down PCBs in contaminated soil FIGURE 16.13

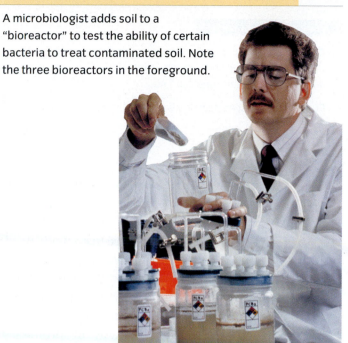

A microbiologist adds soil to a "bioreactor" to test the ability of certain bacteria to treat contaminated soil. Note the three bioreactors in the foreground.

PCBs Polychlorinated biphenyls (PCBs) are a group of 209 industrial chemicals composed of carbon, hydrogen, and chlorine. PCBs were manufactured in the United States between 1929 and 1979 for a wide variety of uses: as cooling fluids in electrical transformers, electrical capacitors, vacuum pumps, and gas-transmission turbines; and in hydraulic fluids, fire retardants, adhesives, lubricants, pesticide extenders, inks, and other materials. Prior to the EPA ban in the 1970s, PCBs were dumped in large quantities into landfills, sewers, and fields. Such improper disposal is one of the reasons PCBs are still a threat today.

The dangers of PCBs first became evident in Japan in 1968, where hundreds of people ate rice bran oil accidentally contaminated with PCBs and consequently experienced serious health problems, including liver and kidney damage. A similar mass poisoning tied to PCBs occurred in Taiwan in 1979. Since then, toxicity tests conducted on animals indicate that PCBs harm the skin, eyes, reproductive organs, and gastrointestinal system. PCBs are endocrine disrupters: They interfere with hormones released by the thyroid gland. Several studies have demonstrated that in utero exposure to PCBs can lead to certain intellectual impairments in children. PCBs may be carcinogenic; they are known to cause liver cancer in rats, and studies in Sweden and the United States have shown a correlation between high PCB concentrations in the body and incidences of certain cancers.

Although high-temperature incineration is one of the most effective ways to destroy PCBs in most solid waste, it is too costly to be used for the removal of PCBs that have leached into soil and water. One way to remove PCBs from soil and water is to extract them with solvents. This method is undesirable because the solvents themselves are hazardous chemicals, and these extraction methods are also costly.

Recently, researchers have discovered several bacteria that degrade PCBs at a fraction of the cost of incineration. Additional research is needed to make the biological degradation of PCBs practical (FIGURE 16.13).

CONCEPT CHECK **STOP**

What is hazardous waste?

What are two sources of dioxins? of PCBs?

Managing Hazardous Waste

LEARNING OBJECTIVES

Compare the Resource Conservation and Recovery Act and the Comprehensive Environmental Response, Compensation, and Liability Act (the Superfund Act).

Explain how green chemistry is related to source reduction.

We have the technology to manage toxic waste in an environmentally responsible way, but it is extremely expensive. Although great strides have been made in educating the public about the problems of hazardous waste, we have only begun to address the issues of hazardous waste disposal. No country currently has an effective hazardous waste management program, but several European countries have led the way by producing smaller amounts of hazardous waste and by using fewer hazardous substances.

CHEMICAL ACCIDENTS

When a chemical accident occurs in the United States, whether at a factory or during the transport of hazardous chemicals, the National Response Center (NRC) is notified. Most chemical accidents reported to the NRC involve oil, gasoline, or other petroleum spills. The remaining accidents involve more than 1000 other hazardous chemicals, such as ammonia, sulfuric acid, and chlorine. In 2005, the NRC database indicated that toxic chemicals were released into the environment during 35,714 separate reported accidents. The state with the greatest number of toxic chemical accidents in 2005 was Texas, which had five significant environmental incidents.

Chemical safety programs have traditionally stressed accident mitigation and adding safety systems to existing procedures. More recently, industry and government agencies have stressed accident prevention through the **principle of inherent safety**, in which industrial processes are redesigned to involve less toxic materials so that dangerous accidents are less likely to occur in the first place.

PUBLIC POLICY AND TOXIC WASTE CLEANUP

Currently, two federal laws dictate how hazardous waste should be managed: (1) the Resource Conservation and Recovery Act, which is concerned with managing hazardous waste being produced now, and (2) the Superfund Act, which provides for the cleanup of abandoned and inactive hazardous waste sites.

The **Resource Conservation and Recovery Act (RCRA)** was passed in 1976 and amended in 1984. Among other things, RCRA instructs the EPA to identify which waste is hazardous and to provide guidelines and standards to states for hazardous waste management programs. RCRA bans hazardous waste from land disposal unless it is treated to meet the EPA's standards of reduced toxicity. In 1992 the EPA initiated a major reform of RCRA to expedite cleanups and streamline the permit system to encourage hazardous waste recycling.

In 1980 the **Comprehensive Environmental Response, Compensation, and Liability Act (CERCLA)**, commonly known as the **Superfund Act**, established a program to tackle the huge challenge of cleaning up abandoned and illegal toxic waste sites across the United States. At many of these sites, hazardous chemicals have migrated deep into the soil and have polluted groundwater. The greatest threat to human health from toxic waste sites comes from drinking water laced with such contaminants.

Cleaning up existing toxic waste: The superfund program
The federal government estimates that the United States has more than 400,000 hazardous waste sites with leaking chemical storage tanks and drums (both above and below ground), pesticide dumps, and piles of mining waste. This estimate does not include the hundreds or thousands of toxic waste sites at military bases and nuclear weapons facilities.

By 2007, 10,753 sites were in the CERCLA inventory, which means the EPA has identified them as qualifying for cleanup (**FIGURE 16.14**). (The count does not include more than 1000 sites that had been cleaned up and removed from the CERCLA inventory since 1980.)

The sites posing the greatest threat to public health and the environment are placed on the **Superfund National Priorities List**, and the federal government will assist in their cleanup. As of 2007, 1556 sites were on the National Priorities List. The five states that led the list in 2007 were New Jersey (115 sites), California (94 sites), Pennsylvania (94 sites), New York (86 sites), and Michigan (65 sites). The average cost of cleaning up a site is $20 million.

One reason for the urgency about cleaning up the sites on the National Priorities List is their locations. With the growth of cities and their suburbs, residential developments now surround many of the dumps. Because the federal government cannot clean up every old dump in the United States, the current landowner, prior owners, and anyone who has dumped waste on or transported waste to a particular site may be liable for cleanup costs.

Although critics decry the slow pace and high cost of cleaning up Superfund sites, the existence of

Cleaning up hazardous waste FIGURE 16.14

A Toxic waste in deteriorating drums at a site near Washington, DC. The metal drums in which much of the waste is stored have corroded and started to leak. Old toxic waste dumps are commonplace around the United States.

B Cleanup of a hazardous waste site near Minneapolis, Minnesota. Removal and destruction of the wastes are complicated by the fact that usually nobody knows what chemicals are present.

CERCLA is a deterrent to further polluting. Companies that produce hazardous waste are now fully aware of the costs of liability and cleanup and are more likely to properly dispose of their hazardous wastes.

MANAGING TOXIC WASTE PRODUCTION

The Superfund Act deals only with hazardous waste produced in the past, not the large amount of toxic waste produced today. There are three ways to manage hazardous waste: (1) source reduction, (2) conversion to less hazardous materials, and (3) long-term storage.

As with municipal solid waste, the most effective approach is source reduction—that is, using less hazardous or nonhazardous materials in industrial processes. Source reduction relies on the increasingly important field of **green chemistry**.

For example, chlorinated solvents are widely used in electronics, dry cleaning, foam insulation, and industrial cleaning. To accomplish source reduction, it is sometimes possible to substitute a less hazardous water-based solvent for the toxic chlorinated one. Substantial source reduction of chlorinated solvents can also be accomplished by reducing solvent emissions. Installing solvent-saving devices benefits the environment and also saves money because smaller amounts of chlorinated solvents must be purchased. No matter how efficient source reduction becomes, however, it will never entirely eliminate hazardous waste.

> **green chemistry**
> A subdiscipline of chemistry in which commercially important chemical processes are redesigned to significantly reduce environmental harm.

The second-best way to deal with hazardous waste is to reduce its toxicity by chemical, physical, or biological means, depending on the nature of the waste. High-temperature incineration, for example, reduces dangerous compounds such as pesticides, PCBs, and organic solvents to safe products such as water and carbon dioxide. The resulting ash is hazardous and must be disposed of in a landfill designed for hazardous materials. Incineration using a *plasma torch* produces such high temperatures (up to 10,000°C, five times higher than conventional incinerators) that hazardous waste is

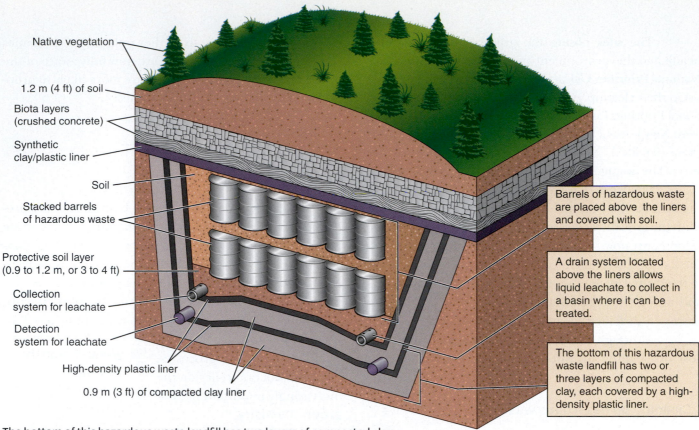

Native vegetation

1.2 m (4 ft) of soil

Biota layers
(crushed concrete)

Synthetic
clay/plastic liner

Soil

Stacked barrels
of hazardous waste

Protective soil layer
(0.9 to 1.2 m, or 3 to 4 ft)

Collection
system for leachate

Detection
system for leachate

High-density plastic liner

0.9 m (3 ft) of compacted clay liner

Barrels of hazardous waste
are placed above the liners
and covered with soil.

A drain system located
above the liners allows
liquid leachate to collect in
a basin where it can be
treated.

The bottom of this hazardous
waste landfill has two or
three layers of compacted
clay, each covered by a high-
density plastic liner.

The bottom of this hazardous waste landfill has two layers of compacted clay,
each covered by a high-density plastic liner. (Some hazardous waste landfills have three
layers of compacted clay.) A drain system located above the plastic and clay liners allows
liquid leachate to collect in a basin where it can be treated, and a leak detection system is
installed between the clay liners. Barrels of hazardous waste are placed above the liners
and covered with soil.

Cutaway view through a hazardous waste landfill FIGURE 16.15

almost completely converted to nontoxic gases, such as carbon dioxide and nitrogen.

Hazardous waste that is not completely detoxified must be placed in long-term storage. Hazardous waste landfills are subject to strict environmental criteria and design features. They are located as far as possible from aquifers, streams, wetlands, and residences. Such a landfill includes several layers of compacted clay and high-density plastic liners at the bottom of the landfill to prevent leaching of hazardous substances into surface water and groundwater (**FIGURE 16.15**). Leachate is collected and treated to remove contaminants. The entire facility and nearby groundwater deposits are carefully monitored to make sure there is no leakage.

Some toxic liquid waste, such as explosives and pesticides, is disposed of by *deep-well injection*. In this tech-

nique, the liquid waste is injected through pipes that extend hundreds of meters into an injection zone located between two impermeable areas. In such geologic formations, the waste is not likely to migrate into aquifers that could be used for drinking or irrigation.

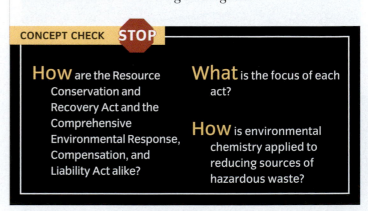

CONCEPT CHECK STOP

How are the Resource Conservation and Recovery Act and the Comprehensive Environmental Response, Compensation, and Liability Act alike?

What is the focus of each act?

How is environmental chemistry applied to reducing sources of hazardous waste?

HIGH-TECH WASTE

In the United States and other highly developed countries, the average computer is replaced every 18 to 24 months, not because it is broken but because rapid technological developments and new generations of software make it obsolete. Old computers may still be in working order, but they have no resale value and are even difficult to give away. As a result, they sit in warehouses, garages, and basements—or are frequently thrown away with the trash and end up in landfills or incinerators. According in the Grass Roots Recycling Network, there were 300 million to 600 million obsolete personal computers in the United States by 2004 (latest data available).

This disposal represents a huge waste of the high-quality plastics and metals (aluminum, copper, tin, nickel, palladium, silver, and gold) that make up computers. Computers also contain the toxic heavy metals lead, cadmium, mercury, and chromium, which could potentially leach from landfills into soil and groundwater. Most computers contain 1.4 to 3.6 kg (3 to 8 lbs) of lead, for example. Several states have passed legislation requiring businesses and residents to **e-cycle** consumer electronics—that is, recycle PCs, color monitors, cell phones, and color televisions.

Currently, about 10 percent of discarded computer and electronic components are e-cycled in the United States. Although some companies handle obsolete computers in the United States (see photograph), many U.S. computers to be recycled are shipped overseas to developing countries such as India, Pakistan, and China. There the computers are disassembled, often using methods that are potentially dangerous to the workers taking them apart. For example, circuit boards are often burned to obtain the small amount of gold in them, and burning releases hazardous fumes into the air.

Some highly developed countries have been more progressive than the United States in dealing with their computer waste. The European Union implemented a Waste Electrical and Electronic Equipment plan in 2005 to recover, recycle, and dispose of electronic waste and remove some of the most hazardous chemicals. Similarly, Japan and other industrialized nations have put in place such policies.

Obsolete computer equipment

The computer monitors at this Texas electronic recycler are being disassembled. Functional tubes will be exported to Thailand, where they will be used to manufacture inexpensive televisions. Broken tubes will be recycled or disposed of in the United States.

SUMMARY

1 Solid Waste

1. **Municipal solid waste** consists of solid materials discarded by homes, office buildings, stores, restaurants, schools, hospitals, prisons, libraries, and other facilities. **Nonmunicipal solid waste** consists of solid waste generated by industry, agriculture, and mining.

2. Use of **sanitary landfills** is the most common method of solid waste disposal, involving compacting and burying waste under a shallow layer of soil. Layers of compacted clay and plastic sheets prevent **leachate** (liquid waste) from seeping into groundwater. Problems with sanitary landfills include the potential for methane gas to seep out and cause explosions, the accidental leaking of toxic leachate, a lack of existing landfill space, and resistance to new landfills near homes and businesses.

3. A **mass burn incinerator** is a large furnace that burns all solid waste except for unburnable items such as refrigerators. Problems associated with incineration of solid waste include the potential for air pollution, difficulties in disposing of the toxic ash produced, the high costs of the process, and difficulties in choosing incinerator sites.

4. In composting, yard waste, food scraps, and other organic wastes are transformed by microbial action into a material that, when added to soil, improves its condition.

2 Reducing Solid Waste

1. In **source reduction**, products are designed and manufactured in ways that decrease the volume of solid waste and the amount of hazardous waste in the solid waste stream.

2. The volume of solid waste produced can be decreased through source reduction, reuse of products, and recycling of materials. Recycling conserves natural resources and is more environmentally benign than landfill disposal but requires a market for the recycled goods.

3. **Integrated waste management** is a combination of the best waste management techniques into a consolidated program to deal effectively with solid waste.

3 Hazardous Waste

1. **Hazardous waste** is a discarded chemical that threatens human health or the environment. Hazardous chemicals may be solids, liquids, or gases and include a variety of acids, dioxins, abandoned explosives, heavy metals, infectious waste, nerve gas, organic solvents, PCBs, pesticides, and radioactive substances.

2. **Dioxins** are hazardous chemicals formed as unwanted byproducts during the combustion of many chlorine compounds. **Polychlorinated biphenyls (PCBs)** are hazardous, oily, industrial chemicals composed of carbon, hydrogen, and chlorine.

4 Managing Hazardous Waste

1. The **Resource Conservation and Recovery Act (RCRA)** instructs the EPA to identify hazardous waste and to provide guidelines and standards for states' hazardous waste management programs. The **Comprehensive Environmental Response, Compensation, and Liability Act (CERCLA)**, or **Superfund Act**, established a program whose goal is to clean up abandoned and illegal toxic waste sites across the United States.

2. The most effective approach to managing hazardous waste is source reduction, reducing the amount and toxicity of hazardous materials used in industrial processes. Source reduction relies on **green chemistry**, a subdiscipline of chemistry in which commercially important chemical processes are redesigned to reduce environmental harm.

KEY TERMS

CRITICAL AND CREATIVE THINKING QUESTIONS

1. Compare the advantages and disadvantages of disposing of waste in sanitary landfills and by incineration.

2. How could source reduction efforts reduce the volume of waste that arises from abandoned automobiles?

3. List what you think are the best ways to treat each of the following types of solid waste, and explain the benefits of the processes you recommend: paper, plastic, glass, metals, food waste, and yard waste.

4. What are dioxins, and how are they produced? What harm do they cause?

5. Suppose hazardous chemicals were suspected to be leaking from an old dump near your home. Outline the steps you would take to (1) have the site evaluated to determine whether there is a danger and (2) mobilize the local community to get the site cleaned up.

6. What are the goals, strengths, and weaknesses of the Superfund Act?

7. What is integrated waste management? Why must a sanitary landfill always be included in any integrated waste management plan?

In an effort to reduce municipal solid waste, many communities have required customers to pay for garbage collection according to the amount of garbage they generate, an approach termed unit pricing or "pay as you throw." The figure to the right illustrates the effects of unit pricing in San Jose, California, on garbage sent to landfills and on wastes diverted through recycling and through separation of yard wastes.

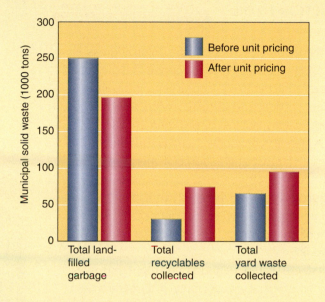

8. How did the implementation of unit pricing in San Jose affect the amount of garbage sent to landfills?

9. How did the implementation of unit pricing affect the quantity of materials recycled or of yard wastes collected?

10. Which component of municipal solid wastes pictured in the graph was most affected by the implementation of unit pricing? Why do you suppose this was the result?

What is happening in this picture ?

- This clothing market in Tanzania sells used clothing from Europe and America. How does this practice affect the volume of solid waste produced?

- These clothes, which come from nonprofit charities such as Goodwill, are sold in local marketplaces. How does this practice affect African clothing industries?

- Africa now exports about $1.6 billion of its textiles and clothing to the United States each year. However, Africa's main competition for the U.S. market is China and other Asian countries. Explain this global connection.

Nonrenewable Energy Resources

17

ADDICTION TO OIL

VIEW THIS IN ACTION
in your WileyPLUS course

In his January 2006 State of the Union address, President George W. Bush warned that the United States is "addicted to oil" (see larger photo). This statement echoes similar claims made over the past several decades, not just by environmentalists but by business leaders, academics, and world leaders. Oil, the world's largest single source of energy, is cheap (relative to most other energy sources), easy to transport and use, and extremely versatile. Oil also has been fairly abundant, with large deposits on every continent except Antarctica. So why the concern with this "addiction"?

First, the United States imports more than two-thirds of the oil that we use, much of it from a small number of countries. We have to rely on these other countries to maintain our current lifestyle, and threats to oil supply are often interpreted as national threats. Second, oil drilling is hazardous and energy intensive, as are transportation and refining, and burning oil (usually as diesel or gasoline) releases a variety of pollutants.

Finally, our addiction cannot continue indefinitely. Some estimates of available oil suggest that we are reaching "peak oil," the point at which the maximum amount of oil is being pumped from underground. Once we pass the peak, the amount of oil available worldwide drops each year, while prices go up. Several of the world's largest oil fields passed their peak productivity in 2005 and 2006.

By 2008, U.S. consumers felt the crunch of depleting supplies as gas prices rose dramatically (see inset). As global supplies continue to tighten, experts predict that the United States will have to scramble to establish energy conservation measures that will weaken our addiction to oil.

Energy Consumption

LEARNING OBJECTIVE

Compare per person energy consumption in highly developed and developing countries.

Human society depends on energy. We use it to warm our homes in winter and cool them in summer; to grow, store, and cook our food; to light our homes; to extract and process natural resources for manufacturing items we use daily; and to power various forms of transportation. Many of the conveniences of modern living depend on a ready supply of energy.

A conspicuous difference in per person energy consumption exists between highly developed and developing nations (**FIGURE 17.1**). As you might expect, highly developed nations consume much more energy per person than developing nations—approximately eight times as much. In the United States, industry uses 42 percent of the nation's total energy, buildings such as homes and offices consume 33 percent, and transportation uses 25 percent.

World energy consumption has increased every year since 1982, with most of the increase occurring in developing countries. From 2003 to 2004, for example, energy consumption increased worldwide by about 2.5 percent, most of it in China and India (**FIGURE 17.2**). One of the goals of developing countries is to improve the standard of living. One way to achieve this is through economic development, a process usually accompanied by a rise in per person energy consumption. Furthermore, the world's energy requirements will continue to increase during the 21st century, as the human population becomes larger, particularly in developing countries.

In contrast, the population in highly developed nations is more stable, and many energy experts think that the per person energy consumption in such countries may be at or near saturation. Additional energy demands in highly developed nations may be met by increasing the energy efficiency of appliances, automobiles, and home insulation (see Chapter 18).

Energy consumption   FIGURE 17.1

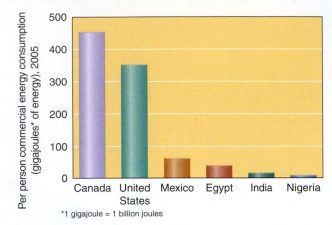

*1 gigajoule = 1 billion joules

A Annual per person commercial energy consumption in selected countries.

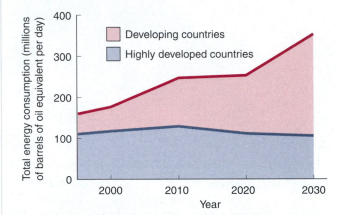

B Projected total energy consumption, to 2030.

Demand for energy in India   FIGURE 17.2

Power lines criss-cross above buildings in Delhi.

CONCEPT CHECK   STOP

HOW does per person energy consumption compare in highly developed and developing countries?

Coal

LEARNING OBJECTIVES

Distinguish between surface mining and subsurface mining.

Summarize the environmental problems associated with using coal.

Describe two technologies that can be used to make coal a cleaner fuel.

C oal, the most abundant fossil fuel in the world, is found primarily in the Northern Hemisphere (**FIGURE 17.3**). The largest coal deposits are in the United States, Russia, China, Australia, India, Germany, and South Africa. The United States has 25 percent of the world's coal supply in its massive deposits. According to the World Resources Institute, known world coal reserves could last for more than 200 years at the present rate of consumption. Coal resources currently too expensive to develop have the potential to provide enough coal to last for 1,000 or more years at current consumption rates.

Utility companies use coal to produce electricity, and heavy industries use coal for steel production. Coal consumption has surged in recent years, particularly in the rapidly growing economies of China and India (**FIGURE 17.4** on next page).

COAL MINING

The two basic types of coal mines are surface and subsurface (underground) mines. If the coal bed is within 30 m (100 ft) or so of the surface, **surface mining** is usually done. In **strip mining**, one type of surface mining, a trench is dug to extract the coal, which is scraped out of the ground and loaded into railroad cars or trucks. Surface mining is used to obtain approximately 60 percent of the coal mined in the United States.

When the coal is deeper in the ground or runs deep into the ground from an outcrop on a hillside, it is

> **surface mining**
> The extraction of mineral and energy resources near Earth's surface by first removing the soil, subsoil, and overlying rock strata.

Distribution of coal deposits FIGURE 17.3

Data are presented as percentages of the 2005 estimated recoverable reserves—that is, of coal known to exist that can be recovered under present economic conditions with existing technologies. (The map is color-coded with the bar graph.)

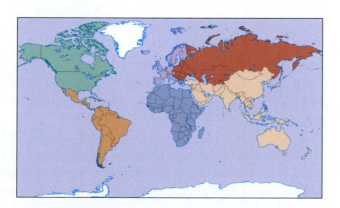

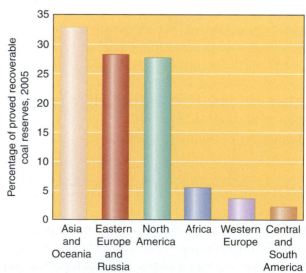

Chinese energy consumption on the rise
FIGURE 17.4

Workers in Shanxi Province load coal onto trucks. Coal provides 65 percent of China's energy. Consumption of coal in China, the highest in the world, may double within 20 years as its economy grows.

Surface coal mine near Douglas, Wyoming
FIGURE 17.5

The overlying vegetation, soil, and rock are stripped away, and then the coal is extracted out of the ground.

> **subsurface mining**
> The extraction of mineral and energy resources from deep underground deposits.

mined underground. **Subsurface mining** accounts for approximately 40 percent of the coal mined in the United States.

Surface mining has several advantages over subsurface mining: It is usually less expensive and safer for miners, and it generally allows more complete removal of coal from the ground. However, surface mining disrupts the land much more extensively than subsurface mining and has the potential to cause serious environmental problems.

ENVIRONMENTAL IMPACTS OF COAL

Coal mining, especially surface mining, has substantial effects on the environment (**FIGURE 17.5**). Prior to the 1977 **Surface Mining Control and Reclamation Act**

(SMCRA), abandoned surface coal mines were usually left as large open pits or trenches. Acid and toxic mineral drainage from such mines, along with the removal of topsoil, which was buried or washed away by erosion, prevented most plants from naturally recolonizing the land. Streams were polluted with sediment and **acid mine drainage**, which is produced when rainwater seeps through iron sulfide minerals exposed in mine wastes (see Chapter 12). Dangerous landslides occurred on hills that were unstable due to the lack of vegetation.

One of the most land-destructive types of surface mining is **mountaintop removal**. According to Environmental Media Services, mountaintop removal has leveled between 15 and 25 percent of the mountaintops in

> **acid mine drainage** Pollution caused when sulfuric acid and dangerous dissolved materials, such as lead, arsenic, and cadmium, wash from coal and metal mines into nearby lakes and streams.

southern West Virginia. The valleys and streams between the mountains are gone as well, filled with mine tailings and debris. At the current rate, half the peaks in that area will be gone by 2020. Mountaintop removal is also occurring in Kentucky, Pennsylvania, Tennessee, and Virginia.

Coal burning generally contributes more of the common air pollutants than burning either oil or natural gas. In the United States, coal-burning electric power plants currently produce one-third of all airborne mercury emissions. Some coal contains sulfur and nitrogen that, when burned, are released into the atmosphere as sulfur oxides (SO_2 and SO_3) and nitrogen oxides (NO, NO_2, and N_2O), many of which form acids when they react with water. These reactions result in **acid deposition**, which is particularly prevalent downwind from coal-burning electric power plants (**FIGURE 17.6**). Acid deposition and forest decline are discussed in greater detail in Chapter 9.

Dead trees enveloped in acid fog on Mt. Mitchell, North Carolina FIGURE 17.6

Forest decline was first documented in Germany and eastern Europe. More recently, it has been observed in eastern North America, particularly at higher elevations. Acid deposition contributes to forest decline.

Burning any fossil fuel releases carbon dioxide (CO_2), a greenhouse gas, into the atmosphere. Currently we are releasing so much CO_2 into the atmosphere that global temperature may be affected (because the increasing concentration of greenhouse gases prevents heat from escaping from the planet). Burning coal causes a more severe CO_2 problem than burning other fossil fuels because coal releases more CO_2 per unit of heat energy produced.

MAKING COAL CLEANER

Sulfur emissions associated with the combustion of coal can be reduced by using **scrubbers**, or desulfurization systems, that clean power plants' exhaust. As polluted air passes through a scrubber, chemicals in the scrubber react with the pollution and cause it to precipitate, or settle out.

Clean coal technologies are methods of burning coal that reduce air pollution. In **fluidized-bed combustion**, crushed coal is mixed with limestone particles in a strong air current during combustion. This clean coal technology produces fewer nitrogen oxides and removes sulfur from the coal. It produces more heat from a given amount of coal, thereby reducing CO_2 emissions per unit of electricity produced.

> ■ **fluidized-bed combustion**
> A clean-coal technology in which crushed coal is mixed with limestone to neutralize acidic compounds produced during combustion.

In the United States several large power plants are testing fluidized-bed combustion, and a few small plants are already using this technology. **The Clean Air Act Amendments of 1990** provide incentives for utility companies to convert to clean coal technologies.

CONCEPT CHECK STOP

Which type of coal mining—surface or subsurface mining—is more land intensive?

What are the environmental impacts of mining and burning coal?

Oil and Natural Gas

Oil and natural gas supplied approximately 63 percent of the energy used in the United States in 2005. In comparison, other U.S. energy sources include coal (23 percent), nuclear power (8.1 percent), and hydropower (3.1 percent). Globally in 2004, oil and natural gas provided 60.6 percent of the world's energy (**FIGURE 17.7**).

Petroleum, or **crude oil**, is a liquid composed of hundreds of hydrocarbon compounds. During petroleum refining, the compounds are separated into different products—such as gases, jet fuel, heating oil, diesel, and asphalt—based on their different boiling points (**FIGURE 17.8**). Oil is also used to produce **petrochemicals**, compounds used to make products such as fertilizers, plastics, paints, pesticides, medicines, and synthetic fibers.

World commercial energy sources
FIGURE 17.7

Note the overwhelming importance of oil, coal, and natural gas as commercial energy sources. "Alternatives" include geothermal, solar, wind, and wood electric power.

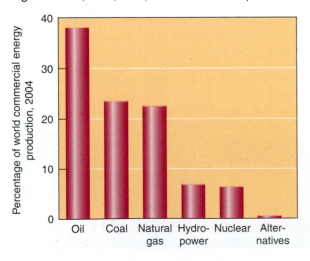

Process Diagram

Petroleum refining FIGURE 17.8

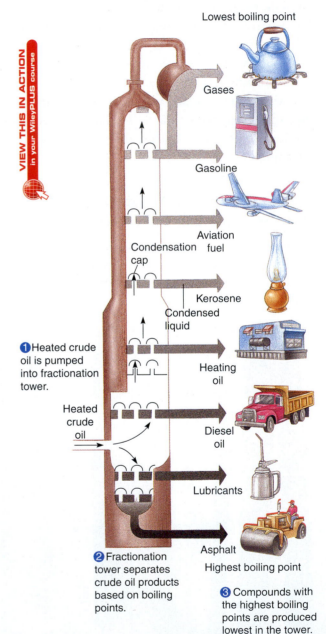

VIEW THIS IN ACTION in your WileyPLUS course

PETROLEUM PRODUCTS

FRACTIONATION TOWER

4 Compounds with lowest boiling points rise highest in the tower.

Lowest boiling point

Gases

Gasoline

Condensation cap

Aviation fuel

Kerosene

Condensed liquid

1 Heated crude oil is pumped into fractionation tower.

Heating oil

Heated crude oil

Diesel oil

Lubricants

2 Fractionation tower separates crude oil products based on boiling points.

Asphalt

Highest boiling point

3 Compounds with the highest boiling points are produced lowest in the tower.

Compared to petroleum, natural gas contains only a few hydrocarbons: methane and smaller amounts of ethane, propane, and butane. Propane and butane are separated from the natural gas, stored in pressurized tanks as a liquid called **liquefied petroleum gas**, and used primarily in rural areas as fuel for heating and cooking. Methane is used to heat residential and commercial buildings, to generate electricity in power plants, and for a variety of purposes in the organic chemistry industry.

Natural gas use is increasing in three main areas: electricity generation, transportation, and commercial cooling. An increasingly popular use of natural gas is **cogeneration**, a clean and efficient process in which natural gas is used to produce both electricity and steam; the heat of the exhaust gases provides energy to make steam for water and space heating (see Chapter 18). Cogeneration systems that use natural gas provide electricity cleanly and efficiently.

As a fuel for trucks, buses, and cars, natural gas offers significant environmental advantages over gasoline or diesel: Natural gas vehicles emit 80 percent to 93 percent fewer hydrocarbons, 90 percent less carbon monoxide, 90 percent fewer toxic emissions, and almost no soot. As of 2008, the United States had more than 150,000 vehicles running on compressed natural gas. The city of Los Angeles has the largest fleet of natural gas–powered transit buses in North America.

Natural gas efficiently fuels residential and commercial air-cooling systems. One example is the use of natural gas in a desiccant-based (air-drying) cooling system, which is ideal for restaurants and supermarkets, where humidity control is as important as temperature control.

The main disadvantage of natural gas is that deposits are often located far from where the energy is used. Because it is a gas and less dense than a liquid, natural gas costs four times more to transport through pipelines than crude oil. To transport natural gas over long distances, it must first be compressed to form liquefied natural gas (LNG) and then carried on specially constructed refrigerated ships. After LNG arrives at its destination, it must be returned to the gaseous state at regasification plants before being piped to where it will be used. Currently, the United States has only four such plants, which severely restricts the importation of natural gas from other countries. American energy companies claim that the United States needs at least 40 regasification plants to keep costs down for natural gas and to meet increasing demands.

RESERVES OF OIL AND NATURAL GAS

Oil and natural gas deposits exist on every continent, but their distribution is uneven. More than half of the world's total estimated reserves are situated in the Persian Gulf region, which includes Iran, Iraq, Kuwait, Oman, Qatar, Saudi Arabia, Syria, United Arab Emirates, and Yemen (**FIGURE 17.9**). Major oil fields also exist in Venezuela, Mexico, Russia, Kazakhstan, Libya, and the United States (in Alaska and the Gulf of Mexico; **FIGURE 17.10**).

Almost half of the world's proved recoverable reserves of natural gas are located in two countries, Russia and Iran. The United States has more deposits of natural gas than western Europe.

Distribution of oil deposits FIGURE 17.9

Data are presented as regional percentages of the 2006 world estimates of crude oil reserves. (The map is color-coded with the bar graph.)

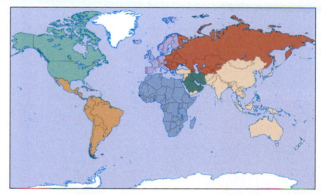

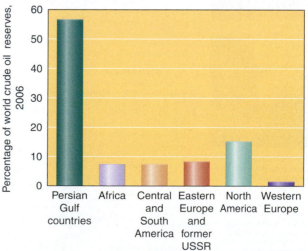

Global energy supply and consumption FIGURE 17.10

Oil supplies are distributed unequally around the world, and energy consumption varies dramatically between regions.

▶ **ENERGY CONSUMPTION**
The use and availability of primary energy resources are unequally distributed around the globe. More than 86 percent of energy consumed globally is from nonrenewable fossil fuels—coal, oil, and natural gas. Consumption of these fuels is greatest in industrialized nations, with the United States using up nearly one-quarter. Developing countries, especially those in sub-Saharan Africa, rely on more traditional sources of energy, such as firewood and dung.

HYDROPOWER
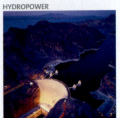

NUCLEAR
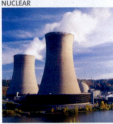

SOLAR

WIND

GEOTHERMAL

ALTERNATIVE ENERGIES
Hydropower provides nearly 18 percent of the world's electricity, but it is limited to countries with adequate water resources, and it poses threats to local watersheds. **Nuclear energy** makes up 17 percent of Earth's electricity, but few countries have adopted it because of potential environmental risks and waste disposal issues. **Solar** and **wind energy** are inexhaustible and are the focus of new energy technologies and research. **Geothermal energy** is efficient but limited to countries with ready sources of hot groundwater, such as Iceland.

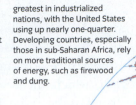

Annual energy consumption, in trillions of British thermal units (BTUs)

- More than 25,000
- 10,001–25,000
- 1001–10,000
- 101–1000
- 10–100
- Less than 10
- No data available

Major energy deposit
- Coal
- Natural gas
- Oil
- Oil transit chokepoint

▼ **RENEWABLE ENERGY**
Renewable sources of energy—geothermal, solar, and wind— make up a small percentage of the world's energy supply. They have a significant impact, however, on local and regional energy supplies, especially for electricity, in places such as the United States, Japan, and Germany. These sources of energy can be regenerated or renewed in a relatively short time, whereas fossil fuels form over geologic time

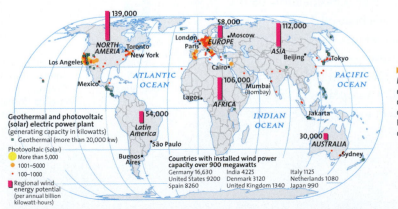

Geothermal and photovoltaic (solar) electric power plant
(generating capacity in kilowatts)
- Geothermal (more than 20,000 kw)

Photovoltaic (Solar)
- More than 5,000
- 1001–5000
- 100–1000
- Regional wind energy potential (per annual billion kilowatt-hours)

Countries with installed wind power capacity over 900 megawatts

Germany 16,630	India 4225	Italy 1125
United States 9200	Denmark 3120	Netherlands 1080
Spain 8260	United Kingdom 1340	Japan 990

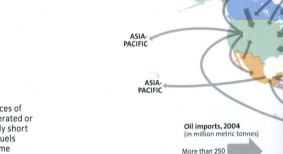

Oil imports, 2004
(in million metric tonnes)
- More than 250
- 175–250
- 75–174
- Less than 75

▶ **FLOW OF OIL WORLDWIDE**
Major oil reserves are clustered in a handful of countries, more than half of which are in the Middle East, whereas the greatest demand for oil is in the United States, Europe, Japan, and China. Other major oil exporters include the Russian Federation, Norway, Venezuela, and Mexico.

Energy enables us to cook our food, heat our homes, move about our planet, and run industry. Every day the world uses some 320 billion kilowatt-hours of energy—equivalent to each person burning 22 lightbulbs nonstop, and over the next century, demand may increase threefold. Consumption is not uniform around the globe. People in industrialized counties consume far greater amounts of energy than those in developing countries. The world's energy supply is still fossil-fuel based, despite advances in alternative energy sources. To meet demand, many countries must import fuels, making the trade of energy a critical, often volatile global political issue. Instability where most oil is found—the Persian Gulf, Nigeria, and Venezuela—make this global economic power line fragile. Insatiable demand where most energy is consumed—the United States, Japan, China, India, and Germany—makes national economies increasingly dependent. Further-more, extraction and use of fossil fuel have serious environmental effects, such as air pollution and climate warming. The challenge for the future? Reducing reliance on fossil fuels, developing alternative energies to meet demand, and mediating the trade-offs between the environment and energy.

GeoBytes

LACK OF ACCESS
More than 2 billion people, mostly in the developing world, do not have access to electricity. Increasingly, small-scale wind and solar projects bring power to poor rural areas.

WINDS OF CHANGE
Worldwide, wind supplies less than 1 percent of electric power, but it is the fastest-growing source, especially in Europe. Denmark gets 20 percent of its electricity from wind.

POWER OF THE SUN
Near Leipzig, Germany, some 33,000 photovoltaic panels produce up to 5 megawatts of power. It is one of the world's largest solar arrays.

GOING NUCLEAR
France gets 78 percent of its electricity from nuclear power. Developing nations, such as China and India, are building new reactors to reduce pollution and meet soaring energy demands.

GROWING PAINS
China is fueling its economic growth with huge quantities of coal, and it suffers from energy-related environmental problems. China is second only to the United States in greenhouse gas emissions that contribute to global warming.

◄ **WORLD OIL SUPPLY**
The world's hunger for oil is insatiable, but the supply is finite and unequally distributed, making it one of the world's most valuable commodities. It is the leading source of energy worldwide, and in industrialized countries it accounts for more than one-third of all energy consumed. Pressure on the world's oil supply continues to mount as both industrialized and developing countries grow more dependent on it to meet increasing energy needs.

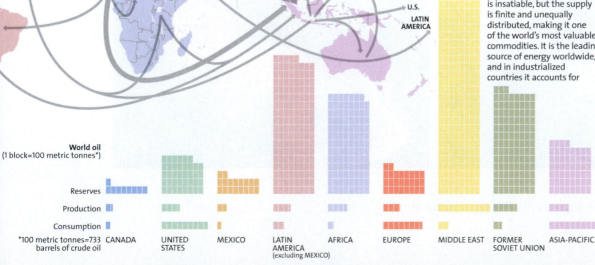

World oil
(1 block=100 metric tonnes*)

Reserves
Production
Consumption

*100 metric tonnes=733 barrels of crude oil

CANADA — UNITED STATES — MEXICO — LATIN AMERICA (excluding MEXICO) — AFRICA — EUROPE — MIDDLE EAST — FORMER SOVIET UNION — ASIA-PACIFIC

ASIA-PACIFIC
U.S. & MEXICO
ASIA-PACIFIC
U.S. LATIN AMERICA

NATIONAL GEOGRAPHIC

Large oil deposits probably exist under the **continental shelves**, the relatively flat underwater areas that surround continents, and in deep-water areas adjacent to the continental shelves. Despite problems like storms at sea and the potential for oil spills, many countries engage in offshore drilling. As many as 18 billion barrels (756 billion gallons) of oil and natural gas may exist in the deep water of the Gulf of Mexico, just off the continental shelf from Texas to Alabama (**Figure 17.11**). Continental shelves off the coasts of western Africa and Brazil are also promising potential sources of oil. Environmentalists and coastal industries such as fishing generally oppose opening the continental shelves for oil and natural gas exploration because of the threat of a major oil spill.

How long will oil and natural gas supplies last?

We cannot predict how many reserves will be discovered, whether technological breakthroughs will allow us to extract more fuel from each deposit, or whether world consumption of oil and natural gas will increase, remain the same, or decrease. Even with technological advances, the most optimistic predictions are for global oil production to peak around 2035. Natural gas is more plentiful than oil. Experts estimate that, at current rates of consumption, readily recoverable reserves of natural gas will keep production rising for at least 10 years after conventional supplies of petroleum have begun to decline.

ENVIRONMENTAL IMPACTS OF OIL AND NATURAL GAS

The environmental problems associated with oil and natural gas result from burning and transporting them. As with coal, the burning of oil and natural gas produces CO_2 that can contribute to global climate warming. Every gallon of gasoline your car or truck burns releases about 9 kg (20 lb) of CO_2 into the atmosphere. Burning oil also leads to acid deposition and the formation of photochemical smog. Oil combustion produces almost no sulfur oxides, but it does produce nitrogen oxides. In fact, gasoline combustion contributes about half the nitrogen oxides released into the atmosphere by human activities. (Coal combustion contributes the rest.)

Natural gas, on the other hand, is a relatively clean, efficient source of energy that contains almost no sulfur and releases far less CO_2, fewer hydrocarbons, and almost no particulate matter compared to oil and coal.

One risk of oil and natural gas production relates to their transport, often over long distances by pipelines or ocean tankers. A serious spill along the route creates an environmental crisis, particularly in aquatic ecosystems, where an oil slick can travel great distances.

Rapid expansion of oil exploration in the Gulf of Mexico, 1961–2001 **Figure 17.11**

Active oil leases give drilling companies the right to obtain oil in a given location.

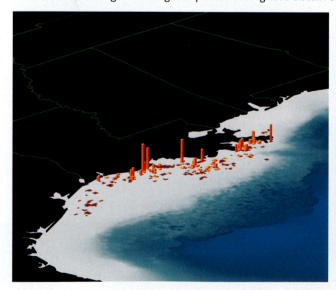 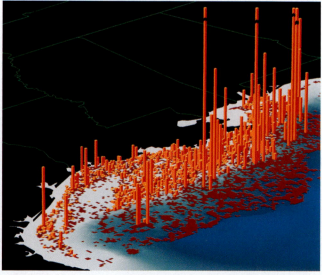

The largest oil spill in the United States In 1989 the supertanker *Exxon Valdez* hit Bligh Reef and spilled 260,000 barrels (10.9 million gallons) of crude oil into Prince William Sound along the coast of Alaska, creating the largest oil spill in U.S. history (**FIGURE 17.12**). According to the U.S. Fish and Wildlife Service and the Alaska Department of Environmental Conservation, more than 30,000 birds (sea ducks, loons, cormorants, bald eagles, and other species) and between 3,500 and 5,500 sea otters died as a result of the spill. The area's killer whale and harbor seal populations declined, salmon migration was disrupted, and the fishing season in the area was halted that year.

Although Exxon declared the cleanup "complete" in late 1989, it left behind contaminated shorelines; continued damage to some species of birds, fishes, and mammals (**TABLE 17.1**); and a reduced commercial salmon catch, among other problems.

Visualizing

The *Exxon Valdez* oil spill FIGURE 17.12

A An aerial view of the massive oil slick at the southwest end of Prince William Sound, Alaska, 1989.

B Workers try to clean the rocky shoreline of Eleanor Island, Alaska, several months after the spill.

C The extent of the spill (black arrows). Water currents caused it to spread rapidly for hundreds of kilometers.

Nearshore coastal wildlife status, Prince William Sound, Alaska, 2005 TABLE 17.1	
Populations that are recovering	**Populations that are not yet recovering**
Clams	Common loons
Intertidal communities	Cormorants (3 species)
Killer whales	Harbor seals
Marbled murrelets	Harlequin ducks
Mussels	Pacific herring
Sea otters	Pigeon guillemots

One positive outcome of the disaster was passage of the **Oil Pollution Act** of 1990. This legislation establishes liability for damages to natural resources resulting from a catastrophic oil spill, including a trust fund that pays to clean up spills when the responsible party cannot, and it requires, by 2015, double hulls on all oil tankers that enter U.S. waters.

Earth's largest oil spill

The world's most massive oil spill occurred in 1991 during the Persian Gulf War, when about 6 million barrels (250 million gallons) of crude oil—more than 20 times the amount of the *Exxon Valdez* spill—were deliberately dumped into the Persian Gulf. Many oil wells were set on fire, and lakes of oil spilled into the desert around the burning oil wells. In 2001 Kuwait began a massive remediation project to clean up its oil-contaminated desert. Progress is slow, and it may take a century or more for the area to completely recover.

CONCEPT CHECK **STOP**

How long are reserves of oil and natural gas projected to last?

What are three environmental problems associated with using oil and natural gas as energy resources?

Nuclear Energy

LEARNING OBJECTIVES

Define *nuclear energy* and describe a typical nuclear power reactor.

Discuss the pros and cons of electric power produced by nuclear energy versus coal.

Describe safety issues associated with nuclear power plants and risks associated with the storage of radioactive wastes.

All atoms are composed of positively charged protons, negatively charged electrons, and electrically neutral neutrons (**FIGURE 17.13**). Protons and neutrons, which have approximately the same mass, are clustered in the center of an atom, making up its nucleus. Electrons, which possess little mass in comparison with protons and neutrons, orbit the nucleus in distinct regions.

As a way to obtain energy, nuclear processes are fundamentally different from the combustion that produces energy from fossil fuels. Combustion is a chemical reaction. In chemical reactions, atoms of one element do not change into atoms of another element, nor does any of their mass (matter) change into energy. The energy re-

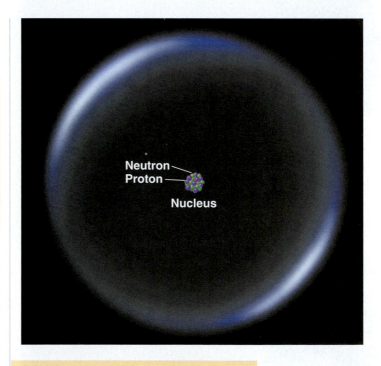

Neutron
Proton
Nucleus

Atomic structure **FIGURE 17.13**

An atom contains a nucleus made of protons and neutrons. Circling the nucleus is a "cloud" of electrons.

leased in chemical reactions comes from changes in the chemical bonds that hold together the atoms. Chemical bonds are associations between electrons, and chemical reactions involve the rearrangement of electrons.

nuclear energy
The energy released by nuclear fission or fusion.

fission The splitting of an atomic nucleus into two smaller fragments, accompanied by the release of a large amount of energy.

In contrast, **nuclear energy** involves changes in the nuclei of atoms; small amounts of matter from the nucleus are converted into large amounts of energy. Nuclear reactions produce 100,000 times more energy per atom than is available from a chemical bond between two atoms.

There are two different nuclear reactions that release energy: fission and fusion. In nuclear **fission**, the process nuclear power plants use, energy is released when a single neutron crashes into a large atom of one element, such as uranium, and splits it into two smaller atoms of different elements (**FIGURE 17.14**). In **fusion**, the process that powers the sun and other stars, two small atoms are combined, forming one larger atom of a different element.

CONVENTIONAL NUCLEAR FISSION

Uranium ore, the mineral fuel used in conventional nuclear power plants, is a nonrenewable resource present in limited amounts in sedimentary rock in Earth's crust. Approximately 11 percent of the world's uranium deposits are located in the United States.

Uranium ore contains three isotopes: U-238 (which makes up 99.28 percent of uranium), U-235 (0.71 percent), and U-234 (less than 0.01 percent). Because U-235, the isotope used in conventional fission reactions, is such a minor part of uranium ore, uranium ore must be refined after mining to increase the concentration of U-235 to about 3 percent. This refining process, called **enrichment**, requires a great deal of energy.

enrichment
The process by which uranium ore is refined after mining to increase the concentration of fissionable U-235.

Nuclear fission FIGURE 17.14

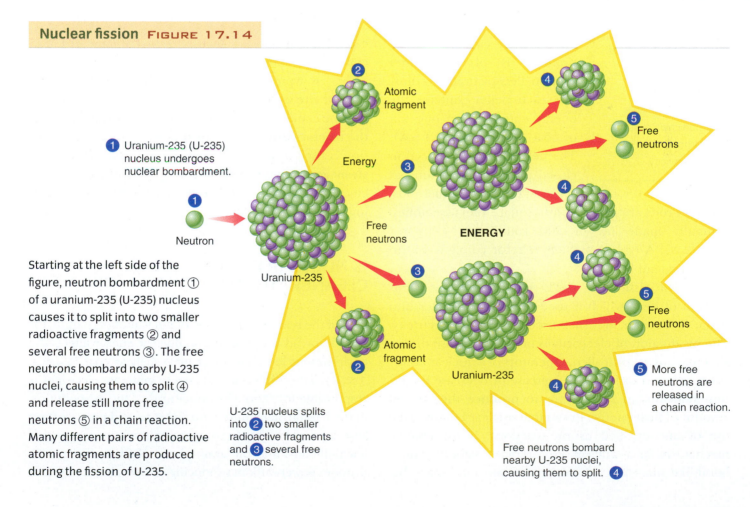

Starting at the left side of the figure, neutron bombardment ① of a uranium-235 (U-235) nucleus causes it to split into two smaller radioactive fragments ② and several free neutrons ③. The free neutrons bombard nearby U-235 nuclei, causing them to split ④ and release still more free neutrons ⑤ in a chain reaction. Many different pairs of radioactive atomic fragments are produced during the fission of U-235.

① Uranium-235 (U-235) nucleus undergoes nuclear bombardment.

Neutron

Uranium-235

Atomic fragment

Energy

Free neutrons

ENERGY

Free neutrons

Atomic fragment

Uranium-235

Free neutrons

U-235 nucleus splits into ② two smaller radioactive fragments and ③ several free neutrons.

Free neutrons bombard nearby U-235 nuclei, causing them to split. ④

⑤ More free neutrons are released in a chain reaction.

Uranium fuel FIGURE 17.15

A Uranium dioxide pellets, held in a gloved hand, contain about 3 percent uranium-235, the fission fuel in a nuclear reactor. Each pellet contains the energy equivalent of 1 ton of coal.

B The uranium pellets are loaded into long fuel rods, which are grouped into square fuel assemblies like the one being inspected here by technicians.

> **nuclear reactor**
> A device that initiates and maintains a controlled nuclear fission chain reaction to produce energy for electricity.

After enrichment, the uranium is processed into small pellets of uranium dioxide; each pellet contains the energy equivalent of one ton of coal (**FIGURE 17.15A**). The pellets are then placed in closed pipes, often as long as 3.7 m (12 ft), called **fuel rods**. The fuel rods are grouped into square **fuel assemblies**, generally made up of 200 rods each (**FIGURE 17.15B**). A typical **nuclear reactor** contains 150 to 250 fuel assemblies.

The fission of U-235 releases an enormous amount of heat, which is used to transform water into steam. The steam, in turn, is used to generate electricity. The production of electricity is possible because the fission reaction is controlled. Operators of a nuclear power plant can start or stop and increase or decrease the fission reactions in the reactor to produce the desired amount of heat energy. Recall that nuclear bombs make use of uncontrolled fission reactions. If the control mechanism in a nuclear power plant were to fail, a bomblike nuclear explosion could not take place be-

cause nuclear fuel has such a low percentage of U-235 compared to bomb-grade material.

A typical nuclear power plant has four main parts: the reactor core, the steam generator, the turbine, and the condenser (**FIGURE 17.16**). Fission occurs in the **reactor core**, and the heat produced by nuclear fission is used to produce steam from liquid water in the **steam generator**. The **turbine** uses the steam to generate electricity, and the **condenser** cools the steam, converting it back to a liquid.

IS NUCLEAR ENERGY CLEANER THAN COAL?

One of the reasons supporters of nuclear energy argue for its widespread adoption is that nuclear energy affects the environment less than fossil fuels such as coal (**TABLE 17.2**). The combustion of coal to generate electricity is responsible for more than one-third of the air pollution in the United States and contributes to acid precipitation and climate warming. In comparison, nuclear energy emits few pollutants into the atmosphere. Nuclear energy also provides power without producing climate-altering CO_2.

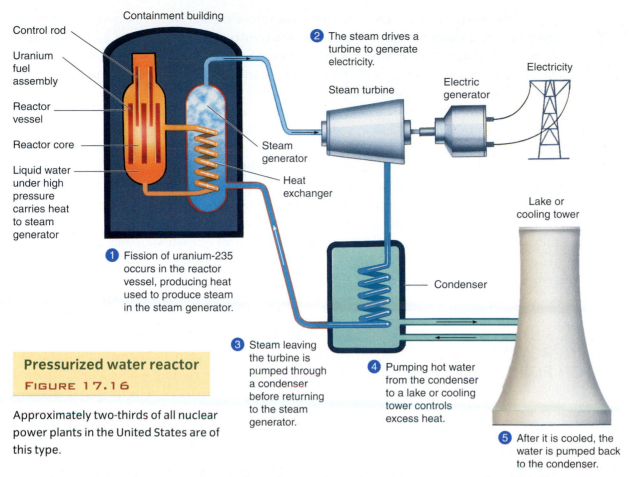

Control rod

Uranium fuel assembly

Reactor vessel

Reactor core

Liquid water under high pressure carries heat to steam generator

Containment building

Steam generator

Heat exchanger

② The steam drives a turbine to generate electricity.

Steam turbine

Electric generator

Electricity

① Fission of uranium-235 occurs in the reactor vessel, producing heat used to produce steam in the steam generator.

③ Steam leaving the turbine is pumped through a condenser before returning to the steam generator.

④ Pumping hot water from the condenser to a lake or cooling tower controls excess heat.

Condenser

Lake or cooling tower

⑤ After it is cooled, the water is pumped back to the condenser.

Pressurized water reactor

FIGURE 17.16

Approximately two-thirds of all nuclear power plants in the United States are of this type.

Comparison of environmental impacts of 1000-MWe coal and conventional nuclear power plants*
TABLE 17.2

Impact	Coal	Nuclear (conventional fission)
Land use	17,000 acres	1900 acres
Daily fuel requirement	9000 tons (of coal)/day	3 kg (of enriched uranium)/day
Availability of fuel, based on present economics	A few hundred years	100 years, maybe longer
Air pollution	Moderate to severe, depending on pollution controls	Low
Climate change risk (from CO_2 emissions)	Severe	No risk
Radioactive emissions, routine	1 curie	28,000 curies
Water pollution	Often severe	Potentially severe at nuclear waste disposal sites
Risk from catastrophic accidents	Short-term local risk	Long-term risk over large areas
Link to nuclear weapons	No	Yes
Annual occupational deaths	0.5 to 5	0.1 to 1

*Impacts include extraction, processing, transportation, and conversion. Assumes that coal is strip-mined. (A 1000-MWe utility, at a 60 percent load factor, produces enough electricity for a city of 1 million people.)

spent fuel Used fuel elements that were irradiated in a nuclear reactor.

However, nuclear energy generates radioactive waste in the form of **spent fuel**. Nuclear power plants also produce radioactive coolant fluids and gases in the reactor. These radioactive wastes are extremely dangerous, and the hazards of their health and environmental impacts require that special measures be taken to ensure their safe storage and disposal.

Opponents of nuclear energy contend that the replacement of coal-burning power plants with nuclear power plants does not significantly reduce the threat of climate warming because only 15 percent of greenhouse gases come from power plants. Auto emissions and industrial processes produce most greenhouse gases, and neither is affected by nuclear power.

CAN NUCLEAR ENERGY DECREASE OUR RELIANCE ON FOREIGN OIL?

International crises, such as the Gulf War of the early 1990s and the Iraq War in the 2000s, occasionally threaten the supply of oil to the United States. Some supporters of nuclear energy assert that our dependence on foreign oil would be reduced if all oil-burning power plants were converted to nuclear plants.

However, oil is responsible for generating only about 3 percent of the electricity in the United States. Replacing electricity generated by oil with electricity generated by nuclear power would do little to lessen our dependence on foreign oil because we would still need oil for heating buildings and for gasoline. Technological advances could change nuclear power's potential contribution in the future.

If electric heat pumps and electric motor vehicles become more common, however, nuclear power plants have the potential to heat buildings and to power automobiles, thus decreasing our reliance on foreign oil.

SAFETY AND ACCIDENTS IN NUCLEAR POWER PLANTS

Although conventional nuclear power plants cannot explode like atomic bombs, accidents do happen in which dangerous levels of radiation are released into the environment and result in human casualties. At high temperatures, the metal encasing uranium fuel can melt, releasing radiation; this is called a **meltdown**. Also, the water used in a nuclear reactor to transfer heat can boil away during an accident, contaminating the atmosphere with radioactivity.

The nuclear industry considers the probability that a major accident will occur low, but public perception of the risk is high for several reasons. Nuclear power risks are involuntary and potentially catastrophic. In addition, many people are distrustful of the nuclear industry. The consequences of such accidents are drastic and life threatening, both immediately and long after the accident has occurred.

Three Mile Island
The most serious nuclear reactor accident in the United States occurred in 1979 at the Three Mile Island power plant in Pennsylvania as a result of human error after the cooling system failed. A partial meltdown of the reactor core took place. Had there been a complete meltdown of the fuel assembly, dangerous radioactivity would have been emitted into the surrounding countryside. Fortunately, the containment building kept almost all the radioactivity released by the core material from escaping. Although a small amount of radiation entered the environment, there were no substantial environmental damages and no immediate human casualties. A study conducted within a 10-mile radius around the plant 10 years after the accident concluded that cancer rates were in the normal range and found no association between cancer rates and radiation from the accident. Numerous other studies have failed to link abnormal health problems (other than increased stress) to the accident.

In the aftermath of the accident, public wariness prompted construction delays and cancellations of several new nuclear power plants across the United States. The accident at Three Mile Island reduced the complacency that was commonplace in the nuclear industry. New safety regulations were put in place, including more frequent safety inspections, new risk assessments, and improved emergency and evacuation plans for nuclear power plants and surrounding communities. The Institute of Nuclear Power Operations was created to promote safety and improve the training of nuclear operators.

Global Locator

Chornobyl, Ukraine FIGURE 17.17

The arrow indicates the site of the explosion. The upper part of the reactor was completely destroyed.

Chornobyl The world's worst nuclear power plant accident took place in 1986 at the Chornobyl plant, located in the former Soviet republic of the Ukraine. One or possibly two explosions ripped apart a nuclear reactor and expelled large quantities of radioactive material into the atmosphere (FIGURE 17.17). The effects of this accident were not confined to the area immediately surrounding the power plant: Significant amounts of radioisotopes quickly spread across large portions of Europe. The Chornobyl accident affected and will continue to affect many nations.

Although cleanup in the immediate vicinity of Chornobyl is finished, the people in Ukraine face many long-term problems. Ultimately, more than 170,000 people had to permanently abandon their homes. Much of the farmland and forests are so contaminated that they cannot be used for more than a century. Inhabitants of many areas of Ukraine cannot drink the water or consume locally produced milk, meat, fish, fruits, or vegetables. Mothers do not nurse their babies because their milk is contaminated by radioactivity. The frequency of birth defects and mental retardation in newborns has in-

creased in affected areas, and children exposed to the Chornobyl fallout experienced increased incidences of leukemia, thyroid cancer, and abnormalities of the immune system (FIGURE 17.18).

THE LINK BETWEEN NUCLEAR ENERGY AND NUCLEAR WEAPONS

Fission is involved in both the production of electricity by nuclear energy and the destructive power of nuclear weapons. Countries that own nuclear power plants have access to the fuel needed for nuclear weapons (by reprocessing spent fuel to make plutonium). Responsible world leaders are concerned about the consequences of terrorist groups and states of concern (such as Iran and North Korea) building nuclear weapons. These concerns have caused many people to shun nuclear energy and to seek alternatives that are not so intimately connected with nuclear weapons.

There are several hundred metric tons of weapons-grade plutonium worldwide. Storing plutonium is a security nightmare because it takes only several kilograms to make a nuclear bomb as powerful as the ones that destroyed the Japanese cities of Nagasaki and Hiroshima in World War II. However, since the 2001 terrorist attacks in the United States, the security of plutonium stockpiles and nuclear power plants has been increased substantially to reduce the chance that terrorist groups could steal plutonium and enriched uranium and use them to make nuclear weapons.

Health consequences of Chornobyl
FIGURE 17.18

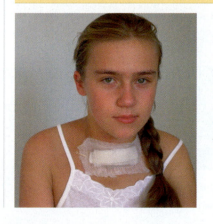

This 14-year-old is recovering from thyroid cancer. A significant (25-fold) increase in thyroid cancer in children and adolescents occurred within a few years after the accident.

RADIOACTIVE WASTES

Radioactive wastes are classified as either "low level" or "high level." **Low-level radioactive wastes** include glassware, tools, paper, clothing, and other items contaminated by radioactivity. They are produced by nuclear power plants, university research labs, nuclear medicine departments in hospitals, and industries. Four sites—located in Washington State, South Carolina, Nevada, and Utah—currently store most of the country's low-level radioactive wastes.

High-level radioactive wastes produced during nuclear fission include the reactor metals (fuel rods and assemblies), coolant fluids, and air or other gases found in the reactor. High-level radioactive wastes are also generated during the reprocessing of spent fuel. Produced by nuclear power plants and nuclear weapons facilities, high-level radioactive wastes are among the most dangerous human-made hazardous wastes.

As the radioisotopes in spent fuel decay, they produce considerable heat and are extremely toxic to organisms; they remain radioactive for thousands of years, and their dangerous level of radioactivity requires special handling. Secure storage of these materials must be guaranteed for thousands of years, until the materials decay sufficiently to be safe. The safe disposal of radioactive wastes is one of the main difficulties that must be overcome if we are to realize the potential of nuclear energy in the 21st century.

What are the best sites for the long-term storage of high-level radioactive wastes? Many scientists recommend storing the wastes in stable rock formations deep in the ground. People's reluctance to have radioactive wastes stored near their homes complicates the selection of these sites. Meanwhile, radioactive wastes continue to accumulate. Many commercially operated nuclear power plants store their spent fuel in huge indoor pools of water or in storage casks on-site. However, none of these plants was designed for long-term storage of spent fuel (**FIGURE 17.19**).

> **low-level radioactive wastes** Solids, liquids, or gases that give off small amounts of ionizing radiation.

> **high-level radioactive wastes** Radioactive solids, liquids, or gases that initially give off large amounts of ionizing radiation.

Storage casks for spent fuel FIGURE 17.19

A On-site storage casks at the Prairie Island nuclear power plant in Minnesota. Each cask holds 40 spent fuel assemblies (17.6 tons).

Cask length: 5.2 m (17 ft)

Protective cover
Lid with metallic seals
Storage cask (neutron shield)
Canister of steel
Spent fuel rods

B Details of a storage cask. Each cask, designed to last at least 40 years, is monitored and will be replaced if leakage occurs.

Yucca Mountain In 1982 the passage of the **Nuclear Waste Policy Act** put the burden of developing permanent sites for civilian and military radioactive wastes on the federal government and required the first site to be operational by 1998. (The deadline has since been postponed to 2010 at the earliest.)

In a 1987 amendment to the Nuclear Waste Policy Act, Congress identified Yucca Mountain in Nevada as the only candidate for a permanent underground storage site for high-level radioactive wastes from commercially operated power plants (see What a Scientist Sees). Since 1983 the U.S. Department of Energy has spent billions of dollars conducting feasibility studies on Yucca Mountain's geology. Results indicate that the site is safe, at least from volcanic eruptions and earthquakes. In 2002 Congress finally approved the choice of Yucca Mountain as the U.S. nuclear-waste repository, despite controversy and opposition from the state of Nevada.

Transporting high-level wastes from nuclear reactors and weapons sites is a major concern of opponents of the Yucca Mountain site. A typical shipment would travel an average of 3700 km (2300 miles), and 43 states would have these dangerous materials passing through them on their way to Yucca Mountain.

Whether or not nuclear waste is eventually stored in Yucca Mountain, the scientific community generally agrees that storage of high-level radioactive waste in deep underground repositories is the best long-term option. Using an underground waste facility is far safer than storing high-level nuclear waste as we do now; storing this waste at many different commercial nuclear reactors poses a greater risk of terrorist attacks, theft, and, possibly, human health problems.

Yucca Mountain

A This tunnel provides access to nuclear waste storage at Yucca Mountain, an arid, sparsely populated area of Nevada.

B An environmental scientist thinks about the huge complex of interconnected tunnels located in dense volcanic rock 305 m (1000 ft) beneath the mountain crest. Canisters containing high-level radioactive waste can be stored in the tunnels.

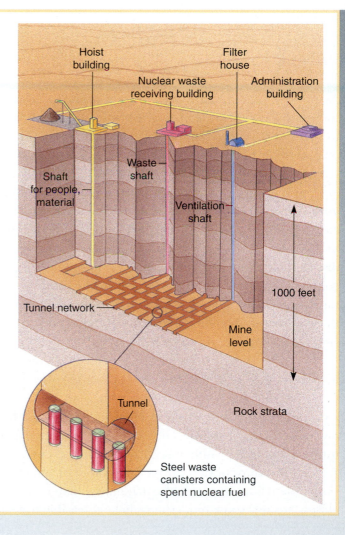

What a Scientist Sees

Over the past three decades, Soviet (and now Russian) practices for radioactive waste disposal have often violated international standards:

- Billions of gallons of liquid radioactive wastes were pumped directly underground, without being stored in protective containers. Russian officials claim that layers of clay and shale at the sites prevent leakage but admit that more leaks than expected have occurred.

- Highly radioactive wastes were dumped into the ocean in amounts double those of dumped wastes from 12 other nuclear nations combined.

Murmansk nuclear waste site.

- Both underground injection and underwater dumping of radioactive wastes continue because Russia lacks alternatives for nuclear waste processing. Potential health and environmental hazards associated with these wastes are unknown because so little data exist for these types of long-term storage.

DECOMMISSIONING NUCLEAR POWER PLANTS

As nuclear power plants age, certain critical sections, such as the reactor vessel, become brittle or corroded. At the end of their operational usefulness, nuclear power plants are not simply abandoned or demolished because many parts have become contaminated with radioactivity.

When a nuclear power plant is closed, it undergoes **decommissioning**. The International Atomic Energy Agency (IAEA) defines three options for decommissioning: storage, entombment, and immediate dismantling. If an old plant is put into **storage**, the utility company guards it for 50 to 100 years while some of the radioactive materials decay, making it safer to dismantle the plant later. Accidental leaks during the storage period are an ongoing concern.

Most experts do not consider **entombment**, permanently encasing the entire power plant in concrete, a viable option because the tomb would have to remain intact for at least 1000 years. Accidental leaks would probably occur during that time, and we cannot guarantee that future generations would inspect and maintain the site.

The third option for the retirement of a nuclear power plant is to dismantle the plant immediately after it closes. Advances in robotics may make it feasible to tear down sections of old plants that are too "hot" (radioactive) for workers to safely dismantle. As the plant is torn down, small sections of it are transported to a permanent storage site.

According to the IAEA, 107 nuclear power plants worldwide were permanently retired as of 2004 (23 of them in the United States), and many nuclear power plants are nearing retirement age. During the 21st century, we may find that we are paying more in our utility bills to decommission old plants than we are to construct new ones.

CONCEPT CHECK   STOP

How does a nuclear reactor produce electricity?

What are the environmental effects of generating electricity through coal combustion or through conventional nuclear fission?

What were some of the short-term effects of the nuclear power plant accident at Chornobyl in Ukraine? What are some long-term effects?

THE ARCTIC NATIONAL WILDLIFE REFUGE

In 1960 Congress declared a section of northeastern Alaska protected because of its distinctive wildlife. In 1980 Congress expanded this wilderness area to form the Arctic National Wildlife Refuge (ANWR; see **FIGURE A**). The proposed opening of ANWR to oil exploration has been an ongoing environment-versus-economy conflict since the refuge's inception. On one side are those who seek to protect rare and fragile natural environments; on the other side are those whose higher priority is the development of some of the last major U.S. oil supplies.

The refuge, called "America's Serengeti," is home to many animal species, including polar bears, arctic foxes, peregrine falcons, musk oxen, Dall sheep, wolverines, and snow geese. It is the calving area for a large migrating herd of caribou (see **FIGURE B**). Although it is biologically rich, the tundra is an extremely fragile ecosystem, in part because of its harsh climate. Arctic organisms are particularly vulnerable to human activities.

In the mid-1990s, prodevelopment interests became more vocal, partly because in 1994, for the first time in its history, the United States imported more than

A Located in the northeastern part of Alaska, the Arctic National Wildlife Refuge is situated close to the Trans-Alaska Pipeline, which begins at Prudhoe Bay and extends south to Valdez. The National Petroleum Reserve–Alaska is also shown.

half the oil it used. Although the Department of the Interior concluded that oil drilling in the wildlife refuge would harm the area's ecosystem, both the Senate and the House of Representatives passed measures to allow it. (President Clinton vetoed the bill.) In 2001, President George W. Bush announced his support for opening the refuge to oil drilling, but after a contentious debate in Congress in 2005, the Senate voted against doing so.

Supporters cite economic considerations as the main reason for drilling for oil in the refuge. Development of domestic oil would improve the balance of trade and make the United States less dependent on foreign countries for our oil. Oil companies are eager to develop this particular site because it is near Prudhoe Bay, where large oil deposits are already being tapped.

Conservationists think oil exploration poses permanent threats to the delicate balance of nature in the Alaskan wilderness, in exchange for a temporary oil supply. They suggest that the money spent drilling for oil would be better used for research into alternative, renewable energy sources and energy conservation—a more permanent solution to the energy problem.

B Members of the caribou herd whose calving grounds are on the Arctic National Wildlife Refuge.

SUMMARY

1 Energy Consumption

1. The per person energy consumption in highly developed nations is eight times higher than that in developing nations.

2 Coal

1. **Surface mining** is the extraction of mineral and energy resources near Earth's surface by first removing the soil, subsoil, and overlying rock strata. **Subsurface mining** is the extraction of mineral and energy resources from deep underground deposits. Surface mining is less expensive and safer but causes more serious environmental problems than subsurface mining.

2. Coal mining can lead to landslides and can pollute streams with sediment and **acid mine drainage**, pollution caused when sulfuric acid and dangerous dissolved materials such as lead, arsenic, and cadmium wash from coal and metal mines into nearby lakes and streams. In **mountaintop removal** a huge shovel gradually removes an entire mountaintop to reach the coal located below. Burning coal releases more CO_2 and contributes more extensively to global climate warming than burning other fossil fuels. The combustion of coal contributes to **acid deposition**, a type of air pollution in which acid falls from the atmosphere to the surface as precipitation or as dry acid particles.

3. Power plants can make coal a cleaner fuel by installing **scrubbers** to clean the power plants' exhaust. **Fluidized-bed combustion** is a clean-coal technology in which crushed coal is mixed with limestone to neutralize the acidic sulfur compounds produced during combustion.

3 Oil and Natural Gas

1. More than half of the world's total estimated oil and natural gas reserves are located in the Persian Gulf region.

2. A serious spill along an oil or gas transportation route creates an environmental crisis, particularly in aquatic ecosystems. The burning of oil and natural gas produces CO_2 that can contribute to global climate warming. Burning oil also leads to acid deposition by producing nitrogen oxides. Natural gas contains almost no sulfur and produces much less CO_2 and other pollutants compared to oil and coal.

4 Nuclear Energy

1. **Nuclear energy** is the energy released by nuclear **fission** or fusion. A **nuclear reactor** is a device that initiates and maintains a controlled nuclear fission chain reaction to produce energy for electricity. A typical reactor contains a **reactor core**, where nuclear fission occurs; a **steam generator**; a **turbine**; and a **condenser**.

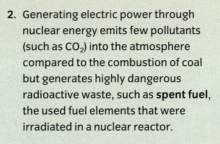

2. Generating electric power through nuclear energy emits few pollutants (such as CO_2) into the atmosphere compared to the combustion of coal but generates highly dangerous radioactive waste, such as **spent fuel**, the used fuel elements that were irradiated in a nuclear reactor.

3. Accidents at nuclear power plants can release dangerous levels of radiation into the environment and result in human casualties. The safe storage of radioactive wastes is another concern associated with nuclear energy. **Low-level radioactive wastes** are radioactive solids, liquids, or gases that give off small amounts of ionizing radiation. **High-level radioactive wastes** are radioactive solids, liquids, or gases that initially give off large amounts of ionizing radiation. Radioactive wastes must be isolated securely for thousands of years. One option for the retirement of an aging nuclear power plant is to decommission it by dismantling it after it closes.

KEY TERMS

CRITICAL AND CREATIVE THINKING QUESTIONS

1. Distinguish among coal, oil, natural gas, and nuclear energy, and compare the environmental impacts of each.

2. What economic priorities and environmental concerns might be shared by coal-mining regions in West Virginia and in China?

3. Some environmental analysts think that the latest war in Iraq was related in part to gaining control over the supply of Iraqi oil. Do you think this is plausible? Explain why or why not.

4. Do you think oil drilling should be permitted in the Arctic National Wildlife Refuge? Why or why not?

5. Which major consumer of oil is most vulnerable to disruption in the event of another energy crisis: electric power generation, motor vehicles, heating and air conditioning, or industry? Why?

6. Can we prevent catastrophic accidents at nuclear power plants in the future? Why or why not?

7. Are you in favor of the United States developing additional nuclear power plants to provide electricity in the 21st century? Why or why not?

8. On the basis of what you have learned about coal, oil, and natural gas, which fossil fuel do you think the United States should exploit during the next 20 years? Explain your answer.

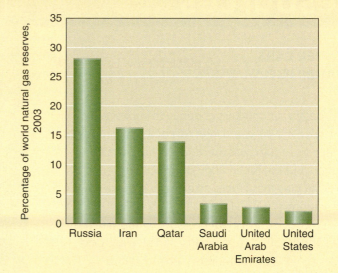

9. Examine the bar graph, which shows the six countries with the greatest natural gas deposits. Is most of the world's natural gas located in North and South America, or in Europe and the Persian Gulf countries?

10. Some scholars think the Industrial Revolution may have been concentrated in Europe and North America because coal is located there. Explain the connection between coal and the Industrial Revolution.

What is happening in this picture ?

- This Gulf of Mexico oil platform, ripped from its mooring by Hurricane Katrina, drifted ashore in Alabama.

- What impact do you suspect Hurricanes Katrina and Rita had on U.S. oil production in late 2005?

- What effect might such natural disasters have on oil supplies and prices?

CLEANER CARS, CLEANER FUELS

VIEW THIS IN ACTION
in your WileyPLUS course

Today's automakers continue to improve the gasoline-powered internal combustion engine so that cars require less fuel and produce less air pollution. But the technology still mainly depends on CO_2-producing fossil fuels, so it doesn't offer a long-term solution to air pollution.

Automakers are under pressure to improve the fuel economy of SUVs, minivans, and pickup trucks; the George W. Bush administration mandated improvements in fuel mileage every year through 2011. Diesel engines, which offer excellent fuel economy but produce more air pollution than gas engines, are a logical choice for these vehicles if their emissions can be cleaned up. Automakers are designing diesel engines that produce less nitrogen oxides and soot; vehicles with these "clean diesel" engines are just now becoming available to consumers.

Hybrid automobiles that use a combination of gasoline or diesel and electricity are a promising technology that is available today. Hybrids are particularly well suited for urban settings. In city driving, vehicles such as the plug-in hybrid electric delivery van used by the *New York Times* (see larger photo) get remarkable fuel mileage and produce relatively little air pollution. Hybrid vehicles remain expensive, but prices are dropping in response to large consumer demand.

Liquid hydrogen is an extremely clean fuel (the only waste product is water). Some energy-efficient vehicles, such as the German bus in the inset photo, incorporate fuel cells that combine stored hydrogen with oxygen from the air to produce electricity. Automakers' prototypes for vehicles powered by fuel cells remain costly, and the range (distance traveled per load of fuel) of these vehicles is still low. Honda, General Motors, and Toyota hope to have fuel cell vehicles on the market by around 2010.

440

Direct Solar Energy

LEARNING OBJECTIVES

Distinguish between active and passive solar heating and describe how each is used.

Contrast the advantages and disadvantages of photovoltaic solar cells and solar thermal electric generation in converting solar energy into electricity.

Define *fuel cell*.

The sun produces a tremendous amount of energy, and most of it dissipates into space. Only a small portion is radiated to Earth. Solar energy is different from fossil and nuclear fuels because it is always available; we will run out of solar energy only when the sun's nuclear fire burns out. To make solar energy useful, we must collect it.

HEATING BUILDINGS AND WATER

In **active solar heating**, a series of collection devices mounted on a roof or in a field gather solar energy. The most common solar collection device is a panel or plate of black metal, which absorbs the sun's energy (**FIGURE 18.1**). Active solar heating is used primarily for heating water, either for household use or for swimming pools. The heat absorbed by the solar collector is transferred to a fluid inside the panel, which is then pumped to the heat exchanger, where the heat is transferred to water that will be stored in the hot water tank. Solar domestic water heating can provide a family's hot water needs year-round. Because more than 8 percent of energy consumed in the

> **active solar heating** A system of putting the sun's energy to use in which collectors absorb the solar energy and pumps or fans distribute the collected heat.

Active solar water heating FIGURE 18.1

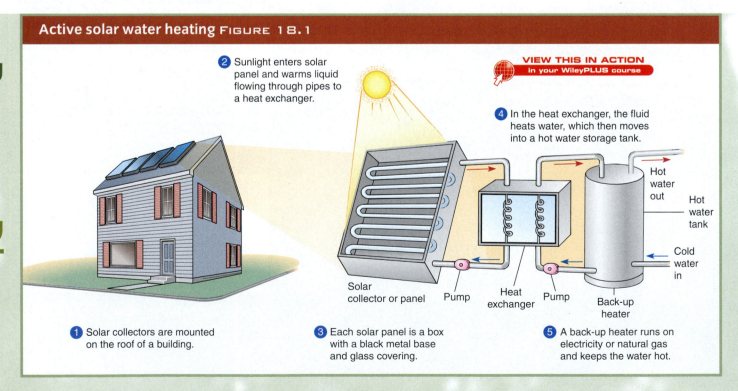

2 Sunlight enters solar panel and warms liquid flowing through pipes to a heat exchanger.

VIEW THIS IN ACTION in your WileyPLUS course

4 In the heat exchanger, the fluid heats water, which then moves into a hot water storage tank.

Solar collector or panel

Pump

Heat exchanger

Pump

Back-up heater

Hot water out

Hot water tank

Cold water in

1 Solar collectors are mounted on the roof of a building.

3 Each solar panel is a box with a black metal base and glass covering.

5 A back-up heater runs on electricity or natural gas and keeps the water hot.

Process Diagram

A Behind the greenhouse glass of this passive solar home, rooms remain at steady temperatures during winter.

Summer sun

VIEW THIS IN ACTION
in your WileyPLUS course

Winter sun

Overhang blocks summer sun.

Vent allows hot air to escape (in summer).

Attic and north-facing wall are heavily insulated.

South-facing double-paned glass allows winter light to enter directly into the room. Double panes reduce heat loss on cold nights.

Insulated drapes or window shades prevent heat loss at night (in winter).

Warm air escapes (in summer).

Thick adobe or stone walls and floor store heat (in winter).

B Several passive designs are incorporated into this home.

United States goes toward heating water, active solar heating could potentially supply a significant amount of the nation's energy demand.

Active solar energy is not used for space heating as commonly as it is used for heating water, but it may become more important when diminishing supplies of fossil fuels force gas and oil prices higher.

In **passive solar heating**, solar energy heats buildings without the need for pumps or fans to distribute the heat. Certain design features are incorporated into a passive solar heating system to warm buildings in winter and help them remain cool in summer (**FIGURE 18.2**). In the Northern Hemisphere, large south-facing windows receive more total sunlight during the day than windows facing other directions. Sunlight entering through the windows provides heat, which is then stored in floors and walls made of concrete or stone, or in containers of water. This stored heat is transmitted throughout the building naturally by **convection**, the circulation that occurs because warm air rises and cooler air sinks.

Buildings with passive solar heating systems must be well insulated so that accumulated heat doesn't escape. Depending on a building's design and location, passive heating can save as much as 50 percent of heating costs. Currently, about 7 percent of new homes built in the United States have passive solar heating features.

> ■ **passive solar heating** A system of putting the sun's energy to use that does not require mechanical devices to distribute the collected heat.

PHOTOVOLTAIC SOLAR CELLS

Photovoltaic (PV) solar cells can convert sunlight directly into electricity (see What a Scientist Sees). They are usually arranged on large panels that absorb sunlight even on cloudy or rainy days.

PV cells generate electricity with no pollution and minimal maintenance. They can be used on any scale, from small portable modules attached to camping lanterns to large, multimegawatt power plants, and can power satellites, uncrewed airplanes, highway signals, wristwatches, and calculators. PV cells' widespread use to generate electricity is currently limited by their low efficiency at converting solar energy to electricity and by the amount of land needed to hold the number of solar panels required for large-scale use. Several thousand acres of today's PV panels would be required to produce the electricity generated by a single large conventional power plant.

In remote areas not served by electric power plants, such as the rural areas of developing countries, it is more economical to use PV cells for electricity than to extend power lines. Photovoltaics generate energy that can pump water, refrigerate vaccines, grind grain, charge batteries, and supply rural homes with lighting. According to the Institute for Sustainable Power, more than 1 million households in developing countries in Asia, Latin America, and Africa have installed PV solar cells on the roofs of their homes. A PV panel the size of two pizza boxes

> **photovoltaic (PV) solar cell** A wafer or thin film of solid-state materials, such as silicon or gallium arsenide, that is treated with certain metals in such a way that the film generates electricity when solar energy is absorbed.

What a Scientist Sees

Photovoltaic Cells

A A student seeing the roof of the Intercultural Center of Georgetown University probably knows that it has arrays of photovoltaic (PV) cells to collect solar energy. The PV system supplies about 10 percent of the school's electricity.

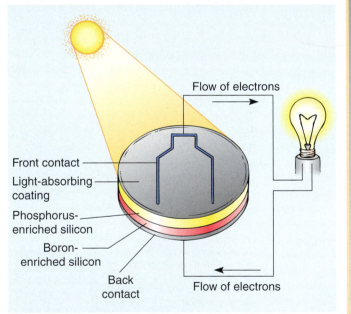

B A scientist looking at those arrays knows that photovoltaic cells contain silicon and other materials. Sunlight excites electrons, which are ejected from silicon atoms. Useful electricity is generated when the ejected electrons flow out of the PV cells through a wire.

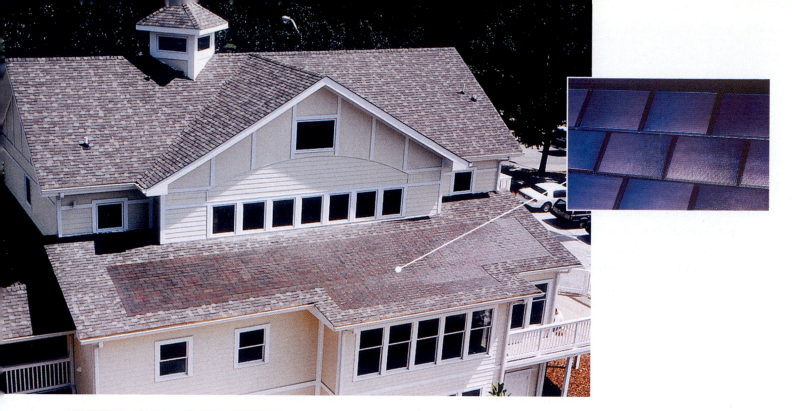

Solar shingles FIGURE 18.3

These thin-film solar cells look much like conventional roofing materials.

supplies a rural household with enough electricity for five lights, a radio, and a television.

Utility companies can purchase PV devices in modular units, which can become operational in a short period. Rather than committing a billion dollars or more and a decade or more to build a new power plant, they can increase generating capacity in small increments. The PV units can provide the additional energy, for example, to power irrigation pumps on hot, sunny days.

The cost of manufacturing PV modules has steadily declined over the past 25 years, from an average factory price of almost $90 per watt in 1975 to about $4.80 per watt in 2007. The cost of producing electricity from PVs has steadily declined from 1970 to the present. Despite this progress, in 2006 the cost was still about $.20 to $.25 per kilowatt-hour. TABLE 18.1 compares the costs of generating electricity using different energy sources, including photovoltaics.

Future technological progress may make PVs economically competitive with electricity produced using conventional energy sources. The production of "thin-film" solar cells (FIGURE 18.3), which are much less expensive to manufacture than standard PVs, has de-

creased costs for PVs across the board. More than 120,000 Japanese homes have installed PV solar-energy roofing in the past few years. The Million Solar Roofs initiative, sponsored by the U.S. government, plans to have solar roofing on 1 million buildings by 2010. Another technological advance that shows promise is dye-sensitized solar cells, which can be produced at about one-fifth the cost of conventional silicon panels.

Generating costs of electric power plants, 2006 TABLE 18.1

Energy source	Generating costs (cents per kilowatt-hour)*
Hydropower	4–10
Biomass	6–9
Geothermal	3–8
Wind	4–7
Solar thermal	5–13
Photovoltaics	20–25
Natural gas	5–7
Coal	4–6
Nuclear power	2–12

*Electricity production and consumption are measured in kilowatt-hours (kWh). As an example, one 50-watt light bulb that is on for 20 hours uses 1 kilowatt-hour of electricity (50 × 20 = 1000 watt-hours = 1 kWh).

SOLAR THERMAL ELECTRIC GENERATION

Systems that concentrate solar energy to heat fluids have long been used for buildings and industrial processes. In **solar thermal electric generation**, electricity is produced by systems that collect sunlight and concentrate it using a combination of mirrors or lenses to heat a working fluid to high temperatures.

In one such system, computer-guided trough-shaped mirrors track the sun for optimum efficiency, center sunlight on nearby oil-filled pipes, and heat the oil to 390°C (735°F) (**FIGURE 18.4**). The hot oil is circulated to a water storage system and used to boil water into super-heated steam, which turns a turbine to generate electricity.

Solar thermal systems often have a backup—usually natural gas—to generate electricity at night and during cloudy days when solar power isn't operating. The world's largest solar thermal system of this type currently operates in the Mojave Desert in southern California.

Solar thermal energy systems are inherently more efficient than other solar technologies because

> ■ **solar thermal electric generation**
> A means of producing electricity in which the sun's energy is concentrated using mirrors or lenses onto a fluid-filled pipe; the heated fluid is used to generate electricity.

they concentrate the sun's energy. With improved engineering, manufacturing, and construction methods, solar thermal energy is becoming cost-competitive with fossil fuels (see Table 18.1). In addition, the environmental benefits of solar thermal plants are significant: These plants don't produce air pollution or contribute to acid rain or global climate change.

SOLAR-GENERATED HYDROGEN

Increasingly, people think of hydrogen as the fuel of the future, as it is abundant as well as easily produced. Electricity generated by photovoltaics or wind energy can split water into the gases oxygen and hydrogen, though this process isn't yet economical. Hydrogen can also be produced using conventional energy sources such as fossil fuels and nuclear power. However, using fossil fuels or nuclear energy to create hydrogen results in the same serious environmental and security problems we discussed in Chapter 17. We therefore limit our discussion to

Solar thermal electric generation FIGURE 18.4

A

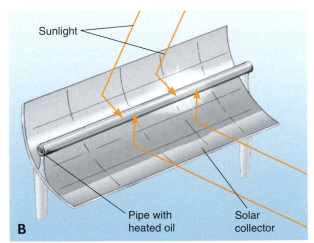

A A solar thermal plant in California uses troughs to focus sunlight on a fluid-filled tube, as shown in **B**. The heated oil is pumped to a water tank where it generates steam used to produce electricity. For simplicity, arrows show sunlight converging on several points; sunlight actually converges on the pipe throughout its length.

Sunlight

Pipe with heated oil

Solar collector

B

hydrogen fuel production using solar electricity, which is sustainable but not yet cost-efficient.

Hydrogen is a clean fuel; it produces water and heat as it burns and produces no sulfur oxides, carbon monoxide, hydrocarbon particulates, or CO_2 emissions. It does produce some nitrogen oxides, though in amounts fairly easy to control. Hydrogen has the potential to provide energy for transportation (in the form of hydrogen-powered automobiles) as well as for heating buildings and producing electricity.

It may seem wasteful to use electricity generated from solar energy to make hydrogen that will then be used to generate electricity. However, the electricity generated by existing photovoltaic cells must be used immediately, whereas hydrogen offers a convenient way to store solar energy as chemical energy. It can be transported by pipeline, possibly less expensively than electricity is transported by wire.

Production of hydrogen from PV electricity currently has a relatively low efficiency (perhaps 10 percent), which means that very little of the solar energy absorbed by the PV cells is actually converted into the chemical energy of hydrogen fuel. Low efficiency translates into high costs. Scientists are working to improve this efficiency and make solar-generated hydrogen fuel commercially viable.

We face other challenges besides high costs if we are to replace gasoline with hydrogen as a transportation fuel. First, we would need to develop a complex infrastructure (such as hydrogen pipelines) to provide hydrogen to service stations. Because hydrogen is extremely volatile, it must be stored, handled, and transported very carefully. Another challenge is developing **fuel cells** for motor vehicles that are inexpensive, safe, and allow the vehicle to drive a long distance without the need to refuel. A fuel cell is an electrochemical cell similar to a battery (**FIGURE 18.5**). Fuel cells represent the most promising way to use hydrogen.

fuel cell A device that directly converts chemical energy into electricity; the fuel cell requires hydrogen and oxygen from the air.

Fuel cells produce power as long as they are supplied with fuel, whereas batteries store a fixed amount of energy. The major automakers are now developing automobiles and buses powered by hydrogen fuel cells (see the chapter introduction).

Fuel cells in a laboratory **FIGURE 18.5**

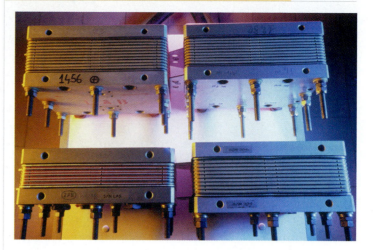

A These fuel cells combine hydrogen and oxygen to create electricity.

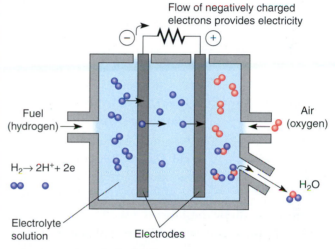

Flow of negatively charged electrons provides electricity

Fuel (hydrogen) →

$H_2 \rightarrow 2H^+ + 2e$

Air (oxygen) ←

H_2O

Electrolyte solution

Electrodes

B Cross-section of a fuel cell.

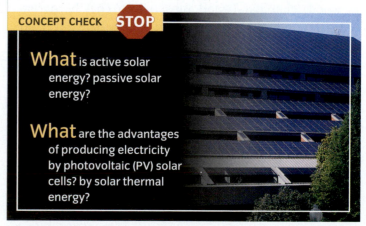

CONCEPT CHECK **STOP**

What is active solar energy? passive solar energy?

What are the advantages of producing electricity by photovoltaic (PV) solar cells? by solar thermal energy?

Indirect Solar Energy

LEARNING OBJECTIVES

Define *biomass* and outline how it is used as a source of energy.

Compare the potential of wind energy and hydropower.

Some renewable energy sources indirectly use the sun's energy. Combustion of *biomass* (organic matter) is an example of indirect solar energy because plants use solar energy for photosynthesis and store the energy in biomass. Windmills, or wind turbines, use *wind energy* to generate electricity. The damming of rivers and streams to generate electricity is a type of *hydropower*—the energy of flowing water.

BIOMASS ENERGY

Biomass, one of the oldest fuels known to humans, consists of materials such as wood, fast-growing plant and algal crops, crop wastes, sawdust and wood chips, and animal wastes. Biomass contains chemical energy that comes from the sun's radiant energy, which photosynthetic organisms use to form organic molecules. Biomass is a renewable form of energy if managed properly.

> **biomass** Plant and animal material used as fuel.

Biomass fuel, which may be a solid, liquid, or gas, is burned to release its energy. Solid biomass fuels such as wood, charcoal (wood turned into coal by partial burning), animal dung, and peat (partly decayed plant matter found in bogs and swamps) supply a substantial portion of the world's energy. At least half of the human population relies on biomass as their main source of energy. In developing countries, wood is the primary fuel for cooking and heat (**FIGURE 18.6**).

It is possible to convert biomass, particularly animal wastes, into **biogas**. Biogas, which is usually composed of a mixture of gases (mostly methane), is like natural gas. It is a clean fuel—its combustion produces fewer pollutants than either coal or biomass. In India and China, several million family-sized **biogas digesters** use microbial decomposition of household and agricultural wastes to produce biogas for cooking and lighting (**FIGURE 18.7**). When biogas conversion is complete, the solid remains are removed from the digester and used as fertilizer.

Biogas has the potential to power fuel cells to generate electricity. A pilot program at Boston's main sewage treatment plant began producing electricity from biogas in 1997. Sewage sludge in large biogas digesters produces methane, which is then burned in a methane fuel cell to produce enough electricity for 150 homes. Like the hydrogen fuel cells discussed earlier in the chapter, methane fuel cells produce relatively few pollutants.

Biomass FIGURE 18.6

Firewood is the major energy source for most of the developing world. Photographed in Nepal.

NEPAL

Global Locator

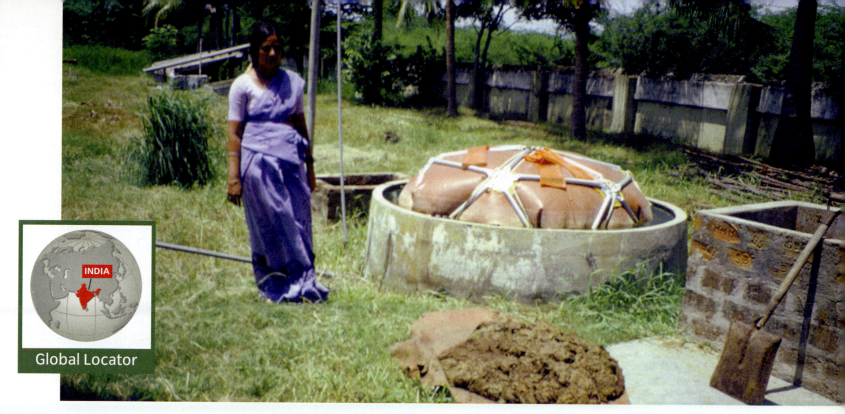

This small-scale biogas digester is being evaluated at a research center. Animal manure placed in the digester decomposes, releasing methane gas that can be used as cooking fuel.

Biogas digester in India FIGURE 18.7

Biomass can also be converted into liquid fuels, especially **methanol** (methyl alcohol) and **ethanol** (ethyl alcohol), which can be used in internal combustion engines. In many parts of the world, automotive fuels must contain 10 percent or more ethanol, and the Indy Car racing series now runs entirely on ethanol. *Biodiesel,* made from plant or animal oils, is becoming popular as an alternative fuel for diesel engines in trucks, farm equipment, and boats. The oil is often refined from waste oil produced at restaurants (such as the oil used to make french fries); biodiesel burns much cleaner than diesel fuel.

Some U.S. energy companies convert sugarcane, corn, or wood crops to alcohol; others are interested in the commercial conversion of agricultural and municipal wastes into ethanol. Currently, the profitability of ethanol is possible only because of government **subsidies** that reduce ethanol's cost. However, as more automakers introduce ethanol-friendly vehicles, these subsidies may cease to be necessary.

Biomass is attractive as a source of energy—and popular with U.S. politicians—because it reduces dependence on fossil fuels. Also, biomass often makes use of waste products, thereby reducing our waste disposal problem. Although it is not completely free of the pollution problems of fossil fuels, biomass combustion produces levels of sulfur and ash that are lower than those that coal produces.

Some problems associated with obtaining energy from biomass include the use of land and water that might otherwise be dedicated to agriculture. This shift toward energy production might decrease food production, contributing to higher food prices.

Also, as mentioned earlier, at least half of the world's population relies on biomass as its main source of energy. Unfortunately, in many areas people burn wood faster than they replant trees. Intensive use of wood for energy has resulted in severe damage to the environment, including soil erosion, deforestation and desertification, air pollution, and degradation of water supplies.

Excessive use of crop biomass can also harm soil quality. Crop residues, such as cornstalks, are increasingly being used for energy. Crop residues left in and on the ground prevent erosion by holding the soil in place; their removal would eventually deplete the soil of minerals and reduce its productivity.

WIND ENERGY

Wind results from the sun warming the atmosphere. **Wind energy** is an indirect form of solar energy: The radiant energy of the sun is transformed into mechanical energy through the movement of air molecules. Wind is sporadic over much of Earth's surface, varying in direction and magnitude. Like direct solar energy, wind energy is a highly dispersed form of energy. Harnessing wind energy to generate electricity has great potential.

> ■ **wind energy**
> Electric energy obtained from surface air currents caused by the solar warming of air.

New wind turbines are huge—100 m (328 ft) tall—and have long blades designed to harness wind energy efficiently (**FIGURE 18.8A** and **18.8B**). As turbines have become larger and more efficient, costs for wind power have declined rapidly—from $.40 per kilowatt-hour in 1980 to a current cost of $.04 to $.07 per kilowatt-hour. Wind power is cost-competitive with many forms of conventional energy. Advances such as turbines that use variable-speed operation may make wind energy an important global source of electricity during the first half of the 21st century.

During the 1990s and early 2000s, wind became the world's fastest-growing source of energy (**FIGURE 18.8C**). Germany currently leads the world as the top producer of wind energy, accounting for almost half of the total European production. Denmark currently generates 21 percent of its electricity using wind energy. Much of this power is generated offshore because ocean winds are strong. Other leading wind energy countries include Spain, the United States, and India.

Harnessing wind energy is most profitable in rural areas that receive fairly continual winds, such as islands, coastal areas, mountain passes, and grasslands. The world's largest concentration of wind turbines is currently located in the Tehachapi Pass at the southern end of the Sierra Nevada mountain range in California.

In the continental United States, some of the best locations for large-scale electricity generation from wind energy are on the Great Plains. The 10 states with the greatest wind energy potential, according to the American Wind Energy Association, are North Dakota, Texas, Kansas, South Dakota, Montana, Nebraska, Wyoming, Oklahoma, Minnesota, and Iowa. In fact, if we developed the wind energy in North Dakota, Texas, and Kansas to their full potential, we could supply enough electricity to meet the current needs of the entire United States! Wind power projects are under way in these and many other states.

Currently, wind energy is captured and placed into regional electricity grids. Deploying wind energy on a national scale—for example, wind energy produced in Texas and used in New York City—requires the development of new technologies for storing and distributing energy.

Wind produces no waste and is a clean source of energy. It produces no emissions of sulfur dioxide, carbon dioxide, or nitrogen oxides. Every kilowatt-hour of electricity generated by wind power rather than fossil fuels prevents 1 to 2 lb of the greenhouse gas CO_2 from entering the atmosphere.

Although the use of wind power doesn't cause major environmental problems, one concern is reported bird and bat kills. The California Energy Commission estimated that several hundred birds turned up dead in the vicinity of the 7,000 turbines at Altamont Pass in California during a two-year study; most had collided with the turbines. Studies later determined that Altamont Pass is a major bird migration pathway. Wind power sites have implemented technical "fixes" such as painted blades and anti-perching devices, or have shut down operations during peak bird migration periods. Developers of future wind farm sites currently conduct voluntary wildlife studies and try to locate sites away from bird and bat routes.

Not all people welcome wind power projects with open arms. The Maple Ridge Wind Farm in upstate New York has almost 200 windmills. Many local residents appreciate the extra money that wind-farm leases provide to the local economy. However, others think the wind turbines ruin their view of the Adirondack Mountains. A similar dispute is occurring over the proposed Nantucket Sound wind farm off the coast of Massachusetts. This 130-turbine wind farm, if built, will be the first offshore wind project in the United States. Massachusetts has some of the highest electricity costs in the nation, and proponents of the wind farm point out that wind energy will help lower energy costs. However, many people are concerned that the wind farm will adversely affect the local economy because tourists will find it offensive.

A California wind farm Wind energy represents a promising alternative that could diminish our dependence on fossil fuels.

B Wind turbine design This basic wind turbine design has a horizontal axis (horizontal refers to the orientation of the drive shaft). Airflow causes the turbine's blades to turn 15 to 60 revolutions per minute (rpm). As the blades turn, gears within the turbine spin the drive shaft. This spinning powers the generator, which sends electricity through underground cables to a nearby utility. (The tower isn't drawn to scale and is much taller than depicted.)

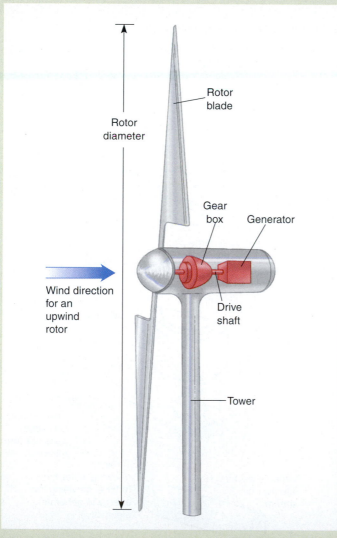

Rotor blade

Rotor diameter

Gear box   Generator

Wind direction for an upwind rotor

Drive shaft

Tower

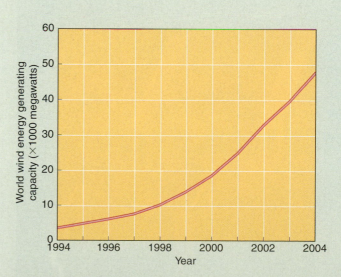

C Trends in wind energy use Global wind energy has shown record growth in recent years.

NATIONAL GEOGRAPHIC

HYDROPOWER

The sun's energy drives the hydrologic cycle, which includes precipitation, evaporation, transpiration, and drainage and runoff (see Figure 5.9). As water flows from higher elevations back to sea level through rivers and streams, dams can harness and make use of its energy. The potential energy of water held back by a dam is converted to kinetic energy as the water turns turbines to generate electricity (**FIGURE 18.9**). **Hydropower**, which is more concentrated than solar energy, is more efficient than any other energy source for producing electricity; about 90 percent of available hydropower energy is converted into consumable electricity.

hydropower A form of renewable energy that relies on flowing or falling water to generate electricity.

Hydropower generates approximately 19 percent of the world's electricity, making it the form of solar energy in greatest use. The 10 countries with the greatest hydroelectric production are, in decreasing order, Canada, the United States, Brazil, China, Russia, Norway, Japan, India, Sweden, and France. In the United States, approximately 2,200 hydropower plants produce between 8 and 12 percent of the country's electricity, the most of any renewable energy source. Highly developed countries have already built dams at most of their potential sites. In many developing nations—particularly in undeveloped, unexploited parts of Africa and South America—hydropower represents a great potential source of electricity.

Hydroelectric power FIGURE 18.9

A Water generates electricity as it moves through the Dalles Dam, which spans the Columbia River between Oregon and Washington.

B A controlled flow of water released down the penstock turns a turbine, which generates electricity.

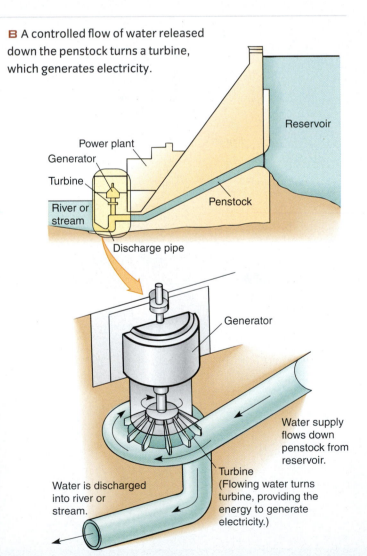

Traditional hydropower technology is only suited for large dams with rapidly flowing water, such as those found on the Columbia River and its tributaries in the Pacific Northwest and British Columbia. Therefore, only about 3 percent of existing U.S. dams generate electricity. New designs allow modern turbines to harness electricity from large, slow-moving rivers or from streams with small flow capacities. These new technologies have the potential to increase the amount of hydroelectric power generated by existing dams.

Building a dam changes the natural flow of a river: Water backs up, flooding large areas of land and forming a reservoir, which destroys plant and animal habitats. Native fishes are particularly susceptible to dams because the original river ecosystem is so altered. The migration of spawning fish is also altered (see Chapter 10). Below the dam, the once-powerful river is reduced to a relative trickle. The natural beauty of the countryside is affected, and certain forms of wilderness recreation are made impossible or less enjoyable, although the dams permit water sports in the reservoir.

With at least 200 large dams around the world, earthquakes may occur during and after the filling of the reservoir behind the dam. The larger the reservoir and the faster it is filled, the greater the intensity of seismic activity. An area need not be seismically active to have earthquakes induced by reservoirs.

If a dam breaks, people and property downstream may be affected. In addition, waterborne diseases may spread through the population. **Schistosomiasis** is a tropical disease caused by a parasitic worm. As much as half the population of Egypt suffers from this disease, largely as a result of the Aswan Dam, built on the Nile River in 1902 to control flooding and used since 1960 to provide electric power. The large reservoir behind the dam provides habitat for the worm, which spends part of its life cycle in the water. Humans are infected by the worm through bathing, swimming, walking barefoot along water banks, and drinking infected water.

In arid regions, the creation of a reservoir results in greater evaporation because it has a larger surface area in contact with the air than the stream or river did. As a result, serious water loss and increased salinity of the remaining water may occur.

When an area behind a dam is flooded, the trees and other vegetation die and are decomposed, releasing the large quantities of carbon that were tied up in organic molecules in the plant bodies as carbon dioxide and methane. These gases absorb infrared radiation and are therefore associated with global climate change.

The construction of large dams involuntarily displaces people from lands flooded by reservoirs. The *Three Gorges Project*, the largest dam ever built, is currently under construction on the Yangtze River in China. This dam, scheduled for completion in 2009 or 2010, will create a reservoir 632 km (412 mi) long. The tops of as many as 100 mountains will become small islands, fragmenting habitat and threatening 57 endangered species. The reservoir will displace almost 2 million people, the largest number for any dam project.

The environmental and social impacts of a dam may not be acceptable to the people living in a particular area. Laws prevent or restrict the building of dams in certain locations. In the United States, the Wild and Scenic Rivers Act prevents the hydroelectric development of certain rivers, although this law protects less than 1 percent of the nation's total river system. Other countries, such as Norway and Sweden, have similar laws.

Dams cost a great deal to build but are relatively inexpensive to operate. A dam has a limited life span, usually 50 to 100 years, because over time the reservoir fills in with silt until it cannot hold enough water to generate electricity. This trapped silt, which is rich in nutrients, is prevented from enriching agricultural lands downstream. For example, Egypt must rely on heavy applications of chemical fertilizer downstream from the Aswan Dam to maintain the fertility of the Nile River Valley.

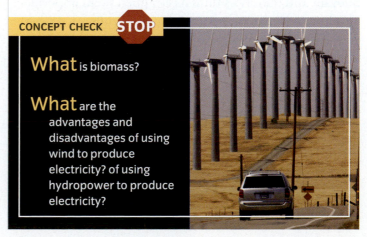

CONCEPT CHECK **STOP**

What is biomass?

What are the advantages and disadvantages of using wind to produce electricity? of using hydropower to produce electricity?

Other Renewable Energy Sources

Geothermal energy and tidal energy are renewable energy sources that are not influenced by solar energy. *Geothermal energy* is the naturally occurring heat within Earth. This heat is used for space heating and to generate electricity. *Tidal energy*, caused by the changes in water level between high and low tides, is exploited to generate electricity on a limited scale.

GEOTHERMAL ENERGY

Geothermal energy, the natural heat within Earth, arises from Earth's core, from friction along continental plate boundaries, and from the decay of radioactive elements. The amount of geothermal energy is enormous. Scientists estimate that just 1 percent of the heat contained in the uppermost 10 km of Earth's crust is equivalent to 500 times the energy contained in all of Earth's oil and natural gas resources.

> **geothermal energy** Energy from Earth's interior, used for space heating or generation of electricity.

Geothermal energy is typically associated with volcanism. Large underground reservoirs of heat exist in areas of geologically recent volcanism. As groundwater in these areas travels downward and is heated, it becomes buoyant and then rises until it is trapped by an impermeable layer in Earth's crust, forming a **hydrothermal reservoir**. Hydrothermal reservoirs contain hot water and possibly steam, depending on the temperature and pressure of the fluid. Some of the hot water or steam may escape to the surface, creating hot springs or geysers. Hot springs have been used for thousands of years for bathing, cooking, and heating buildings. Drilling a well brings the hot fluid from a hydrothermal reservoir to the surface, where a power station may use it to supply heat directly to consumers or to generate electricity (**FIGURE 18.10A**). The electricity these power stations generate is inexpensive and reliable.

The United States is the world's largest producer of geothermal electricity. Electric power is currently produced at 17 different geothermal fields in California, Nevada, Utah, and Hawaii. The world's largest geothermal power plant—The Geysers in northern California—provides electricity for 1.7 million homes. Other important producers of geothermal energy include the Philippines, Italy, Japan, Mexico, Indonesia, and Iceland (**FIGURE 18.10B**).

Iceland, a country with minimal oil and natural gas resources, is located on the mid-Atlantic ridge, a boundary between two continental plates. Iceland is therefore an island of intense volcanic activity with considerable geothermal resources. Iceland uses geothermal energy to generate electricity to heat two-thirds of its homes. In addition, most of the fruits and vegetables required by the people of Iceland are grown in geothermally heated greenhouses.

Is geothermal energy renewable? As a source of heat for geothermal energy, the planet is inexhaustible on a human time scale. However, the water used to transfer the heat to the surface isn't inexhaustible. Some geothermal applications recirculate all the water back into the underground reservoir, ensuring many decades of heat extraction from a given reservoir.

Geothermal energy is considered environmentally benign because it emits only a fraction of the air pollutants released by conventional fossil fuel–based energy technologies. The most common environmental hazard is the emission of hydrogen sulfide (H_2S) gas, which comes from the very low levels of dissolved minerals and salts found in the steam or hot water. Hydrogen sulfide smells like rotten eggs and is toxic to humans in high concentrations. A lesser concern is that the surrounding land may subside, or sink, as the water from hot springs and their connecting underground reservoirs is removed.

Scientists are studying how to economically extract some of the vast amount of geothermal energy stored in hot, dry rock. Such a technology could greatly expand the extent and use of geothermal resources.

Geothermal power plant FIGURE 18.10

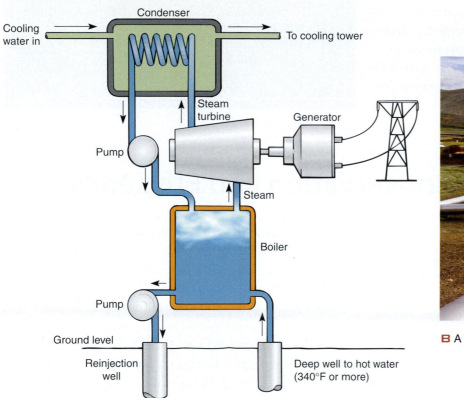

A Steam separated from hot water pumped from underground turns a turbine and generates electricity.

B A geothermal power plant in Iceland.

Heating and cooling buildings with geothermal energy

Geothermal heat pumps (GHPs) take advantage of the difference in temperature between Earth's surface and subsurface (at depths from 1 m to about 100 m). In an underground arrangement of pipes containing circulating fluids, GHPs extract natural heat in winter, when Earth acts as a heat source, and transfer excess heat underground in summer, when Earth acts as a heat sink. Geothermal heating systems can be modified to provide supplemental hot water.

Though GHPs have been available for many years, they aren't widely used because their installation is expensive. However, with the growth of *green architecture* (see the case study at the end of this chapter) and rising fuel costs, commercial and residential use of GHPs is on the rise. The system's benefits include low operating costs—which may be half those of conventional systems—and high efficiency.

TIDAL ENERGY

Tides, or the rise and fall of the surface waters of the ocean and seas that occur twice each day, are the result of the gravitational pull of the moon and the sun. A dam built across a bay can harness the energy of large tides to generate electricity. As the tide falls, water flowing back to the ocean over the dam's spillway turns a turbine and generates electricity through **tidal energy**.

Power plants using tidal power are in operation in France, Russia, China, and Canada. Tidal energy can't become a significant resource worldwide because few areas have large enough differences in water level between high and low tides to make power generation feasible. The most promising locations for tidal power in North America include the Bay of Fundy in Nova Scotia, Passamaquoddy Bay in Maine, Puget Sound in Washington, and Cook Inlet in Alaska.

Other problems associated with tidal energy include the high cost of building a tidal power station and potential environmental problems associated with tidal energy in **estuaries**, coastal areas where river currents meet ocean tides. Fishes and countless invertebrates migrate to estuaries to spawn. Building a dam across the mouth of an estuary would prevent these animals from reaching their breeding habitats.

CONCEPT CHECK **STOP**

What are the pros and cons of using geothermal energy to produce electricity?

What are the pros and cons of using tidal power to produce electricity?

Energy Solutions: Conservation and Efficiency

LEARNING OBJECTIVES

Distinguish between energy conservation and energy efficiency and give examples of each.

Define cogeneration.

Human requirements for energy will continue to increase, if only because the human population is growing. In addition, energy consumption continues to increase as developing countries raise their standards of living. We must therefore place a high priority not only on developing alternative sources of energy but on **energy conservation** and **energy efficiency**.

To illustrate the difference between energy conservation and energy efficiency, let's consider automobile gasoline consumption. Energy conservation measures to reduce gasoline consumption would include carpooling and lowering driving speeds, whereas energy efficiency measures would include designing and manufacturing more fuel-efficient automobiles. Conservation and efficiency accomplish the same goal—saving energy.

Many energy experts consider energy conservation and energy efficiency the most promising energy "sources" available because they save energy for future use and buy us time to explore new energy alternatives. Developing technologies for energy conservation and efficiency costs less than developing new sources or supplies of energy; the technologies also improve the economy's productivity. The adoption of energy-efficient technologies generates new business opportunities, including the research, development, manufacture, and marketing of those technologies.

Energy-efficient technologies and greater efforts at conservation would also provide important environmental benefits by reducing air pollution. CO_2 emissions that contribute to global climate change, acid precipitation, and other environmental problems are related to large quantities of energy production and consumption.

ENERGY CONSUMPTION TRENDS AND ECONOMICS

Although the U.S. economy has become more energy efficient, total energy consumption has increased in recent years, partly due to an increase in population. The per person energy consumption in developing nations is substantially less than that in industrialized countries (see Figure 17.1A), yet the greatest per person increase in energy consumption today is occurring in these developing nations. The rising energy demand in developing nations is caused by increases in economic development and population, as well as the use of older, less expensive, and less energy-efficient technologies.

Developing countries are forced to balance developing their economies with controlling environmental degradation. At

energy conservation

Using less energy—by reducing energy use and waste, for example.

energy efficiency

Using less energy to accomplish a given task—by using new technology, for example.

first glance, these two goals appear mutually exclusive. However, both can be realized by adopting the new technologies now being developed in industrialized nations to achieve greater energy efficiency.

ENERGY-EFFICIENT TECHNOLOGIES

The development of more efficient appliances, automobiles, buildings, and industrial processes has helped reduce energy consumption in highly developed countries. Compact fluorescent light bulbs produce light of comparable quality to that of incandescent light bulbs but require only 25 percent of the energy and last up to 15 times longer. Although relatively expensive, the energy-efficient bulbs more than pay for themselves in energy savings. Standard long-tube fluorescent bulbs have become more efficient. New condensing furnaces require approximately 30 percent less fuel than conventional gas furnaces. "Superinsulated" buildings use 70 to 90 percent less heat energy than buildings insulated using standard methods (**FIGURE 18.11**).

The **National Appliance Energy Conservation Act (NAECA)** sets national appliance efficiency standards for refrigerators, freezers, washing machines, clothes dryers, dishwashers, room air conditioners, and ranges and ovens (including microwaves). For example, refrigerators built today consume 75 percent less energy than comparable models built in the mid-1970s.

Energy costs often account for 30 percent of a company's operating budget. It makes good economic sense for businesses in older buildings to invest in energy improvements, which often pay for themselves in a few years. Energy improvements may be as simple as fine-tuning existing heating, ventilation, and air conditioning systems or as major as replacing all the windows and lights. Either way, both the environment and the company's bottom line can benefit.

Superinsulated buildings FIGURE 18.11

A Some of the characteristics of a superinsulated home, which is so well insulated and airtight it doesn't require a furnace in winter. Heat from the inhabitants, light bulbs, and appliances provides almost all the necessary heat.

B A superinsulated office building in Toronto, Canada, has south-facing windows with insulating glass. The building is so well insulated it uses no furnace.

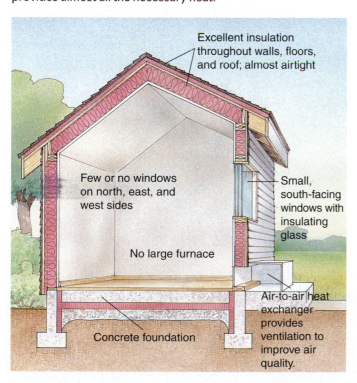

Excellent insulation throughout walls, floors, and roof; almost airtight

Few or no windows on north, east, and west sides

Small, south-facing windows with insulating glass

No large furnace

Air-to-air heat exchanger provides ventilation to improve air quality.

Concrete foundation

The Daihatsu prototype UFE (ultra fuel economy) III hybrid can achieve 169 miles per gallon.

Automobile efficiency has improved dramatically since the mid-1970s as a result of the use of lighter materials and designs that reduce air drag. The average fuel efficiency of new passenger cars doubled between the mid-1970s and the mid-1980s. It declined after that, as larger vehicles became more popular, but high gasoline prices are reversing the trend: Sales of compact and subcompact cars increased in 2006, and sales of trucks and SUVs dropped. Further, significant gains could easily be made using current technology. Automobiles with fuel efficiencies of 60 to 65 mpg could be routinely manufactured within the next decade or so, and manufacturers are developing even more efficient models for the future (FIGURE 18.12).

Cogeneration One energy technology that has a bright future is **cogeneration**, or **combined heat and power (CHP)**. Cogeneration involves the production of two useful forms of energy from the same fuel. A CHP system generates electricity, and then the steam produced during this process is used rather than wasted. The system's overall conversion efficiency (that is, the ratio of useful energy produced to fuel energy used) is high because some of what is usually waste heat is incorporated into the process.

Cogeneration can be cost-effective on both small and large scales. Modular CHP systems enable hospitals, hotels, restaurants, factories, and other businesses to harness steam that would otherwise be wasted to heat buildings, cook food, or operate machinery before it cools and gets pumped back into the boiler as water (FIGURE 18.13). Larger CHP systems can produce electricity for local utilities.

> **cogeneration (CHP)** An energy technology that involves recycling "waste" heat.

In this example of a cogeneration system, fuel combustion generates electricity in a generator. The electricity produced is used in-house or sold to a local utility. The waste heat (leftover hot gases or steam) is recovered for useful purposes, such as industrial processes, heating of buildings, hot water heating, and generation of additional electricity.

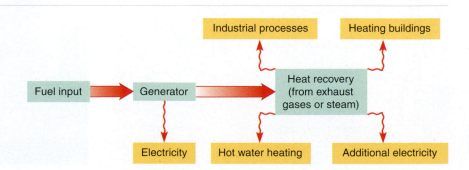

Environmentally aware homeowners have often paid a large economic price when installing renewable energy sources such as solar panels or wind-driven generators. Such systems can be expensive to purchase and install, and they provide electricity only intermittently, when weather conditions permit. During unfavorable conditions, residents must purchase electricity from their local utility.

In a growing trend, some utility companies are permitting homeowners who produce their own energy to "net meter" their electricity. Any excess energy homeowners generate is supplied to the utility's power grid, and the homeowners' electric meters run backward. Net metering offsets energy costs over a billing period by essentially providing individuals the full retail price for the electricity they generate. By itself, net metering doesn't turn residential energy generation into an inexpensive prospect, but it certainly makes it more affordable, encouraging the additional development of renewable energy for homes. Allowed to "bank" their energy and withdraw it later, consumers save money while contributing to a cleaner environment.

Electric meters can run backward for homeowners participating in net metering.

ELECTRIC POWER COMPANIES AND ENERGY EFFICIENCY

Changes in the regulations that govern electric utilities allow these companies to make more money by generating less electricity. Such programs provide incentives for energy conservation and thereby reduce power plant emissions that contribute to environmental problems.

Electric utilities can often avoid the massive expenses of building new power plants or purchasing additional power by helping electricity consumers save energy. Some utilities support energy conservation and efficiency by offering cash awards to consumers who install energy-efficient technologies. Other utilities give customers energy-efficient compact fluorescent light bulbs, air conditioners, or other appliances. They then charge slightly higher rates or a small leasing fee, but the greater efficiency results in savings for both the utility company and the consumer.

The utility company makes more money from selling less electricity because it does not have to invest in additional power generation to meet increased demand. The consumer saves because the efficient light bulbs or appliances use less energy, which more than offsets the higher rates.

According to the American Council for an Energy-Efficient Economy, U.S. electric power plants are themselves an important target for improved energy efficiency. Much heat is lost during the generation of electricity. If all that wasted energy were harnessed—by cogeneration, for example—it could be used productively, thereby conserving energy.

Another way to increase energy efficiency would be to improve our electric grids because about 10 percent of electricity is lost during transmission. To accomplish this, some energy experts envision that future electricity will be generated far from population centers, converted to supercooled hydrogen, and transported through underground superconducting pipelines. The technology to build such conduits has not yet been developed.

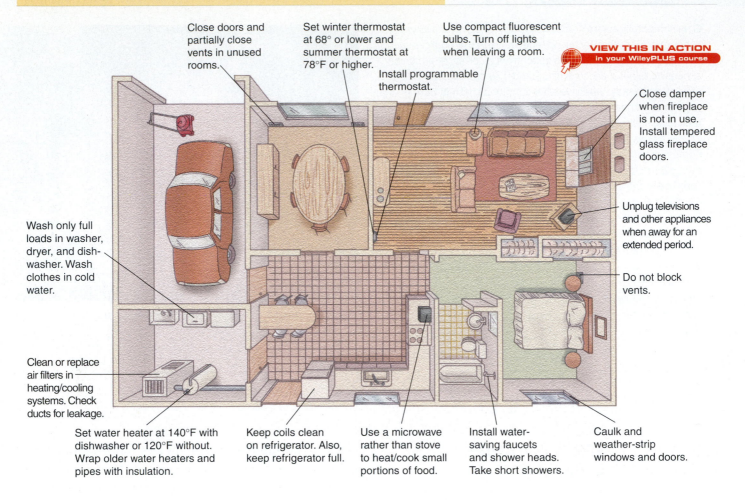

Close doors and partially close vents in unused rooms.

Set winter thermostat at 68° or lower and summer thermostat at 78°F or higher. Install programmable thermostat.

Use compact fluorescent bulbs. Turn off lights when leaving a room.

VIEW THIS IN ACTION in your WileyPLUS course

Close damper when fireplace is not in use. Install tempered glass fireplace doors.

Wash only full loads in washer, dryer, and dish-washer. Wash clothes in cold water.

Unplug televisions and other appliances when away for an extended period.

Do not block vents.

Clean or replace air filters in heating/cooling systems. Check ducts for leakage.

Set water heater at 140°F with dishwasher or 120°F without. Wrap older water heaters and pipes with insulation.

Keep coils clean on refrigerator. Also, keep refrigerator full.

Use a microwave rather than stove to heat/cook small portions of food.

Install water-saving faucets and shower heads. Take short showers.

Caulk and weather-strip windows and doors.

ENERGY CONSERVATION AT HOME

The average U.S. household spends several thousand dollars each year on utility bills. This cost could be reduced considerably with investment in energy-efficient technologies (FIGURE 18.14). Although a more energy-efficient house might cost more up front, depending on the technologies employed, the improvements usually pay for themselves in two or three years. Energy efficiency has become an essential element of design codes nationwide.

Some energy-saving improvements, such as thicker wall insulation, are easiest to install while a home is being built. Other improvements can be made to older homes to reduce heating and cooling costs and enhance energy efficiency. Examples include installing thicker attic insulation, installing storm windows and doors, caulking cracks around windows and doors, replacing inefficient furnaces and refrigerators, and adding heat pumps.

News articles often provide these and other suggestions. Most local utility companies will also perform an energy audit on a home for little or no charge.

CONCEPT CHECK STOP

What is the difference between energy conservation and energy efficiency?

What is cogeneration?

GREEN ARCHITECTURE

A recent addition to the Midtown Manhattan skyline represents a new standard in green architecture in New York City (**FIGURE A**). Completed in 2006, the Hearst Tower, home of Hearst Publishing, was designed to achieve a degree of energy efficiency 26 percent higher than that of standard office buildings. The building's innovations earned it the city's first Gold LEED (Leadership in Energy and Environmental Design) certification from the U.S. Green Building Council (USGBC) (**FIGURE B**).

The Hearst Tower's striking "diagrid" frame, which incorporates a system of steel and glass triangles, floods the interior with natural light while using approximately 2,000 tons less steel than a conventional frame. Most of the steel (90 percent) used in the 46-story, 856,000-square-foot building is recycled.

Other energy-saving designs within the Hearst Tower boost the efficiency of the building's heating and cooling systems. Windows are coated to reduce solar radiation, and heating and air conditioning equipment cools and ventilates using only outside air for three-quarters of the year. In addition, embedded polyethylene within the atrium's limestone floor circulates water for both cooling and heating. Unique to the building's design, a 10-story waterfall called the "Icefall" chills the atrium. The Icefall's water supply comes from rainwater collected at the roof and drawn into a 53,000-liter (14,000-gallon) basement reclamation tank. This water is also used to irrigate the building's plants.

Natural light is enhanced by the placement of few internal walls and only low partitions. Sensors turn off lights in empty rooms and control the amount of artificial light provided based on the natural light being received.

Finally, building health is protected by several interior features. Low-vapor paints and low-toxicity sealants coat surfaces, while furniture and carpeting make use of low-toxicity recycled content and materials obtained from sustainable forests.

A The Hearst Tower in Manhattan offers important contributions to green architecture.

B The Hearst Tower's Gold LEED certification is the first for a New York City high-rise.

SUMMARY

1 Direct Solar Energy

1. **Active solar heating** is a system of putting the sun's energy to use in which a series of collectors absorb the solar energy, and pumps or fans distribute the collected heat. Active solar heating is used primarily for heating water. **Passive solar heating** is a system of putting the sun's energy to use that does not require mechanical devices to distribute the collected heat in order to heat buildings.

2. A **photovoltaic (PV) solar cell** is a wafer or thin film of solid-state materials, such as silicon or gallium arsenide, that is treated with certain metals in such a way that it generates electricity when solar energy is absorbed. PVs generate electricity with no pollution and minimal maintenance but are limited by their low efficiency and by the amount of land needed for their large-scale use. **Solar thermal electric generation** is a means of producing electricity in which the sun's energy is concentrated by mirrors or lenses onto a fluid-filled pipe; the heated fluid is used to generate electricity. Solar thermal energy systems are efficient and provide significant environmental benefits; but they are only now becoming more cost-competitive with fossil fuels.

3. A **fuel cell** is a device that directly converts chemical energy into electricity; a fuel cell requires hydrogen fuel and oxygen (from the air).

2 Indirect Solar Energy

1. **Biomass** is plant and animal material used as fuel. Biomass fuel—materials such as wood, fast-growing plant and algal crops, crop wastes, sawdust and wood chips, and animal wastes—is burned to release energy.

2. **Wind energy** is electric energy obtained from surface air currents caused by the solar warming of air. Restricted primarily to rural areas that receive fairly continual winds, wind power is a clean and cost-effective source of energy, but wind turbines can kill birds and bats. **Hydropower** is a form of renewable energy that relies on flowing or falling water to generate electricity. Hydropower is highly efficient, but the dams built in traditional hydropower projects can greatly disrupt the natural environment and displace local residents.

3 Other Renewable Energy Sources

1. **Geothermal energy** is the use of energy from Earth's interior for either space heating or generation of electricity. **Tidal energy** is a form of renewable energy that relies on the ebb and flow of the tides to generate electricity.

4 Energy Solutions: Conservation and Efficiency

1. **Energy conservation** is using less energy—for example, by reducing energy use and waste. **Energy efficiency** is using less energy to accomplish a given task—for example, with new technology. Examples of energy conservation measures that reduce gasoline consumption include carpooling and lowering driving speeds; energy efficiency measures include designing and manufacturing more fuel-efficient automobiles.

2. **Cogeneration** is an energy technology that involves recycling "waste" heat. It produces two useful forms of energy from the same fuel: It generates electricity and captures and uses the steam released.

KEY TERMS

CRITICAL AND CREATIVE THINKING QUESTIONS

1. Do you think that a vehicle fuel efficiency of 60 miles per gallon could be reasonably achieved at present? Explain your answer.

2. Japan wants to make use of solar power, but it does not have extensive tracts of land for building large solar power plants. Which solar technology do you think is best suited to Japan's needs? Why?

3. Explain the following statement: Unlike fossil fuels, solar energy is not resource limited but is technology limited.

4. One advantage of the various forms of renewable energy, such as solar thermal and wind energy, is that they cause no net increase in atmospheric carbon dioxide. Is this true for biomass? Why or why not?

5. Give an example of how one or more of the alternative energy sources discussed in this chapter could have a negative effect on each of these aspects of ecosystems: soil preservation, natural water flow, foods used by wild plant and animal populations, and preservation of the diversity of organisms found in an area.

6. Explain how energy conservation and efficiency are major "sources" of energy.

7. Evaluate which forms of energy, other than fossil fuels and nuclear power, have the greatest potential where you live.

8. List energy conservation measures you could adopt in each of the following aspects of your life: washing laundry, lighting, bathing, cooking, buying a car, and driving a car.

The map below shows the average daily total of solar energy (on an annual basis) received on a solar collector that tilts to compensate for latitude.

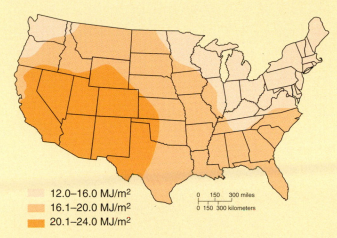

12.0–16.0 MJ/m^2
16.1–20.0 MJ/m^2
20.1–24.0 MJ/m^2

0   150   300 miles
0   150   300 kilometers

9. Based on the map, which state would you rather live in, Maine or Montana, if you wanted to heat your house using solar energy?

10. Which region of the United States is the best area for year-round solar energy collection?

What is happening in this picture ?

These hoses suck methane, which is used as fuel, from decomposing trash in California.

What type of renewable energy is being provided?

What would be some advantages and disadvantages of this type of energy production?

GLOSSARY

acid deposition   A type of air pollution that includes sulfuric and nitric acids in precipitation as well as dry acid particles that settle out of the air.

acid mine drainage   Pollution caused when sulfuric acid and dangerous dissolved materials, such as lead, arsenic, and cadmium, wash from coal and metal mines into nearby lakes and streams.

active solar heating   A system of putting the sun's energy to use in which a series of collectors absorb the solar energy, and pumps or fans distribute the collected heat.

acute toxicity   Adverse effects that occur within a short period after high-level exposure to a toxicant.

age structure   The number and proportion of people at each age in a population.

air pollution   Various chemicals (gases, liquids, or solids) present in the atmosphere in high enough levels to harm humans, other organisms, or materials.

aquaculture   The growing of aquatic organisms (fishes, shellfish, and seaweeds) for human consumption.

aquifer depletion   The removal of groundwater faster than it can be recharged by precipitation or melting snow.

artificial eutrophication   Overnourishment of an aquatic ecosystem by nutrients such as nitrates and phosphates, due to human activities such as agriculture and discharge from sewage treatment plants.

atmosphere   The gaseous envelope surrounding Earth.

benthic environment   The ocean floor, which extends from the intertidal zone to the deep ocean trenches.

biocentric preservationist   A person who believes in protecting nature from human interference because all forms of life deserve respect and consideration.

biochemical oxygen demand (BOD)   The amount of oxygen that microorganisms need to decompose biological wastes into carbon dioxide, water, and minerals.

biodiversity hotspots   Relatively small areas of land that contain an exceptional number of endemic species and are at high risk from human activities.

biological control   A method of pest control that involves the use of naturally occurring disease organisms, parasites, or predators to control pests.

biological diversity   The number and variety of Earth's organisms.

biological magnification   The increase in toxicant concentrations as the toxicant passes through successive levels of the food chain.

biomass   Plant and animal material used as fuel.

biome   A large, relatively distinct terrestrial region with similar climate, soil, plants, and animals regardless of where it occurs in the world.

biosphere   The layer of Earth containing all living organisms.

biotic potential   The maximum rate at which a population could increase under ideal conditions.

boreal forest   A region of coniferous forest (such as pine, spruce, and fir) in the Northern Hemisphere; located just south of the tundra.

broad-spectrum pesticide   A pesticide that kills a variety of organisms, including beneficial organisms, in addition to the target pest.

bycatch   The fishes, marine mammals, sea turtles, seabirds, and other animals caught unintentionally in a commercial fishing catch.

carbon management   Ways to separate and capture the CO_2 produced during the combustion of fossil fuels and then sequester (store) it.

carcinogen   Any substance (chemical, radiation, or virus) that causes cancer.

carrying capacity   The maximum population that can be sustained by a given environment or by the world as a whole.

chaparral   A biome with mild, moist winters and hot, dry summers; vegetation is typically small-leafed evergreen shrubs and small trees.

chlorofluorocarbons (CFCs)   Human-made organic compounds containing chlorine and fluorine; now banned because they attack the stratospheric ozone layer.

chronic toxicity   Adverse effects that occur after a long period of low-level exposure to a toxicant.

clear-cutting   A logging practice in which all the trees in a stand of forest are cut, leaving just the stumps.

climate   The average weather conditions that occur in a place over a period of years.

cogeneration (CHP)   An energy technology that involves recycling "waste" heat.

command and control regulation   Pollution control laws that work by setting limits on levels of pollution.

community   A natural association that consists of all the populations of different species that live and interact together within an area at the same time.

compact development   The design of cities in which tall, multiple-unit residential buildings are close to shopping and jobs, and all are connected by public transportation.

competition   The interaction among organisms that vie for the same resources in an ecosystem (such as food or living space).

conservation biology   The scientific study of how humans impact organisms and of the development of ways to protect biological diversity.

conservation easement   A legal agreement that protects privately owned forest, rangeland, or other property from development for a specified number of years.

conservation tillage   A method of cultivation in which residues from previous crops are left in the soil, partially covering it and helping to hold it in place until the newly planted seeds are established.

consumption overpopulation   A situation in which each individual in a population consumes too large a share of resources.

contour plowing   Plowing that matches the natural contour of the land.

Coriolis effect The tendency of moving air or water to be deflected from its path and swerve to the right in the Northern Hemisphere and to the left in the Southern Hemisphere.

cost-benefit diagram A diagram that helps policymakers make decisions about costs of a particular action and benefits that would occur if that action were implemented.

crop rotation The planting of a series of different crops in the same field over a period of years.

decommissioning The dismantling of a nuclear power plant after it closes.

deep ecology worldview A worldview based on harmony with nature, a spiritual respect for life, and the belief that humans and all other species have an equal worth.

deforestation The temporary or permanent clearance of large expanses of forest for agriculture or other uses.

degradation (of land) Natural or human-induced processes that decrease the future ability of the land to support crops or livestock.

demographic transition The process whereby a country moves from relatively high birth and death rates to relatively low birth and death rates.

demographics The applied branch of sociology that deals with population statistics.

desert A biome in which the lack of precipitation limits plant growth; deserts are found in both temperate and tropical regions.

desertification Degradation of once-fertile rangeland or tropical dry forest into nonproductive desert.

dose–response curve In toxicology, a graph that shows the effect of different doses on a population of test organisms.

dust dome A dome of heated air that surrounds an urban area and contains a lot of air pollution.

ecological niche The totality of an organism's adaptations, its use of resources, and the lifestyle to which it is fitted.

ecological succession The process of community development over time, which involves species in one stage being replaced by different species.

ecology The study of the interactions among organisms and between organisms and their abiotic environment.

economic development An expansion in a government's economy, viewed by many as the best way to raise the standard of living.

ecosystem A community and its physical environment.

ecosystem services Important environmental benefits, such as clean air to breathe, clean water to drink, and fertile soil in which to grow crops, that the natural environment provides.

El Niño–Southern Oscillation (ENSO) A periodic, large-scale warming of surface waters of the tropical eastern Pacific Ocean that temporarily alters both ocean and atmospheric circulation patterns.

endangered species A species that faces threats that may cause it to become extinct within a short period.

endemic species Organisms that are native to or confined to a particular region.

energy conservation Using less energy—by reducing energy use and waste, for example.

energy efficiency Using less energy to accomplish a given task—by using new technology, for example.

energy flow The passage of energy in a one-way direction through an ecosystem.

enhanced greenhouse effect The additional warming produced by increased levels of CO_2 and other gases that absorb infrared radiation.

enrichment The process by which uranium ore is refined after mining to increase the concentration of fissionable U-235.

environmental ethics A field of applied ethics that considers the moral basis of environmental responsibility.

environmental justice The right of every citizen to adequate protection from environmental hazards.

environmental science The interdisciplinary study of humanity's relationship with other organisms and the physical environment.

environmental sustainability The ability to meet humanity's current needs without compromising the ability of future generations to meet their needs.

environmental worldview A worldview based on how the environment works, our place in the environment, and right and wrong environmental behaviors.

estuary A coastal body of water, partly surrounded by land, with access to the open ocean and a large supply of fresh water from a river.

evolution The cumulative genetic changes in populations that occur during successive generations.

exponential population growth The accelerating population growth that occurs when optimal conditions allow a constant reproductive rate.

external cost A harmful environmental or social cost that is borne by people not directly involved in buying or selling a product.

extinction The elimination of a species from Earth.

first law of thermodynamics Energy cannot be created or destroyed, although it can change from one form to another.

fission The splitting of an atomic nucleus into two smaller fragments, accompanied by the release of a large amount of energy.

flowing-water ecosystem A freshwater ecosystem such as a river or stream in which the water flows in a current.

fluidized-bed combustion A clean-coal technology in which crushed coal is mixed with limestone to neutralize acidic compounds produced during combustion.

food insecurity The condition in which people live with chronic hunger and malnutrition.

forest decline A gradual deterioration and eventual death of many trees in a forest.

freshwater wetlands Lands that shallow fresh water covers for at least part of the year; wetlands have a characteristic soil and water-tolerant vegetation.

fuel cell A device that directly converts chemical energy into electricity; the fuel cell requires hydrogen and oxygen from the air.

full cost accounting The process of evaluating and presenting to decision makers the relative benefits and costs of various alternatives.

genetic engineering The manipulation of genes (for example, taking a specific gene from one species and placing it into an unrelated species) to produce a particular trait.

genetic resistance Any inherited characteristic that decreases the effect of a given agent (like a pesticide) on an organism (like a pest).

geothermal energy The use of energy from Earth's interior for either space heating or generation of electricity.

germplasm Any plant or animal material that may be used in breeding.

green chemistry A subdiscipline of chemistry in which commercially important chemical processes are redesigned to significantly reduce environmental harm.

greenhouse gases The gases that absorb infrared radiation; include water vapor, carbon dioxide, methane, and certain other gases.

groundwater The supply of fresh water under Earth's surface that is stored in underground aquifers.

growth rate (r) The rate of change (increase or decrease) of a population's size, expressed in percent per year.

gyres Large, circular ocean current systems that often encompass an entire ocean basin.

habitat fragmentation The breakup of large areas of habitat into small, isolated patches.

hazardous waste Any discarded chemical that threatens human health or the environment.

high-level radioactive wastes Radioactive solids, liquids, or gases that initially give off large amounts of ionizing radiation.

highly developed countries Countries with complex industrialized bases, low rates of population growth, and high per person incomes.

hydropower A form of renewable energy that relies on flowing or falling water to generate electricity.

incentive-based regulation Pollution control laws that work by establishing emission targets and providing industries with incentives to reduce emissions.

industrialized agriculture Modern agricultural methods, which require a large capital input and less land and labor than traditional methods.

infant mortality rate The number of infant deaths under age one per 1000 live births.

infrared radiation Electromagnetic radiation with wavelengths longer than visible light but shorter than microwaves; perceived as invisible waves of heat energy.

integrated pest management (IPM) A combination of pest control methods that, if used in the proper order and at the proper times, keep the size of a pest population low enough to prevent substantial economic loss.

integrated waste management A combination of the best waste management techniques into a consolidated program to deal effectively with solid waste.

intertidal zone The area of shoreline between low and high tides.

invasive species Foreign species that spread rapidly in a new area if free of predators, parasites, or resource limitations that may have controlled their population in their native habitat.

landscape A region that includes several interacting ecosystems.

less developed countries Countries with a low level of industrialization, a very high fertility rate, a very high infant mortality rate, and a very low per person income relative to highly developed countries.

low-level radioactive wastes Solids, liquids, or gases that give off small amounts of ionizing radiation.

marginal cost of pollution abatement The added cost of reducing one unit of a given type of pollution.

marginal cost of pollution The added cost of an additional unit of pollution.

mass burn incinerator A large furnace that burns all solid waste except for unburnable items such as refrigerators.

microirrigation A type of irrigation that conserves water by piping it to crops through sealed systems.

minerals Elements or compounds of elements that occur naturally in Earth's crust.

moderately developed countries Countries with a medium level of industrialization and per person incomes lower than those of highly developed countries.

monoculture Ecological simplification in which only one type of plant is cultivated over a large area.

municipal solid waste Solid materials discarded by homes, offices, stores, restaurants, schools, hospitals, prisons, libraries, and other commercial and institutional facilities.

national income accounts A measure of the total income of a nation's goods and services for a given year.

natural capital Earth's resources and processes that sustain living organisms, including humans; includes minerals, forests, soils, groundwater, clean air, wildlife, and fisheries.

natural selection The tendency of better-adapted individuals—those with a combination of genetic traits better suited to environmental conditions—to survive and reproduce, increasing their proportion in the population.

neritic province The part of the pelagic environment that overlies the ocean floor from the shoreline to a depth of 200 m (650 ft).

nonmunicipal solid waste Solid waste generated by industry, agriculture, and mining.

nonpoint source pollution Pollutants that enter bodies of water over large areas rather than being concentrated at a single point of entry.

nonrenewable resources Natural resources that are present in limited supplies and are depleted as they are used.

nuclear energy The energy released by nuclear fission or fusion.

nuclear reactor A device that initiates and maintains a controlled nuclear fission chain reaction to produce energy for electricity.

nutrient cycling The pathway of various nutrient minerals or elements from the environment through organisms and back to the environment.

oceanic province The part of the pelagic environment that overlies the ocean floor at depths greater than 200 m (650 ft).

optimum amount of pollution The amount of pollution that is economically most desirable.

overburden Soil and rock overlying a useful mineral deposit.

overgrazing When too many grazing animals consume the plants in a particular area, leaving the vegetation destroyed and unable to recover.

overnutrition A type of malnutrition in which there is an overconsumption of calories that leaves the body susceptible to disease.

ozone thinning The removal of ozone from the stratosphere by human-produced chemicals or natural processes.

passive solar heating A system of putting the sun's energy to use without requiring mechanical devices to distribute the collected heat.

pathogen An agent (usually a microorganism) that causes disease.

people overpopulation A situation in which there are too many people in a given geographic area.

persistent organic pollutants (POPs) Persistent toxicants that bioaccumulate in organisms and travel through air and water to contaminate sites far from their source.

pesticide Any toxic chemical used to kill pests.

pheromone A natural substance produced by animals to stimulate a response in other members of the same species.

photochemical smog A brownish orange haze formed by chemical reactions involving sunlight, nitrogen oxides, and hydrocarbons.

photovoltaic solar cell (PV) A wafer or thin film of solid state materials, such as silicon or gallium arsenide, that are treated with certain metals in such a way that the film generates electricity when solar energy is absorbed.

plate tectonics The study of the processes by which the lithospheric plates move over the asthenosphere.

point source pollution Water pollution that can be traced to a specific spot.

population A group of organisms of the same species that live together in the same area at the same time.

population ecology The branch of biology that deals with the number of individuals of a particular species found in an area and why those numbers increase or decrease over time.

poverty A condition in which people are unable to meet their basic needs for food, clothing, shelter, education, or health.

precautionary principle Making decisions about adopting a new technology or chemical product by assigning the burden of proof of its safety to its developers.

predation The consumption of one species (the prey) by another (the predator).

primary air pollutants Harmful chemicals that enter directly into the atmosphere from either human activities or natural processes.

primary treatment Treating wastewater by removing suspended and floating particles by mechanical processes.

rangeland Land that isn't intensively managed and is used for grazing livestock.

renewable resources Resources that are replaced by natural processes and that can be used forever, provided they are not overexploited in the short term.

replacement-level fertility The number of children a couple must produce to "replace" themselves.

restoration ecology The study of the historical condition of a human-damaged ecosystem, with the goal of returning it as close as possible to its former state.

risk The probability of harm (such as injury, disease, death, or environmental damage) occurring under certain circumstances.

risk assessment The use of statistical methods to quantify risks so they can be compared and contrasted.

runoff The movement of fresh water from precipitation and snowmelt to rivers, lakes, wetlands, and the ocean.

salinization The gradual accumulation of salt in a soil, often as a result of improper irrigation methods.

saltwater intrusion The movement of seawater into a freshwater aquifer near the coast.

sanitary landfill The most common method of disposal of solid waste, by compacting it and burying it under a shallow layer of soil.

savanna A tropical grassland with widely scattered trees or clumps of trees.

scientific method The way a scientist approaches a problem, by formulating a hypothesis and then testing it.

second law of thermodynamics When energy is converted from one form to another, some of it is degraded into heat, a less usable form that disperses into the environment.

secondary air pollutants Harmful chemicals that form in the atmosphere when primary air pollutants react chemically with each other or with natural components of the atmosphere.

secondary treatment Treating wastewater biologically to decompose suspended organic material; secondary treatment reduces the water's biochemical oxygen demand.

sewage Wastewater from drains or sewers (from toilets, washing machines, and showers); includes human wastes, soaps, and detergents.

shelterbelt A row of trees planted as a windbreak to reduce soil erosion of agricultural land.

sick building syndrome Eye irritations, nausea, headaches, respiratory infections, depression, and fatigue caused by indoor air pollution.

smelting The process in which ore is melted at high temperatures to separate impurities from the molten metal.

soil The uppermost layer of Earth's crust, which supports terrestrial plants, animals, and microorganisms.

soil erosion The wearing away or removal of soil from the land.

soil horizons The horizontal layers into which many soils are organized, from the surface to the underlying parent material.

solar thermal electric generation A means of producing electricity in which the sun's energy is concentrated by mirrors or lenses onto a fluid-filled pipe; the heated fluid is used to generate electricity.

source reduction An aspect of waste management in which products are designed and manufactured in ways that decrease the amount of solid and hazardous waste in the solid waste stream.

species richness The number of different species in a community.

spent fuel Used fuel elements that were irradiated in a nuclear reactor.

spoil bank A hill of loose rock created when the overburden from a new trench is put into the already excavated trench during strip mining.

standing-water ecosystem A body of fresh water surrounded by land and whose water does not flow; a lake or a pond.

subsistence agriculture Traditional agricultural methods, which are dependent on labor and a large amount of land to produce enough food to feed oneself and one's family.

subsurface mining The extraction of mineral and energy resources from deep underground deposits.

surface mining The extraction of mineral and energy resources near Earth's surface by first removing the soil, subsoil, and overlying rock strata.

surface water Precipitation that remains on the surface of the land and does not seep down through the soil.

sustainable agriculture Agricultural methods that maintain soil productivity and a healthy ecological balance while having minimal long-term impacts.

sustainable consumption The use of goods and services that satisfy basic human needs and improve the quality of life but that also minimize resource use.

sustainable development Economic growth that meets the needs of the present without compromising the ability of future generations to meet their needs.

sustainable forestry The use and management of forest ecosystems in an environmentally balanced and enduring way.

sustainable soil use The wise use of soil resources, without a reduction in the amount or fertility of soil, so it is productive for future generations.

sustainable water use The wise use of water resources, without harming the essential functioning of the hydrologic cycle or the ecosystems on which present and future humans depend.

symbiosis Any intimate relationship or association between members of two or more species; includes mutualism, commensalism, and parasitism.

temperate deciduous forest A forest biome that occurs in temperate areas where annual precipitation ranges from about 75 cm to 126 cm (30 to 50 in).

temperate grassland A grassland with hot summers, cold winters, and less rainfall than is found in the temperate deciduous forest biome.

temperate rain forest A coniferous biome with cool weather, dense fog, and high precipitation.

temperature inversion A layer of cold air temporarily trapped near the ground by a warmer, upper layer.

tertiary treatment Advanced wastewater treatment methods that are sometimes employed after primary and secondary treatments.

threatened species A species whose population has declined to the point that it may be at risk of extinction.

total fertility rate (TFR) The average number of children born to each woman.

toxicology The study of toxicants, chemicals with adverse effects on health.

tropical rain forest A lush, species-rich forest biome that occurs where the climate is warm and moist throughout the year.

tundra The treeless biome in the far north that consists of boggy plains covered by lichens and small plants such as mosses; it has harsh, very cold winters and extremely short summers.

ultraviolet (UV) radiation That part of the electromagnetic spectrum with wavelengths just shorter than visible light; can be lethal to organisms at high levels of exposure.

undernutrition A type of malnutrition in which there is an underconsumption of calories or nutrients that leaves the body weakened and susceptible to disease.

urban heat island Local heat buildup in an area of high population.

urbanization The process whereby people move from rural areas to densely populated cities.

utilitarian conservationist A person who values natural resources because of their usefulness to us but uses them sensibly and carefully.

water pollution Any physical or chemical change in water that adversely affects the health of humans and other organisms.

Western worldview A worldview based on human superiority and dominance over nature, the unrestricted use of natural resources, and increased economic growth to manage an expanding industrial base.

wilderness A protected area of land in which no human development is permitted.

wildlife corridor A protected zone that connects isolated unlogged or undeveloped areas.

wind energy Electric energy obtained from surface air currents caused by the solar warming of air.

zero population growth The state in which population remains the same size because the birth rate equals the death rate.

Chapter 1

Figure 1–2: Data from Population Reference Bureau; **Graph in Creative and Critical Thinking Questions:** Data from *Science* Vol. 315 (February 16, 2007), page 913.

Chapter 2

First section on sustainable development: *Our Common Future.* World Commission on Environment and Development. Cary, NC: Oxford University Press (1987), p. 22; **Eight principles of deep ecology:** Naess, A. and D. Rothenberg. *Ecology, Community and Lifestyle.* Cambridge, UK: Cambridge University Press (2001). Reprinted with permission of Cambridge University Press; **Five recommendations in section on overall plan for sustainable living:** Brown, L. *Plan B 2.0: Rescuing a Planet under Stress and a Civilization in Trouble.* New York: W.W. Norton (2006); **Graphs in Creative and Critical Thinking Questions:** Data from U.S. National Climate Assessment.

Chapter 3

Page 54, excerpt from the "Wilderness Essay": a letter written by Wallace Stegner to David Pesonen of the U. of California's Wildland Research Center; **fourth section, beginning on p. 62, Environmental Economics:** Main source: Levin, J. "The Economy and the Environment: Revising the National Accounts." *IMF Survey* (June 4, 1990).

Chapter 4

Chapter introduction inset, Guilette, E.A., M.M. Meza, M.G. Aguilar, A.D. Soto, and I.E. Garcia. "An Anthropological Approach to the Evaluation of Preschool Children Exposed to Pesticides in Mexico." *Environmental Health Perspectives* (May 1998); **Figure 4.1:** Data from National Safety Council; art design from Health feature in *National Geographic* (August 2006); **Figure 4.2:** Adapted from *Science and Judgment in Risk Assessment.* Washington, D.C.: National Academy Press (1994). Data in graph from "CDC's Second National Report on Human Exposure to Environmental Chemicals" (2003); **Table 4.2:** Adapted from U.S. Centers for Disease Control and Prevention, National Center for Chronic Disease Prevention and Health Promotion, Division of Adolescent and School Health, Youth Risk Behavior Survey; **Figure 4.8B:** Data from Grier, J.W. "Ban of DDT and Subsequent Recovery of Reproduction in Bald Eagles" Copyright 1982, American Association for the Advancement of Science; **Figure 4.9B:** Data from Woodwell, G.M., C.F. Worster, Jr. and P.A. Isaacson. "DDT Residues in an East Coast Estuary: A Case of Biological Concentration of a Persistent Insecticide. *Science* Vol 156 (May 12, 1967); **Table 4.4:** Josten, M.D. and J.L. Wood. *World of Chemistry*, 2nd edition. Philadelphia: Saunders College Publishing, (1996).

Chapter 5

Figures 5.8, 5.9, 5.10, 5.12, and 5.13: Values are from Schlesinger, W.H. *Biogeochemistry: An Analysis of Global Change*, 2nd edition. Academic Press, San Diego (1997) and based on several sources; **What a Scientist Sees: Resource Partitioning:** Adapted from MacArthur, R.H. "Population Ecology of Some Warblers of Northeastern Coniferous Forests." *Ecology*, Vol. 39 (1958).

Chapter 6

Figure 6.1: Based on data from the World Wildlife Fund; Figure 6.2: Based on Holdridge, L. *Life Zone Ecology.* Tropical Science Center, San Jose, Costa Rica (1967); **Data for all climate graphs (in Figures 6.4–6.12) from** www.worldclimate.com; **Figure 6.17:** Adapted from Figure 14.11 on p. 428 in B.W. Murck, B.J. Skinner, and D. Mackenzie. *Visualizing Geology*, Hoboken, NJ: John Wiley & Sons, Inc. (2008); **Figures 6.18B and 6.18C: Adapted** from Figure 15.13 on p. 244 and Figure 15.19 on p. 247, respectively, in S.A. Alters and B. Alters *Biology: Understanding Life*, Hoboken, NJ: John Wiley & Sons, Inc. (2006).

Chapter 7

Figure 7.5B: After V.C. Scheffer. "The Rise and Fall of a Reindeer Herd." *Sci. Month.*, Vol. 73 (1951); **Figures 7.6, 7.13, 7.14, and 7.15 and graph for Critical and Creative Thinking Question:** Data from Population Reference Bureau; **Figure 7.8:** Data from *World Population Prospects, The 2004 Revision*, United Nations Population Division; **What a Scientist Sees: Education and Fertility:** Data from U.S. Census Bureau; **Table 7-1:** "Urban Agglomerations 2005," U.N. Department of Economic and Social Affairs, Population Division.

Chapter 8

Figure 8.5: Adapted from Figure 6.11 in A.F. Arbogast *Discovering Physical Geography*/Hoboken, NJ: John Wiley & Sons, Inc. (2007); **What a Scientist Sees: Air Pollution from Natural Sources and Figure 8.8A:** EPA; Figure 8.14: Air Quality Planning and Standards, Office of Air and Radiation, EPA.

Chapter 9

Figure 9.5: Global Land-Ocean Temperature Index, Goddard Institute of Space Studies, NASA; **Figure 9.6:** Dave Keeling and Tim Whorf, Scripps Institution of Oceanography, La Jolla, California; **Figure 9.12:** Richard Feely and Dana Greeley, NOAA Pacific Marine Environmental Laboratory, as shown on page 111 in *National Geographic* (November 2007); **Graph in Case Study:** U.S. Department of Energy Carbon Dioxide Information Analysis Center; **Map in Creative and Critical Thinking Questions:** U.S. Global Change Research Program.

Chapter 10

Figure 10.2: Adapted from Figure 2.8 on p. 28 in Strahler, A. and A. Strahler. *Physical Geography: Science and Systems of the Human Environment.* Hoboken, NJ: John Wiley & Sons, Inc. (2002); **Figure 10.5:** *Control of Water Pollution from Urban Runoff.* Paris: Organization for Economic Cooperation and Development (1986); **Figure 10.8:** U.S. Geological Survey; **Figure 10.16:** Adapted from Joesten, M.D., and J.L. Wood. *World of Chemistry*, 2nd edition. Philadelphia: Saunders College Publishing (1996).

Chapter 11

Figure 11.1B: Adapted from Figure 12.5A on p. 347 in Murck, B.W., B.J. Skinner, and D. Mackenzie. *Visualizing Geology*, Hoboken, NJ: John Wiley & Sons, Inc. (2008); **Figure 11.2:** Adapted from Figure 6.31 on p. 148 in A.F. Arbogast. *Discovering Physical Geography.* Hoboken, NJ: John Wiley & Sons, Inc. (2007); **Figure 11.5:** Adapted from Figure 14-1 in Karleskint, G. *Introduction to Marine Biology.* Philadelphia: Harcourt College Publishers (1998); **Figure 11.6A:** Adapted from Marangos, J.E., M.P. Crosby, and J.W. McManus. "Coral Reefs and Biodiversity: A Critical and Threatened Relationship." *Oceanography*, Vol. 9 (1996); **Figure 11.11A:** After Halpern, B.S. et al. "A Global Map of Human Impact on Marine Ecosystems." *Science*, Vol. 319, No. 5865, pp. 948-952 (February 15, 2008); **What a Scientist Sees: Ocean Warming and Coral**

Bleaching, Figure B: Adapted from National Assessment Synthesis Team, *Climate Change Impacts on the United States: The Potential Consequences of Climate Variability and Change* (Report for the U.S. Global Change Research Program). Cambridge, UK: Cambridge University Press (2001); **Figure 11.15B:** From www.noaa.gov; **graph in Critical and Creative Thinking Questions:** Adapted from *Vital Signs 2006-2007*, page 27, Worldwatch Institute (2006), and based on data from the FAO.

Chapter 12

Figure 12.7: Adapted from Joesten, M.D., and J.L. Wood. *World of Chemistry*, Second Edition. Philadelphia: Saunders College Publishing (1996); **Figure 12.17:** *One Planet, Many People: Atlas of Our Changing Environment*. United Nations Environment Program (2005)—Part A based on graph on page 30 (FAO 2000), Part B based on table on page 29 (*World Atlas of Desertification*).

Chapter 13

Table 13.1: U.S. Dept. of Interior, U.S. Dept. of Agriculture, and U.S. Dept. of Defense; **Table 13.2:** National Park Service; **Table 13.3:** From Box 1.1 in Noss, R.F., M.A. O'Connell, and D.D. Murphy. *The Science of Conservation Planning: Habitat Conservation Under the Endangered Species Act*, Island Press: World Wildlife Fund (1997). Reprinted with permission; **Figure 13.1:** Adapted from the map, "Federal Lands in the Fifty States," produced by the Cartographic Division of the National Geographic Society (October 1996); **Figure 13.3:** Values are from Schlesinger, W.H. *Biogeochemistry: An Analysis of Global Change*, 2nd edition. Academic Press, San Diego (1997) and based on several sources; **Figure 13.6A:** Based on Goode Base Map (S.R. Eyre 1968); **Figure 13.7:** Data obtained from Marsh, W.M. and J.M. Grossa, Jr., *Environmental Geography*, 2nd edition, John Wiley & Sons, Inc. (2002); **graph in Critical and Creative Thinking Questions:** World Resources Institute and U.N. Food and Agricultural Organization.

Chapter 14

Figure 14.2: Based on data from U.N. Food and Agricultural Organization and UNICEF; **Figure 14.3:** Adapted from FAO data, as reported in Brown, L.R. et al. *Vital Signs 2006-2007*. New York: W.W. Norton & Company (2006); **Figure 14.4:** Adapted from G.H. Heichel, "Agricultural Production and Energy Resources." *American Scientist*, Vol. 64 (January/February 1976); **Figure 14.8:** Data from USDA; **Figure 14.15:** From "International Travel and Health. Vaccination Requirements and Health Advice: Situation as of January 1, 1992." Geneva: World Health Organization (1992). Reproduced by permission of the World Health Organization; **Figure 14.18:** Adapted from Gardner, G., "IPM and the War on Pests." *World Watch*, Vol. 9, No. 2 (March/April 1996). Redrawn by permission of Worldwatch Institute; **What a Scientist Sees: Pesticide Use and New Pest Species (graph):** Adapted from Debach, P. *Biological Control by Natural Enemies*. New York: Cambridge University Press (1974); **graph in Critical and Creative Thinking Questions:** Data from UNESCO/FAO.

Chapter 15

Table 15.1: Adapted from page 527 of *Climate Change Impacts on the United States*, A report of the National Assessment Synthesis Team, U.S. Global Change Research Program, Cambridge University Press (2001); **Table 15.2:** U.S. Fish and Wildlife Service; **Data in first section on number of species:** Data from Reaka-Kudla, M.L., D.E. Wilson, and E.O. Wilson. *Biodiversity II*. Washington, D.C.: Joseph Henry Press (1997); **Figure 15.1B:** After M.L. Cody and J.M. Diamond, eds., *Ecology and Evolution of Communities*. Harvard University, Cambridge, (1975); **What a Scientist Sees: Where Is Declining Biological Diversity the Greatest Problem:** Map by Conservation International; **quote in Critical and Creative Thinking Questions:** Aldo Leopold, *A Sand County Almanac*. Oxford University Press, Inc. (1991); **graph in Critical and Creative Thinking Questions:** Data from The Nature Conservancy.

Chapter 16

Figures 16.2 and 16.3: Data from EPA; **Figure 16.4:** Data from Ecobalance, Inc., and Integrated Waste Services Association; **Figure 16.9A:** Adapted from T. Zeller, Jr., "Recycling: The Big Picture," *National Geographic* (January 2008); **Figure 16.5:** Adapted from Rocky Mountain Arsenal Remediation Venture Office; **graph in Critical and Creative Thinking Questions:** EPA.

Chapter 17

Unless noted otherwise, all energy facts cited in this chapter were obtained from the Energy Information Administration (EIA), the statistical agency of the U.S. Department of Energy (DOE). **Table 17.1:** Integral Consulting. "2005 Assessment of Lingering Oil and Resource Injuries from the *Exxon Valdez* Oil Spill (Restoration Project 040776)"; **Table 17.2:** Adapted from two sources: Hinrichs, R.A. and M. Kleinbach. *Energy: Its Use and the Environment*, 3rd edition. Philadelphia: Harcourt College Publishers (2002) and *Science for Democratic Action*, Vol. 8, No. 3 (May 2000); **Figures 17.1A, 17.6, and 17.8:** EIA (DOE); **Figure 17.1B:** Adapted from Figure 13-3a in Harris, J.M. *Environmental and Natural Resource Economics: A Contemporary Approach*, second edition. Houghton Mifflin (2006) and based on data from the World Bank; **p. 427, graph in Critical and Creative Thinking Questions:** EIA (DOE).

Chapter 18

Unless noted otherwise, all energy facts cited in this chapter were obtained from the Energy Information Administration (EIA), the statistical agency of the U.S. Department of Energy (DOE). **Table 18.1:** Data from Kammen, D.M. "The Rise of Renewable Energy." *Scientific American* (September 2006); **Figure 18.5B:** Adapted from Hinrichs, R.A. *Energy: Its Use and the Environment*, 2nd ed. Philadelphia: Saunders College Publishing (1996); **Figure 18.8C:** From *Vital Signs 2005*, Worldwatch Institute, W.W. Norton (2005) and based on data from BTM Consult, American Wind Energy Association, and European Wind Energy Association; **map in Critical and Creative Thinking Questions:** DOE.

Chapter 1

Page 2: NG Image Collection; page 5 (top): Earth Imaging/Stone/Getty Images; page 5 (bottom): NG Image Collection; page 5: NG Image Collection; page 5 (top): Earth Imaging/Stone/Getty Images; page 5: Raymond Gehman/National Geographic Society; page 5: Jim Richardson/National Geographic Society; page 5: Ben Osborne/Stone/Getty Images; page 5: AP/ Wide World Photos; page 6 (left): Peter Menzel; page 6 (right): Peter Menzel; page 7 (left): Peter Menzel; page 8: Ali Abbas/©Corbis; page 8: NG Maps; page 10: Macduff Everton/Iconica/Getty Images; page 13 (top): John Anderson/ NG Image Collection; page 13 (center): Raymond Gehman/NG Image Collection; page 13 (bottom): Sam Kittner/NG Image Collection; page 14 (top left): Gene Carl Feldman/Sea WIFs/NASA; page 14 (top left): Lara Hansen, Adam Markham/WWF/NG Maps; page 14 (top right): Photodisc/Getty Images; page 14 (top right): Global Forest Watch/WRI/NG Maps; page 14 (bottom left): Caroline Rogers/USGS; page 14 (bottom left): NOAA/ NMFS/UNEP-WCMC/NG Maps; page 14 (bottom right): Bruce Dale/ National Geographic Society; page 14 (bottom right): USDA Global Desertification Map/National Geographic Society; page 15 (top): National/ Naval Ice Center/NG Maps; page 15 (center): NG Maps; page 15 (bottom): NASA Earth Observatory; page 16: Raymond Gehman/National Geographic Society; page 17: Jim Richardson/National Geographic Society; page 19 (left): Nicole Duplaix/National Geographic Society; page 19 (bottom): Courtesy of Matthew Price; page 20 (bottom): Ben Osborne/Stone/Getty Images; page 20 (top): Jim Richardson/National Geographic Society; page 22: Raymond Gehman/National Geographic Society; page 23: AP/Wide World Photos; page 24 (left): John Anderson/NG Image Collection; page 24 (right): AP/Wide World Photos; page 25 (bottom): David H. Wells/CORBIS.

Chapter 2

Page 26: Sam Abell/National Geographic Society; page 27 (top): William Thomas Cain/Getty Images; page 27: Mark Edwards/Peter Arnold, Inc.; page 27: Jenny Hager/The Image Works; page 27: Donna DeCesare; page 27: Tim Laman/National Geographic Society; page 29 (left): Peter Essick/National Geographic Society; page 29 (right): NG Image Collection; page 29: NG Maps; page 30 (top): William Thomas Cain/Getty Images; page 31: Richard Nowitz/NG Image Collection; page 32: Minnesota Historical Image Collection/CORBIS; page 32 (right): Mark Edwards/Peter Arnold, Inc.; page 32: NG Maps; page 33 (left): Sam Abell/National Geographic Society; page 33 (top right): Secret Sea Visions/Peter Arnold, Inc.; page 33 (bottom right): Alan and Sandy Cary/Peter Arnold, Inc.; page 34: Priit Vesilind/National Geographic Society; page 35: Jenny Hager/The Image Works; page 37 (top left): Donna DeCesare; page 37 (center left): Emory Kristof/NG Image Collection; page 37 (bottom left): Bertrand Rieger/Hemis/©Corbis; page 37 (center right): Rafael Macia/Photo Researchers, Inc.; page 37 (bottom right): Christian Shallert/©Corbis; page 38: Jodi Cobb/National Geographic Society; page 38: NG Maps; page 40: Nick Nichols/NG Image Collection; page 41: Fred Hoogervorst/Panos Pictures; page 43: Peter Treanor/Alamy; page 44: Tim Laman/National Geographic Society; page 45: Jenny Hager/The Image Works; page 47: Liane Cary/Age Fotostock.

Chapter 3

Page 48: James P. Blair/NG Image Collection; page 49: John Schwieder/ Alamy; page 49: Courtesy Library of Congress; page 49: Rick Poley/Visuals Unlimited; page 49 (bottom): Randy Olson/NG Image Collection; page 49 (bottom): Gerd Ludwig/NG Image Collection; page 50 (left): USDA/NG Image Collection; page 50 (right): John Schwieder/Alamy; page 51: Courtesy Library of Congress; page 52: Bettman/Corbis Images; page 53 (top): Courtesy National Archives; page 53 (bottom): Eugene Fisher/NG Image Collection; page 54: Photo by Tom Coleman/Courtesy Aldo Leopold Foundation, Baraboo, WI.; page 55: Erich Hartmann/Magnum Photos, Inc.; page 56: Todd Gipstein/National Geographic Society; page 57: Roger Rosentreter/ iStockphoto; page 58 (top): Bob Sacha/National Geographic Society; page 58 (bottom): Michael Nichols/National Geographic Society; page 60: Tim Laman/National Geographic Society; page 61: Rick Poley/Visuals Unlimited; page 63 (top left): Creatas/SUPERSTOCK; page 63 (top right): age foto-stock/SUPERSTOCK; page 63 (bottom): Randy Olson/NG Image Collection; page 63: NG Maps; page 64: Tyrone Turner/National Geographic Society; page 67 (bottom): Randy Olson/NG Image Collection; page 68 (top): Gerd Ludwig/NG Image Collection; page 68: NG Maps; page 68 (bottom): Gerd Ludwig/NG Image Collection; page 69: USDA/NG Image Collection; page 69: Todd Gipstein/National Geographic Society; page 69: Tim Laman/ National Geographic Society; page 70: ©Steve Greenberg; page 71: Reuters Newmedia Inc./Corbis Images.

Chapter 4

Page 72: David Alan Harvey/NG Image Collection; page 73: Sisse Brimberg/NG Image Collection; page 73: Carsten Koall/Getty Images; page 73: Andy Crump/Photo Researchers, Inc.; page 73: Wolfgang Flamish/ zefa/©Corbis; page 73: Andy Levin/Photo Researchers; page 73: Stuart Bauer; page 75: Sisse Brimberg/NG Image Collection; page 76: Scott Camazine/ Alamy; page 76: Sisse Brimberg/NG Image Collection; page 78: SPL/Photo Researchers, Inc.; page 79 (top): Courtesy Millipore Corp; page 79 (bottom): Bruce Dale/NG Image Collection; page 79 (bottom): NG Maps; page 80: Justin Guariglia/NG Image Collection; page 81: Carsten Koall/Getty Images; page 82 (top): Roy Toft/National Geographic Society; page 82 (bottom): Rod Planck/Photo Researchers, Inc.; page 84: Andy Crump/Photo Researchers, Inc.; page 85: Wolfgang Flamish/zefa/©Corbis; page 87: Courtesy Centers for Disease Control ; page 88 (top): Karen Kasmauski/NG Image Collection; page 89: Daniel LeClair/Reuters/Landov; page 90: Andy Levin/Photo Researchers; page 91: Raymond Gehman/NG Image Collection; page 92: Stuart Bauer; page 93: Roy Toft/National Geographic Society; page 93: Sisse Brimberg/NG Image Collection; page 93: Carsten Koall/Getty Images; page 95 (bottom): James L. Amos/NG Image Collection.

Chapter 5

Page 96: Harmut Schwarzbach/Peter Arnold, Inc.; page 96 (inset): Robert Caputo/NG Image Collection; page 97: Taylor S. Kennedy/NG Image Collection; page 97 (bottom right): Norbert Rosing/NG Image Collection; page 97 (right): Dr Jeremy Burgess/Photo Researchers; page 97: Jason Edwards/NG Image Collection; page 97 (left): Mattias Klum/NG Image Collection; page 97: ©AP/Wide World Photos; page 99 (top): Norbert Rosing/NG Image Collection; page 99 (center): Geroge F. Mobley/NG Image Collection; page 100 (right): NG Image Collection; page 100 (left): Taylor S. Kennedy/NG Image Collection; page 101 (left): Romeo Gacad/AFP/Getty Images; page 101 (right): Max Rossi/Reuters/Corbis; page 102: Raymond Gehman/NG Image Collection; page 103 (top left): John Eastcott and Yva Momatiuk/NG Image Collection; page 103 (bottom left): Roy Taft/NG Image Collection; page 103 (top right): Tim Laman/NG Image Collection; page 103 (bottom right): Norbert Rosing/NG Image Collection; page 106: Roy Taft/NG Image Collection; page 110 (right): Dr Jeremy Burgess/Photo Researchers; page 110 (left): Dr. Robert Calentine/Visuals Unlimited; page 114: Jason Edwards/NG Image Collection; page 115: Rob and Ann Simpson/Visuals Unlimited; page 117 (top left): Jack Jeffrey Photography; page 117 (top right): Minden Pictures/Getty Images; page 117 (bottom left): Tim Laman/NG Image Collection; page 117 (bottom right): Darlyne A. Murawski/NG Image Collection; page 118: Maria Stenzel/NG

Image Collection; page 119: Chris Johns/NG Image Collection; page 119: NG Maps; page 119: NG Maps; page 120 (left): Mattias Klum/NG Image Collection; page 120 (right): Robert Sisson/NG Image Collection; page 121 (left): Joel Sartore/NG Image Collection; page 121 (right): SUPERSTOCK; page 122: ©AP/Wide World Photos; page 123: NG Image Collection; page 123 (bottom center): Dr Jeremy Burgess/Photo Researchers; page 123: Macduff Everton/NG Image Collection; page 124: Joel Sartore/NG Image Collection; page 125: Darlyne A. Murawski/NG Image Collection.

Chapter 6

Page 126: Raymond Gehman/NG Image Collection; page 127: Paul Nicklen/NG Image Collection; page 127: Kathleen Revis/NG Image Collection; page 127: O. Louis Mazzatenta/NG Image Collection; page 127: Wolfgang Kaehler; page 127: ©AP/Wide World Photos; page 129 (left): Michael Melford/NG Image Collection; page 129 (right): Todd Gipstein/NG Image Collection; page 130: Paul Davis/The Global Land Cover Facility, University of Maryland Institute for Advanced Computer Studies; page 130: Brand X/SUPERSTOCK; page 130: Hemis. Fr/SUPERSTOCK; page 130: Photodisc/SUPERSTOCK; page 130: age fotostock/SUPERSTOCK; page 130: Polka Dot Images/SUPERSTOCK; page 130: Michael McCoy/Photo Researchers, Inc.; page 130: Corbis/SUPERSTOCK; page 130: age fotostock/SUPERSTOCK; page 131: age fotostock/SuperStock, Inc.; page 131: James Steinberg/Photo Researchers, Inc.; page 131: Corbis/SUPERSTOCK; page 131: Imagemore/SUPERSTOCK; page 131: Steve Coleman/SUPERSTOCK; page 131: Ernest Manewal/SUPERSTOCK; page 132: Paul Nicklen/NG Image Collection; page 133: Maria Stenzel/NG Image Collection; page 134: Sam Abell/NG Image Collection; page 135: James P. Blair/NG Image Collection; page 136: Robert Hynes/NG Image Collection; page 137: Ken Lucas/Visuals Unlimited; page 138: Steve McCurry/NG Image Collection; page 139: Raymond Gehman/NG Image Collection; page 140: Mitsuaki Iwago/NG Image Collection; page 141: Medford Taylor/NG Image Collection; page 143: Kathleen Revis/NG Image Collection; page 144: Frans Lanting/NG Image Collection; page 145: John Eastcott and Momatiuk/NG Image Collection; page 146 (top): Skip Brown/NG Image Collection; page 146 (bottom): Joe Stancampiano/NG Image Collection; page 146: NG Maps; page 148: Nicole Duplaix/NG Image Collection; page 149 (top right): NG Maps; page 149: Ralph Lee Hopkins/NG Image Collection; page 149: ©J. Dunning/VIREO; page 149: ©Gerald & Buff Corsi/Visuals Unlimited; page 149: ©Fritz Polking/Visuals Unlimited; page 149: ©Joe & Mary Ann McDonald/Visuals Unlimited; page 149: Tierbild Okapia/Photo Researchers, Inc.; page 149: Eric Hosking/Photo Researchers, Inc.; page 151: O. Louis Mazzatenta/NG Image Collection; page 152 (top): Wolfgang Kaehler; page 152 (center left): Glenn N. Oliver/Visuals Unlimited; page 152 (center right): Wolfgang Kaehler; page 154 (top): ©AP/Wide World Photos; page 154 (bottom): ©AP/Wide World Photos; page 155: Nicole Duplaix/NG Image Collection; page 155 (bottom): Joe Stancampiano/NG Image Collection; page 157 (top): Joseph T. Collins/Photo Researchers; page 157 (bottom): Jodi Cobb/NG Image Collection.

Chapter 7

Page 158: Lou Linwei/Alamy; page 158: Sidney Hastings/NG Image Collection; page 159: Yvona Momatiuk and John Eastcott/Photo Researchers; page 159: NG Maps; page 159 (right): Karen Kasmauski/NG Image Collection; page 159: Karen Kasmauski/NG Image Collection; page 159: Urban Redevelopment Authority of Pittsburgh; page 159 (left): Getty Images; page 160 (left): Wolcott Henry/NG Image Collection; page 160 (right): David Pluth/NG Image Collection; page 162: CNRI/Science Photo Library/Photo Researchers; page 163: Michael Abbey/Photo Researchers; page 164: Yvona Momatiuk and John Eastcott/Photo Researchers; page 166: Karen Kasmauski/NG Image Collection; page 167: James P. Blair/NG Image Collection; page 167: NG Maps; page 168: Jodi Cobb/NG Image Collection; page 168: NGS Maps; page 168 (bottom): NG Maps; page 168: NG Maps; page 169 (bottom right): NG Maps; page 171 (left): Annie Griffiths Belt/NG Image Collection; page 171 (right): Karen Kasmauski/NG Image Collection; page 175 (left): Pablo Corral Vega/NG Image

Collection; page 175 (right): Larry C. Price; page 176: Kenneth Macleish/NG Image Collection; page 176: NG Maps; page 177: Sean Spague/Alamy; page 177: NG Maps; page 178: Karen Kasmauski/NG Image Collection; page 179: Karen Kasmauski/NG Image Collection; page 180: Sean Spague/Alamy; page 181: Richard Nowitz/NG Image Collection; page 182: mediacolor s/Alamy; page 183: Urban Redevelopment Authority of Pittsburgh; page 185: Stuart Franklin/NG Image Collection; page 186 (left): Getty Images; page 186 (right): Corbis Images; page 187: Yvona Momatiuk and John Eastcott/Photo Researchers; page 187: NG Maps; page 187 (right): Karen Kasmauski/NG Image Collection; page 188: Richard Nowitz/NG Image Collection; page 188: Karen Kasmauski/NG Image Collection; page 189: Lynn Johnson/NG Image Collection.

Chapter 8

Page 190: Norbert Rosing/NG Image Collection; page 191 (top left): Roger Harris/Photo Researchers, Inc.; page 191: Mario Tama/Getty Images; page 191: James P. Blair/NG Image Collection; page 191: Guy Starling/NG Image Collection; page 191: Joerg Boethling/Peter Arnold, Inc.; page 191: Raymond Gehman/NG Image Collection; page 192 (top left): Roger Harris/Photo Researchers, Inc.; page 193 (left): Kenneth Garrett/NG Image Collection; page 193 (right): Jack Finch/Photo Researchers; page 198: Sarah Leen/NG Image Collection; page 199: David Cavagnaro/Peter Arnold, Inc.; page 200 (top): Mario Tama/Getty Images; page 200 (bottom): Emory Kristof/NG Image Collection; page 203: Frank Lukasseck/©Corbis; page 204: James P. Blair/NG Image Collection; page 206 (right): John D. Cunningham/Visuals Unlimited; page 206 (left): John D. Cunningham/Visuals Unlimited; page 208: Michael S. Yamashita/NG Image Collection; page 208: NG Maps; page 210: Joerg Boethling/Peter Arnold, Inc.; page 212: Raymond Gehman/NG Image Collection; page 213: Guy Starling/NG Image Collection; page 215: Peter Menzel Photography.

Chapter 9

Page 216: National Snow and Ice Data Center, W. O. Field, B. F. Molnia; page 217: National Snow and Ice Data Center, W. O. Field, B. F. Molnia; page 217: SuperStock, Inc.; page 217: James A. Sugar/National Geographic Society; page 217: David Austen/NG Image Collection; page 217: Al Petteway/National Geographic Society; page 217: Maria Stanzel/NG Image Collection; page 218: Maria Stanzel/NG Image Collection; page 221: SuperStock, Inc.; page 221 (left): Maria Stanzel/NG Image Collection; page 223: James A. Sugar/National Geographic Society; page 225: Stock Trek/PhotoDisc Green/Getty; page 225: NG Maps; page 226: Michael Nichols/NG Image Collection; page 226: NG Maps; page 227 (top left): FLPA/Alamy; page 227 (top right): Thomas R. Fletcher/Alamy; page 227 (center): Peter Scoones/Photo Researchers, Inc.; page 227 (bottom): NG Maps; page 229: David R. Frazier Photolibrary/Alamy; page 229: Peter Scoones/Photo Researchers, Inc.; page 231: NASA Media Services; page 232: David Austen/NG Image Collection; page 235 (left): Al Petteway/National Geographic Society; page 235 (top right): Christian Mehr; page 235 (bottom right): Christian Mehr; page 236: Mike Dobel/Alamy; page 238: Christian Mehr; page 239: Thomas Nebbia/NG Image Collection.

Chapter 10

Page 240: Macduff Everton/National Geographic Society; page 240: T. Dietrich/Alamy; page 241: Jodi Cobb/NG Image Collection; page 241: ©AP/Wide World Photos; page 241 (right): ©Raymond G. Barnes; page 241: Steve Winter/NG Image Collection; page 241: Bruce Dale/NG Image Collection; page 242: Jodi Cobb/NG Image Collection; page 244: Jodi Cobb/NG Image Collection; page 245: ©AP/Wide World Photos; page 246: Courtesy NASA; page 248: Dan Lamont, Seattle, Washington; page 249: NASA/USGS; page 249: NG Maps; page 250 (top left): NG Maps; page 250 (bottom left): NG Maps; page 250 (bottom right): NG Maps; page 250 (center): NG Maps; page 251: NG Maps; page 252: NG Maps; page 252: Dan Lamont, Seattle, Washington; page 253 (left): Courtesy U. S. Dept. of Energy; page 253 (right): ©Raymond G. Barnes; page 255 (left): James L. Stanfield/NG Image

Collection; page 255 (right): ©AP/Wide World Photos; page 258 (left): Rich Buzzelli/Tom Stack & Associates; page 258 (right): W.A. Banaszewski/Visuals Unlimited; page 259: Steve Winter/NG Image Collection; page 261: W.A. Banaszewski/Visuals Unlimited; page 264: Ted Shreshinsky/Corbis Images; page 266 (left): Steve McCurry/NG Image Collection; page 266 (right): ©AP/Wide World Photos; page 267: NG Maps; page 267: Bruce Dale/NG Image Collection; page 268: Ted Shreshinsky/Corbis Images; page 269: Sandra Baker/Getty.

Chapter 11

Page 270: Brian J. Skerry/NG Image Collection; page 270: Jose Cort/Courtesy NOAA; page 271: Norbert Wu/Peter Arnold, Inc.; page 271: Science VU/Visuals Unlimited; page 271: Courtesy Dave Matilla/NOAA; page 271: Bill Curtsinger/NG Image Collection; page 273 (bottom): Courtesy NASA; page 277 (top left): Burt Curtsinger/NG Image Collection; page 277 (top right): Stuart Westmorland/Photo Researchers; page 277 (bottom left): Alex Kerstitch/Visuals Unlimited; page 277 (bottom right): Paul Zahl/NG Image Collection; page 278: Tim Layman/NG Image Collection; page 279 (top): Raul Touzon/NG Image Collection; page 279 (bottom): Norbert Wu/Peter Arnold, Inc.; page 280: Hugh Rose/Danita Delimont; page 280 (bottom): Joe Stancampiano/NG Image Collection; page 281: Dante Fenolio/Photo Researchers, Inc.; page 281: Tim Layman/NG Image Collection; page 283 (top left): Courtesy B. Halpern, K. Selkoe, C. Kappel, F. Micheli; National Center for Ecological Analysis and Synthesis; page 285: Brian J. Skerry/NG Image Collection; page 286: Flip Nicklin/Getty Images; page 286: Flip Nicklin/Getty Images; page 287: Peter W. Glynn/NG Image Collection; page 288: Science VU/Visuals Unlimited; page 288: Science VU/Visuals Unlimited; page 290: Courtesy Dave Matilla/NOAA; page 291: Bill Curtsinger/NG Image Collection; page 292: Flip Nicklin/Getty Images; page 293: Brian J. Skerry/NG Image Collection.

Chapter 12

Page 294: Pat Armstrong/Visuals Unlimited; page 294 (bottom left inset): Photo by Jack Waters www.telliquah.com; page 295: Bruce F. Molnia/Terraphotographics/BPS; page 295: R. Ashley/Visuals Unlimited; page 295: Courtesy U.S. Dept. of Agriculture; page 295: Grant Heilman/Grant Heilman Photography; page 301: Mikhail Pozhenko/iStockphoto; page 301: iStockphoto; page 301: Matthew Ragen/iStockphoto; page 301: iStockphoto; page 301: Mike Clarke/iStockphoto; page 301: Christoph Ermel/iStockphoto; page 301: Andrew Johnson/iStockphoto; page 301: Dave White/iStockphoto; page 301: Jack Cobben/iStockphoto; page 301: Matt Meadows/Alamy; page 301: Vladimir Melnik/iStockphoto; page 301: Wesley VanDinter/iStockphoto; page 301: Predrag Novakovic/iStockphoto; page 301: Nikki Lowry/iStockphoto; page 301: Don Wilkie/iStockphoto; page 301: Jozsef Szasz-Fabian/iStockphoto; page 302 (top): Bruce F. Molnia/Terraphotographics/BPS; page 305 (left): David Hiser/Stone/Getty Images; page 305 (right): R. Ashley/Visuals Unlimited; page 306 (bottom): William Felger/Grant Heilman Photography; page 307: Leste Lefkowitz/Corbis Images; page 308: Photo by Ray Weil, courtesy Martin Rabenhorst; page 309: Courtesy U.S. Dept. of Agriculture; page 311 (left): Grant Heilman/Grant Heilman Photography; page 311 (right): Courtesy U. S. Department of Agriculture; page 313 (left): Courtesy U. S. Department of Agriculture; page 313 (right): Blaine Harrington III/Alamy; page 314 (left): Kevin Flaming/Corbis Images; page 314: NGS Maps; page 317: Melissa Farlow/NG Image Collection; page 357 (top center): Grant Heilman/Grant Heilman Photography, iStockphoto.

Chapter 13

Page 318: SUPERSTOCK; page 318 (bottom left inset): SUPERSTOCK; page 319: Sam Abell/NG Image Collection; page 319: Peter Essick/NG Image Collection; page 319: Steve McCurry/NG Image Collection; page 319 (top left): (c)Larry Ulrich; page 319: NG Maps; page 319 (right): James P. Blair/NG Image Collection; page 321: Sam Abell/NG Image Collection; page 323: Kirtley-Perkins/Visuals Unlimited; page 325: Stephen Sharnoff/NG Image Collection; page 326: Courtesy Forest Stewardship Council—US; page 327:

Peter Essick/NG Image Collection; page 329 (center left): NRSC/Photo Researchers, Inc.; page 329 (bottom left): Pegasus/Visuals Unlimited; page 329 (center right): Marin Hervey/Natural History Photographic Agency; page 331: Steve McCurry/NG Image Collection; page 332: NGS Maps ; page 333: Steve Smith/SUPERSTOCK; page 335 (top left): (c)Larry Ulrich; page 335 (top right): Tom Bean/Stone/Getty Images; page 335 (bottom): Jim Parkin/Alamy; page 336: Tom Murphy/NG Image Collection; page 337 (left): James Schwabel/Alamy; page 337 (right): Riich Reid/NG Image Collection; page 338: Chris Luneski/Alamy; page 340: Pacific Stock/SUPERSTOCK; page 340: Corbis/SUPERSTOCK; page 340: Prisma/SUPERSTOCK; page 340: Corbis/SUPERSTOCK; page 340: Edward Parker/Alamy; page 340: age foto-stock/SUPERSTOCK; page 340: NG Maps; page 340: NG Maps; page 341: age fotostock/SuperStock, Inc.; page 341: Digital Vision/SUPERSTOCK; page 341: Ingram Publishing/SUPERSTOCK; page 341: Prisma/SUPER-STOCK; page 341: SuperStock, Inc.; page 341: Pacific Stock/SUPERSTOCK; page 341: NG Maps; page 342 (top): Bruce Dale/NG Image Collection; page 342 (bottom): James P. Blair/NG Image Collection; page 345: Corbis Sygma, High Country News/©Steve Greenberg.

Chapter 14

Page 346: Sisse Brimberg/NG Image Collection; page 346 (bottom left inset): H. Donnezan/Photo Researchers, Inc.; page 347: Paul Souders/©Corbis; page 347: David Keith Jones/Alamy; page 347: Sarah Leen/NG Image Collection; page 347: Jim Richardson/NG Image Collection; page 347: Courtesy Max Badgley; page 347: Michael S. Quinton/NG Image Collection; page 349 (left): Paul Souders/©Corbis; page 349 (center): Reuters Newmedia, Inc./Corbis Images; page 349 (right): Bill Wisser/Digital Railroad, Inc; page 351: Biosphoto/NouN/Peter Arnold, Inc.; page 352: David Keith Jones/Alamy; page 353: Sarah Leen/NG Image Collection; page 354: Photo by Agri-Graphics, Ltd, courtesy Winifred Hoffman; page 357 (top left): Courtesy NASA; page 357 (top center): Lawrence Migdale/Photo Researchers, Inc.; page 357 (top right): John Elk III/Alamy; page 357 (left): ©AP/Wide World Photos; page 357 (right): Courtesy Peggy Greb/USDA; page 359: ©AP/Wide World Photos; page 360: Jim Richardson/NG Image Collection; page 362: Bill Bachman/Alamy; page 363: Darlyne A. Murawski/NG Image Collection; page 364 (top): Lawrence Migdale/Photo Researchers, Inc.; page 364 (bottom): Courtesy Max Badgley; page 365: Nigel Cattlin/Alamy; page 366: Bill Bachman/Alamy; page 367: Michael S. Quinton/NG Image Collection; page 367: David Keith Jones/Alamy; page 368: Nigel Cattlin/Alamy; page 369: Cary Wolinsky/NG Image Collection.

Chapter 15

Page 370: George Grall/NG Image Collection; page 370 (bottom left inset): Frans Lanting/Minden Pictures, Inc.; page 371: George Grall/NG Image Collection; page 371: Joel Sartore/NG Image Collection; page 371: Bryan & Cherry Alexander Photography/Alamy; page 371: Terry Whitaker/Alamy; page 371: John Burcham/NG Image Collection; page 372: NG Image Collection; page 373: David Cavagnaro; page 374: George Grall/NG Image Collection; page 375: Doug Wechsler; page 376: David Cavagnaro; page 378: Joel Sartore/NG Image Collection; page 379: Joel Sartore/NG Image Collection; page 381 (top left): Frans Lanting/Minden Pictures, Inc.; page 381 (top right): W. Robert Moore/NG Image Collection; page 381 (bottom): Chris Johns/NG Image Collection; page 383: Andrew Sacks/Time Life Pictures/Getty Images; page 384: Dan McCoy/Science Faction/Getty Images; page 386 (left): University of Wisconsin - Madison Arboretum; page 386 (right): Virginia Kline/University of Wisconsin - Madison Arboretum; page 387 (left): Steve Nexbitt, Florida Fish and Wildlife Conservation Commission; page 387 (center): Richard O. Bierregaard, Jr.; page 387: Bryan & Cherry Alexander Photography/Alamy; page 389: Melissa Farlow/NG Image Collection; page 390 (top): Terry Whitaker/Alamy; page 391 (bottom): Joel Sartore/NG Image Collection; page 391 (top): John Burcham/NG Image Collection; page 392: Dan McCoy/Science Faction/Getty Images; page 393: Michael Nichols/NG Image Collection, (left) Steve Nexbitt,

Florida Fish and Wildlife Conservation Commission, Melissa Farlow/NG Image Collection.

Chapter 16

Page 394: Joel Sartore/NG Image Collection; page 395: Inga Spence/Visuals Unlimited; page 395 (center): Shari Lewis/©AP/Wide World Photos; page 395: Andy Levin/Photo Researchers, Inc.; page 395 (bottom): Gary Milburn/Tom Stack & Associates; page 395: Peter Essick/NG Image Collection; page 396: Ramin Talaie/©Corbis; page 399: Inga Spence/Visuals Unlimited; page 400: Jose Azel/Aurora Photos; page 401; page 402: Jose Azel/Aurora Photos; page 403: FOR BETTER OR FOR WORSE ©2006 Lynn Johnston Productions. Dist. by Universal Press Syndicate. Reprinted with permission. All rights reserved; page 405 (left): Frances Roberts/Alamy; page 405 (right): Shari Lewis/©AP/Wide World Photos; page 407: Andy Levin/Photo Researchers, Inc.; page 407: NG Maps; page 408: NG Image Collection; page 409 (left): Chromosohm/Photo Researchers, Inc.; page 409 (right): DOE/Science Source; page 411 (center): Courtesy USDA; page 411 (bottom): Gary Milburn/Tom Stack & Associates; page 413: Peter Essick/NG Image Collection; page 414: Ramin Talaie/©Corbis; page 415: Ulrich Doering/Alamy.

Chapter 17

Page 416: Darren McCollester/Getty Images; page 416: Mark Tantrum/Alamy; page 417: Alexander Pischel/Alamy; page 417: Peter Essick/NG Image Collection; page 417 (bottom): Charles Mason/Black Star; page 417 (left): Courtesy Westinghouse Electric Corp., Commercial Nuclear Fuel Division; page 417: Steven J. Kazlowski/Alamy; page 418: Alexander Pischel/Alamy; page 420 (left): Peter Essick/NG Image Collection; page 420 (right): Kristin Finnegan Photography; page 421: John Shaw Photography; page 424: Jim Richardson/NG Image Collection; page 424: Mark Burnett/Photo Researchers, Inc.; page 424: Courtesy National Renewable Energy Lab; page 424: John Mead/Photo Researchers, Inc.; page 424: Photo Researchers, Inc.; page 424: U.S. Department of Energy/USGS Mineral Resources Program;

page 424: The Geothermal Energy Association/Barry Soloman/Michigan Technological University/Sandia National Laboratories/Windpower Monthly News Magazine; page 424: NG Image Collection/BP Statistical View of World Energy; page 425: Peter Essick/NG Image Collection;page 427 (top): Chris Wilkins/AFP/Getty Images; page 427 (bottom): Charles Mason/Black Star; page 428 (top): Chris Wilkins/AFP/Getty Images; page 430 (left): Courtesy Westinghouse Electric Corp., Commercial Nuclear Fuel Division; page 430: Jayson Mellom Photography; page 433: NGS Maps; page 433 (top): Novosti; page 433 (bottom): Caroline Penn/Corbis Images; page 434 (left): Peter Essick/Aurora Photos; page 436: Fedoseyev Lev/ITAR-TASS/Landov LLC; page 437: Steven J. Kazlowski/Alamy; page 438 (top): Kristin Finnegan Photography; page 439 (bottom left inset): Peter Cosgrove/©AP/Wide World Photos.

Chapter 18

Page 440: Mario Tama/Getty Images; page 440 (bottom left inset): Sarah Leen/NG Image Collection; page 441: Hank Morgan/Getty Images; page 441: Steve McCurry/NG Image Collection; page 441: Phil Degginger/Alamy; page 441: Courtesy Ontario Hydro; page 441: Emmanuel Dunand/Getty Images; page 443: Otis Imboden/NG Image Collection, Courtesy of Uni-Solar. Energy Conservation Devices, Discover Magazine; inset, United Solar Systems, Discover Magazine.; page 445: Courtesy of Uni-Solar. Energy Conservation Devices, Discover Magazine; inset, United Solar Systems, Discover Magazine; page 446: Hank Morgan/Getty Images; page 447: Pasquale Sorrentino/Photo Researchers, Inc.; page 448: Steve McCurry/NG Image Collection; page 448: NGS Maps; page 449: Prof. David Hall/Photo Researchers, Inc.; page 449: NGS Maps; page 451: Justin Sullivan/Getty Images; page 452: Corbis/SUPERSTOCK; page 453: Justin Sullivan/Getty Images; page 455: Phil Degginger/Alamy; page 457: Courtesy Ontario Hydro; page 458: Iain Materton/Alamy; page 459: Todd Gipstein/NG Image Collection; page 461: Nick Wood/Alamy; page 461 (bottom): Emmanuel Dunand/Getty Images; page 463: Sarah Leen/NG Image Collection.

INDEX

Note: *f* indicates figures; *t* indicates tables

SECOND EDITION

VISUALIZING
ENVIRONMENTAL
SCIENCE